Springer Earth System Sciences

More information about this series at http://www.springer.com/series/10178

Roy H. Behnke · Michael Mortimore
Editors

The End of Desertification?

Disputing Environmental Change in the Drylands

Springer

Editors
Roy H. Behnke
Odessa Centre
Great Wolford, Shipston-on-Stour
UK

Michael Mortimore
Drylands Research
Milborne Port
UK

ISSN 2197-9596 ISSN 2197-960X (electronic)
Springer Earth System Sciences
ISBN 978-3-642-16013-4 ISBN 978-3-642-16014-1 (eBook)
DOI 10.1007/978-3-642-16014-1

Library of Congress Control Number: 2015952050

Springer Heidelberg New York Dordrecht London

Printed on acid-free paper

Springer-Verlag GmbH Berlin Heidelberg is part of Springer Science+Business Media
(www.springer.com)

Acknowledgments

In May 2013, the contributors to this volume were invited to attend a meeting in rural Somerset, UK, to present early drafts of their chapters and discuss the organization of this book. This meeting was instrumental in sustaining enthusiasm for what became a long process leading to eventual publication. Two organizations—the International Institute for Environment and Development (IIED), in UK, and the Feinstein International Center of Tufts University, in USA—provided the funding that made our meeting possible. In particular, we thank Ced Hesse and Camilla Toulmin of IIED and Helen Young of Tufts for their personal encouragement and generous institutional support. The meeting was hosted by the Odessa Centre Ltd., directed by Carol Kerven, who also recorded the results of our meeting and later edited a number of the chapters in the book. Cara Kerven and Tristam Barrett provided the logistical support and practical hospitality that made the meeting a pleasant as well as intellectually gratifying experience. Countless scholars, informants, and assistants have contributed in different ways to the work recorded in the chapters of this book, whose completion represents a truly collaborative effort. To all these individuals and institutions we owe our deepest thanks.

Contents

Chapter 1
Introduction: The End of Desertification?

Roy Behnke and Michael Mortimore

Abstract The opening chapters of this book examine something that never occurred but was widely believed to have existed—the late 20th century desertification crisis in the Sahel. Recent advances in climatology and changing weather patterns have effectively terminated further scientific debate about the existence of widespread Sahelian desertification, providing us with an opportunity to take stock and draw lessons. The logical and empirical shortcomings of the concept of desertification have been known for decades but the idea has been institutionalized at the global level and is remarkably resilient. The middle section of this book presents new reasons for concluding that the concept of desertification is no longer analytically useful and that we should instead struggle to better define and measure dryland degradation. The closing chapters of the book provide case studies from around the world that examine the use and relevance of the desertification concept. Despite an increasingly sophisticated understanding of dryland environments and societies, the uses now being made of the desertification concept in parts of Asia exhibit many of the shortcomings of earlier work done in Africa. It took scientists more than three decades to transform a perceived desertification crisis in the Sahel into a non-event. This book is an effort to critically examine that experience and accelerate the learning process in other parts of the world.

Keywords Desertification · Dryland degradation · UNCCD · Sahel · China

In *The End of Nomadism?* Humphrey and Sneath (1999) questioned the future of pastoralism in Inner Asia and challenged the analytical categories routinely used to characterize the world's mobile pastoral systems. The question in the title of the present book draws attention to the shortcomings of another, related scientific

R. Behnke (✉)
Odessa Centre Ltd., 2 the Ridgeway, Great Wolford CV36 5NN, UK
e-mail: roy_behnke@msn.com

M. Mortimore
Drylands Research, Cutters Cottage, 8 Glovers Close, Milborne Port, Sherborne DT9 5ER, UK
e-mail: mike@mikemortimore.co.uk

R.H. Behnke and M. Mortimore (eds.), *The End of Desertification?*
Springer Earth System Sciences, DOI 10.1007/978-3-642-16014-1_1

concept, that of desertification, which has become a political tool of global importance even as the scientific basis for its use grows weaker. We shall argue that the concept of desertification has ceased to be analytically useful and distorts our understanding of social-environmental systems and their sustainable development, particularly in poor countries with variable rainfall and persistent poverty. In the interests of better policy and governance, we need to reconsider the scientific justification and the institutionalized promotion of attempts to combat desertification.

There are three parts to our indictment:

Part I of this book examines something that never occurred but was widely believed to have existed—the late 20th century desertification crisis in the Sahel—the affair that ushered in the modern 'Age of Desertification' (Davis, Chap. 8). Science advances by failing as well as succeeding, and understanding the aetiology of a misconception can help to prevent its repetition. Recent advances in climatology and changing weather patterns in the Sahel have effectively terminated further scientific debate about the existence of widespread desertification in that region in the late 20th century, providing us with an opportunity to take stock and draw lessons. The chapters in Part I also exemplify the methodological sophistication and care with which current research investigates social, political, economic and environmental change in the Sahel, including the politicized nature of the desertification concept (Chap. 2), the changing paradigms in understanding social-environmental systems (Chap. 3), the relations between degradation and conflict (Chap. 4), re-greening through on-farm afforestation (Chap. 5), the combination of earth satellite data with ground observations (Chap. 6) and adaptation or resilience of agro-ecosystems (Chap. 7). This work has dispelled the very notion of widespread, catastrophic environmental degradation in the region, altered the recommended approach to environmental management, and underscored the need for scientifically informed governance.

The logical and empirical shortcomings of the concept of desertification have been known for decades (Chaps. 2 and 3; Mortimore 1989; Thomas and Middleton 1994; Warren and Agnew 1988; Swift 1996) but the idea has been institutionalized at the global level and is remarkably resilient. Part II of this book presents new reasons for concluding that the concept of desertification is no longer analytically useful and that we should instead struggle to better define and operationalize the admittedly difficult concept of dryland degradation.[1] The main planks in our argument are the pre-scientific and now indefensible set of ideas that generated the term 'desertification' (Chap. 8), the undefined and unquantifiable nature of the

[1]Drylands receive relatively low precipitation in the form of rainfall or snow and have an aridity index of <0.65. The aridity index is a measure of the ratio between average annual precipitation and total annual potential evapotranspiration. Drylands can be subdivided into: hyper-arid deserts (<0.05 index of aridity), arid (0.05–0.20 index of aridity), semi-arid (0.2–0.5 index of aridity) and dry sub-humid (0.5–0.65 index of aridity). A further defining characteristic of many (but not all) drylands is a strongly seasonal and sharply variable distribution of precipitation, both within and between rainy seasons (UN Environment Management Group 2011).

concept on a global or regional scale (Chap. 9), the need and opportunity to subsume desertification within a larger framework of global change (Chap. 10), and our current understanding of rangeland dynamics, which contradicts important aspects of desertification theory (Chap. 11).

Part III presents regional case studies that document dryland degradation and assess the efficacy of state policies to control degradation in different semi-arid environments. Chapters 12–14 and 16 examine the environmental impact of livestock production and crop agriculture in East Africa, South America and the Mediterranean. Chapter 15 looks at how journalists in three countries depict pastoralists, who are often viewed as one of the main agents of desertification. The final three chapters of the book, Chap. 17–19, document a worrying disconnection between policy formulation and environmental research on dryland issues in Central Asia and China. The use now being made of the concept of desertification in parts of Asia exhibits uncomfortable parallels to the history of the concept in the Sahel, suggesting that a reappraisal of Sahelian desertification is of more than regional or historical interest.

A final overview, Chap. 20, evaluates the contents of this book in terms of the 'drylands development paradigm', a recent and authoritative guide to investigating the complexities of dryland environments and societies.

1.1 Part I: Desertification in the Sahel: The Meaning of a Non-event

Given the complexity, imprecision and methodological opacity of desertification, the concept has lost its analytical utility for science and practice. Our argument begins in the Sahel, where a large literature has developed surrounding the widespread use of the concept, where the term itself was popularized, and where mainstream desertification orthodoxy has now been effectively dismantled by new scientific knowledge.

1.1.1 Crisis Versus Evolution

If desertification denotes an environmental crisis consisting of irreversible degradation on a sub-continental scale, then the most significant thing about desertification in the Sahel is that it never happened. Localized, even severe land degradation certainly exists in the region. Widespread, long-term degradation of some components of the environment may also be occurring (Herrmann and Sop on trees, Chap. 5), or not (Hiernaux et al. Chap. 6 on herbaceous and shrub vegetation in the northern Sahel). But the care with which researchers now work indicates the subtlety of these long-term environmental trends (Miehe et al. 2010) and, on occasion, the debatable nature of their existence (Hein and de Ridder 2006; Prince et al. 1998,

2007). While degradation is certainly a reality in the Sahel at some localities or with respect to certain components of the environment, there is no evidence of a catastrophic regional environmental crisis: 'Existing data do not support the claim that the African Sahel is a desertification hotspot' (Lepers et al. 2005: 122).

In the last century there was a Sahelian crisis, but not one of desertification. Beginning in the 1970s, a prolonged and severe drought caused tremendous human suffering. This drought was initially seen as 'a catalyst which exposed the deleterious effects of long-term degradation by people' (Thomas and Middleton 1994: 27). Recent shifts in rainfall patterns have witnessed the re-greening of much of the Sahel (Seaquist et al. 2009; Bégué et al. 2011; Olsson et al. 2005; Anyamba and Tucker 2005; Herrmann et al. 2005), and in retrospect it now seems that the mostly reversible effects of a persistent drought were simply confused with long-term degradation.

1.1.2 Blame and Response

Destructive land use practices by peasants and pastoralists—primarily primitive farming techniques and overgrazing—were blamed for causing Sahelian drought and desertification. In the late 1970s a renowned expert on the Sahel listed the causes of desertification as: overgrazing, overcultivation, uprooting woody species, borehole drilling and livestock vaccination campaigns (Le Houérou 1977); the UNCOD in 1977 identified overcultivation, overgrazing, deforestation and mismanagement of irrigated cropland (Thomas and Middleton 1994). It was argued that by removing vegetation, traditional agriculture exacerbated erosion and increased the reflectivity of the earth's surface, permanently shifting the regional climate towards more arid conditions (Charney et al. 1975). This argument was eventually refuted by climatological research summarized and discussed in detail by Giannini in Chap. 10. Climatological research demonstrated that the immediate cause of the great Sahel droughts had not even been terrestrial in origin, but resulted from variations in sea surface temperatures, and to the extent that these oceanic temperature anomalies had terrestrial causes, these could be traced to the changing composition of the air pollutants emitted in the northern hemisphere. Residents of the Sahel had not, as previously asserted, been the agents of their own misery.

The conclusive scientific demise of the Sahelian desertification paradigm only came about in the last decade or so, but serious reservations about the standard narrative emerged in the late 1980s (Mortimore 1989; Toulmin and Brock, Chap. 2). In the intervening decades this counter narrative coalesced into a compelling alternative explanation of the evolution of Sahelian landscapes and society (Mortimore, Chap. 3), and—in some parts of the Sahel—provided the rationale for innovative technical and political responses to the challenge of combining conservation with economic development (Boubacar, Chap. 7).

1.1.3 Institutionalization

Of the many things that the history of desertification in the Sahel can teach us, a lesson about the interaction between science and public policy is possibly the most important. The protracted Sahel drought that began in the early 1970s promoted widespread concern about desertification and sparked a global political response. As discussed in Chap. 2 by Camilla Toulmin and Karen Brock, desertification has for nearly half a century been a formal institutional process, a topic of occasional journalistic attention and the subject of an international convention and of environmental policy in individual nation states. Commitment to the concept of desertification goes beyond the scientific community and this has, in retrospect, complicated the attempt to objectively assess its status.

As a global institutional phenomenon, desertification has gone through two main phases. In the first phase, the United Nations Conference on Desertification (UNCOD), was convened in Nairobi in 1977 in the wake of the Sahel drought of the early 1970s, and gave rise to the nonbinding PACD or Plan of Action to Combat Desertification, coordinated by the United Nations Environment Program (UNEP). In the second phase, the PACD, which was supposed to run until 2000 but was widely perceived to be ineffective, was supplanted at a second UN conference, the Conference on Environment and Development (UNCED) or 'Earth Summit' held in Rio de Janeiro in 1992. The Rio conference set in motion negotiations to create a binding international convention on desertification, which eventually came into force in 1994 and continues to operate, the UNCCD or the Convention to Combat Desertification (Toulmin and Brock, Chap. 2).

In neither of these two time periods—from 1977–92 or from 1992 to the present—has the relationship between science and the international desertification bureaucracy been close or productive. In the initial phase under PACD, UNEP essentially ran an advocacy campaign that exaggerated the extent, severity and threat posed by desertification, making the official version of desertification a target for informed scientific criticism (Thomas and Middleton 1994; Helldén 1991; Tucker et al. 1991):

> UNEP limited the scope of knowledge communicated to policy makers through its exclusive choice of experts and having final say on how knowledge was presented….-favouring the policy domain over the science domain….The casualty of this trade-off was the credibility of scientific knowledge (Grainger 2009: 417).

As a result, "when the United Nations finally created the Convention to Combat Desertification (UNCCD) in 1994, policy was seriously disconnected from science" (Toulmin and Brock, Chap. 2, citing Andersson et al. 2011: 306).

If scientists require clarity in the concepts they employ, the politicians and administrators who create and manage large institutions have other, very pragmatic requirements. In the search for money and support, they need a problem that is dramatic enough to command immediate attention, simple enough to be quickly grasped, and general enough to satisfy diverse interest groups; they need what

Jeremy Swift has called a development narrative—a powerful story line with clear, broadly applicable policy implications and urgent funding needs (Swift 1996).

In Chap. 2 Toulmin and Brock explain how such a crisis narrative for the Sahel was assembled and subsequently challenged. The mainstream interpretation of desertification in the UN era initially replicated earlier colonial concerns about the rapid expansion of the Sahara as a result of native agricultural practices (Stebbing 1935, 1937). These predictions of environmental disaster were sufficiently apocalyptic to mobilize money, and the proposed solutions were technical enough to avoid the appearance of political involvement by international or bilateral development agencies in the affairs of newly independent African states. By the early 1990s scientific critiques were effective in blunting some of the more egregious elements of the crisis narrative, such as inflated estimates of the geographical expansion of the Sahara or the confounding of drought with permanent land degradation (Tucker et al. 1991; Helldén 1991; Olsson 1993). Following the Earth Summit in 1992, the definition of desertification adopted by the UNCCD reflected these adjustments. Dropping the desert metaphor and the insistence on either human causation or permanence, the official definition was sufficiently vague to encompass all kinds of environmental decline—including changes that did not mimic the onset of desert conditions or were only transitory—by almost any imaginable cause—natural or human:

> Desertification is land degradation in arid, semi-arid and sub-humid areas resulting from various factors, including climatic variations and human activities. (United Nations Convention to Combat Desertification, Bonn, Germany: Convention text).

In the decades since ratification, the broad UNCCD definition of desertification has permitted the perceived purpose of the Convention to evolve in step with changing policy concerns and shifts in development theory. In the 1990s official thinking on desertification began to reflect the fashion for participatory or bottom-up development, and to distance itself (though not entirely consistently, as discussed by Toulmin and Brock in Chap. 2) from the older rhetoric of coercion and centralization (Stiles 1995). Post-2000, desertification has been depicted as a potential security threat, reflecting the emergence of terrorism, state failure and forced migration as international policy concerns (Benjaminsen, Chap. 4). With the increasing prominence of climate change, desertification has recently been reframed as one part of land degradation 'in any climatic zone—not just in drylands' and linked to wider issues of resilience and adaptation in response to global warming (Reed and Stringer 2015: 27). Finally, as the only legally binding international agreement with 'a dual focus on environmental and developmental concerns', the UNCCD has also been depicted as a useful mechanism for promoting pro-poor economic growth (Middleton et al. 2011: 2).

In sum, the institutionalization of desertification within the UN system has fostered the conviction that the concept must be relevant to something important. Unconstrained by a precise definition of the phenomenon or by a credible system of scientific oversight (Grainger 2009) and driven by competition for international funding (Toulmin and Brock, Chap. 2), what this relevance might be has been open to multiple interpretations.

1.1.4 Learning from the Sahel

Part I of this book provides an opportunity to reassess the utility of the concept of desertification from the perspective of contemporary research on the Sahel. Because it acted as the catalyst for the modern 'Age of Desertification' (Davis, Chap. 8), the Sahel has been subjected to unusually high levels of scientific scrutiny for a prolonged period of time. The focus of our argument is not, however, strictly on the Sahel itself but on learning from a protracted period of interaction between science and environmental policy, in order to make sense of what is happening now in parts of the world that may not be as intensively studied. It took science more than three decades of work on the Sahel to transform a perceived desertification crisis into a non-event. This book is an effort to critically examine that experience and accelerate the learning process.

Michael Mortimore (Chap. 3) provides an overview of what we now know about the state of the Sahel environment. He argues that there is little evidence across a broad range of indicators—forest cover, rangeland condition, soil fertility or overall biological productivity—for an irreversible decline in the Sahelian environment since the middle of the 20th century. There is, on the other hand, ample evidence for high and occasionally destructive levels of environmental variability, a reality that residents of the Sahel have adapted to with remarkable success despite poverty, political marginality and rapid population growth.

A reappraisal of the relationship between Sahelian society and its environment is therefore required, what Mortimore characterizes as a shift from a desertification to a resilience paradigm. In the desertification paradigm, humans cause irreversible degradation by disturbing the inherent stability of their natural environment. In response, technical solutions are externally imposed on local communities, which are perceived to be the source of the problem, and biophysical variables are used to measure success or failure. Alternatively, in the resilience paradigm, local resource users adapt to a variable environment that they cannot control, and their responses provide temporary solutions to a shifting set of problems and opportunities. Mortimore closes with a plea for a decentralized approach to development that respects the ingenuity and effectiveness of these responses and seeks to build on them.

Yamba Boubacar (Chap. 7) provides a concrete example of how Niger adopted and then implemented just such a decentralized approach. Innovation in Niger's case was a response to multiple crises—an extended drought (1972–84) that undermined the livelihoods and destroyed the working capital of rural households, accompanied by government indebtedness and currency devaluation. In response, in the 1980s government abandoned a centralized forestry strategy that was modelled on colonial policy and 'switched from protecting forest reserves to the overall management and use of plant resources'. The technical centrepiece of this effort was a new method for clearing farmers' fields that encouraged the retention and natural regeneration of indigenous trees. Equally essential, however, were political changes at the national level, new land laws that recognized the property rights of rural people, and programmes to enhance the capacity of local institutions to manage natural resources.

Instead of a purely environmental crisis, the real challenge is to manage linked processes of political, social and environmental change. Chapters 5 and 6 provide examples of the methodological care with which current research addresses this challenge in the Sahel.

Stefanie Herrmann and Tene Sop (Chap. 5) document trends in Sahelian woody vegetation from multiple perspectives: satellite data, ethnobotanical studies, historical vegetation surveys, and the analysis of the size class distributions of current tree populations. While recent remote sensing studies suggest a regreening of the Sahel and a general increase in bio-productivity, the field studies reviewed in this chapter document the opposite trend for woody vegetation, including the loss of large trees, trees that require wetter conditions, or species that are economically valuable to local inhabitants. In contrast to the general trend, the good news about trees comes from farmers' fields where they are more intensively managed and have become more dense and diverse than before the great droughts, a consequence of the political processes detailed by Boubacar in Chap. 7:

> As farmer-managed natural regeneration has shown, human population growth and agricultural intensification do not have to entail losses in vegetation productivity and environmental degradation …. On the contrary, where sound management is practised, farmers' fields stand out by improved vegetation cover and diversity, while unprotected woodlands and unmanaged fields continue to degrade (Chap. 5).

Pierre Hiernaux, Cecile Dardel, Laurent Kergoat and Eric Mougin (Chap. 6) present an equally complex picture based on decades of ecosystem monitoring at two study sites, a sparsely populated pastoral area in Mali and a more densely settled agro-pastoral region in Niger. In the more arid pastoral study site, vegetation dynamics were erratic, complex, and unpredictable—largely driven by fluctuations in rainfall and unresponsive to the impact of grazing. At the more humid agro-pastoral site, the expansion of cropland and high concentrations of livestock increased the spatial heterogeneity of the ecosystem, pushed rangeland resilience to its limits, and left a long-lasting imprint on the landscape, an outcome that some observers might interpret as 'degradation'. Long-term monitoring in specific locations—Mali versus Niger, pastoral versus agro-pastoral areas—demonstrates the inadequacy of sweeping generalizations about the Sahel; ecological change is as varied and locally specific as the heterogeneous social and physical environments in which it takes place.

Tor Benjaminsen (Chap. 4) reveals, nonetheless, a continuing appetite in some quarters for broad generalizations—in this instance an alleged connection between increased conflict levels and a scarcity of natural resources. Benjaminsen argues that lack of evidence for widespread desertification in the Sahel makes it an unlikely cause of conflict. In two case studies from Mali, he demonstrates that there was also no correlation between the timing of episodes of conflict and drought periods, making it unlikely that drought directly provoked increased levels of conflict. There were, on the other hand, compelling alternative political and historical explanations: state-sponsored agricultural encroachment on pastoral grazing resources, rent-seeking by government officials, the embezzlement of emergency aid, the temporary retreat of state control over rural areas, and the perceived para-military suppression of pastoral regions.

Stepping back from the individual chapters, a common message emerges from Part I. In considerable measure, science grows by making mistakes, acknowledging them, and moving on. With respect to the science on desertification in the Sahel, this process of rejection, invention and rejuvenation has been complicated because desertification is not simply a scientific concept. It also provides a basis for public policy and for the funding of global institutions, and concepts that are useful for this purpose often do not provide a convenient platform for the conduct of impartial science. Publicists, administrators and politicians active in the public arena thrive on crises and on unequivocal prescriptions for addressing crises. Careful science tends to deflate and moderate this rhetoric: Crisis degrades into evolution, environmental collapse becomes a problematic process of environmental change, local resource users are transformed from environmental villains into potential collaborators in conservation and development. There would seem to be an inherent tension, well illustrated by the history of ideas about desertification in the Sahel, between science that is flamboyant enough to be politically influential and yet cautious and complex enough to withstand scrutiny. Walking this tightrope is a recurrent challenge for research conducted in the service of public policy.

1.2 Part II: Challenging the Desertification Model

Several papers in this collection raise new questions about the continued relevance of the concept of desertification, both for the Sahel and for drylands more generally.

1.2.1 If You Can't Define It You Can't Measure It, and if You Can't Measure It You Can't Map It

Diana Davis (Chap. 8) and Stephen Prince (Chap. 9) address the same question from very different perspectives—'What is desertification?' Davis answers this question historically; Prince argues that the concept remains vague and cannot be defined operationally. Taken in combination, these chapters suggest that, despite a plethora of definitions and decades of discussion, we still do not know—and may never know—exactly what we mean by desertification.

The concept of desertification is not a 20th century invention; it was a product of early European scientific and prescientific thinking that survived virtually unchanged into the 20th century. Desertification is a variant of desiccation theory, 'the idea that deforestation causes the climate to dry out and diminishes rainfall', a theory that was widely accepted in European scientific and philosophical circles in the 18th and 19th centuries (Chap. 8). It was the French in their North African colonies who first applied desiccation theory to semi-arid environments and invented the concept and term 'desertification'—the creation or expansion of deserts by people. In North Africa the French colonial authorities invoked desertification to explain the impoverished condition of the natives in what appeared to be a degraded landscape. Avoidable environmental mismanagement by the natives helped rationalize the need for French imperial rule, for the reallocation of land to European settlers, and for the criminalization of native agricultural practices (Davis 2007). From their North African possessions, French administrators and scientists carried the desertification concept across the Sahara into West Africa. Chapter 8 traces these historical linkages, step by step, showing how colonial experts eventually provided the basis in the 1950s for UNESCO's Arid Zone Program, which in turn provided the background for the UN Conference on Desertification convened in 1977, and ultimately for the UNCCD, ratified in 1996.

As Davis makes clear in Chap. 8, the mid-20th century image or metaphor of desertification was perfectly clear, widely shared, and centuries old: It referred to what Bovill (1921) called the encroaching Sahara, a human-created desert where a desert did not naturally occur, 'a living environment becoming irreversibly sterile and barren' (Nicholson 2002: 51). The problem with this metaphor was that it had become scientifically indefensible by the early 1990s (Toulmin and Brock, Chap. 2). Driven by political expediency, the response to this impasse was not to abandon the concept as untenable, but to expand it. The negotiations that created the UNCCD were dominated by differences between developed and developing countries concerning the purpose of the new Convention (Corell 1999). Political support for the Convention 'rested on the diplomatic artifice of ambiguity' predicated on 'a consensus definition of the problem agreed in negotiations at UNCCD …[that] formalized the ambiguous roles of environmental degradation and drought [in the desertification concept]' (Grainger 2009: 419). If artful compromise trumped scientific clarity in the construction of the Convention, it continued to do so with its implementation. Writing in 2009, Grainger's estimation of the chances for improving scientific inputs into the official desertification regime was poor: 'Social order between developed and developing country Parties [is maintained] by means of [a] … combination of narratives compatible with both sets of discourses….Better scientific inputs could puncture the ambiguity on which the political viability of the CCD depends' (Grainger 2009: 425). As late as 2012, the UNCCD remained the only 'Rio Convention' without a functioning or proposed independent intergovernmental scientific body 'to guide policymakers' considerations of scientific evidence that is credible, relevant and legitimate' (Thomas et al. 2012: 123).

In Chap. 9 Stephen Prince documents the lingering scientific consequences of this political process: our inability to quantify and map either the global extent or severity of desertification. In the mid-1990s desertification was characterized as a 'blanket term for a whole range of specific biological, chemical and physical changes in the environment' whose 'breadth rendered its actual use impractical, for example in terms of attempts to quantify desertification' (Thomas and Middleton 1994: 9). In Chap. 9 Prince argues that little has changed: Desertification remains 'a nebulous, all-encompassing concept' that 'is not a single phenomenon … and is therefore incapable of simple measurement' (Chap. 9). As a result, existing global maps of desertification are often based on subjective assessments by experts and cannot be replicated or used for monitoring. Moreover, there is no general agreement between different global map sets and there are no strong correlations between the levels of desertification depicted in the maps and the environmental variables that are generally thought to cause degradation: 'Consequently, unsupported statements about the extent and severity of desertification abound,' (Chap. 9). Prince concludes that 'Progress will be difficult unless the unitary concept of desertification is abandoned' in favour of measuring aspects of land degradation—changes in vegetative biomass, the responsiveness of the vegetation to rainfall, or rates of soil erosion, for example—that are each susceptible to quantification (Chap. 9). These scientific advances are, however, dependent on institutional reform, including 'independent organizations at national, regional and global scales to undertake routine monitoring' and 'technically credible leadership, free from external pressures … to address the issues of degradation at international-global scales and to harmonize policy-relevant research, monitoring and interpretation' (Prince, Chap. 9). As ever, desertification research remains enmeshed in global politics.

1.2.2 From Regional Desertification to Global Climate Change

Given the centuries of desiccation theory that preceded the 'Age of Desertification,' it may have been 'intuitive,' as Alessandra Gianinni notes in Chap. 10, for researchers in the 1970s to look to regional land use to explain regional climate change. At that time the attractiveness of a desiccationist interpretation was also reinforced by the emergence of a new mechanism that purported to explain how desiccation worked: the albedo effect. According to this theory, by reducing vegetation cover, desertification altered the reflectivity of the earth's surface and caused drought, which further reduced the vegetation cover, causing more drought—a degenerative biophysical feedback process (Charney 1975; Charney et al. 1975). At the bottom of this spiral of causation were the usual suspects—expanding numbers of local land users who were denuding the landscape—which set the stage for the standard remedies—population control, afforestation (Xue and Shukla 1996), and the reduction of livestock numbers (Otterman 1974). Desiccation theory had, in its essential features, been updated for the 20th century.

As Gianinni observes in Chap. 10, since the early work of Charney and his associates, regional models of land-atmosphere interactions have become increasingly sophisticated with the inclusion of more variables (such as soil moisture) and a more realistic depiction of these variables (Taylor et al. 2002). As a result, 'It has become increasingly clear that the processes by which anthropogenic land use change can reduce precipitation are physically plausible, and consistent with a natural positive feedback response of the land surface to precipitation' (Gianinni, Chap. 10). What the regional land-surface models could never explain, however, was the magnitude of the Sahel droughts, which were simply too big to be caused by observable levels of land use change in the region. A resolution to this impasse only occurred when climatologists turned to simulating the timing rather than the amplitude of rainfall fluctuations, and did so by looking at the Sahel climate from a planetary rather than a regional perspective.

As Gianinni explains, over the last century in the Sahel, alternating periods of high and low rainfall—such as the wet 1950s and 1960s followed by persistently dry conditions at the end of the 1960s, or the regreening currently taking place—are all correlated with global variations in sea surface temperatures. Dry periods in the Sahel are the consequence of elevated sea surface temperatures in the tropical oceans combined with cooler conditions in the north Atlantic (Palmer 1986; Folland et al. 1986). In the case of the great Sahel droughts at the end of the 20th century, the warming of the equatorial Indian Ocean and south Atlantic was caused by greenhouse gas emissions, while the cooling of the northern oceans was, aside from natural variation, caused by sulphate aerosol pollutants from the northern hemisphere (Gianinni, Chap. 10). Air pollution from greenhouse gasses continues, and so does the warming of the world's oceans; pollution from sulphate aerosols increased until the mid-1980s and has declined since the introduction of legislation aimed at controlling acid rain in North America and Europe. Both the great Sahel drought and current regreening trends reflect these pollution-driven shifts in oceanic temperature:

> As warming of the tropical oceans began to emerge in the 1970s, the reduced warming of the North Atlantic ... starved the African continent of the moisture needed to trigger deep convection, causing persistent drought. Now that the North Atlantic has reversed its trend and is warming, a wetter Sahel has become possible, as manifest in a "partial recovery" of the rains since the mid-1990s (Gianinni, Chap. 10).

Droughts in the Sahel are primarily caused by oceanic temperatures, but may be suppressed or amplified by land use changes in the Sahel itself. This conclusion shifts the primary blame for the great 20th century Sahel droughts from local to global anthropogenic influence:

> In response to our best estimates of global sea surface temperature, state-of-the-art atmospheric models require no information on human-induced land cover/land use change to reproduce Sahelian drought. These simulations lead to the conclusion that drought was caused by large-scale, if subtle, shifts in oceanic temperatures, not by local anthropogenic pressure on the environment (Gianinni, Chap. 10).

1.2.3 *Beware the Charismatic Story*

Managed grazing systems—i.e., pastoralism broadly conceived—occupy about a quarter of the earth's land surface and are the most widespread form of land use on the planet, and dryland biomes—savannahs, grasslands, shrublands and deserts—contribute over three quarters of the world's total grazing area (Asner et al. 2004). Given the global importance of drylands and of grazing in drylands, if the concept of desertification is to have any utility, it must contribute to a better understanding of the impact of extensive livestock production on these environments. Reflecting this imperative, seven chapters in this book—Chaps. 6, 11–18—focus on rangelands and various forms of pastoralism and livestock farming.

In Chap. 11 Lynn Huntsinger re-examines the 'founding narrative' of the rangeland management profession in America—the widely accepted story of how human greed and uncontrolled resource exploitation degraded the 'pristine' rangeland ecosystems of the American West. Huntsinger argues that this 'declensionist' reading of Western history as a process of decline has been reinforced by an equally declensionist interpretation of Western range ecology, and that both are flawed.

Huntsinger limits her social and biological critique to a re-examination of the American experience, but her analysis has wider implications. In the 1970s when desertification was institutionalized globally, classical American rangeland management of the sort discussed by Huntsinger was influential far beyond the borders of the United States. In its classical form, the discipline was predicated on Clementsian theories of plant succession leading to climax vegetation, concepts of rangeland condition that identified good and bad plants, a paramilitary commitment to fire suppression, and an authoritarian approach to resource management. If these ideas have now been found inappropriate in their homeland, as Huntsinger suggests is the case, then a reappraisal of their global significance is overdue.

Effective story telling has been part of the problem. Stories, Huntsinger argues, are dangerous because they are such a powerful form of communication. The theories of rangeland ecology that dominated North American and much of the rest of the world for most of the 20th century were essentially stories—powerful simplifying narratives with moral implications. These stories were persuasive because they contained a considerable element of truth, but they were fundamentally misleading because their clear story lines excluded the complications, qualifications and exceptions that litter more balanced but less exciting accounts of reality. By the late 1980s, however, complications, qualifications and exceptions finally overwhelmed the simple Clementsian succession story, resulting in a more accurate and effective model of vegetation change—the states and transition format based on nonequilibrium theory. Alas, as Huntsinger emphasizes, this was an explanatory framework without a plot:

> Unfortunately, the "multiple stable states" model of vegetation change does not have the same simple, moralistic and appealing story as linear succession—there really isn't a beginning, end, or moral lesson. There is a site, it rains and things change or it doesn't rain

> and things change. Changes may be permanent. Rainfall is unpredictable and not influenced by human actors. We need to watch, experiment, and record to learn what is going on.
>
> The role of grazing in this scenario is one of many interacting dramatic changes, and changes without directionality. Removal of grazing may have little or no impact, and threshold dynamics may result in multiple stable states that have little to no relation to a "climax" or previously identified "potential" vegetation. There is no simple story here.

The state and transition format may be scientifically robust but it creates, in Huntsinger's words, a 'narrative vacuum,' a natural world populated by plants that lack a 'moral compass', plants without a human story to tell. In North America, Huntsinger argues that commercial interests purveying a mixture of story-telling and pseudo-science have filled the narrative vacuum created by the retreat of the Clementsian consensus.

If science is to regain the initiative and influence public policy, it must provide an interpretation of rangeland dynamics that is empirically adequate and yet simple enough to engage non-scientists. Drawing on research that provides the background to the present book, the following section argues that range science is moving towards such a synthesis.

1.2.4 Contemporary Range Science and the Desertification Paradigm

The following discussion examines the intellectual equipment—the concepts, theories and some of the field evidence—that can now be brought to bear on questions about the impact of extensive livestock production on rangeland environments. There are at least three practical questions that a non-specialist policy maker would want answered about the relationship between livestock and desertification:

- When livestock change vegetation, what kinds of changes should be characterized as 'degradation' or desertification?
- Under what circumstances is livestock grazing likely to cause changes in rangeland vegetation?
- How can rangelands be managed to limit degradation?

Contemporary range science provides answers to these questions that are different from those that prevailed in the 1970s and 1980s when desertification was first institutionalized in the UN system. In part these differences reflect scientific progress. Equally important, however, are the changes that have happened on the rangelands in the succeeding decades, particularly to pastoralists in developing areas. Subsistence-oriented livestock producers have been incorporated into commercial markets, increasingly adopting husbandry practices dependent upon purchased inputs and orientated to the production of commodities for sale. In some areas traditional migratory systems of livestock production are also in decline, for a variety of reasons including conversion of key pastoral resource areas to other forms of land use, the fragmentation and privatization of communal lands, the importance of markets and services that anchor families to settlements, and the availability of new technologies—from water development, to feed supplementation or disease prevention—that render some kinds of movement unnecessary. These developments have blurred one of the standard dichotomies of desertification orthodoxy—the contrast between the primitive nomad and the modern, sedentary livestock producer. Taken in combination with advances in our understanding of rangeland ecology, these changes suggest new answers to the practical resource management questions enumerated above, answers that fundamentally challenge the concept of desertification.

The following three sections of this introduction outline some contemporary answers to our three management questions. We focus on grazing and rangeland issues because they have been central to the desertification debate since its inception, and clearly illustrate the relationship between science and environmental policy. It is important to note, however, that livelihood systems based on cropping or on mixed crops and livestock, routinely support larger human and animal populations than specialized livestock production and deserve consideration in their own right. With increasing market integration, dryland farming, like pastoralism, is also undergoing a transition to different and potentially higher levels of accepted risk. The land changes associated with these processes have been characterized as desertification in reports and publications targeted at policy makers and general readers. Equally, however, some recent policy-oriented analyses simply forgo the use of the term desertification, preferring to employ less prescriptive frameworks such as soil fertility management, resilience and agricultural intensification. While not reviewed here, these trends are fully discussed in a number of recent UN publications and development reports (Mortimore et al. 2009; UN Environment Management Group 2011; UNCCD/UNDP 2011; Davies et al. 2012).

1.2.4.1 When Is Vegetation Change Degradation?

'Desertification' connotes harmful environmental change, which implies an ability to distinguish between harmful and beneficial changes. In the 1970s, late succession grasses, especially perennials, represented both good livestock feed and the natural climax vegetation, and loss of such plants was routinely construed as evidence of degradation. Contemporary research suggests a more complicated situation in which the history of grazing in a region has an important influence on degradation processes (Milchunas et al. 1988). In areas that were not exposed to high levels of ungulate grazing prior to the introduction of livestock, such as parts of the south-western United States or Australia, degradation is often an unambiguous progression in which the later stages of environmental disruption become ever more expensive or time-consuming to reverse, or are irreversible (Schlesinger et al. 1990; Ludwig et al. 2004). But the situation is not always this clear-cut.

In much of Eurasia, Africa and North America, grazing by large hooved mammals—either domestic or wild—has contributed to creating the existing vegetation, and it would be misleading to characterize grazing as an unnatural disturbance. In these circumstances, the attributes that might be used to identify rangeland condition—particular kinds of plants, high levels of plant productivity, livestock productivity or biodiversity—are not consistently linked to one another, or to grazing intensity. Three case studies from regions with a long evolutionary history of grazing—the rangelands of South Africa, the North American prairie, and the Mediterranean Basin—illustrate the complexities of environmental change in these environments.

In South Africa, a nation-wide study concluded that heavy grazing altered the composition of plant communities by increasing the proportion of annuals and favouring opportunistic, grazing-resistant species with low palatability and forage quality—weedy, grazing-induced degradation from the perspective of climax vegetation. Despite these changes in species composition, at none of the very heavily utilized communal rangeland sites was there a decline in species richness—degradation defined in terms of the loss of plant biodiversity (Rutherford and Powrie 2013).

The maintenance of an imagined 'climax' vegetation may also conflict with wildlife biodiversity. In the United States the most rapidly declining kinds of birds are those associated with rangelands, and these declines coincided with nationwide improvements in rangeland condition and health, as defined by Clementsian succession theory (Fuhlendorf et al. 2012). Many of the bird species that suffered were those that required disturbed ground and low vegetation structure caused by heavy grazing or fire. In terms of the habitat requirements of North American semi-arid bird populations, there was no single livestock stocking rate or type of vegetation that met the needs of all species. Instead what was beneficial was a diversity of different habitats—a high level of local, botanical variability caused by different levels of grazing intensity (Fuhlendorf et al. 2006, 2012; Fuhlendorf and Engle 2001).

In a wide-ranging review, Seligman and Perevolotsky concluded that 'grazing by domestic ruminants is seldom irreversibly destructive to landscape values' for winter rainfall regimes of the Mediterranean Basin (Seligman and Perevolotsky 1994: 93–4; Perevolotsky and Seligman 1998). Animal output per unit area was also higher under heavy grazing and primary production was not reduced (Crespo 1985; Gutman et al. 1990). As in South Africa, the herbaceous plant community formed under very intensive grazing was no less diverse or rich in species than less disturbed vegetation, but the composition was substantially altered (Hadar et al. 1999).

The preceding cases illustrate the difficulties of identifying botanical indicators of degradation in environments long exposed to grazing. In South Africa and the Mediterranean Basin heavy grazing produced changes in the composition of pastures as grazing-tolerant plants replaced ones that were grazing intolerant, but with no loss in diversity or species richness, and, at least in the Mediterranean, with no loss of livestock productivity. In North America, intense but localized grazing that would traditionally constitute overgrazing was essential to the survival of certain species of wildlife. Cases like this suggest that degradation cannot be identified with particular kinds of plants or assessed according to a single metric—be it biodiversity, primary or secondary production. A more realistic measure of environmental resilience may instead be the maintenance of diverse landscapes capable of sustaining different production systems, species and environmental services. Instead of asking how drylands can be managed to limit desertification, we might instead ask how they can be managed to maximize this heterogeneity.

1.2.4.2 Under What Circumstances Is Grazing Likely to Cause Vegetation Change?

The threat posed by spreading deserts or 'desert edge displacement' (Veron et al. 2006) was a central component of the early desertification literature (Lamprey 1988; Aubréville 1949), as was the perception that drylands were fragile environments especially prone to degradation.

The non-equilibrium theory of grazing systems dramatically reversed this assessment (Ellis and Swift 1988). Part of the resilience of semi-arid rangelands can be attributed to the characteristics of the plants themselves. Some semi-arid grasses are robust to grazing because the evolutionary adaptations that promote survival to drought also equip them to evade, resist or tolerate large herbivores (Coughenour 1985). More important, however, is the response of livestock populations to severe droughts, which occur frequently in regions of low rainfall. Following droughts, herd sizes recover slowly and the vegetation has an opportunity to rebound in the absence of significant grazing pressure (Ellis and Swift 1988).

In areas subject to low and erratic rainfall, recurrent livestock population crashes also suppress livestock numbers over the long term. As a result, in the absence of feed supplementation, herbivore biomass per unit of plant biomass—i.e., the herbivore load relative to primary production—increases exponentially with average annual rainfall (Oesterheld et al. 1998; McNaughton et al. 1989; Fritz and Duncan 1994). Subjected to both more constant and increased grazing pressure, vegetation in high rainfall areas is more exposed than in low rainfall areas to the effects of grazing (Coppock, Chap. 12).

This conclusion is supported by a global meta-analysis that discovered no instances of widespread or zonal vegetation change—i.e. change on a spatial scale that might qualify as desertification—in non-equilibrium rangelands with sufficiently high levels of rainfall variability (von Wehrden et al. 2012). The same pattern emerges when grazing systems are examined along a rainfall gradient in a single region. As Hiernaux et al. explain in Chap. 6, grazing leaves little imprint on the vegetation in more arid areas of Mali but has a significant impact under more humid conditions in Niger.

Mongolia provides another well-researched example. In Mongolia, grazing pressure was of minor importance and weather conditions determined vegetation characteristics at the arid end of the rainfall gradient, but differences in grazing pressure had an increasingly marked effect on vegetation under more humid conditions (Fernandez-Gimenez and Allen-Diaz 1999, 2001; Zemmrich et al. 2010), and degradation became more of a concern as the level and reliability of rainfall increased (Wesche et al. 2010; Stumpp et al. 2005; Khishigbayar et al. 2015).

Similar results emerged from a long-term study in Senegal that substituted a temporal for a spatial rainfall continuum (Miehe et al. 2010). In an area that had experienced shifts in weather conditions, fluctuations in precipitation drove vegetation change and masked the effects of grazing during a run of dry years with variable rainfall. However, in a wetter phase when rainfall variability had declined, grazing caused changes in the species composition of the vegetation that might be construed as degradation (Miehe et al. 2010). Contrary to the anxieties about 'desertification' that typically emerge in times of drought, in Senegal the risk of grazing-induced vegetation change coincided with periods of higher rainfall.

There are exceptions to the association between aridity and reduced grazing impacts. Some of these exceptions can be attributed to the susceptibility of certain types of vegetation to grazing (Oliva et al. Chap. 13; Todd and Hoffman 2009; Hacker et al. 2006). Problems also occur when mobility is too successful at buffering migratory livestock populations from feed shortages. This possibility arises when mobile herds have access to large areas of grazing that are insensitive to heavy use during periods of feed scarcity—extensive marshes or floodplains, for example. The reliable food supplies available in these areas can mitigate the effects of an erratic climate and sustain livestock populations that are large enough to damage vegetation that is vulnerable to grazing in adjacent areas (Illius and O'Connor 1999).

By far the most common exceptions to the nonequilibrium model are caused by differences in the way livestock are managed (Briske et al. 2003; O'Connor 1995). With access to feed supplementation, water development and markets, commercial ranchers have an enhanced ability to suppress fluctuations in livestock populations caused by variable weather conditions, removing the periods of low grazing pressure that buffer arid rangelands from the impact of livestock. Under these conditions, episodic fluctuations in rainfall may produce the most visible vegetation responses, 'but these changes are unlikely to be directional provided there is no long-term rainfall trend. In contrast, the response of a species to heavy grazing is … generally unidirectional … and although year-by-year changes due to grazing are small, they are cumulative (O'Connor 1995: 59). While the immediate effects of grazing may not be dramatic, managers with sufficient resources at their disposal can override the climatic constraints that would otherwise protect arid rangelands from livestock-induced degradation (Ellis and Lee 2003).

Most livestock owners would like to reduce their exposure to climatic variability and there is increasing evidence that producers in developing areas are now acquiring industrial inputs that make this possible. The most important of these inputs are purchased or home-grown feed supplements, the construction of artificial water points, and enhanced disease prevention—all of which suppress livestock population crashes and increase livestock productivity. Unless these interventions are combined with management practices that ensure the periodic resting of rangelands, they also increase the risk of long-term resource degradation (Vetter and Bond 2012; Vetter 2005; Hary et al. 1996; Miehe et al. 2010; Müller et al. 2007; Li et al. Chap. 18).

Arid rangelands are not the fragile, degradation-prone environments of the orthodox desertification paradigm, but they are currently under stress, not from the nomadic menace of the older narrative, but from intensification and commercialization processes that are at once economically attractive and environmentally challenging.

1.2.4.3 New Approaches to Managing Extensive Grazing Systems

Since the 18th century, proponents of desertification have identified nomadism with environmental destruction (Davis, Chap. 8). By the 20th century, the standard antidote to nomadism, modelled on commercial ranching in developed countries and promoted by numerous donor-financed livestock development projects, was settled ranching and rotational grazing.

Designed to move livestock on a predetermined schedule through a sequence of fenced paddocks, rotational grazing schemes were promoted as the modern, orderly, and scientifically endorsed alternative to the presumed randomness and environmental destructiveness of migratory movement. Despite the lingering popularity of rotational systems among some commercial ranchers and rangeland professionals, research has repeatedly demonstrated that fenced systems of grazing rest and rotation do not reliably increase plant or livestock production in arid and semi-arid rangelands (Heady 1961; O'Reagain and Turner 1992; Briske et al. 2008). As scientific confidence in rotation grazing systems has eroded, there has been a steady advance in the understanding of the ecological processes that underpin wild animal migrations, which sustain some of the greatest concentrations of animal biomass on earth. Migration also has a measurable impact on the productivity and viability of free-ranging animal populations, both domesticated and wild, and on the state of the resources that they use (Milner-Gulland et al. 2011).

For our understanding of desertification, the significance of these findings lies in a revaluation of the implications of rangeland heterogeneity. By forcing livestock on a predetermined schedule through a sequence of relatively small fenced areas, often at high densities and for short periods of time, rotational systems attempted to suppress selective grazing and distribute animals evenly across a landscape, and made little attempt to exploit the temporal and spatial variability in forage quality and quantity that is an inherent feature of many dryland environments (Fynn 2012). Migratory systems—both in developing (Behnke et al. 2011; Butt 2010; Butt et al. 2009; Moritz et al. 2013) and industrial settings (Huntsinger et al. 2010; Huband et al. 2010; Baena and Casas 2010; Abreu et al. 2010)—take a different approach, seeking to exploit rather than suppress rangeland heterogeneity. These systems encourage livestock to track the best grazing areas available on a seasonal basis, to respond to unpredictable episodic variations in the location of favourable and unfavourable areas, and to exploit the distinctive features of individual localities. Benefits derived from this kind of movement include higher growth rates for mobile herbivore populations (Wang et al. 2006), the creation of grazing-induced lawns of highly nutritious forage (Hempson et al. 2014); mitigation of the effects of drought

(McAllister et al. 2006); higher sustainable stocking rates (Boone and Hobbs 2004); and facilitation of the co-existence of wildlife and livestock (Western et al. 2009).

Different criteria are also relevant for identifying harmful vegetation changes in migratory systems. The grazing gradients that occur around water points and settlements have for decades been central to the argument that pastoralism is a major cause of desertification (Sinclair and Fryxell 1985). Herbaceous plants in the heavily used 'sacrifice zones' adjacent to water points or settlements tend to be short, quick-growing annual species well adapted to defoliation and trampling, but providing small quantities of forage—unattractive grazing weeds in terms of a traditional Clementsian classification. However, these heavily grazed plant communities often provide forage of high nutritional quality (García et al. 2014; Vetter and Bond 2012), albeit for short periods of time (Anderson et al. 2010; Anderson and Hoffman 2007). While the ephemeral nature of these feed sources may render them unsuitable for permanent use, their transitory productivity is well suited to mobile systems that exploit resources during periods of optimal productivity and then move on to new areas and other kinds of resources (Bollig and Schulte 1999; Thomas and Twyman 2005). Increased spatial scale facilitates these migratory systems but is not an absolute requirement. Recent work has extended the principle of heterogeneous resource exploitation to private properties in which movement takes place over shorter distances (Fynn 2012; Bailey and Brown 2011; Hempson et al. 2014).

Mobile livestock production is not suited to all natural environments or to the needs of all producers. In some regions mobility is not a realistic alternative because resources are evenly distributed in space or fluctuations in their productivity are inappropriately timed. In other settings, the pressures of population growth or the advantages of agricultural intensification may render mobility impossible or make it unattractive. At the extensive margins, however, there are and will remain large dryland areas suitable only to low-input livestock production. Contrary to the desertification paradigm, for these marginal drylands, the scientific rationale for the environmental and economic benefits of mobility is stronger than ever.

1.2.5 Summing up the Challenge to Desertification

Nearly a decade ago Veron, Paruelo and Oesterheld undertook an assessment of desertification (2006). Their review noted many of the difficulties highlighted in this book—exaggerated journalistic accounts of the extent of the desertification threat, a plethora of competing definitions, and flawed attempts to measure the extent or severity of the problem. Despite these shortcomings, their appraisal concluded on an optimistic note:

> There are no reasons to believe that desertification ecology faces harder challenges than other disciplines (e.g. the definition of invasive species in invasion ecology, or of endangered species in conservation ecology) (Veron et al. 2006: 760).

According to this passage, desertification studies are like any other scientific sub-discipline: Growing pains are inevitable and temporary setbacks are to be expected and overcome. We are sceptical of this conclusion and the chapters in this book explain why.

When it comes to understanding and controlling desertification, scientists have not been the only interested party. From its inception to the present, desertification research has been targeted at and deeply involved in the formulation of public policy. The idea of desertification that was broadly accepted until the late 20th century labelled a clearly defined process: desert encroachment caused by destructive land use practices and population growth in dryland areas. This vision was sufficiently apocalyptic to capture the interest of policy makers, and had the added attraction of justifying the imposition of imperial control and pandering to European conceptions of their own technical superiority (Davis, Chap. 8). By the late 1980s, however, this portrayal of desertification had become difficult to defend against contrary scientific evidence and was replaced by a UNCCD definition that equated desertification with dryland environmental decline, irrespective of its permanence, cause, or similarity to desert conditions (Toulmin and Brock, Chap. 2 and Mortimore, Chap. 3). This omnibus definition has been effective in holding the UNCCD together institutionally, but the vagueness that made it institutionally effective has also rendered it unworkable as a basis for the objective quantification and mapping of desertification on a global scale (Prince, Chap. 9).

It is also becoming increasingly difficult for observers to attribute dryland environmental change to regional land use alone, as the concept of desertification implies, or to distinguish confidently between the effects of local and global influences. Our evolving understanding of the Sahel drought of the 1970s and 1980s is indicative of this shift in perspective. Unbeknownst to observers at the time, these droughts were a manifestation of global climate change, so much so that the ability of current climate models to replicate this event 'now constitutes a litmus test of our confidence in these models' (Gianinni, Chap. 10). But the Sahel is not an isolated case. Recent changes in the rangeland vegetation of Mongolia, for example, have been attributed to excessive grazing pressure or to rising temperatures associated with climate change, or to an unknown combination of these factors (Khishigbayar et al. 2015). Similar uncertainty surrounds the extent to which bush encroachment is caused by elevated levels of CO_2, changing rainfall patterns associated with global warming or by land use variables, such as the intensity of grazing or the suppression of fire (Buitenwerf et al. 2012; Lohmann et al. 2012; O'Connor et al. 2014). From Patagonia to Central Asia, the case studies in this book frequently cite a new source of uncertainty: the impact of climate change at local or regional levels. The pervasiveness of global change calls into question the concept of desertification as a distinct, regional form of dryland degradation that can be understood or managed in isolation from changes that are now rapidly taking place on a planetary scale.

The concept of desertification also contradicts current understandings of the complexities of dryland degradation. The UNCCD definition of land degradation assumes that a decline in biodiversity is consistently associated with declines in the economic and biological productivity of the land:

> Land degradation is defined by the United Nations Convention to Combat Desertification as the 'reduction or loss of the biological or economic productivity *and* complexity' of terrestrial ecosystems (our emphasis). That no provision is made in this definition for 'productivity *or* complexity' reflects a common negation of the possibility of finding an inverse or part-inverse relationship between the two… (Rutherford and Powrie 2010: 692).

The assumptions that underpin the UNCCD definition of degradation make sense in a simple Clementsian framework in which climax vegetation is assumed to be uniformly good; they make little sense in a state and transition framework in which different measures of degradation such as biodiversity, primary and secondary productivity are free to vary independently of one another, and have been shown to actually do so in numerous field studies. Field research also questions the attempt to identify and maintain uniformly good environments, however 'good' might be defined. Heterogeneous landscapes sustain diverse plants, animal species and environmental services, and the attempt to create homogenous environments can have negative consequences both for conservation and for mobile production systems that exploit heterogeneity.

The problems of evaluating vegetation change in positive or negative terms are indicative of a broader ambiguity. This dilemma corresponds to what Andrew Warren has called the contextual nature of degradation: 'simple, universal systems of judging land degradation … for precise criteria, or criteria based on economic performance are in vain. The evaluation of land degradation cannot be reduced to nutrient budgets, soil depth, soil water holding capacity, to economics or to politics' (Warren 2002: 457). If 'degradation' is situationally dependent, then desertification studies cannot characterize the condition of the earth's dryland environments by applying a uniform system of evaluation. At global and regional scales, a more realistic goal is the construction of maps or data sets, now heavily dependent on remote sensing, that summarize what we know about different measures of change in semi-arid environments: estimates of biomass production or rain use efficiency, changes in land cover, biodiversity or plant species richness, erosion rates, soil carbon levels, etc. (Prince, Chap. 9). Evaluation of the positive or negative implications of these attributes will remain, as the case studies in in this book demonstrate, subject to field documentation and local (sometimes contested) interpretation.

1.3 Part III: Regional and Country Case Studies

The opening chapters of Part III continue the discussion of pastoralism and rangeland issues, but pose additional questions about the broader socio-economic context in which environmental change takes place.

Chapter 13 by Gabriel Oliva, Juan Gaitan and Daniela Ferrante illustrates the caution that is required when generalizing about the impact of livestock on the environment. The area examined in Chap. 13, the Patagonian highlands of Argentina, fits the definition of a low rainfall, highly variable nonequilibrium environment in which grazing might be expected to have a minimal effect on vegetation. In fact, unlike the more arid pastoral areas of the Sahel discussed in Chap. 6, the pastoral exploitation of Patagonia has left a lasting imprint on the environment. The rangelands of Patagonia were slow to collapse following the introduction of domestic sheep at very high stocking densities at the end of the 19th century, but once collapsed have been equally slow to recover their former condition:

> Some vegetation changes documented in Patagonia have resulted in desert-like conditions, and they have not shown clear signs of recovery thereafter, even with careful, conservative management of the land and in a unchanging or slightly positive rainfall scenario (Chap. 13).

The difficulties of managing rangelands 'that change slowly and more or less permanently at a time-scale that is difficult to perceive' (Chap. 13) are compounded by the alternative botanical and economic criteria that can be invoked to identify preferred management options. Officially recommended stocking rates do not quickly regenerate lost grazing capacity, but they are sufficient to prevent further change. In economic terms, these botanically sustainable rates maximize meat output but lead to declines in wool output, which is also an economically important product for local ranchers. Unfortunately, wool production is maximized by the much heavier stocking rates that caused the vegetation to change in the first place. Given their marginal economic situation and the unresponsiveness of the rangelands to destocking, Oliva et al. doubt the willingness and ability of smaller ranchers to adhere to recommended stocking levels.

Layne Coppock (Chap. 12) documents another instance of grazing-induced degradation, the Borana plateau of southern Ethiopia. For a pastoral area, the Borana plateau experiences relatively high levels of rainfall and low levels of inter-annual rainfall variability. In this relatively constant environment, cattle numbers have increased steadily over recent decades. Combined with the suppression of fire and reduced herd mobility, increased grazing pressure has caused erosion and major changes in the vegetation, including a shift from grasslands to shrublands, the spread of exotic plants, and the loss of preferred forage species—degradation as defined both by Coppock and by local pastoralists. Coppock identifies the root cause of this degradation as a rapidly expanding human population with few non-pastoral livelihood opportunities. As long as farming techniques and levels of on-farm investment remain unchanged, cultivation is possible only in favoured parts of the plateau. The Ethiopian economy also has a very limited capacity to gainfully employ rural immigrants from pastoral areas. In the absence of significant non-pastoral sources of income, more people need more livestock, which then make unsustainable demands on the environment—a classic instance of desertification driven by population pressure.

Writing about the southern Bolivian Andes, David Preston (Chap. 14) documents an optimistic alternative to the downward economic and environmental spiral described by Coppock. Despite huge differences in natural environment and social organization, the income diversification, commercial engagement and adaptation described by Preston for the Andes conforms to the resilience paradigm presented by Mortimore in Chap. 3 for the Sahel. In both settings, outmigration has helped households adjust to periods of environmentally induced stress, and provided the income and markets that have made agricultural innovation possible. In Bolivia, outmigration has also led to declines in rural human and livestock populations, relieving pressure on natural resources. As in the Sahel, government officials and development agencies perceived the region in terms of environmental degradation, based on exaggerated estimates of overgrazing and erosion, for which there was little evidence: 'Current soil and vegetation degradation processes observed are largely associated with climatic and geomorphological factors' (Chap. 14). While crops and livestock have certainly left an imprint on the Andean landscape, Preston's analysis suggests that this was a price worth paying, because farming and herding have helped rural households exploit changing economic opportunities and maintain a satisfactory quality of life.

Read in combination with one another, Chaps. 12–14 probe the structural as opposed to the immediate or proximate causes of environmental degradation. Local communities certainly have the capacity to overexploit their natural resources, but what equips some to adapt to changing circumstances and avoid negative outcomes, while others fail? The history of adaptation to drought in the Sahel (Mortimore, Chap. 3) suggests that rural households are adept at discovering pathways that minimize environmental damage and maximize their chances for survival, but the result is not under their control:

> The speed of such strategies, and their flexibility or competence for managing multiple stresses and opportunities, is determined by macroeconomic rather than local factors. At the regional scale, economic integration facilitates adaptive strategies in the remotest villages of the Sahel. It appears that it is economic development, not adaptive capacity, that may be too slow (Mortimore 2010: 137).

From this perspective, different outcomes in Patagonia, the Bolivian Andes and Ethiopia—relative success in resource conservation and economic welfare versus failure—may own less to obvious differences in local circumstances than to the wider economic situation: Bolivian agro-pastoralists have employment and emigration opportunities that are unavailable to Ethiopian Boran; in Patagonia economic incentives undermine resource conservation for poorer ranchers. As exemplified in Patagonia, the desertification paradigm does, on occasion, accurately identify the immediate causes of degradation—on the ground at the local level where the hoe or hoof meets soil and plant. It is less well equipped to address the broader political and economic factors that both motivate and circumscribe these material encounters.

Mark Mulligan, Sophia Burke and Andrew Ogilvie (Chap. 16) present an analysis of social and environmental change in the Mediterranean that has nonetheless succeeded in avoiding these limitations. In the 1990s, European concerns about desertification began, as is often the case, after a multi-decadal drought attracted media and political attention to a simplistic notion of large-scale desert-like degradation. As Mulligan and his co-authors note, desertification is an 'umbrella term'—make of it what you will. With a rich database at their disposal and long-term governmental support, European researchers have exploited the open-ended nature of the desertification concept to build an increasingly sophisticated picture of the interaction between Mediterranean environment and society. To do this, researchers have progressively expanded the kinds of causal processes that they have incorporated into computerized models of environmental change. As described in Chap. 16, these models initially portrayed environmental degradation in simple pressure-response terms—apply increasing pressure to a resource and the environmental response is a problem. A subsequent generation of models incorporated the possibility of negative feedback—people notice a developing problem and act to mitigate the pressures that are causing it—a pressure-response-feedback system. Finally, current modelling—characterized as 'tipping points and complexity'—depicts the compromises between human welfare and environmental conservation that arise in seeking to sustain 'ecosystem service provision whilst maintaining and increasing agricultural production' in a complex industrial economy. The outcome, aptly summarized in the title of Chap. 16, is 'much more than simply "desertification"'. Indeed, the positive contribution of the concept of desertification to this impressive research effort is difficult to identify, aside from the alarming misconceptions about desertification that were instrumental in generating political and financial support in the first place.

Sarah Robinson (Chap. 17) examines land degradation in five Central Asian republics—Kazakhstan, Kyrgyzstan, Tajikistan, Turkmenistan and Uzbekistan—from the late Soviet period to the present. With the exception of the disappearing Aral Sea—an unmitigated and well-documented environmental disaster—Robinson documents a high level of uncertainty regarding the state of the Central Asian environment. Irrespective of this uncertainty, among regional officials and scientists 'the belief that [land degradation] must be getting worse is almost universal' (Robinson, Chap. 17). Why? How does a consensus emerge from a flawed, partial and equivocal body of evidence?

Robinson links the pervasive belief in environmental decline to the institutional arrangements that currently fund environmental science in the region. Generously supported in the Soviet period, scientists must now seek external funding, both to supplement salaries and produce new research. While significant international funding is targeted at combatting land degradation, only a minute fraction of this money goes to support basic environmental research on the extent of the problem. The reasonable conclusion for national scientists to draw from this arrangement is that the donors already know what they want to hear: Degradation exists and we need to stop it.

> Overall, international funding has rarely resulted in improved quality or availability of those primary data which drive the allocation of funds. The ambiguity and complexity of the

evidence for different types of land degradation discussed here are not reflected in the grey literature associated with the dispersal of funds to combat degradation (Robinson, Chap. 17).

In Central Asia, the perception of widespread environmental degradation is what Warren and Agnew, following Thompson and Warburton (1985), have characterized as an institutionalized fact—defined as a fact 'that an institution wanted to believe, one that served its purposes' (Warren and Agnew 1988). In this instance in Central Asia, the 'fact' of degradation is sustained by the congruent interests of a combination of institutions, both national and international.

Strictly national institutions govern the creation of environmental perceptions in China, as described by Hong Jiang in Chap. 19. In China, one governmental organization, the State Forestry Administration (SFA) controls all aspects of desertification work—policy, research, control and monitoring—to the extent that, in Jiang's estimation, 'Desertification control is trapped in the institutional interest of the SFA'. Decades after the great Sahel droughts, the SFA programmes bear an uncanny resemblance to UNEP's earlier efforts to combat desertification in Africa. The main culprits are the same—local people who farm, overgraze and collect too much firewood—as is the top-down, technical response, dominated by officials and experts, while local cultural practices, particularly pastoralism, are disregarded. Even the names are the same: Great Green Walls now march, or are set to march, across both China and West Africa. Another unwelcome similarity is the disconnection between policy and science. Being a forestry organization, the SFA promotes trees. Trees are good, irrespective of their limited ability to survive in arid climates, or the demands they place on limited water supplies. As Jiang shows, there is robust scientific evidence against unrestrained tree planting, but this evidence has little impact on policy because of the dominant institutional position of the SFA. Mindful of the implications of speaking out, environmental scientists will say things in private that they hesitate to say in public.

The social organization of information is also the focus of Chap. 15, Mike Shanahan's analysis of the journalistic portrayal of pastoralists, one of the key protagonists in desertification theory. Shanahan describes how the media depicts pastoralism in Kenya, India and China, and in each country the message is different: 'In Kenya, pastoralists feature mostly in 'bad news' stories of conflict and drought… In China, the media presented pastoralists as the cause of environmental degradation and as (generally happy) beneficiaries of government investment and settlement projects. In India, newspapers portrayed pastoralists … as vulnerable people whose livelihoods were at risk.' (Shanahan, Chap. 15). In no country was there discussion of the environmental rationale behind pastoral mobility or of the economic benefits derived from pastoralism, and in all three countries the media image of pastoralism was self-reinforcing because journalists depended on other journalists as their most common source of information. In all three countries journalists expressed personal opinions at variance with their own public pronouncements, and in China these personal opinions were also at variance with state policies:

> [M]ost journalists (71.5 %) felt that herding did not cause damage to the environment, and more than two-thirds (67.8 %) even felt that herding had a positive effect on the environment. Among these journalists 71.4 % disagreed that herders need to settle instead of herding livestock… As one journalist commented: "Their lives are strongly impacted by the policy to make them settle down for reason of keeping stability, [and this] damages already-vulnerable ecology …." (Shanahan, Chap. 15).

It would appear that professional assessments, both scientific (Jiang, Chap. 19) and journalistic (Shanahan, Chap. 15) are more sceptical of Chinese environmental policy than the public record might suggest.

Wenjun Li, Yanbo Li and Gongbuzeren provide further evidence in Chap. 18 for the plurality of views in China on environmental issues. Their analysis examines the opinions of three groups of stakeholders—the government, academic evaluators, and pastoralists—on the state's rangeland degradation policies, which favour either the confinement or complete removal of livestock from many rangeland areas. Though they express their positions differently, both government and academic observers tend to endorse the environmental effectiveness of these programmes, while recognizing that ecological benefits are achieved at a cost to livestock production and household economic welfare. Herders go further, condemning the programmes as uniformly destructive—environmentally, socially and economically. The official programmes, Li et al. argue, have failed because they do not recognize the interdependent relationship between the rangeland environment and pastoral communities. In destroying pastoral communities and their way of life, government is unknowingly destroying the means to achieve its environmental objectives.

Li et al.'s conclusions pertain directly to Inner Mongolia, but environmental research elsewhere in China suggests a wider relevance. The Quinghai-Tibetan Plateau is a vast, high-altitude grassland and desert that stretches across parts of five autonomous regions or provinces in China and is the source of some of Asia's mightiest rivers—the Mekong, Yellow, Yangzte, upper Indus and Brahmaputra. Official statistics and Chinese state policies depict the Qinghai-Tibetan Plateau, as severely degraded as a result of 'backward' pastoral land use practices, particularly overstocking, and a large body of scientific literature purports to sustain this conclusion. A recent comprehensive literature review nonetheless argues that the extent and magnitude of rangeland degradation on the Plateau is unknown due to poor documentation and subjective monitoring programmes (Harris 2010). A common thread running through the Chinese degradation literature is the assumption that the Tibetan Plateau is a natural ecosystem disturbed by grazing, an assumption that has been challenged by advances in the archaeological interpretation of pollen and charcoal records. Archaeological research suggests that the vegetation of the Tibetan highlands is a pastoral creation that began about 8800 years ago with the conversion of forests into high-yielding pastures through the human use of fire to control woody vegetation and the replacement of wild herbivores with much larger numbers of domestic livestock. According to this interpretation, the present species composition and plant morphology of the area are a response to the selective foraging of free-ranging livestock and are remarkably resilient to grazing pressure

(Miehe et al. 2009). Far from damaging the vegetation, there is evidence that grazing in this environment increases pasture productivity and improves its nutritive quality, ‘results [that] call into question the prevailing opinion on the Tibetan Plateau, that grazing decreases the productivity of the rangelands’ (Klein et al. 2007: 553). At least in Inner Mongolia and on the Tibetan Plateau, it would appear that China’s grasslands need their pastoralists and livestock.

This book begins in Africa and concludes in Asia. In between it documents the diversity of meanings that different people have attached to the concept of desertification. It also describes a sobering continuity.

The official version of desertification that currently prevails among policymakers in Central Asia and China is hardly different from the one that was commonly accepted when UNCOD was convened in Nairobi in 1977. The perceived agents of environmental destruction—local land users—are unchanged, as are the autocratic and technical solutions that are officially advocated, despite contrary scientific evidence. What has changed is the ability of some national governments to enforce their will. For Sahelian states in the late 20th century, desertification was a useful tool for leveraging the international assistance that they needed if they were to exercise control and extract profit from peripheral rural communities. The limited international support that these states received and their own constrained means set limits to what they could do. The local level resilience, adaptation, and environmental recovery that has been subsequently documented, and the awakening of some Sahelian governments to the advantages offered by these processes, are an unintended consequence of the inability of national administrations and international agencies to enforce their will on rural Africa.

But things are different in much of Central and Inner Asia. Quite apart from environmental concerns, the Chinese state has its own political and economic reasons for exerting unqualified control over its peripheral ethnic minorities, and it has ample administrative and financial resources to do so. If flawed environmental policies are part of the means of this control, the concept of desertification could yet be implicated in the wholesale destruction of environmental values and rural livelihoods. It has taken decades to set the record straight in the Sahel; in China, the environmental and human sciences do not have decades to moderate potentially destructive policies. Hopefully, the global case studies, the scientific advances, and the reappraisal of Sahel history reported in this book will alert Asian policy makers to options that they have not previously considered.

A final overview, Chap. 20 by Mark Stafford Smith, assesses the key messages contained in this book. Looking back over the history of the desertification debate, Stafford Smith emphasizes the role of power and narrative—the power of national and international centres relative to remote and often marginal dryland populations, and the role of simplifying narratives like desertification in guiding the interactions between centre and dryland periphery. Stafford Smith then asks ‘What next?’ His answer is both optimistic and a challenge to scientists and others concerned about the future of drylands and the people who live in them: Create a ‘new and appealing narrative of drylands … that is strong enough to rebut (or evade) the apparent inevitability of centralized power dynamics, and to respond to the inexorable realities of global change.' Maybe the most effective response to the desertification narrative is to simply move on.

Acknowledgements We would like to thank Carol Kerven, Mark Stafford Smith, and Cara Kerven for reading and providing helpful comments on the various multiple drafts of this introduction.

References

Abreu, U. G. P., McManus, C., & Santos, S. A. (2010). Cattle ranching, conservation and transhumance in Brazilian pantanal. *Pastoralism: Research, Policy and Practice, 1*(1), 99–114.

Anderson, P. M. L., & Hoffman, M. T. (2007). The impacts of sustained heavy grazing on plant diversity and composition in lowland and upland habitats across the Kamiesberg mountain range in the Succulent Karoo, South Africa. *Journal of Arid Environments, 70*(4), 686–700.

Anderson, P. M. L., Hoffman, M. T., & O'Farrell, P. J. (2010). Above ground perennial plant biomass across an altitudinal and land-use gradient in Namaqualand, South Africa. *South African Journal of Botany, 76*(3), 471–481.

Andersson, E., Brogaard, S., & Olsson, L. (2011). The political ecology of land degradation. *Annual Review of Environment and Resources, 36*, 295–319.

Anyamba, A., & Tucker, C. J. (2005). Analysis of Sahelian vegetation dynamics using NOAA-AVHRR NDVI data from 1981–2003. *Journal of Arid Environments, 63*(3), 596–614.

Asner, G. P., Elmore, A. J., Olander, L. P., Martin, R. E., & Harris, A. T. (2004). Grazing systems, ecosystem responses, and global change. *Annual Review of Environment and Resources, 29*, 261–299.

Aubréville, A. (1949). *Climats, Forêts et Désertification de l'Afrique Tropicale*. Paris: Société d'Éditions Géographiques, Maritimes et Coloniales.

Baena, P. M., & Casas, R. (2010). Past, present and future of *transhumancia* in Spain: Nomadism in a developed country. *Pastoralism: Research, Policy and Practice, 1*(1), 72–90.

Bailey, D. W., & Brown, J. R. (2011). Rotational grazing systems and livestock grazing behavior in shrub-dominated semi-arid and arid rangelands. *Rangeland Ecology and Management, 64* (1), 1–9.

Bégué, A., Vintrou, E., Ruelland, D., Claden, M., & Dessay, N. (2011). Can a 25-year trend in Soudano-Sahelian vegetation dynamics be interpreted in terms of land use change? A remote sensing approach. *Global Environmental Change, 21*(2), 413–420.

Behnke, R. H., Fernandez-Gimenez, M. E., Turner, M. D., & Stammler, F. (2011). Pastoral migration: Mobile systems of livestock husbandry. In E. J. Milner-Gulland, J. M. Fryxell, & A. R. E. Sinclair (Eds.), *Animal migration: A synthesis* (pp. 144–171). Oxford: Oxford University Press.

Bollig, M., & Schulte, A. (1999). Environmental change and pastoral perceptions: Degradation and indigenous knowledge in two African pastoral communities. *Human Ecology, 27*(3), 493–514.

Boone, R. B., & Hobbs, N. T. (2004). Lines around fragments: Effects of fencing on large herbivores. *African Journal of Range and Forage Science, 21*(3), 147–158.

Bovill, E. W. (1921). The encroachment of the Sahara on the Sudan. *Journal of the Royal African Society, 1*, 259–269.

Briske, D. D., Derner, J. D., Brown, J. R., Fuhlendorf, S. D., Teague, W. R., Havstad, K. M., et al. (2008). Rotational grazing on rangelands: Reconciliation of perception and experimental evidence. *Rangeland Ecology and Management, 61*(1), 3–17.

Briske, D. D., Fuhlendorf, S. D., & Smeins, F. E. (2003). Vegetation dynamics on rangelands: A critique of the current paradigms. *Journal of Applied Ecology, 40*(4), 601–614.

Buitenwerf, R., Bond, W. J., Stevens, N., & Trollope, W. S. W. (2012). Increased tree densities in South African savannas: 50 years of data suggests CO_2 as a driver. *Global Change Biology, 18* (2), 675–684.

Butt, B. (2010). Pastoral resource access and utilization: Quantifying the spatial and temporal relationships between livestock mobility, density and biomass availability in southern Kenya. *Land Degradation and Development, 21*(6), 520–539.

Butt, B., Shortridge, A., & WinklerPrins, A. M. (2009). Pastoral herd management, drought coping strategies, and cattle mobility in southern Kenya. *Annals of the Association of American Geographers, 99*(2), 309–334.

Charney, J. G. (1975). Dynamics of deserts and drought in the Sahel. *Quarterly Journal of the Royal Meteorological Society, 101*(428), 193–202.

Charney, J., Stone, P. H., & Quirk, W. J. (1975). Drought in the Sahara: A biogeophysical feedback mechanism. *Science, 187*, 434–435.

Corell, E. (1999). *The negotiable desert: Expert knowledge in the negotiations of the convention to combat desertification, Linköping Studies in Arts and Sciences* (Vol. 19). Linköping, Sweden: Linköping University.

Coughenour, M. B. (1985). Graminoid responses to grazing by large herbivores: adaptations, exaptations, and interacting processes. *Annals of the Missouri Botanical Garden, 1*, 852–863.

Crespo, D. G. (1985). Importance of grazing trials in determining the potential of rainfed Mediterranean pastures. FAO-European Cooperative Network on Pasture and Fodder Crop Production. *Bulletin, 4*, 85–92.

Davies, J., Poulsen, L., Schulte-Herbruggen, B., Mackkinnon, K., Crawhall, N., Henwood, W. D., et al. (2012). *Conserving dryland biodiversity*. Nairobi: IUCN.

Davis, D. K. (2007). Resurrecting the granary of Rome: Environmental history and French colonial expansion in North Africa. Athens: Ohio University Press.

Ellis, J., & Lee, R.-Y. (2003). Collapse of the Kazakstan livestock sector: A catastrophic convergence of ecological degradation, economic transition and climatic change. In C. Kerven (Ed.), *Prospects for pastoralism in Kazakstan and Turkmenistan: From state farms to private flocks* (pp. 52–74). London: RoutledgeCurzon.

Ellis, J. E., & Swift, D. M. (1988). Stability of African pastoral ecosystems: Alternate paradigms and implications for development. *Journal of Range Management Archives, 41*(6), 450–459.

Fernandez-Gimenez, M. E., & Allen-Diaz, B. (1999). Testing a non-equilibrium model of rangeland vegetation dynamics in Mongolia. *Journal of Applied Ecology, 36*(6), 871–885.

Fernandez-Gimenez, M., & Allen-Diaz, B. (2001). Vegetation change along gradients from water sources in three grazed Mongolian ecosystems. *Plant Ecology, 157*(1), 101–118.

Folland, C. K., Palmer, T. N., & Parker, D. E. (1986). Sahel rainfall and worldwide sea temperatures, 1901–85. *Nature, 320*(6063), 602–607.

Fritz, H., & Duncan, P. (1994). On the carrying capacity for large ungulates of African savanna ecosystems. *Proceedings of the Royal Society of London B: Biological Sciences, 256*(1345), 77–82.

Fuhlendorf, S. D., & Engle, D. M. (2001). Restoring heterogeneity on rangelands: Ecosystem management based on evolutionary grazing patterns. *BioScience, 51*(8), 625–632.

Fuhlendorf, S. D., Engle, D. M., Elmore, R. D., Limb, R. F., & Bidwell, T. G. (2012). Conservation of pattern and process: Developing an alternative paradigm of rangeland management. *Rangeland Ecology and Management, 65*(6), 579–589.

Fuhlendorf, S. D., Harrell, W. C., Engle, D. M., Hamilton, R. G., Davis, C. A., & Leslie, D. M, Jr. (2006). Should heterogeneity be the basis for conservation? Grassland bird response to fire and grazing. *Ecological Applications, 16*(5), 1706–1716.

Fynn, R. W. (2012). Functional resource heterogeneity increases livestock and rangeland productivity. *Rangeland Ecology and Management, 65*(4), 319–329.

García, C. A., Schellberg, J., Ewert, F., Brüser, K., Canales-Prati, P., Linstädter, A., et al. (2014). Response of community-aggregated plant functional traits along grazing gradients: Insights from African semi-arid grasslands. *Applied Vegetation Science, 17*(3), 470–481.

Grainger, A. (2009). The role of science in implementing international environmental agreements: The case of desertification. *Land Degradation and Development, 20*(4), 410–430.
Gutman, M., Seligman, N. G., & Noy-Meir, I. (1990). Herbage production of Mediterranean grassland under seasonal and yearlong grazing systems. *Journal of Range Management, 1*, 64–68.
Hacker, R. B., Hodgkinson, K. C., Melville, G. J., Bean, J., & Clipperton, S. P. (2006). Death model for tussock perennial grasses: Thresholds for grazing-induced mortality of mulga Mitchell grass (*Thyridolepis mitchelliana*). *The Rangeland Journal, 28*(2), 105–114.
Hadar, L., Noy-Meir, I., & Perevolotsky, A. (1999). The effect of shrub clearing and grazing on the composition of a Mediterranean plant community: Functional groups versus species. *Journal of Vegetation Science, 10*(5), 673–682.
Harris, R. B. (2010). Rangeland degradation on the Qinghai-Tibetan plateau: a review of the evidence of its magnitude and causes. *Journal of Arid Environments, 74*(1), 1–12.
Hary, I., Schwartz, H. J., Pielert, V. H., & Mosler, C. (1996). Land degradation in African pastoral systems and the destocking controversy. *Ecological Modelling, 86*(2), 227–233.
Heady, H. F. (1961). Continuous vs. specialized grazing systems: a review and application to the California annual type. *Journal of Range Management, 1*, 182–193.
Hein, L., & De Ridder, N. (2006). Desertification in the Sahel: A reinterpretation. *Global Change Biology, 12*(5), 751–758.
Helldén, U. (1991). Desertification: Time for an assessment? *Ambio, 1*, 372–383.
Hempson, G. P., Archibald, S., Bond, W. J., Ellis, R. P., Grant, C. C., Kruger, F. J., et al. (2014). Ecology of grazing lawns in Africa. *Biological Reviews,*. doi:10.1111/brv.12145.
Herrmann, S. M., Anyamba, A., & Tucker, C. J. (2005). Recent trends in vegetation dynamics in the African Sahel and their relationship to climate. *Global Environmental Change, 15*(4), 394–404.
Huband, S., McCracken, D. I., & Mertens, A. (2010). Long and short-distance transhumant pastoralism in Romania: Past and present drivers of change. *Pastoralism: Research, Policy and Practice, 1*, 55–71.
Humphrey, C., & Sneath, D. (1999). *The end of nomadism? Society, state and the environment in inner Asia*. Durham: Duke University Press.
Huntsinger, L., Forero, L. C., & Sulak, A. (2010). Transhumance and pastoralist resilience in the western United States. *Pastoralism: Research, Policy, and Practice, 1*(1), 1–15.
Illius, A. W., & O'Connor, T. G. (1999). On the relevance of nonequilibrium concepts to arid and semiarid grazing systems. *Ecological Applications, 9*(3), 798–813.
Khishigbayar, J., Fernández-Giménez, M. E., Angerer, J. P., Reid, R. S., Chantsallkham, J., Baasandorj, Y., & Zumberelmaa, D. (2015). Mongolian rangelands at a tipping point? Biomass and cover are stable but composition shifts and richness declines after 20 years of grazing and increasing temperatures. *Journal of Arid Environments, 115*, 100–112.
Klein, J. A., Harte, J., & Zhao, X. Q. (2007). Experimental warming, not grazing, decreases rangeland quality on the Tibetan Plateau. *Ecological Applications, 17*(2), 541–557.
Lamprey, H. F. (1988). Report on the desert encroachment reconnaissance in northern Sudan: 21 October to 10 November 1975. Desertification Control Bulletin.
Le Houérou, H. N. (1977). The nature and causes of desertization. In M. Glantz (Ed.), *Desertification: environmental degradation in and around Arid Lands* (pp. 17–38). USA: Westview Press, Boulder, CO.
Lepers, E., Lambin, E. F., Janetos, A. C., DeFries, R., Achard, F., Ramankutty, N., & Scholes, R. J. (2005). A synthesis of information on rapid land-cover change for the period 1981–2000. *BioScience, 55*(2), 115–124.
Lohmann, D., Tietjen, B., Blaum, N., Joubert, D. F., & Jeltsch, F. (2012). Shifting thresholds and changing degradation patterns: climate change effects on the simulated long-term response of a semi-arid savanna to grazing. *Journal of Applied Ecology, 49*(4), 814–823.
Ludwig, J. A., Tongway, D. J., Bastin, G. N., & James, C. D. (2004). Monitoring ecological indicators of rangeland functional integrity and their relation to biodiversity at local to regional scales. *Austral Ecology, 29*(1), 108–120.

McAllister, R. R., Gordon, I. J., Janssen, M. A., & Abel, N. (2006). Pastoralists' responses to variation of rangeland resources in time and space. *Ecological Applications, 16*(2), 572–583.
McNaughton, S. J., Oesterheld, M., Frank, D. A., & Williams, K. J. (1989). Ecosystem-level patterns of primary productivity and herbivory in terrestrial habitats. *Nature, 341*, 142–144.
Middleton, N., Stringer, L., Goudie, A., & Thomas, D. (2011). *The forgotten billion: MDG achievement in the drylands*. New York and Bonn: United Nations Development Programme and UNCCD Secretariat.
Miehe, S., Kluge, J., Von Wehrden, H., & Retzer, V. (2010). Long-term degradation of Sahelian rangeland detected by 27 years of field study in Senegal. *Journal of Applied Ecology, 47*(3), 692–700.
Miehe, G., Miehe, S., Kaiser, K., Reudenbach, C., Behrendes, L., Duo, L., & Schlütz, F. (2009). How old is pastoralism in Tibet? An ecological approach to the making of a Tibetan landscape. *Palaeogeography, Palaeoclimatology, Palaeoecology, 276*(1), 130–147.
Milchunas, D. G., Sala, O. E., & Lauenroth, W. (1988). A generalized model of the effects of grazing by large herbivores on grassland community structure. *American Naturalist, 1*, 87–106.
Milner-Gulland, E. J., Fryxell, J. M., & Sinclair, A. R. E. (Eds.). (2011). *Animal migration: A synthesis*. Oxford: Oxford University Press.
Moritz, M., Scholte, P., Hamilton, I. M., & Kari, S. (2013). Open access, open systems: Pastoral management of common-pool resources in the Chad Basin. *Human Ecology, 41*(3), 351–365.
Mortimore, M. (1989). Adapting to drought. Farmers, famines and desertification in West Africa. Cambridge: Cambridge University Press.
Mortimore, M. (2010). Adapting to drought in the Sahel: Lessons for climate change. *Wiley Interdisciplinary Reviews: Climate Change, 1*(1), 134–143.
Mortimore, M., Anderson, S., Cotula, L., Davies, J., Faccer, K., & Hesse, C. et al. (2009). Dryland opportunities: A new paradigm for people, ecosystems and development. Gland, Switzerland: IUCN; London: International Institute for Environment and Development; Nairobi: United Nations Development Programme.
Müller, B., Frank, K., & Wissel, C. (2007). Relevance of rest periods in non-equilibrium rangeland systems—A modelling analysis. *Agricultural Systems, 92*(1), 295–317.
Nicholson, S. E. (2002). What are the key components of climate as a driver of desertification. In J. Reynolds & D.M. Stafford Smith (Eds.), *Global desertification: Do humans cause deserts* (pp. 41–57). Berlin: Dahlem University Press.
O'Connor, T. G. (1995). Transformation of a savanna grassland by drought and grazing. *African Journal of Range and Forage Science, 12*(2), 53–60.
O'Connor, T. G., Puttick, J. R., & Hoffman, M. T. (2014). Bush encroachment in southern Africa: Changes and causes. *African Journal of Range and Forage Science, 31*(2), 67–88.
Oesterheld, M., DiBella, C. M., & Kerdiles, H. (1998). Relation between NOAA-AVHRR satellite data and stocking rate of rangelands. *Ecological Applications, 8*(1), 207–212.
Olsson, L. (1993). On the causes of famine: Drought, desertification and market failure in the Sudan. *Ambio, 1*, 395–403.
Olsson, L., Eklundh, L., & Ardö, J. (2005). A recent greening of the Sahel—Trends, patterns and potential causes. *Journal of Arid Environments, 63*(3), 556–566.
O'Reagain, P. J., & Turner, J. R. (1992). An evaluation of the empirical basis for grazing management recommendations for rangeland in southern Africa. *Journal of the Grassland Society of Southern Africa, 9*(1), 38–49.
Otterman, J. (1974). Baring high-albedo soils by overgrazing: a hypothesized desertification mechanism. *Science, 186*(4163), 531–533.
Palmer, T. N. (1986). Influence of the Atlantic, Pacific and Indian oceans on Sahel rainfall. *Nature, 322*, 251–253.
Perevolotsky, A., & Seligman, N. G. (1998). Role of grazing in Mediterranean rangeland ecosystems—Inversion of a paradigm. *Bioscience, 48*, 1007–1017.

Prince, S. D., Colstoun, D., Brown, E., & Kravitz, L. L. (1998). Evidence from rain-use efficiencies does not indicate extensive Sahelian desertification. *Global Change Biology, 4*(4), 359–374.

Prince, S. D., Wessels, K. J., Tucker, C. J., & Nicholson, S. E. (2007). Desertification in the Sahel: A reinterpretation of a reinterpretation. *Global Change Biology, 13*(7), 1308–1313.

Reed, M.S., & Stringer, L.C. (2015). Climate change and desertification: anticipating, assessing and adapting to future change in drylands. UNCCD Third Scientific Conference Impulse Report. Cancun, Mexico.

Rutherford, M. C., & Powrie, L. W. (2010). Severely degraded rangeland: implications for plant diversity from a case study in Succulent Karoo, South Africa. *Journal of Arid Environments, 74*(6), 692–701.

Rutherford, M. C., & Powrie, L. W. (2013). Impacts of heavy grazing on plant species richness: A comparison across rangeland biomes of South Africa. *South African Journal of Botany, 87*, 146–156.

Schlesinger, W. H., Reynolds, J. F., Cunningham, G. L., Huenneke, L. F., Jarrell, W. M., Virginia, R. A., & Whitford, W. G. (1990). Biological feedbacks in global desertification. *Science, 247* (4946), 1043–1048.

Seaquist, J. W., Hickler, T., Eklundh, L., Ardö, J., & Heumann, B. W. (2009). Disentangling the effects of climate and people on Sahel vegetation dynamics. *Biogeosciences, 6*(3), 469–477.

Seligman, N., & Perevolotsky, A. (1994). Has intensive grazing by domestic livestock degraded Mediterranean Basin rangelands? In M. Arianoutsou & R. H. Groves (Eds.), *Plant-animal interactions in Mediterranean-type ecosystems* (pp. 93–103). Netherlands: Springer.

Sinclair, A. R. E., & Fryxell, J. M. (1985). The Sahel of Africa: Ecology of a disaster. *Canadian Journal of Zoology, 63*(5), 987–994.

Stebbing, E. P. (1935). The encroaching Sahara: the threat to the West African colonies. *Geographical Journal, 1*, 506–519.

Stebbing, E. P. (1937). The threat of the Sahara. *Journal of the Royal African Society, 1*, 3–35.

Stiles, D. (1995). *Social aspects of sustainable dryland management*. Chichester, England: Wiley.

Stumpp, M., Wesche, K., Retzer, V., & Miehe, G. (2005). Impact of grazing livestock and distance from water source on soil fertility in southern Mongolia. *Mountain Research and Development, 25*(3), 244–251.

Swift, J., (1996). Desertification: Narratives, winners and losers. In M. Leach & R. Mearns (Eds.), *The lie of the land: Challenging received wisdom on the African environment* (pp. 73–90). Martlesham: James Currey Ltd.

Taylor, C. M., Lambin, E. F., Stephenne, N., Harding, R. J., & Essery, R. L. (2002). The influence of land use change on climate in the Sahel. *Journal of Climate, 15*(24), 3615–3629.

Thomas, R. J., Akhtar-Schuster, M., Stringer, L. C., Marques, M. J., Escadafal, R., Abraham, E., & Enne, G. (2012). Fertile ground? Options for a science–policy platform for land. *Environmental Science & Policy, 16*, 122–135.

Thomas, D. S., & Middleton, N. J. (1994). *Desertification: Exploding the myth*. Chichester, England: Wiley.

Thomas, D. S., & Twyman, C. (2005). Equity and justice in climate change adaptation amongst natural-resource-dependent societies. *Global Environmental Change, 15*(2), 115–124.

Thompson, M., & Warburton, M. (1985). Uncertainty on a Himalayan scale. *Mountain Research and Development, 1*, 115–135.

Todd, S. W., & Hoffman, M. T. (2009). A fence-line contrast reveals effects of heavy grazing on plant diversity and community composition in Namaqualand, South Africa. *Plant Ecology, 142* (1), 169–178.

Tucker, C. J., Dregne, H. E., & Newcomb, W. W. (1991). Expansion and contraction of the Sahara Desert from 1980 to 1990. *Science, 253*(5017), 299–301.

UN Environment Management Group. (2011). *Global drylands: A UN system-wide response*. New York: Environment Management Group of the United Nations.

UNCCD/UNDP. (2011). The Forgotten Billion: MDG achievement in the drylands. New York: United Nations Development Programme; Bonn, Germany: United Convention to Combat Desertification.

Veron, S. R., Paruelo, J. M., & Oesterheld, M. (2006). Assessing desertification. *Journal of Arid Environments, 66*(4), 751–763.

Vetter, S. (2005). Rangelands at equilibrium and non-equilibrium: recent developments in the debate. *Journal of Arid Environments, 62*(2), 321–341.

Vetter, S., & Bond, W. J. (2012). Changing predictors of spatial and temporal variability in stocking rates in a severely degraded communal rangeland. *Land Degradation and Development, 23*(2), 190–199.

von Wehrden, H., Hanspach, J., Kaczensky, P., Fischer, J., & Wesche, K. (2012). Global assessment of the non-equilibrium concept in rangelands. *Ecological Applications, 22*(2), 393–399.

Wang, G., Hobbs, N. T., Boone, R. B., Illius, A. W., Gordon, I. J., Gross, J. E., & Hamlin, K. L. (2006). Spatial and temporal variability modify density dependence in populations of large herbivores. *Ecology, 87*(1), 95–102.

Warren, A. (2002). Land degradation is contextual. *Land Degradation and Development, 13*(6), 449–459.

Warren, A., & Agnew, C. (1988). An assessment of desertification and land degradation in arid and semi-arid areas. London: IIED Drylands Programme, Drylands Paper (2).

Wesche, K., Ronnenberg, K., Retzer, V., & Miehe, G. (2010). Effects of large herbivore exclusion on southern Mongolian desert steppes. *Acta Oecologica, 36*(2), 234–241.

Western, D., Groom, R., & Worden, J. (2009). The impact of subdivision and sedentarization of pastoral lands on wildlife in an African savanna ecosystem. *Biological Conservation, 142*(11), 2538–2546.

Xue, Y., & Shukla, J. (1996). The influence of land surface properties on Sahel climate. Part II. Afforestation. *Journal of Climate, 9*(12), 3260–3275.

Zemmrich, A., Manthey, M., Zerbe, S., & Oyunchimeg, D. (2010). Driving environmental factors and the role of grazing in grassland communities: A comparative study along an altitudinal gradient in Western Mongolia. *Journal of Arid Environments, 74*(10), 1271–1280.

Part I
Lessons from the Sahel

Chapter 2
Desertification in the Sahel: Local Practice Meets Global Narrative

Camilla Toulmin and Karen Brock

Abstract For nearly a century, crisis narratives about desertification have dominated policy discourse on the Sahelian drylands. This chapter looks at some of the ways in which these have shaped policy interventions in the drylands over the decades, and how contemporary development thinking offers better options for resilient dryland livelihoods. We argue that solutions to the environmental and economic problems faced by dryland systems—especially in the context of climate change—need to be more firmly rooted in a nuanced understanding of ecological change and the links between climate, vegetation and people. They must also involve a shift in power to local people, recognizing the value of marrying modern science with indigenous knowledge systems. Dryland peoples are more likely to prosper when governments reverse heavy-handed attempts to manage these areas. Greater promise lies with decentralizing power and decision-making to local institutions, and recognizing local tenure rights and systems for securing access to land.

Keywords Drylands · Policy · Discourse · Livelihoods · Decentralization and resilience

For many generations, the people of the Sahel have lived with consistently unreliable rainfall and long-term cycles of periodic drought. This is the nature of their dryland climate; it is the backdrop to their lives and livelihoods, and it has been the setting within which they have developed social and environmental strategies to support their prosperity. But for nearly a century, crisis narratives about desert

C. Toulmin (✉)
International Institute of Environment and Development, 80-86 Gray's Inn Road, London WC1X 8NH, UK
e-mail: Camilla.Toulmin@iied.org

K. Brock
Green Ink, Hawson Farm, Buckfastleigh, Devon TQ11 0HX, UK
e-mail: k.brock@greenink.co.uk

R.H. Behnke and M. Mortimore (eds.), *The End of Desertification?*
Springer Earth System Sciences, DOI 10.1007/978-3-642-16014-1_2

advance, and a broader desertification of the region's resources—forged in the aftermath of drought—have dominated policy discourse on the Sahelian drylands. The attitudes and arguments behind these narratives have remained remarkably tenacious, despite being "stuck within poorly researched data on degradation and desertification" (Hesse et al. 2013: 64) and systematically excluding the views, lived experience and practices of Sahelian people.[1]

Many 'solutions' to the problems of the drylands, mostly led by international aid agencies and governments, have featured large-scale projects with top-down planning, including human resettlement and large-scale irrigation, all aimed at changing and transforming nature. These interventions, and the scientific research they draw on to frame their design, have frequently misunderstood the highly variable and uncertain resource availability that characterizes this region. They have also disregarded the knowledge of Sahelian people and downplayed the importance of the practices and institutions that underpin their livelihoods in the face of this uncertainty. Although the range of interventions has altered over time, the positions of many international development agencies and national governments in the region continue to reflect the idea that mastery over nature is a better option than understanding and living with the uncertain conditions of the Sahel.

Alongside the mainstream narrative, several counter-narratives—which question 'desertification'—have emerged from the social, economic and political as well as the biophysical sciences. We locate ourselves in this tradition, alongside other proponents who argue that the biophysical manifestations of desertification and drought in the Sahel are inseparable from their underlying social, political and economic causes, consequences and remedies (e.g., Stamp 1940; Behnke and Scoones 1992; Sendzimir et al. 2011; Andersson et al. 2011). We concur with the view that a policy focus on technical solutions, to the detriment of social and political considerations, runs the risk of failure (Keeley and Scoones 2003), and that a much better balance between science and the knowledge and experience of Sahelian people should form the basis for future interventions in the drylands (Hesse et al. 2013).

Questioning the mainstream narrative is not a denial of the environmental and economic problems faced by dryland systems; on the contrary, we argue that it is critical to understand the ecology, instability and links between climate, vegetation and people, if solutions are to be found that fit within existing institutions and knowledge systems. Looking forward, this recognition of diversity and the significance of local knowledge is particularly important, as dryland people must continue to adapt to the extreme climatic events and hotter temperatures predicted for their ecosystems. Concerted support will be needed, recognizing the knowledge and agency of local people, offering partnership and shared learning, to maintain the resilience of dryland ecosystems within a wider social and geographic landscape (Hesse et al. 2013).

---

[1] In contrast with the rest of the chapters in this volume, our use of the term 'Sahel' and 'Sahelian' also includes the drylands of eastern Africa, and includes Ethiopia and Kenya.

From this perspective, resilient livelihoods in the Sahel can best be strengthened by interventions that support farmers, pastoralists and fishers through decentralized governmental and nongovernmental institutions that recognize them as citizens with rights to natural resources and valid knowledge that can be harnessed. Local management and control are key attributes in building the resilience, diversity and productivity of pastoral, mixed farming and agricultural livelihood systems in the face of climate change.

In this chapter, we look at how the desertification discourse has shaped policy interventions in the drylands over the decades, and ways in which contemporary development discourse may offer better opportunities to support resilient dryland livelihoods. We highlight key historical shifts in the evolution of the narratives and counter-narratives of desertification discourse, and their relationship with both scientific research and the livelihoods of Sahelian people. We trace the development of the 'received wisdom'—that a combination of inappropriate land use practices, rising population and periodic drought mean that Sahelian dryland ecosystems are being rapidly and permanently degraded by desertification—from its origins in colonial West Africa through the major Sahelian droughts of the 1970s and 1980s. We discuss its subsequent manifestations in the international policy architecture that has framed many interventions in the region.

We go on to examine how policy has responded to other key shifts in the language of development policy—climate change adaptation and mitigation, and poverty reduction—as well as better grounded scientific analysis. We describe a range of development interventions from across dryland Africa associated with different aspects of these narratives on dryland change. We discuss the opportunities presented for successful interventions to strengthen livelihood resilience in the face of climate change.

2.1 Desiccation and Blame—The Crisis Narrative

The ideas behind the term 'desertification' have been traced to the 1920s and 1930s, when colonial administrators, scientists and foresters began to describe the apparent drying out of their West African territories (Davis—this volume, Mortimore 1989; Swift 1996).

Given the long-held concern of the colonial authorities about desiccation, from December 1936 to February 1937, a joint Anglo-French Commission made its way through northern Nigeria and southern Niger to assess "the extent to which desert conditions are being created, and what is causing these conditions". While recognizing that 1936 had been a particularly wet rainy season, they concluded that there was no evidence to support a general desiccation of the region. Among their counter-intuitive findings they discovered that tree cover was reduced in the more southerly, wetter districts, due to higher population density, while the long-established farming areas further north were well-wooded due to farmers keeping numerous trees on their farmlands, and using living hedges as field boundaries. They saw abundant natural regeneration of tree and shrub cover, but also acknowledged that while there was no apparent danger of desiccation, there was "unquestionably an impoverishment of the sylvan conditions of the country … due almost entirely to uncontrolled expansion of shifting cultivation". They proposed a solution entailing "permanent farmlands, properly demarcated, regularly manured, adequately timbered with trees of local economic importance and with an assured supply of water."

Despite the conclusions of the Anglo-French Commission, other views of the encroaching Sahara came to dominate the story of desertification in the 1930s, as discussed in detail by Davis (this volume). This included the work of the British forester Stebbing, who travelled widely in the region in 1937, and concluded that the southern boundary of the Sahara was indeed expanding. Stebbing's view came to form the foundations of a powerful story of a vicious cycle of desertification: human misuse and population increase cause desiccation, which causes the desert to move south; and rainfall becomes more variable as a result of erosion and vegetation degradation. To counter this cycle, Stebbing proposed the extension of government authority over agricultural practices, an approach that resonated well with French efforts to establish a centralized natural resource management bureaucracy in their colonies, particularly through their Eaux et fôrets departments, under which elaborate regulations enforced by a paramilitary force served to constrain the natural-resource use rights of rural people and enrich a small, mostly urban elite (Ribot 2001).

Stebbing's narrative of the causes and effects of desertification, and the type of interventions needed to stop its destructive effect, were early versions of a blueprint from which many top-down approaches to governing dryland natural resources subsequently stemmed. The "desert advance" story has been replicated and reinterpreted many times since the 1930s by actors with different interests in the Sahel. As an early example of a development narrative (Roe 1991), this story of desertification vastly simplified complex dryland ecological systems, and helped to establish consensus for state intervention to manage resources (Krätli and Enson 2013). It was also associated closely with scientists of particular disciplines—in this case, soil scientists, foresters and physical geographers—and perspectives from other disciplines were marginal (Andersson et al. 2011).

Despite its critics, the desertification narrative became an enduring storyline because it was particularly effective in the service of the ruling political classes of the region, both colonial and post-colonial. Key to this success was its framing of desertification as a crisis, not only generating consensus and giving legitimacy to authoritarian interventions, but also allowing the attribution of blame. The narrative casts Sahelian farmers and, to an even greater extent, pastoralists in the role of culprits in the desertification process (Hesse and Odhiambo 2006; Sendalo 2009). In this framing, their poverty and ignorance are responsible for making them the consequent victims of its impacts.

To really take hold, however, a crisis narrative needs a real crisis (Krätli and Enson 2013). Historically rooted in the droughts of the 1920s, the desertification crisis narrative was used less in the 1950s and 1960s. During this period, although there were still periodic droughts (Mortimore 1989), wetter years in the Sahel brought greater yields and fewer pressures on scarce water and pasture (Swift 1996); the preoccupations of colonial government became more keenly focused on soil erosion. But when major droughts began to hit the region in the late 1960s, reaching their peak in 1972–74, the desertification crisis narrative found new resonance.

In addition to a large-scale famine relief effort across much of the region, international preoccupation with desertification in the Sahel grew throughout the 1970s, part of a wider growth in concerns about human impacts on the environment which started with the United Nations Educational, Scientific and Cultural Organization's (UNESCO) first Biosphere Conference in Paris in 1968. The international political and institutional response to the drought was spearheaded by of a set of United Nations (UN) reviews of the state of scientific knowledge about the causes and effects of desertification in drylands across the world (see, for example, Lamprey 1975; UN 1977). This period, culminating in the UN Conference on Desertification (UNCOD) in 1977, can be characterized by the consolidation of what Hajer (1995) describes as a 'discourse coalition'—a group of actors, usually with a bureaucratic base, who share a similar set of stories that account for why a particular phenomenon exists and what should be done about it. In this case, a discourse coalition formed, with its centre around the United Nations Environment Programme (UNEP), receiving support from a range of actors within African governments which together reinforced the storyline of an "apocalyptic vision" of desertification as "an inexorable and almost contagious process" (Swift 1996: 80).

Just as the received wisdom had served the implicit interests of the colonial project of centralized administration over natural resources 40 years previously, so in the 1970s it dovetailed with the agenda of many independent African governments to strengthen control over their dryland regions. This was especially the case for governments with a significant nomadic pastoral population, who were perceived as a threat to national unity because of their ability, through movement with their herds, to evade registration and taxation by the central state (Fratkin 2001; Niamir-Fuller 1998).

The UN General Assembly endorsed the Global Plan of Action to Combat Desertification that emerged from the UNCOD in 1977,[2] and the Desertification Branch of UNEP was charged with its implementation. Proposed activities included establishing a Sahelian green belt that revived Stebbing's proposal of the 1930s, as well as five large-scale transnational projects to manage the region's aquifers, and a regional livestock project aimed at controlling rangeland degradation through control of stocking levels, perceived to be excessive (UNCOD 1978).

Despite these project activities, when UNEP reviewed progress in 1984, it concluded that the land lost to desertification had continued to rise steadily at the same rate that had been reported in 1977 (Mabbutt 1984; UNEP 1984). But questions must be asked about the quality of the data used to support these

[2]Resolution 32/172 of 19 December 1977.

assessments. Swift (1996) notes that, in some cases, national government estimates of population and land use change were based on very limited survey data and heavily reliant on expert opinion or assessment. Despite questions raised about data quality (Berry 1983), these estimates were aggregated and translated into global desertification indicators. These indicators were used to produce figures that became the new headlines for the received wisdom on desertification,[3] and then widely used by a broad range of policy actors trying to raise the profile of environmental change on international policy agendas.

In contrast to these aggregate assessments, a growing number of researchers working in the Sahel were finding that the desertification narrative was not borne out on the ground. Their studies showed that the science behind the desertification narrative was poorly understood, and use of the term often confuses three related but distinct ecological processes: drought, desiccation and land degradation. However, their views, along with those of the farmers and herders they studied, were side-lined (Swift 1996). Many such researchers—from a range of disciplines including anthropology, ecology, economics and range science—went on to try and fill the knowledge gap that arose from a lack of primary research on the processes of dryland degradation in the 1970s and 1980s. Often funded by bilateral aid agencies, such research sought to gain recognition for the diverse people, livelihood systems, and settings across the Sahelian region, and their remarkable ability to survive and prosper in often harsh, uncertain conditions.[4] Nonetheless, Andersson et al. (2011) argue that there was a growing disconnection between the desertification discourse and scientific understanding of land degradation, which meant that "when the United Nations finally created the Convention to Combat Desertification (UNCCD) in 1994, policy was seriously disconnected from science" (2011: 306).

By their very nature, mainstream narratives have their detractors, and desertification is no exception. A further upsurge of research into land degradation and Sahelian livelihood systems during the 1980s and 1990s challenged some of the conceptual foundations of the desertification story, allowing the forging of a significant counter-narrative that framed dryland degradation as the outcome of complex interactions between climate, ecosystems and social systems in inherently dynamic environments (Behnke and Scoones 1992). This strand of thought also emphasized the importance and value of local knowledge and institutions for developing the drylands, framing interventions in support of the adaptive capacities of local people (Rochette 1989).

The Earth Summit at Rio in 1992, and the UN conventions that resulted—on desertification, biological diversity and climate change[5]—aimed to establish a new era of international environmental policy and associated frameworks. The two

[3]"Desertification threatens 35 % of the Earth's land surface and 20 % of its population; 75 % of threatened area and 60 % of the threatened population are already affected" (Swift 1996: 81).

[4]See, for example Haramata, a research bulletin about the drylands published by IIED between 1987 and 2010.

[5]UN Convention to Combat Desertification (1994), UN Convention on Biological Diversity (1992), and UN Framework Convention on Climate Change (1992).

broad perspectives on desertification—the received wisdom describing an inexorable spread driven by poor land management practices, and the counter-narrative emphasizing the long-term rhythms of climate variability and complex social and environmental dynamics—co-existed uneasily while the new institutional architecture was shaped and solidified.

The received wisdom, embedded in existing international and domestic bureaucratic structures, maintained its dominant position. The alternative narrative found resonance in a set of policy spaces outside those officially labelled as dealing with desertification, and amongst a set of actors who were not part of formal bureaucratic processes, including NGOs. Questioning the value of global processes to address essentially local problems of marginalization, its advocates increasingly sought to fit the science of dryland degradation into national and local policy processes on agriculture, poverty reduction and climate change adaptation.

2.2 The Earth Summit and the UN Convention to Combat Desertification

While UNCOD and the Plan of Action to Combat Desertification (PACD) raised the profile of desertification on international environmental agendas during the 1970s and 1980s, the issues of climate change and biodiversity loss were growing in prominence more quickly. Some African states were concerned that this would lead to desertification being ignored, and first proposed a UN convention on desertification in the run-up to the 1992 Earth Summit in Rio (Stringer 2006). Their rallying cry was for resources to support the victims of desertification. Although discouraged by several European countries at the summit—who argued that desertification did not have the same global dimensions as biodiversity and climate change—a commitment was eventually made to establish negotiations on a desertification convention. This decision was seen as a concession to African leaders who needed to show some return from the Rio conference in terms of resources to distribute through their governments.

The process of negotiating the UNCCD opened in May 1993. Burkinabè politician and diplomat the late Hama Arba Diallo was appointed head of a temporary secretariat, and an International Panel of Experts on Desertification[6] was convened. Government representatives were presented with a draft text drawn from documents agreed at Rio and, five negotiation sessions later, the UNCCD international treaty was signed by governments and adopted. Following ratification by 195 states, it entered into force at the end of 1996. Despite lobbying from a number of multilateral organizations which wanted to become the seat of this new convention, an independent Permanent Secretariat was established in Bonn; Diallo became the first Executive Secretary of the UNCCD. A Conference of the Parties

[6]Which included Camilla Toulmin, one of the authors of this chapter.

(CoP), held every two years, oversees implementation and reviews progress, while a science conference meets to review desertification science and local knowledge.

The principal objective of the UNCCD is to "combat desertification and mitigate the effects of drought, particularly in Africa". It defines desertification as "land degradation in arid, semi-arid and dry sub-humid areas resulting from various factors, including climatic variations and human activities" (www.unccd.int). The language of the convention frames local communities as the main managers of drylands, and appears to reject top-down approaches in favour of involving local communities in interventions to combat desertification. Nonetheless, it operates through a series of National Action Programmes, which identify the main causes and consequences of desertification at the country level and define a programme of priority responses very similar to those of the earlier UN-PACD.

While the Permanent Secretariat administers the UNCCD, it has no powers to enforce the convention. The UNCCD has its own funding institution, the Global Mechanism, but this channels existing financial support, rather than establishing and allocating new resources. New funding comes through the Global Environment Facility (GEF), but "desertification must compete with other GEF issues, and is often a low priority" (Stringer 2006: 1). This has led to the UNCCD becoming the "poor relation" among the three Rio conventions (Tollefson and Gilbert 2012: 1); in 2011, for example, GEF funding to UNCCD was only 10 % of that directed to the UN Convention on Biological Diversity (UNCBD). It has proved difficult to raise funds for the diffuse and poorly described set of problems that comprise desertification, and funding for dryland interventions has increasingly tended to be channelled through non-CCD interventions focused on soils and land management, such as the World Bank's Sustainable Land Management programme.

A range of other international organizations have recognized the importance of dryland regions, their people and resources for achieving broader global goals, whether it be biological diversity, the MDGs, or livelihood security (IUCN 2009, 2012; UNDP 2011; UN 2011). For some, the drylands are seen as a great untapped resource, where increased investment would yield substantial returns. For others, the drylands need to be addressed because their current low-income status and marginalization puts at risk global campaigns around the MDGs, or achievement of anti-polio campaigns.

The UNCCD competes for recognition and funding with the UNCBD and the UN Framework Convention on Climate Change by trying to get the attention of global audiences. Current documents emphasize the connection between desertification, conflicts and crises. In a 2014 document, the UNCCD Secretariat asserts that 12 million hectares of land become barren each year, and conflicts over land lead to civil war, sexual violence and genocide (UNCCD 2014a). Despite the arrival of a new, dynamic Executive Secretary in 2014, there remain longstanding questions about whether the international convention model is really the right strategy for addressing drylands development. Furthermore, Stringer (2006: 3) notes that "the UNCCD relies on only a small number of scientists relative to the amount of research done on desertification issues […] so the views of many experts are overlooked". Over a decade ago, Toulmin (2001) observed that the convention "has tied people into a series of CoP performances which demonstrate no linkage with real problems on the ground." By framing land degradation as an issue for action at the global level, the UNCCD lets national governments off the hook for decades of neglect of land, soils and farming systems. It is perhaps too much to expect of a structure created by and answerable to national governments—and therefore accountable at the national level—that it should make a strong case for local people to be at the heart of development.

Below we present two interventions in the Sahel that are related to the UNCCD. The first, the Desert Margins Program (DMP), was conceived when the CGIAR was invited to join the negotiations for the UNCCD in 1993; it was part-funded by the UNCCD through UNEP. The second, the Great Green Wall for the Sahara and the Sahel Initiative (GGWSSI), has been driven by African governments and has also been part-funded by the UNCCD, through the Global Mechanism, with additional support from the European Union (EU). Both illustrate the way that approaches to desertification enshrined in the UNCCD have been translated into interventions, through the new institutional architecture established after Rio.

These two examples illustrate how the international political framework for desertification has been translated into concrete initiatives, and some of the challenges that have ensued from implementation. These include the strong managerial approach to intervention which often overlooks how people actually live in the drylands, and a continued assumption that technology-based solutions to natural resource management problems will trickle down effectively to local people through government and national agricultural research bureaucracies.

2.2.1 *The Desert Margins Program—Putting Research into Practice*[7]

The idea of an international programme to address the problems of Africa's desert margins was tabled in 1993 when CGIAR member, the International Crops Research Institute for the Semi-Arid Tropics (ICRISAT), joined the international discussions to elaborate the UNCCD. By 1997, the idea had crystallized into the Desert Margins Program (DMP). A five-year pilot phase was launched to identify the causes of land degradation in nine countries on the margins of the Sahara and Kalahari deserts. This provided the basis for the full DMP, a multi-partner programme headquartered at ICRISAT in Niamey, Niger implemented by UNEP, and funded by GEF and nine member countries. Three two-year cycles of research-for-development were planned, commencing in 2002.

The DMP was positioned at the overlap of the CCD and the CBD mandates, resting on the assumption that "desertification can be avoided or reversed by adopting technologies and practices that simultaneously strengthen biodiversity and

[7]Unless otherwise indicated, the material in Sect. 2.2.1 is drawn from Koala et al. (2013).

improve farmers' livelihoods" (Koala et al. 2013: 1). Acknowledging the importance of indigenous knowledge, DMP researchers set out to "source traditional methods for regenerating land and managing it sustainably" (Op. cit. p. 3).

DMP involved a network of different funding and implementing organizations that crossed many levels and spaces. In addition to ICRISAT and UNEP, four other CGIAR centres, four international research organizations, and national agricultural research systems from each of the nine countries joined forces in this programme.

The first phase identified benchmark sites and selected key land management and livelihood diversification strategies. These included the establishment of tree nurseries and market gardens, mechanized construction of half-moon planting pits, pasture reseeding, and community-based rangeland monitoring. The second phase used the benchmark sites as hubs for demonstrating these options and training local people in their application, with a declared focus on using participatory methodologies.

Despite claiming considerable achievements—including more than 1500 ha of land under improved management in Burkina Faso, Mali, Niger and Senegal, and a part in shifting perceptions about African drylands among national policymakers—funds for the third phase of the DMP, which would have involved the scaling up of successful approaches beyond the demonstration hubs, were "deferred" after a "change in donor funding policies" (Op. cit. p. iii). This was publicly attributed to poor communication of programme outputs, and a "lack of demonstrated impact from the investments made so far" (Op. cit. p. 76)—a potential problem flagged up by a UNEP technical review of the original project proposal in 2001 (UNEP 2001). Privately, some of those involved also highlighted internal communication problems; top-heavy, bureaucratic processes of programme management; and a tokenistic approach to local people's participation.

2.2.2 The Great Green Wall—Reconstructing a Tenacious Anti-Desertification Project

Championed by former President Obasanjo of Nigeria and proposed to the African Union (AU) Assembly in July 2005, the Great Green Wall for the Sahara initiative originally aimed to arrest the southwards advance of the Sahara desert and improve livelihoods (AUC 2006). The Great Green Wall for the Sahara and the Sahel Initiative (GGWSSI) was formally launched by the AU in 2007 (FAO 2013). The rhetorical roots of the Great Green Wall lie firmly in the received wisdom on desertification. An early concept note for the initiative (AUC 2006: 2) states that "the threat of desertification is real" and that the erosion of livelihoods by the "continuous encroachment of desert areas" is "a source of grave concern".

The idea of a continuous process of forest degradation has a long history in West Africa stretching back to the colonial period, when it was popularized by foresters (e.g., Aubréville 1949; see Davis, this volume), as does the view that "some age-old deleterious farming and cultural practices would need to be addressed with the view to changing them, while more appropriate ones are introduced or promoted where they exist" (AUC 2006: 6).

In the years following the launch, the EU and the Food and Agriculture Organization of the United Nations (FAO) developed country strategies and project portfolios, assessed capacity needs and implemented pilot projects. During this period, the narrative of the project shifted from its original emphasis on large-scale reforestation—literally, a great green wall of trees to "stop the spread of the desert"—to reflect more contemporary objectives, such as climate resilience and carbon sequestration. In 2010, eleven countries signed a convention to create the Great Green Wall Agency. The GGWSSI was re-launched in Addis Ababa in 2011 as a priority project of the Africa–EU partnership, with resources also provided by FAO and the Global Mechanism of the CCD.

The GEF became involved in 2011, with a ministerial consultation to agree on priority areas it could address with its resources. This led to the development, with the World Bank, of the Sahel and West Africa Program in Support of the Great Green Wall Initiative (SAWAP) (World Bank 2013a). This US$1 billion portfolio of projects in 12 countries[8] is implemented by the New Partnership for Africa's Development (NEPAD), through TerrAfrica, a partnership whose members include multilateral agencies, regional organizations and national governments (TerrAfrica et al. 2011). Its headline aim is to expand sustainable land and water management in targeted landscapes and climate-vulnerable areas.

By the time of the 2011 re-launch, much of the promotional and operational language of the GGWSSI had changed. Its headline aim was reported as "tackling the detrimental social, economic and environmental impact of land degradation and desertification in the Sahara and Sahel region, in particular by supporting local community efforts in sustainable management and use of natural resources" (EU 2011: 1) to achieve "a mosaic of actions contributing to a variety of sustainable rural development and natural resource management programmes" (FAO 2013: 1).

The complex institutional architecture of the initiative makes it difficult to find public accounts of the GGWSSI, and assess how successful it has been in the sustainable management and use of natural resources by local communities. A two-page information sheet on GGWSSI produced by the Africa–EU Partnership reports on a 600-ha pilot site in the north of Senegal where reforestation with gum arabic trees started in 2009, and which is being used to "educate local farmers" and supply them with "resilient seeds, technical assistance and a forum to share information with other villages" (Africa-EU partnership nd: 2). But there is little evidence as yet of tangible, local-level outcomes from the first phases of the GGWSSI.

Implementation of the SAWAP is now in progress. Many of the programme components include investment in resilient and carbon-smart natural resource

---

[8]Benin, Burkina Faso, Chad, Ethiopia, Ghana, Mali, Mauritania, Niger, Nigeria, Senegal, Sudan, Togo.

management—actions to address land degradation, biodiversity conservation, climate change and sustainable forest management—and are consistent with the host of national action plans to combat desertification drawn up under the UNCCD. Although there is strong emphasis on strengthening the capacity of in-country institutions, in line with the aim of supporting the implementation of the UNCCD through building effective partnerships between national and international actors, there is relatively little mention of the local institutions of farmers and pastoralists who are actually managing resources on the ground, day to day. Nor is there any sense that the most effective forms of dryland development could be achieved by recognizing and empowering local people and their institutions.

2.3 Looking Beyond Desertification: Other Policy Narratives for Dryland Livelihoods

While the UNCCD has succeeded in maintaining the profile of desertification on international policy agendas, its weaknesses include a failure to create policy synergy with the other Rio conventions, despite the scientific overlap between desertification, climate change and biodiversity loss (Andersson et al. 2011). Consequently, many research and development actors are looking beyond the label of 'desertification' in their search for effective, people-centred solutions to the problems of the drylands.

In the decades following the Rio Summit of 1992, two major growth areas in international development policy—climate change and poverty reduction—have offered new spaces and opportunities for making progress in achieving more sustainable livelihoods in the Sahel. While desertification has not been central to either issue, key ideas from climate change and poverty debates have overlapped with and cross-fertilized the work of many researchers and development practitioners in the drylands.

International policy responses to climate change have generated new buzzwords in development policy (Cornwall and Brock 2006), such as 'adaptation' and 'resilience'. They have also created new modes of intervention in natural resource management, such as the Clean Development Mechanism (CDM) and the REDD+[9] mechanism. Based on the concept of payment for ecosystem services through carbon markets, these mechanisms were intended to open up new ways of promoting sustainable land management by putting a price on soil and forest carbon. This frames the maintenance and 'repair' of nature as a tradable commodity (Fairhead et al. 2012), and offers farmers and herders a crucial potential role in

[9]REDD+ is an effort under the UN Framework Convention on Climate Change to provide financial incentives for developing countries to reduce CO_2 emissions from deforestation and forest degradation, and to conserve, manage and enhance existing forest carbon stocks. The Clean Development Mechanism provides for emission reduction projects that generate Certified Emission Reduction units which can be traded in emission trading schemes.

managing carbon stocks in vegetation and soils, and therefore climate change mitigation. Section 2.3.1 discusses an example of a REDD+ project in Kenya which demonstrates the promise of applying such approaches in the drylands.

International policy responses to poverty, crowned by the project to achieve the Millennium Development Goals (MDGs), include an increased recognition of poverty as a multidimensional phenomenon (e.g., World Bank 2000; Alkire and Foster 2009), comprising far more than lack of income, but also including social exclusion and political marginalization. Since the food price crises of 2007–2008, there has also been a radical reassessment of agriculture's importance in development, with many arguing that in sub-Saharan Africa, where the majority of poor people depend on agriculture for their livelihood, the achievement of the MDGs must hinge on agriculture-led growth (Cabral and Scoones 2007). In the rest of this section, we discuss five cases that illustrate how dryland interventions have been shaped by these broader shifts in narrative: the Kasigau project in Kenya (Sect. 2.3.1), the Ethiopian Productive Safety Net Programme (Sect. 2.3.2), changes in post-drought reforestation in Niger (Sect. 2.3.3), two decades of pastoralist water development in Chad (Sect. 2.3.4) and a devolved fund for climate change adaptation in dryland Kenya (Sect. 2.3.5).

2.3.1 Carbon-Funded Development in Dryland Kenya—The Kasigau Project

The Kasigau project, the first accredited REDD+ project in Africa to issue carbon credits, conserves a 200,000-ha dryland forest in Kenya's coastal region, which is an important wildlife corridor. In 2013, it was one of 15 REDD+ projects in Kenya, only one other of which is located in the drylands. Atela's (2013) study of the Kasigau project, on which this section is based, presents information on project actors and narratives, community and state engagement, land tenure and livelihood impacts gathered through interviews and focus group discussions with participating households, CBO leaders, project staff and government personnel.

The project is a commercial venture, developed by a private US company, Wildlife-Works, specializing in wildlife conservation and ecotourism, which generates carbon through managing a protected forest area, as well as engaging community members in activities meant to reduce pressure on the protected forest. Its activities are designed to reduce greenhouse gas emissions and sequester carbon in soils and vegetation. The project developers have implemented a carbon accounting system and used it to claim carbon payments through the REDD+ mechanism. Ranch shareholders receive some of this revenue, and some goes to fund community projects.

The Kasigau project land was initially classified as government trust land, but was reclassified in the 1970s and allocated to private ranches. Both Somali pastoralists and Taita farming communities held rights in these ranches. By the 1990s, many were dissatisfied by the lack of income-generation opportunities offered by land ownership, and many sold their shares to immigrants, and to Wildlife-Works, which has become a major shareholder in one of the communal ranches. Other shareholders have leased their shares to the project. Prior to the REDD+ project, the ranch areas were not generating significant incomes for their owners, shareholders or neighbouring local communities, and were thus perceived to be 'valueless'. The project design argues that conserving the dryland forest for carbon increases the land's value to local people.

In contrast to most REDD+ projects in Kenya, the Kasigau project is being implemented in a dryland area where poverty levels are high and access to water and other livelihood assets is very limited. The project earned US$180,000/year in carbon revenue for its first five years of conservation activities in the protected areas. One third of this goes into a REDD+ project trust fund, the majority of which has been spent on classroom rehabilitation, bursaries for schoolchildren, and community water projects in almost equal proportions. So although local people do not engage in the management of carbon production and accounting, which is carried out by the project, they can apply to the trust fund for local development resources.

Because of the way the project links carbon benefits to specific local vulnerabilities, such as low-value land productivity, water scarcity and illiteracy, it has been favourably perceived among the Kasigau people, appearing to reverse their long history of exclusion from resources by centralized state-based resource management regimes. The success of this project in its early stages rests partly on the unusual mixture of private and communal land tenure systems that prevails in the area. This has allowed those local people who already have formal tenure rights to become shareholders in the carbon-financed enterprise, and those without land rights to benefit from the project in other ways.

But broader changes in national land tenure systems may have an impact on this early success. Kenya established a National Land Commission (NLC) in 2012, one of a range of reforms introduced under the 2010 constitution, with the task of recommending a National Land Policy to parliament. Since its establishment, there has been an ongoing power struggle at the national level between the NLC and the Ministry of Lands. It remains to be seen exactly what the outcome of this highly politicized process will be, but there is a fear that if the new policies favour individual private ownership over communal land tenure, the livelihood opportunities that currently result from the Kasigau project may be lost.

During interviews and focus group discussions, participating households, leaders of community-based organizations, and project and government staff all identified some encouraging signs of success in the Kasigau project. But successful carbon-financed development interventions in the drylands are relatively rare, and it is important to locate them within a broader national context of climate change policy. Ochieng Odhiambo (2013) examines the intersection between climate change and dryland development narratives in Kenya, noting that while climate change debates have fed a traditional storyline which blames pastoral land use for degradation, there is also now a strong counter-narrative. The latter is drawn from many decades of research in dryland agriculture, and it articulates clearly the variability and resilience of the drylands. This latter narrative has been used to support mobile pastoralism as the best use of marginal drylands, and "views the arid and semi-arid lands as resilient rather than vulnerable, and pastoralists as active producers equipped with traditional knowledge and experience and having in place institutions capable of harnessing the resilience of the arid and semi-arid lands for improved productivity and livelihood security even in the face of climate change" (Ochieng Odhiambo 2013: 22). This approach to the drylands has gained ground within the country's national climate change response strategy, greatly helped by the active engagement of former Minister for Northern Kenya, Mohamed Elmi, himself from a pastoral background, in advocating this shift in approach.

Hesse et al. (2013) identify the importance of resilience as a key concept in understanding how complex socio-ecological systems react to the bio-physical challenges of climate change. Socio-economic factors—such as the operation of markets, social networks and land tenure—are crucial for building or eroding resilience to the biophysical risks of climate change. The next section discusses a similar recognition of socio-economic factors in increasing or diminishing the possibility of poverty reduction.

2.3.2 *Social Protection for the Drylands? Lessons from the Ethiopian Productive Safety Net Programme*

Food aid for drought relief has been an important mode of international intervention in the drylands which, while saving lives, failed to save livelihoods (Hesse et al. 2013). Recent alternatives to this approach include a range of social protection interventions, designed as an integrated set that simultaneously supports livelihoods and strengthens resilience to climate change (Béné et al. 2012). Social protection has not yet been widely applied in the Sahel. In Mali, for example, social safety nets remain very limited in scope, are heavily dependent on external financing, and are poorly targeted. A high proportion of revenue fails to reach the poorest households (Bastagli and Toulmin 2014). Nonetheless, there are lessons to be learned from elsewhere in dryland Africa, such as the experience of extending Ethiopia's Productive Safety Net Programme from the country's highlands to its drylands. The challenges of this process highlight the importance of understanding complex dryland livelihoods.

The Productive Safety Net Programme (PSNP), one of the largest social protection programmes in sub-Saharan Africa, was initiated by the Government of Ethiopia in 2004. With the support of a group of donor agencies, implementation began in 2005, targeting 5 million chronically food-insecure people, mostly in highland agricultural regions. The PSNP has now spread to the predominantly dryland, pastoral regions, reaching 8 million beneficiaries in about 1.5 million households (Béné et al. 2012).

Under the PSNP, chronically food-insecure households receive support for several months of the year, and for a period of up to five years, building their resilience until they are better able to cope with moderate shocks. Drought is one of the most commonly reported household-level shocks in both highland and lowland Ethiopia (Van Domelen et al. 2010).

Targeted households receive cash and food in two ways: either unconditionally (direct support), or in exchange for contributions to labour-intensive public works that are selected through a public consultation process. A World Bank evaluation of lessons learned from the first phase of implementation notes a strong tendency to select natural resource management projects that could "reverse the severe degradation of watersheds, a major constraint on food production" (Van Domelen et al. 2010: 70). Public works activities included building soil and water conservation structures, water harvesting, sustainable farming, and rehabilitating and reclaiming marginal lands.

The first phase of the PSNP was predominantly focused on the densely populated highland agricultural zone of the country, and was designed to respond to the risks and vulnerabilities faced by people living there. Dryland people, whose livelihoods include pastoralism, face different risks and experience different vulnerabilities from farmers in highland ecosystems. Research on the expansion of the PSNP to the dryland Afar and Somali regions highlights the complexity of their livelihoods, pointing out that, "contrary to received wisdom, only a small proportion of lowland populations pursue a purely 'pastoralist' livelihood in the way of keeping and moving livestock across the rangeland to access fodder and water" (Sabates-Wheeler et al. 2011: 11). Livelihood strategies include a combination of pastoralism and agriculture, participation in cross-border markets, and the commercial harvest of natural resources. The opening of the pastoral lowlands, particularly for plantation agriculture and irrigation schemes, is an important recent development, posing major challenges to pastoral livelihoods.

Although from the outset the PSNP design envisaged the need for a "pastoral PSNP", with "public works and transfer payment mechanisms tailored to the needs of pastoralists" (Van Domelen et al. 2010: 26), there is little evidence that the complexity of dryland livelihoods has been well understood. A process to design and implement a pilot programme to tailor the PSNP to pastoral livelihoods was launched in 2006, and the PSNP was extended to the Afar Region in 2006 and to the Somali Region in 2008. But the implementation manual was the same as that used in the highland regions (Sabates-Wheeler et al. 2011). This suggests a lack of understanding of the need for a different design to accommodate the constraints and complexities of a dryland existence. The programme's instruments need to be re-designed better to respond to the problems of a fluctuating food-insecure population, and the tendency of the poorest people in dryland areas to spend much of their time in peri-urban settings.

2.3.3 Farmers and Dryland Re-Greening in Niger

Much of the new agricultural development discourse, shaped by the push and pull between different ideas and agencies, favours large-scale, irrigated farming, and market-driven processes of agricultural modernization. But these approaches have seldom been successfully applied to the small-scale, rainfed crop farming that is at the centre of many dryland livelihood strategies. Krätli and Enson (2013) argue that a systems approach is needed, covering a wider landscape, to illuminate the larger-scale functions of livestock-farming systems at the regional level, which are key to understanding how farmers and herders have managed the efficient cycling of nutrients in the Sahel. Equally, Hesse et al. (2013) highlight the importance of learning from many success stories of dryland crop production in the Sahel to understand the factors that help maintain sustainable land management under variable market, social and climatic conditions. Two decades of positive change in land management in Niger have occurred despite such variable conditions (Boubacar, Chap. 7).

In the last 30 years of the 20th century, drought and famine resulted in livestock losses and human migration on an immense scale in Niger. High levels of soil erosion and tree mortality fuelled the story of a relentless advance of the Sahara desert. Since the mid-1990s this story has been challenged by the reforestation of more than 5 million hectares in Maradi and Zinder Regions, estimated to involve 200 million trees and to have had a positive influence on the livelihoods of 4.5 million people (Sendzimir et al. 2011).

Most regions of the Sahel have experienced increased rainfall since the late 1980s, but few have witnessed the extent of re-greening evident in the Maradi–Zinder region, or the rates of increase in tree density since the drought of 1984–85. By examining the interaction of biophysical, livelihood and governance parameters underlying these changes, the analysis shows that forest decline was reversed when certain interventions were able to change vicious cycles to virtuous ones.

International response to the intense Sahelian drought of the early 1970s led to internationally funded projects for food security and reforestation, but tree planting schemes failed to reforest any significant area. The first successful experiments in reforestation were prompted in the early 1980s by the identification of a deep root system that allowed "shrubs" to regrow into trees when pruned in a particular way. An international NGO, Serving in Mission, experimented with and formalized these insights as a set of practices now known as 'farmer-managed natural regeneration'. They were influenced by a donor-prompted review of existing agricultural practices, undertaken after the widespread failure of post-drought reforestation interventions.

When another severe drought in 1984–85 devastated the livelihoods of hundreds of thousands of farmers, Serving in Mission offered a food-for-work programme, with labour invested in farmer-managed natural regeneration. When the rains returned, enough farmers continued to apply this approach that news of its promising results spread throughout the Maradi–Zinder region.

Growth in these methods coincided with national political changes triggered by the death of Niger's president in 1987, which led to a decade of political vacuum. While it deprived the Maradi–Zinder region of central government resources, it also meant that forestry officers—"a quasi-paramilitary force that on occasion could perversely interpret the law so as to extract prohibitive fines" (Sendzimir et al. 2011: 6)—no longer determined tree management on farmers' fields.

New-found opportunities for farmer autonomy were reinforced by an increasing donor emphasis on self-reliance, partly in response to the failure of earlier interventions. An International Fund for Agricultural Development (IFAD) project supported the development of oversight committees to enable local people to decide how and when to apply farmer-managed natural regeneration, aided by NGO outreach campaigns. From the late 1980s, IFAD also worked with regional offices of the agriculture and environment ministries to support the introduction of farming innovations, and also the establishment of monitoring and learning organizations.

Sendzimir et al. conclude that "systems analysis also shows why there was no single silver bullet to restore the integrity of agro-ecosystems or the communities that relied on them. A number of interventions at different scales and at different times combined to foster successful woodland regeneration…. Interventions catalysed the shift from vicious to virtuous circles, and it is this multi-layered pattern of reinforcement that has sustained re-greening…. The failure of single-issue policies becomes more understandable when one recognizes that the pattern of interactions was more important to the sustained success of re-greening than any single factor or process" (Op cit: 17).

This case serves as powerful illustration of the possibilities for change that lie beyond the boundaries of single projects, programmes and policies, but are nonetheless influenced by each. Discussing this case alongside several other well-documented accounts of resilient agro-sylvo-pastoral systems in the Sahel, Hesse et al. (2013) conclude that there are many different ways to support dryland crop and livestock farming, including strengthening customary land tenure, protecting livestock corridors, identifying and sharing innovations in crop and forage management to generate high-value residues for fodder, sharing soil and water conservation practice, and strengthening local seed networks for circulating local varieties more resilient to drought.

Nonetheless, the contrast between this agenda and current policy approaches to agriculture in the Sahel remains stark. The World Bank's 2013 draft regional approach to the Sahel (World Bank 2013b) places the expansion of irrigation at the top of its list of priorities for the region, making no direct mention of dryland crop farming, and illustrating that considerable work remains to be done to shift the narrative in key agencies. There is no evidence of the World Bank having learned any lessons from previous dryland interventions, especially poor performance in large scale irrigation schemes. Nor is there any acknowledgement that local people, their knowledge and institutions constitute a great asset in developing more resilient livelihood systems.

2.3.4 Pastoral Water Development in Chad

In many sub-Saharan African countries, pastoralism is still portrayed in public opinion as an outdated practice rather than a major contributor to national economies; Shanahan (2013), for example, notes that pastoralists remain invisible in the press, unless there is a story to be told about conflict between pastoralists and farmers. Yet a recent evaluation of two decades of work on pastoral water development funded by the Agence Française de Développement (AFD) in pastoral areas of Chad (Krätli et al. 2013), illuminates what can be done to overcome such persistent stereotypes.

Between 1993 and 2013, the AFD implemented a series of water sector interventions in pastoral areas of Chad, comprising 11 projects in three regions and representing two decades of uninterrupted support for the pastoral economy. The evaluation of these interventions found that, although they were not without problems, they represent an innovative approach to pastoral development, firmly based on the new understandings of pastoralism that emerged in the 1980s and 1990s. The approach has been widely praised for its support for water infrastructure, set within a political and institutional arena within which to manage conflict.

The received wisdom on pastoralism, embedded in orthodox desertification narratives, labels it as illogical, archaic, subsistence-based, with pastoralists and their hungry livestock as the major culprits in generating irreversible desertification (Herskovits 1926; Murdock 1959; FAO 1980). The first AFD projects were launched in an era when the Government of Chad was still following the colonial policy of trying to settle (and thereby control) nomadic people, and to encourage them to take up different forms of livestock rearing, especially ranching. The basic principle of the early AFD projects—which were innovative, especially in Francophone West Africa—was to recognize the importance of pastoral mobility, and find ways to safeguard it. This principle was derived from research evidence which in 1993 was still marginal and countercultural: that livestock mobility was a key strategy in the successful and sustainable exploitation of inherently unpredictable Sahelian rangelands (Behnke et al. 1993).

Applying this principle to the provision of water meant shifting from a sectoral approach—which for 40 years had restricted livestock movement by ignoring traditional water use rights—to a pastoral water development approach, which uses water as an entry point for interventions aimed at supporting pastoral mobility. The end goal of AFD's pastoral water development intervention was not water provision per se, but instead, "water provision as a means of safeguarding pastoral systems, governing pastoral spaces and achieving peace, and a strategic decision to operate on a scale that encompasses the main north–south transhumance routes" (Krätli et al. 2013: iv).

So, although the AFD projects resulted, for example, in the rehabilitation of wells, the construction of seasonal ponds, the marking of transhumance corridors and creation of new water points, they did so using methods that prioritized social consensus, and developing tools for dialogue. They also worked on the basis of free access to pastoral water, to be managed by a set of local institutions. In a region where water-related conflicts between pastoralists and farmers are not uncommon, the evaluation notes that "what is most striking is that in 20 years there have been no violent conflicts over the works funded" (Op cit: iv). This case study demonstrates the importance of investing in both physical infrastructure, and intangible institutions which, though invisible to the eye, a critical to building effective systems for managing dryland resources.

2.3.5 A Devolved Fund for Climate Change Adaptation in Kenya

A long history of marginalization has left many dryland areas with weak institutions to plan and govern resources, ineffective services, and higher levels of poverty than other regions. Climate change will exacerbate these existing causes of inequality. A pilot project in Kenya's Isiolo County (IIED 2014) has developed and tested a model for climate change adaptation in a dryland environment which is characterized by high mobility and variability. It has established a devolved County Adaptation Fund (CAF) to finance investments in public goods, which have been prioritized by communities as critical to climate change adaptation.

Kenya's 2010 constitution grants county governments authority over social and economic development within their county, to be planned according to local priorities. This legal framework for decentralized governance has provided the opportunity to develop the Isiolo CAF.

The first phase of the process was funded by grants from the UK Department for International Development and the Catholic Organization for Relief and Development Aid, and has been supported by the Government of Kenya. Before establishing the CAF, capacity-development and institution-building initiatives were implemented, an important part of which was building a common understanding of dryland ecology, livelihoods and climate change. This was achieved through workshops for community members and government personnel on climate and ecology; community-led resilience assessments, to understand factors that strengthen or undermine capacity to address climate variability; and the production of digital resource maps for planning (Hesse 2014).

The CAF is managed by adaptation planning committees, appointed following an information campaign and public selection process. Together with government planners and local organizations, the committees conduct participatory livelihood and resilience assessments. They then use the findings to design investments that will promote climate resilient growth and adaptive livelihoods, based on agreed criteria.

In its first round, the CAF invested in: the rehabilitation, fencing and construction of dams, wells and tanks, and accompanying water governance activities; funding the operational costs of traditional range management institutions; rehabilitation of a livestock laboratory; a vaccination programme and livestock survey; and a radio station transmitter. The second round followed the pattern of the first, but also saw increased engagement with county government, legislature, and donors—with strong interest in the workings of the fund expressed by groups from dryland areas in other countries.

To date, the CAF has been run as a pilot for local level adaptation planning, with revisions to the procedure manual for the second round based on lessons from the first. But broader lessons are relevant for the successful adaptation of this bottom-up community-led approach to planning.

While bottom-up development planning initiatives like the Isiolo CAF have important technical lessons, they are also fundamentally political processes in support of devolution. As such, they need continual support from existing government institutions and local communities, and ways found to demonstrate clearly the benefits of community-led approaches. This demands leadership from a team of people from the community who understand the issues and process, and who command respect. In Isiolo, this was initially provided by government staff and local organizations, and is now increasingly being provided by local people who are members of the adaptation planning committees.

Development partners and implementing organizations need to strike the right balance between ensuring genuine control by local communities over decision making, and guaranteeing sound financial management in a context of high risk. This means good communication, transparency, and timely and regular provision of information to government and development partners.

The final aim of this process is to mainstream this kind of approach into local government, by building local ownership of the process, and seeking every opportunity to integrate leadership and accountability with local institutions.

These examples show how success can be achieved in dryland areas, when proper attention is given to local people, institutions and knowledge systems. In some cases, the introduction of new measures has been key, such as establishment of new markets, as in the dry forests of Kasigau, Kenya, and of safety nets in Ethiopia. In other cases, it has been the removal of government constraint that has unleashed energy and innovation, such as in Niger and Chad. In the case of northern Kenya, it has been the political recognition of local voices and priorities as central to decision-making and allocation of resources. The new Constitution of 2010 has been central to achieving this turn-around, which reverses many decades of neglect, and offers new space for local people to take charge of their own decision-making structures.

2.4 Conclusions: A New Profile for the Sahel

A new paradigm for drylands development has been in the making for more than a decade, drawing on evidence of what works, in a range of settings. While recognizing that there is no ideal recipe for unqualified success, its key elements involve a shift in power to local people, and removing attempts at bureaucratic control. This shift in the politics of development includes marrying modern science with indigenous knowledge systems, decentralizing power and decision-making to local institutions, and recognizing local tenure rights and systems for securing access to land. Recent debate around how to support adaptation to climate change, and build resilience to climatic shocks, has focused on the importance of community-based adaptation (e.g., Bryan and Behrman 2013). Alongside the decentralization of decision-making, there have been trials with local funding models, aiming to demonstrate that 'good governance' at local levels offers an alternative system for delivering funds, better tailored to local circumstances, than control and allocations made at national level.

This new paradigm has also been supported by landmark work on pastoral ecology from the early 1990s. This helped shift thinking away from classical analysis based on the concept of equilibrium towards seeing dryland grazing systems as inherently unstable, driven by fluctuations in rainfall, and with no assumption of some state of equilibrium to which they tend to return (Westoby et al. 1989; Behnke et al. 1993). Drawing on complexity theory, revisions to pastoral ecology have led to very different prescriptions for policy and project interventions. Such changes in approach need to become embedded in university courses and training programmes to ensure the next generation of professionals and government administrators are properly prepared for the reality of the systems they study and administer. A start has been made on this with a series of training materials developed, tested and implemented in West and East Africa, which have been adapted into a 2–3 week course on pastoralism that is being taught at undergraduate and postgraduate level at the universities of Jigjiga, Bule Hora and Samara in Ethiopia (C. Hesse, pers. comm.).

Looking back over 35 years of engagement in drylands research and development, there has been a constant tug-of-war between those who advocate a rooted decentralized approach, drawing on local knowledge and perspectives, and those who assert their power and expertise through planning large-scale schemes and mobilizing big investments. At times, the bottom-up paradigm seems to be in the ascendant, with vocal champions in key positions able to argue the case. At other times, the top-down planners seize back the initiative, using a crisis narrative—such as the urgency of addressing climate change—to retake the initiative and regain control of budgets. In the most recent bout of wrestling, the World Bank and EU have put more than US$8 billion on the table for major investments in irrigated agriculture and livestock development in the Sahel, to address both security and climate change concerns (World Bank 2013c). Plans are to develop 'growth poles' which will offer prospects for economic development and jobs for people who might otherwise be tempted to take up arms. The big official funding agencies have been joined in this modernization agenda by a range of new actors, such as agribusinesses seeking land, and investors in large-scale infrastructure, such as dams. These new actors are much less ready to listen to hard-won evidence for local approaches working best. They want to establish control, make big investments, and see results.

Reversing this imbalance in knowledge, political power and funding is thus an ongoing political battle. Scientists whose work challenges the equilibrium models held so dear by proponents of ranching and sedentarization continue to engage in policy debates to shift attitudes. NGOs and researchers help demonstrate the rationale for indigenous knowledge and land use practice by learning from, communicating the experiences of, and advocating on behalf of farmers and pastoralists.

But the power of interests often pushes in favour of large scale investments, regardless of suitability, and heavy-handed management by government when local people are much better placed to make sound decisions about land and water use.

In some cases, central government has found it increasingly difficult to maintain its administrative hold on distant dryland regions, and consequently has relinquished effective bureaucratic control. This welcome reversal of centralized power was at the root of the re-greening of the Sahel in the 1980s, when the government of Niger ceded control to local communities in part due to limited resources, and in part through a formal process of establishing decentralized government (Boubacar, Chap. 7). Elsewhere, a shift in the politics and power can help dryland regions and their populations achieve greater voice, as has been the case in Kenya. Following the 2003 elections, in which the ruling party lost power for the first time since independence, a new Ministry for the Development of Northern Kenya was established, and drylands development was made an integral part of Kenya's medium-term development plan 'Vision 2030' (Government of the Republic of Kenya 2007). A new constitution was adopted in 2010, which pledged greater powers to decentralized local government, and two years later was launched the national policy for the sustainable development of northern Kenya and other arid lands, bringing to an end their historic marginalization.

But offering clear evidence of the benefits of more participatory approaches and the use of indigenous knowledge and practice is not the only factor at work in shaping the balance of power in dryland regions. There is an innate political and bureaucratic urge to master and control. Governments are usually keen to maintain or re-establish an administrative or military presence, especially in border lands, and there are growing interests keen to push big schemes that can mobilize large-scale resources, especially for irrigation.

Setting dryland regions within the broader global picture, there is growing recognition that we must re-design our economic system if we are to live within the resources of our single planet. Agreement in 2015 of a set of Sustainable Development Goals is the clearest global manifestation of this at present, alongside the goal of agreeing a set of targets to limit greenhouse gas emissions, and thereby keep projected climate change within what is hoped will be non-catastrophic levels. While we might seek to master nature, there are limits we must understand and respect if we are to maintain a prosperous society for all. For generations, local communities in the drylands have known the importance of living within their means, but it is only gradually that their expertise has been called upon by those designing policy and practice.

The picture of dryland futures has become further complicated in the last decade by a crisscrossing of new opportunities and growing conflicts. In sub-Saharan Africa, groundwater reserves appear much larger than had been thought (MacDonald et al. 2012), offering new possibilities for livestock and cropping. Solar energy has transformed options for millions of people 'off-grid', and could generate large-scale supplies if challenges in technology and transmission can be resolved. The search for oil, gas and other minerals has increased interest and investment in formerly neglected regions. Mobile phones have been a tremendous boon in linking formerly marginal areas into mainstream communication, markets and information sources. Under favourable circumstances, urbanization and market development provide an extension of livelihood options for rich and poor alike. But at the same time, growing conflict and occupation of large desert spaces by militant groups—in Libya, Mali, Niger, northern Nigeria, northeast Kenya and Somalia—are driving a stronger security interest in managing these large arid areas. A strong military presence now blocks the transit of goods, movement of herds and management of water and grazing in several former pastoral zones, and herders themselves have become caught in the crossfire.

UNCCD (2014b) makes an explicit link between land degradation, poverty, migration and conflict. It argues that land degradation, if unchecked, will bring further extremism, radicalization and resource-driven crises, as people seek ways to assure their survival. In its latest strategy, UNCCD states that "our failure to act is bringing issues that were once imagined to be domestic problems into matters of global instability" (UNCCD 2014b: 4). Decentralized management of land and resources, which recognizes the role of Sahelian farmers and herders in decisions that shape their livelihoods, is being squeezed out in favour of top-down security concerns. While climate change is adding greater uncertainty to rainfall and temperature patterns, and setting new parameters within which people must organize their lives. Let us try to learn from evidence and experience, and put dryland people, their knowledge, institutions and priorities at the centre of plans for these variable and uncertain regions.

References

Africa-EU Partnership. (n.d.) Reducing vulnerability to climate change, land degradation and drought: The Great Green Wall for the Sahara Initiative. http://www.africa-eu-partnership.org/sites/default/files/success-story-files/ggwssi_en_fin.pdf. Accessed March 22, 2014.

Alkire, S., & Foster, J. (2009). *Counting and multidimensional poverty measures*. OPHI Working Paper Series 7. Oxford: Oxford Poverty and Human Development Initiative.

Andersson, E., Brogaard, S., & Olsson, L. (2011). The political ecology of land degradation. *Annual Review of Environment and Resources, 36*, 295–319.

Atela, J. O. (2013). *Governing REDD+: Global framings versus practical evidence from the Kasigau Corridor REDD+ Project, Kenya*. STEPS Working Paper 55. Brighton: STEPS Centre.

Aubréville, A. (1949). Climats, forêts et désertification de l'Afrique tropicale. Société d'Edition de Géographie Maritime et Coloniale, Paris.

AUC. (2006). *The Green Wall for Sahara Initiative: A concept note*. Addis Ababa: Commission of the African Union, Rural Economy and Agriculture Department.

Bastagli, F., & Toulmin, C. (2014). *Mali: Economic factors behind the crisis*. EP/EXPO/B/DEVE/2013/13. Brussels: European Parliament, Directorate-General for External Policies, Policy Department.

Behnke, R., & Scoones, I. (1992). *Rethinking range ecology: Implications for rangeland management in Africa*. Dryland Networks Programme Issue Paper No. 33. London: International Institute for Environment and Development and Overseas Development Institute.

Behnke, R., Scoones, I., & Kerven, C. (1993). *Range ecology at disequilibrium: New models of variability and pastoral adaptation in African savannas*. London: Overseas Development Institute.

Béné, C., Devereux, S., Sabates-Wheeler, R. (2012). *Shocks and social protection in the Horn of Africa: Analysis from the Productive Safety Net Programme in Ethiopia*. IDS Working Paper 395. Brighton: Institute of Development Studies.

Berry, L. (1983). *Assessment of progress in the implementation of the Plan of Action to Combat Desertification in the Sudano-Sahelian Region 1977–1984*. New York: UN Sudano-Sahelian Office.

Bryan, E., & Behrman, J. (2013). *Community-based adaptation to climate change*. CAPRi Working Paper 109. Washington, DC: International Food Policy Research Institute.

Cabral, L., & Scoones, I. (2007). *Donor policy narratives: What role for agriculture? Future agricultures policy brief 16*. Brighton: Institute of Development Studies.

Cornwall, A., & Brock, K. (2006). The new buzzwords. In P. Utting (Ed.), *Reclaiming development agendas: Knowledge, power and international policy making* (pp. 43–72). Basingstoke: Palgrave Macmillan and United Nations Research Institute for Sustainable Development.

EU. (2011). Great Green Wall for Sahara and Sahel—Combat desertification, improving food security and climate change adaptation. Delegation of the European Union to the African Union. http://eeas.europa.eu/delegations/african_union/press_corner/all_news/news/2011/20110929_01_en.htm. Accessed May 22, 2014.

Fairhead, J., Leach, M., & Scoones, I. (2012). Green grabbing: A new appropriation of nature? *Journal of Peasant Studies, 39*(2), 237–261.

FAO. (1980). *Natural resources and the human environment for food and agriculture*. Environment Paper 1. Rome: Food and Agriculture Organization of the United Nations.

FAO. (2013). Forging innovative partnerships for the implementation of the Great Green Wall for the Sahara and the Sahel Initiative. Rome: Food and Agriculture Organization of the United Nations. http://www.fao.org/partnerships/great-green-wall/news-and-events/news-detail/en/c/210633/. Accessed May 22, 2014.

Fratkin, E. (2001). East African pastoralism in transition: Maasai, Borana, and Rendille Cases. *African Studies Review, 44*(3), 1–25.

Government of the Republic of Kenya. (2007). *Kenya: Vision 2030*. Nairobi: Government of the Republic of Kenya.

Hajer, M. (1995). *The politics of environmental discourse: Ecological modernization and the policy process*. Oxford: Clarendon Press.

Herskovits, M. J. (1926). The cattle complex in East Africa. *American Anthropologist, 28*(1), 230–272.

Hesse, C. (2014). Harnessing pastoral knowledge for climate change adaptation in the drylands: Participatory digital mapping in the drylands of Kenya and Tanzania. International Institute for Environment and Development, London and GEODATA Institute, University of Southampton.

Hesse, C., & Odhiambo, M. (2006). Strengthening pastoralists' voice in shaping policies for sustainable poverty reduction in ASAL regions of East Africa. In *Paper Presented at Conference on Pastoralism and Poverty Reduction in East Africa: A Policy Research Conference*. Nairobi, Kenya, June 27–28, 2006.

Hesse, C., Anderson, S., Cotula, L., Skinner, J., & Toulmin, C. (2013). *Managing the boom and bust. Supporting climate resilient livelihoods in the Sahel*. IIED Issue Paper. London: International Institute for Environment and Development.

IFAD. (n.d.) *IFAD's strategy in Niger*. Rome: International Fund for Agricultural Development. http://operations.ifad.org/web/ifad/operations/country/home/tags/niger. Accessed February 15, 2015.

IIED. (2014). *Isiolo County Adaptation Fund: Activities, costs and impacts after the 1st investment round*. Kenya National Drought Management Authority Project Report. London: International Institute for Environment and Development.

IUCN. (2009). *Dryland opportunities: A new paradigm for people, ecosystems and development*. Gland, Switzerland: International Union for the Conservation of Nature.
IUCN. (2012). *Conserving dryland biodiversity*. Nairobi: International Union for the Conservation of Nature.
Keeley, J., & Scoones, I. (2003). *Understanding environmental policy processes: Cases from Africa*. London: Earthscan.
Koala, S., Tabo, R., Gandah, M., Sessay, M., O'Connell, E., Chater, S., et al. (2013). *Biodiversity and livelihoods: New ways forward for Africa's desert margins*. Niamey: International Crops Research Institute for the Semi-Arid Tropics.
Krätli, S., & Enson S.-J. (2013). *A review of global public policy narratives on the drylands*. Climate Change Group Working Paper. London: International Institute for Environment and Development.
Krätli, S., Monimart, M., Jallo, B., Swift, J., & Hesse, C. (2013). *Evaluation of AFD Group interventions in pastoral water development in Chad over the last 20 years*. Final report. London: International Institute for Environment and Development.
Lamprey, H. F. (1975). Report on the desert encroachment reconnaissance in Northern Sudan, 21 October to 10 November 1975. *Desertification Control Bulletin, 17*, 1–7.
Mabbutt, J. (1984). A new global assessment of the status and trends of desertification. *Environmental Conservation, 11*(2).
MacDonald, A. M., Bonsor, H. C., Dochartaigh, B. E. O., & Taylor, R. G. (2012). British Geological Survey—Quantitative maps of groundwater resources in Africa. *Environmental Research Letters, 7*(2), 546–556.
Mortimore, M. (1989). *Adapting to drought: farmers, famines and desertification in West Africa*. Cambridge: Cambridge University Press.
Murdock, G. P. (1959). Africa: Its peoples and their culture history. New York: McGraw-Hill.
Niamir-Fuller, M. (1998). The resilience of pastoral herding in Sahelian Africa. In F. Berkes & C. Folke (Eds.), *Linking social and ecological systems: Management practices and social mechanisms for building resilience* (pp. 250–284). Cambridge: Cambridge University Press.
Ochieng Odhiambo, M. (2013). *The unrelenting persistence of certain narratives: An analysis of changing policy narratives about the ASALs in Kenya*. Position paper prepared for the global workshop of the New Perspectives on Climate Resilient Drylands Development Project.
Ribot, J. (2001). *Science, use rights and exclusion: A history of forestry in Francophone West Africa*. Drylands Issue Paper E104. Edinburgh: International Institute for Environment and Development.
Rochette, R. M. (1989). *Le Sahel en lutte contre la désertification*. Eschborn: Gesellschaft für Technische Zusammenarbeit.
Roe, E. (1991). "Development narratives" or making the best of blueprint development. *World Development, 19*(4), 287–300.
Sabates-Wheeler, R., Lind, J., & Hoddinott, J. (2011). Implementing social protection in pastoralist areas: How local distribution structures moderate PSNP outcomes in Ethiopia. Available at SSRN: http://ssrn.com/abstract=1997339 or http://dx.doi.org/10.2139/ssrn.1997339.
Sendalo, D. S. C. (2009). *A review of land tenure policy implications on pastoralism in Tanzania*. Dar-es-Salaam: Department of Livestock Research, Training and Extension. http://www.unpei.org/sites/default/files/e_library_documents/A_Review_of_Land_Tenure_Policy_Implications_on_Pastoralism_in_Tanzania.pdf.
Sendzimir, J., Reij, C. P., & Magnuszewski, P. (2011). Rebuilding resilience in the Sahel: Regreening in the Maradi and Zinder regions of Niger. *Ecology and Society, 16*(3), 1. doi:10.5751/ES-04198-160301.
Shanahan, M. (2013). *Media perceptions and portrayals of pastoralists in Kenya, India and China*. IIED Gatekeeper 154. London: International Institute for Environment and Development.
Stamp, L. D. (1940). The southern margin of the Sahara: Comments on some recent studies on the question of desiccation in West Africa. *Geographical Journal, 85*, 506–524.

Stebbing, E. P. (1937). *The forests of West African and the Sahara: A study of modern conditions*. London: Chambers.

Stringer, L. (2006). The UN Convention to Combat Desertification. http://www.scidev.net/global/desert-science/policy-brief/the-un-convention-to-combat-desertification.html. Accessed May 27, 2014.

Swift, J. (1996). Desertification: Narratives, winners and losers. In M. Leach & R. Mearns (Eds.), *The lie of the land: Challenging received wisdom on the African environment*. Oxford: James Currey.

TerrAfrica, World Bank, GEF. (2011). Sahel and West Africa program in support of the Great Green Wall Initiative—To expand sustainable land and water management in targeted landscapes and climate-vulnerable areas. Washington, DC: World Bank.

Tollefson, J., & Gilbert, N. (2012). Earth Summit: Rio report card. *Nature, 486*, 20–23. doi:10.1038/486020a.

Toulmin, C. (2001). *Lessons from the theatre: Should this be the final curtain call for the Convention to Combat Desertification?*. London: International Institute for Environment and Development.

UN. (1977). *Desertification: Its causes and consequences*. Oxford: Pergamon Press.

UNCCD. (2014a). *Desertification: The invisible frontline*. Bonn: Secretariat of the UN Convention to Combat Desertification.

UNCCD. (2014b). *The UNCCD: laying the groundwork for future security (2014–2015)*. Bonn: Secretariat of the UN Convention to Combat Desertification.

UNCOD. (1978). *Round-up, plan of action and resolutions*. New York: United Nations. http://www.ciesin.org/docs/002-478/002-478.html.

UNDP. (2011). *The forgotten billion: MDG achievement in the drylands*. New York: United Nations Development Programme and UN Convention to Combat Desertification.

UNEP. (1984). General assessment of progress in the implementation of the Plan of Action to Combat Desertification 1978–1984: Report of the Executive Director. Governing Council, twelfth Session, UNEP/GC.12/9. Nairobi: UN Environment Programme.

UNEP. (2001). *Desert Margins Program*. Project review from UNEP. Nairobi: United Nations Environment Programme.

United Nations. (2011). *Global drylands: A UN system-wide response*. New York: United Nations Environment Management Group.

Van Domelen, J., Coll-Black, S. et al. (2010). *Designing and implementing a rural safety net in a low income setting: Lessons learned from Ethiopia's Productive Safety Net Program 2005–2009*. Addis Ababa: Productive Safety Net Program. http://siteresources.worldbank.org/SAFETYNETSANDTRANSFERS/Resources/EthiopiaPSNPLessonsLearnedLite.pdf. Accessed May 23, 2014.

Westoby, M., Walker, B., & Noy-Meir, I. (1989). Opportunistic management of rangelands not at equilibrium. *Journal of Range Management, 42*, 266–274.

World Bank. (2000). *World development report 2000/2001: Attacking poverty*. Washington, DC: The World Bank and New York: Oxford University Press.

World Bank. (2013a). *Sahel and West Africa Program (SAWAP) in support of the great green wall*. [Working Paper 82293]. Washington, DC: The World Bank. http://documents.worldbank.org/curated/en/2013/10/18456494/sahel-west-africa-program-sawap-support-great-green-wall. Accessed June 18, 2014.

World Bank. (2013b). *Regional approach to the Sahel*. Unpublished draft policy document. Washington, DC: The World Bank.

World Bank. (2013c). http://www.worldbank.org/en/news/press-release/2013/11/04/global-leaders-pledge-billions-of-dollars-for-africas-Sahel. Accessed December 1, 2014.

Chapter 3
Changing Paradigms for People-Centred Development in the Sahel

Michael Mortimore

Abstract Key paradigm changes are discernible in the science of desertification or land degradation in the West African Sahel. These have been characterized as a shift from a 'desertification' to a 'resilience' paradigm. The first of these positions, based on an equilibrial model of social-ecological systems, refers to a range of measurable changes which are unilinear with a strong likelihood of irreversibility. The second is based on a disequilibrial model which begins with evidence for sustainable practices under local knowledge and management of variability. Whereas the desertification paradigm attempts to correct misuse through exogenous interventions, designed to transform natural resource management by improved technologies, the resilience paradigm, based on optimizing endogenous capacities, offers an evolutionary trajectory of development adapted to the constraints of small-scale farming and pastoralism. Guided by this distinction, the complexity and variability of the social-environmental systems in the Sahel are disaggregated under the following headings: bio-productivity changes as reflected in remote sensing; rangeland management; deforestation (the expansion of cultivation, burning, fuel-wood cutting and the transition to sustainable practice); and soil nutrient management. Against the backdrop of rapid demographic change, poverty reduction and adaptive capacity are briefly reviewed. An 'escape from Malthus' is being attempted through a smallholder intensification pathway in situ and an income diversification pathway that extends to non-local opportunities. Governments, donors, and other agencies can use these as a platform for policies and interventions.

Keywords Adaptation · Desertification · Dryland farming · Sahel · Variability

---

M. Mortimore (✉)
Drylands Research, Cutters Cottage, 8 Glovers Close, Milborne Port, Sherborne DT9 5ER, UK
e-mail: mike@mikemortimore.co.uk

R.H. Behnke and M. Mortimore (eds.), *The End of Desertification?*
Springer Earth System Sciences, DOI 10.1007/978-3-642-16014-1_3

3.1 Change and Complexity in the Sahel

A conception of drylands as a major global biome has now emerged as a policy priority, against a background of the Millennium/Sustainable Development Goals, recognizing that in view of the drylands' spatial extent (41 % of global land surface) and demographic weight (35 % of global human population), the Goals will not be achieved without making substantial progress in the drylands. The global drylands have recently been unwrapped in a series of policy-oriented studies published by international organizations (IUCN 2009; UNDP/UNCCD 2011; UN/EMG 2011; Davies et al. [IUCN] 2012). All place emphasis on natural resource (or ecosystem) management and all draw attention to Sahelian local experience and knowledge.

Relations between society and environment in the West African Sahel region are intimately bound up with the variable rainfall, in space as well as time. They are highly complex in terms of the heterogeneous and interactive nature of social and environmental systems. And they are dynamic, in the sense that subtle and multi-dimensional changes—some slow, some fast—have taken place in the past, and still continue. This ecological and social diversity overlays the deceptive regularity of the moisture gradient from southern (sub-humid) through semi-arid and arid to northern (hyper-arid) biomes.

In terms of the theme of this book, this characterization presents a major challenge to development policy, debate and practice. For a long time, debate has been dominated by global diagnostics such as desert advance, irreversible land degradation, or deforestation. The persistence (and influence) of what we term here the 'desertification paradigm' in scientific, policy and international discourse on the Sahel has hitherto tended to prefer diagnostic-prescriptive framings imposed 'top-down' by governments and external actors. Yet while research studies at micro-scale have presented nuanced analyses of environmental intimacy, complexity and dynamics in specific social-ecological systems, they have not adequately informed the design of policies and interventions at macro-scale, which tend to reflect the viewpoint of the state, informed by science rather than local people (Scott 1998).[1]

3.2 The Desertification Paradigm

Desert advance in regions bordering the southern Sahara is an idea that goes back to early 20th century perceptions (Davis, this volume; Mortimore 1998): 'desséchement' (Hubert 1920); desiccation (Bovill 1921), desert advance (Stebbing

[1]The literature is too large to review. For examples, Raynaut et al. (1997) set out the interactions between social and agro-ecological systems in the Sahel region, Mortimore and Adams (1999) show the complexity of these relations at micro-scale, and Cour and Snrech (1994) identified the long-term regional dynamics and predicted future trajectories of change up to 2020.

1935), desertification (Aubréville 1949), progressive aridity and 'ensablement' (Chevalier 1950). 'Misuse' of the land by local people (Timberlake 1986) has transformed the less arid into more arid woodland and thus to move the latitudinal vegetation zones southwards over time, with negative implications for productivity. Estimates were made of the rate of desert advance (USAID 1972; Lamprey 1975; Ibrahim 1978). The idea is not yet dead (UNEP 2007).

The seeming culpability of humans was endorsed by the United Nations Conference on Desertification (UNCOD 1977: 3) and according to the (then) Director General of UNEP, six million hectares of land were lost 'totally' each year and, in a further 21 million hectares, productive capacity was 'reduced to the point of zero economic productivity' (Tolba 1986). However, under fire from critics, the inclusion of natural causes (rainfall and drought) and a preference for the less emotive expression 'land degradation' resulted in the definition published in the first edition of the *World Atlas of Desertification* (Thomas and Middleton 1992) being adopted at its inception by the UN Convention to Combat Desertification (UNCCD 1993):

> Desertification is land degradation in arid, semi-arid and sub-humid areas resulting from various factors, including climatic variations and human activities

which included soil erosion caused by wind and/or water; deterioration of the physical, chemical, and biological or economic properties of soil; and long-term loss of natural vegetation (UN Convention to Combat Desertification, Part 1 (Introduction), Article 1, Use of terms).

Popular perceptions are still framed by the first edition of the *Atlas,* and its successor, the second edition (Middleton and Thomas 1997). A fundamental challenge is encountered in separating natural-induced from human-induced processes or causes (Thomas and Middleton 1992, p. 11; Reynolds and Stafford Smith 2001).

The assessment of global land degradation employed in both editions of the *World Atlas of Desertification* relied heavily on GLASOD (Global Assessment of Human-induced Soil Degradation), a survey of homogeneous landscape units compiled from subjective reports by 250 soil scientists (Oldeman and Hakkeling 1990). This assessment concluded that soil degradation affected 224 million hectares of drylands, at four levels of severity (light, moderate, strong and extreme) (Table 3.1). Degradation includes wind and water erosion, chemical, and physical

Table 3.1 GLASOD data on the Sahel (millions of hectares)

	Water erosion	Wind erosion	Soil nutrient depletion	Physical deterioration	Human induced soil degradation
Light	97.5	156.2	25.6	18.7	109.8
Moderate	24.7	99.5	8.8	8.7	80.3
Strong	18.2	4.9	5.0	3.1	30.8
Extreme	2.2	0.8	0	0	3.1

Source Middleton and Thomas (1997)

Note Figures relate to the continental Sahel and include Sudan and parts of eastern Africa (total 802.3 million hectares)

deterioration, superimposed on a global vegetation index (GVI) derived from satellite radiometer data (averaged from 1983 to 1990).

The expert assessments were delivered at a point in time, rather than being based on change in key variables over a known period. The severity categories thus indicate states rather than rates of change. The 'global maps are designed to give an overall, if exaggerated, impression of the scale of the soil degradation problem world-wide' (*ibid* 1997: 17). GLASOD defined soil degradation as 'human-induced phenomena which lower the current and/or future capacity of the soil to support human life' (*ibid,* 1: 11). According to Middleton and Thomas (1997: 71), the main causes of soil degradation in the Sahel are: overgrazing (118.8 million hectares), agriculture (34.8), "overexploitation" (54.2), and deforestation (16.3 million hectares).

Over a period of eight decades (1920s–1990s), desertification was consolidated, in an influential expert orthodoxy, as a product of misuse of the land by indigenous farmers and pastoralists, factoring in a changeable climate and underlying population growth (Eckholm and Brown 1977). The implicit basis of the desertification paradigm (WRI 2005: Chap. 22) is the idea of an ecosystem at equilibrium in which a perturbation is normally followed by a natural readjustment back to a stable state. Within this frame, the 'advancing Sahara' hypothesis implied a risk of an irreversible form of land degradation which (if caused by humans) called for urgent interventions with the aim of arresting unsustainable practices and restoring an ecosystem to its baseline state.

The UNCCD has consistently addressed this theory of change in the global drylands, and its current goal is

> to forge a global partnership to reverse and prevent desertification/land degradation and to mitigate the effects of drought in affected [arid and sub-humid] areas in order to support poverty reduction and environmental sustainability (www.unccd.int).

The desertification paradigm adopted a diagnostic of human agency (Garcia and Escudero 1982; Garcia and Spitz 1986) in a crisis scenario, a prioritization of scientific (as opposed to local) knowledge, a prescription of urgent technical interventions to transform land use practices, and (to achieve this) the mobilization of resources on a grand scale. Its institutionalization in the form of the UNCCD and funding channels for 'desertification control' raised the profile of counter-measures at the expense of independent scientific analysis. However, science and practice are now achieving better convergence in some areas of natural resource management in drylands.[2]

[2]Reynolds et al. (2007). Work on 'desertification indicators' commissioned from an independent panel of scientists by the UNCCD for the Buenos Aires Conference of the Parties (2009) is published in 'Special Issue on Understanding Dryland Degradation Trends' *Land Degradation and Development* (2011), vol 22/2. Subsequently work on economic issues is ongoing.

3.3 Questioning the Dominant Paradigm: Resilience

In contrast to the 'desertification paradigm', which asserts degradation in the natural ecosystem, and its consequences for the social system, and advocates changes in practice based on science, an alternative view focuses on the evidence of adaptive capacities based on local knowledge of small-scale farming and pastoralism. This view points to the failure of many development interventions and yet to the 'success stories' of transitions to sustainable practices based on full participation and building on local experience (Safriel et al. 2005). Given the environmental variability with which farmers and pastoralists have to contend, their 'repertoire' of livelihood strategies and technologies enables them to adapt to change not necessarily by returning to the status quo (as predicted in an equilibrium model), but to manage state-changes in the social-environmental system (a non-equilibrial model) (Hesse et al. 2013).

Originating in ecological science, the idea of resilience in this usage recognizes that the natural ecosystems in drylands are not characteristically at equilibrium. As their productivity depends directly on variable rainfall, they are better understood as in a state of non-equilibrium and uncertainty. Extreme events—droughts, or floods —may lead to species loss and replacement. But the ecosystem retains its structure and functional integrity (Holling 1973, 2001)—it is said to be unstable but resilient.

The resilience model has been applied to nomadic pastoralism where herd mobility is the principal (but not the only) response to variability in the natural environment (Benhke et al. 1993). Evidence from interviewing Sahelian farmers suggests that such a view fits the strategies employed to manage their livelihoods under conditions of uncertainty. Such flexibility is driven in the first place by drought (Adams and Mortimore 1997) and expressed in an inventory of adaptive strategies for dealing with rainfall variability (Table 3.2): in agronomy, crop selection, wild food sources, timber and non-timber forest products for sale, manufacturing, processing, services and labour hiring within the community or migration on a seasonal or longer term basis for earning incomes further afield. For pastoralists the essence of adaptation is still herd mobility, in response to spatial and temporal variability in pastures and fodder. The advantages of adaptive range management, spatial mobility, herd management and breeding is apparent in every study of Sahelian pastoralism from the past (e.g., Stenning 1959; Dupire 1962) to the present (e.g., Krätli 2008).

In an extension of the model, it is contended that resilient adaptation is also driven by the need to maintain productive resources—soils, woodland, animals—in the longer term (Table 3.3). Managing variability is not always easy to distinguish from sustainable use of natural resources: the first is episodic, the second longer term ('combating desertification'). But adaptation in the longer term includes a measure of labour intensification, which is a response both to necessity (as 'saturation' gradually overwhelms 'free land' in the economy of colonization) and to growing market opportunities for staple and niche foods and animal products (in an economy of urbanization). These changes amount to investments—at a micro-scale —in smallholder intensification.

Table 3.2 An inventory of adaptive strategies to variability

Elements of the social-ecological system	Activity change	Beyond the household	Outside the area
Crop farming (rainfed or irrigated)	Re-planting Buy borrow or beg seed Reduce fertilization Change crop mixtures (drought resistant or tolerant varieties) Increase or reduce weeding Cultivate more wetlands Make do with less farm labour	Transfer farm labour to another system Hiring out own labour sell assets (land)	Hiring out own labour through migration to urban or more humid areas
Livestock breeding/keeping	Optimize herd mobility to access spatial-temporal variability of pasture Manage animal health Sell surplus bulls and vulnerable cows Make do with less labour Purchase water	Plant and grow crops (agro-pastoralists) Herding contracts with farmers' livestock Sell breeding/producing stock	Hiring out own labour through migration to urban areas Transhumance
Fishing (where surface water)	More or less depending on the resource	Absorb farm labour (at harvest time)	Set up camps near seasonal floods Urban sales
Forestry/tree management	Protect regenerating useful trees on farms Increase trimming and lopping for fuelwood Plant shade and fruit trees, tree lots, boundaries	Absorb additional labour cutting/selling fuelwood	Sell fuelwood to traders from other areas
Harvesting non-timber forest/wild products	Conserve knowledge of wild foods Increase intensity of collection Extend collecting from more to less preferred edible plant parts Process and cook wild foods	Absorb additional labour	Sell high value wild products to traders

(continued)

Table 3.2 (continued)

Elements of the social-ecological system	Activity change	Beyond the household	Outside the area
Manufacturing local materials	harvest raw materials, especially dum palm fronds increase output of marketable products	absorb additional labour making/selling mats, ropes, baskets	sell manufactures to traders
Service provision	Increase sale of ready foods, confectioneries Visit more local markets	Begging, patron-client relationships	Visit markets outside the area Unskilled employment in towns (e.g., guards) Begging in urban centres
Paid employment	Send more people to find jobs elsewhere		
Trading	Bulking local products for export	Absorb additional labour	Transport and sell livestock or other commodities in urban centres

Sources Various

A recent study by the IUCN argues that the key to sustainable dryland development is to reverse the status of drylands as 'investment deserts' (IUCN 2009). This theme is taken up in an analysis of dryland investment opportunities in a UN System-wide Response (UN 2011), and reiterated in a new report from UNDP-UNCCD (2011). Responses to perceived land degradation, at the smallholder level, are constrained by capital scarcity and for this reason consist of incremental, small-scale processes, whether assisted by donors and governments or autonomously financed using private savings. A large part of the value added may come from family labour, analytically indistinguishable from current operations, though with time an increasing access to financial capital is visible in drylands as in more favoured biomes.

Adaptation can go wrong. Among such strategies there lurk possibilities of maladaptation or dysfunctional adaptation, which leads to increased vulnerability to climate variability or change, including for other stakeholders (Magnan et al., in prep.). Maladaptation is a real possibility where external actors misjudge the impact of interventions by the state and its partners (Scott 1998); or on a local scale, where rights to resources are contested between stakeholders, for example between farmers and graziers. The resilience model accords higher status to local knowledge, practice and capacity in managed dryland ecosystems.

Evidence of such incremental growth into greater resilience in farming ecosystems of the West African Sahel is provided in meso- and micro-scale studies, for example: Senegal (Faye et al. 2001), Burkina Faso (Mazzucato and Niemeijer 2000; Reij and Thiombiano 2003), Niger (Mortimore et al. 2001) and northern Nigeria (Mortimore 1993; Mortimore and Adams 1997). Resilience does not mean

Table 3.3 Adaptation to land degradation

Elements of the social-ecological system	Incremental adaptations through autonomous investments
Crop farming (rainfed or irrigated)	Increase irrigated wetland cultivation (engineering, water management, negotiated use rights) Obtain livestock Increase nutrient cycling through livestock or composting Increase weeding frequency Intercrop grain with legume (N-fixing) crops Reduce plant spacing Soil and water conservation (ridging, terracing) Grazing/manuring contracts with herders Purchase fertilizers (incl. manure) Protect natural regeneration of on-farm trees
Livestock breeding or keeping	Breed for optimal performance in variable conditions (not for maximum milk) Increase mobility/transhumance where better grazing opportunities allow Cut-and-carry feeding during growing season Maximize use of crop residues (own, purchased or contracted) Purchase feedstock
Forestry/tree management	Protect regenerating trees on farmland, boundaries and cattle tracks Community reservation of woodlands and controlled grazing Community managed tree lots
Harvesting non-timber forest/wild products	Community protection of valuable natural resources Community control of exported scarce natural products, esp. fuelwood
Manufacturing local materials	Community protection of valuable natural resources
Paid employment	Send more children to school for qualifications Extend education to higher ages Send more people to find jobs Diversify employment sources Use migrant remittances for productive investments
Trading	Invest in buying and selling

Sources Various

universal improvements in well-being. In the Kano region of northern Nigeria, signs of pressure, some negative, have accompanied the evolution of the social-ecological system (Maconachie 2007). Nor does it offer certainty in predicting the future. Development indicators across a range of householder activities are needed to assess resilience and adaptive capacity. In the Sahel as a whole, a degree of resilience is implied in the evidence that average food production per capita has not fallen below (though it has struggled to keep up with) growth in the human population, despite very high rates of human fertility.[3]

[3]FAOSTAT, Rome: Food and Agriculture Organization of the UN.

In order to progress this argument, in the following sections, the desertification debate is disaggregated into discourses on bioproductivity (using remote sensing indicators), rangeland management, deforestation (agricultural expansion and biodiversity, burning, and fuelwood cutting), soil nutrient conservation and fertility, and demographic issues (poverty and adaptation).

3.4 Bio-Productivity: The View from Space?

Declining bio-productivity, often claimed to be irreversible, has been central to claims of desertification from the beginning. But empirical, in situ measurements of productivity based on field surveys are limited to small areas and when they are scaled up to larger areas, ecosystems, countries or region, reliability has to be thrown to fortune especially where change over time is targeted. On the other hand, earth satellite data—repeated at regular intervals—offer a compatible basis for estimating environmental change, using the key parts of the spectrum as proxy indicators of biological productivity. Applying this principle, using the Normalized Difference Vegetation Index (NDVI), or 'greenness index', to the African Sahel, has produced surprising counter-evidence to the view of progressive degradation. A strongly significant increase was observed between 1982 and 2006 (Fig. 3.1), a trend that is ongoing and has been confirmed in numerous studies (Knauer et al. 2014).[4] The methodology and documentation of these analyses are reviewed by Herrmann and Sop (this volume), Prince et al. (this volume) and Mbow et al. (2015).

Following the Sahel Drought (1972–74), a second major drought cycle (1984–85), and a decline in average annual rainfall of 25 % or more (from the 1960s to the 1990s), the region became (and remains) an archetype of dryland challenges. Many early interventions in natural resource management (e.g., afforestation) failed. Too often they were imposed 'top-down' and only inadequately addressed the systemic context (e.g., rainfall variability, mobility in pastoral systems, flexibility in farming systems). Development planning grew naturally from humanitarian relief and assumptions of technology transfer and authoritarian government.

These regional data confirm early findings on the oscillation of the desert edge since the 1980s (Tucker et al. 1991), and confirm a positive correlation between greening and rainfall. Average annual rainfall recovered from an all-time low in the drought cycle of 1984–85 (Eklundh and Olsson 2003; Herrmann et al. 2005; Olsson et al. 2005). The consensus on this recovery—though incomplete and highly variable, between years and seasons (Anyamba and Tucker 2005; Nicholson 2005)

[4]Remote sensing data has wide-ranging applications including estimating plant parameters (photosynthetically active radiation, leaf area index, plant biomass and primary productivity, evapotranspiration and water content), the assessment of phenology, and crop yield assessment as well as the analysis of vegetation trends since 1982 and of evidence for climate versus management agency (Knauer et al. 2014).

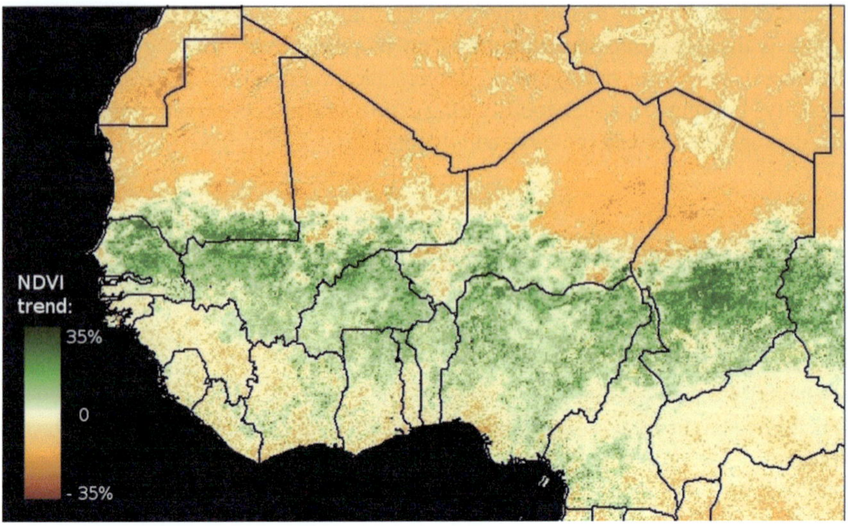

Fig. 3.1 NDVI trend in the West African Sahel, 1982–2006. Monthly NDVI and cumulative rainfall of the same plus the two preceding. *Source* Hermann in IUCN (2009). Technical Note: Linear trends in the NDVI (Normalized Difference Vegetation Index) are shown in percentages. Trends were computed from monthly 8 km resolution AVHRR NDVI time series produced by the GIMMS group, NASA Goddard Space Flight Center, USA, extended from previous work reported in Herrmann et al. (2005) Representations of the 'greenness index' vary in detail depending on the years included and differences in data processing methodology.

confirms the possibility of resilience in Sahelian vegetation in the long term (Hiernaux, this volume; Fig. 3.1). Global trends also appear to be positive (Bai et al. 2008), perhaps reflecting enhanced CO_2 emissions, which would increase biomass production where or when rainfall is not the limiting factor. If this hypothesis is substantiated, the biological productivity of the Sahelian ecosystems, together with the flora, fauna and human communities they support, are globally integrated, in parallel with the linkages discovered between ocean temperatures and Sahelian rainfall (Gianinni, this volume).

The relations between rainfall and vegetation greening are directly affected by the seasonal movements of the West African Monsoon, which vary from year to year, in terms of several parameters including monthly distribution, soil nutrients and moisture, and rainfall use efficiency (RUE: in agriculture, 'crop per drop').[5] These considerations illustrate the complexity and variability of rainfall-greening relationships (Figs. 3.2 and 3.3).

[5]This is believed to increase with rainfall until a point is reached where other factors (such as soil nutrients) reduce the RUE of total rainfall. In Senegal, according to official statistics, millet and groundnut yields per mm of rainfall RUE increased from 1960 to 1995 against a decline in the rainfall (Faye et al. 2001).

Fig. 3.2 An abandoned millet harvest in north-east Kano State, Nigeria, November 1973 (second year of the Great Sahel Drought).

Fig. 3.3 Bringing in a successful harvest (millet, sorghum, cowpea), eastern Kano State, Nigeria, September 1996 (Note on-farm trees. A year of above-average rainfall).

The positive correlation found between rainfall trends and trends in the greenness index does not necessarily invalidate the thesis of degradation resulting from human activities (including deforestation, grazing, or cultivation). In the regional pattern, regressive relationships are also found locally, possibly reflecting management impact, or soil type. It is likely that some forms of degradation (such as loss of biodiversity, invasion of unwanted species, or declining crop yields) were not picked up by the remote sensing products used or were disguised by the positive effects of increasing rainfall (Knauer et al. 2014)—'net' greening. Much more needs to be done to disentangle the effects of management from those of climate. "What these findings imply in terms of vegetation changes on the ground, however, remains mostly unaddressed in these satellite-based studies" (Herrmann and Sop, this volume). "Many uncertainties exist in regression models" (Mbow et al. 2015).

The use of data from earth satellites offers the possibility of detecting degradation (or sustainability) at landscape scale and monitoring its progression, not only with repeated remote sensing, but by means of integrated social and economic indicators (Nkonya et al. 2010). Remotely sensed data is also useful for monitoring interventions at this scale, and over a long period, such as soil and water conservation in the central plateau of Burkina Faso (stone lines and *zaï* planting holes); and protected natural regeneration in Maradi and Zinder regions of Niger (Reij et al. 2005; Sendzimir et al. 2011). Determining agency for such transformations at a landscape scale is not a simple matter of attributing 'success' to a single intervention. Multiple drivers of change are responsible, including local and science-based knowledge. In Maradi and Zinder, 500,000 ha are claimed to have benefited from protected natural regeneration.

The use of satellite data to estimate bio-productivity has three major drawbacks. First, plant biomass may offer only an imperfect proxy for economic value: for example where invasive plants, useless for grazing, affect the observed values; where standing stocks of timber which grow incrementally may be under-estimated; or where fallow cycles are interspersed with cultivation. On Sahelian farms, biomass production per year (including grain, residues, weeds and compost)—produced in only 3–4 months—may be comparable to that in bush fallows (Mortimore and Turner 2005); so extending agricultural clearance may not be accurately represented as degradation. Second, there is little understanding of what plant formations yield a given NDVI value at a micro-scale, impeding the assessment of change. Third, the history of land use in a given pixel is not known and would be costly to find out; but necessary for the complete understanding of the trends observed. This is relevant to understanding local exceptions to the overall greening trend.

Herrmann (this volume) uses collective recall of pastoralists to build a profile of tree degradation in Senegal; Hiernaux (this volume) finds vegetational resilience through long-term monitoring of grazing land in Burkina Faso and western Niger; and Hof et al. (2006), using ground measurements in conjunction with remote sensing in a grazing reserve in northwest Nigeria, found no evidence of widespread land degradation. Thus, more research is needed to align remotely sensed trends with in situ land use histories.[6]

3.5 Rangeland: Blame the Rain?

Received wisdom from observers of the Sahel Drought of 1969–74 had it that desertification was primarily caused by over-exploitation of natural ecosystems (Mensching and Ibrahim 1977; Garcia 1981; Dregne 1983). By this logic, grazing increasing numbers of animals on a fixed amount of land encounters a bio-productivity threshold and thereafter destroys or modifies pasture communities, resulting in reduced carrying capacity. The dramatic increase in animal mortality and loss of condition that accompanies or follows major drought events in Sahelian rangelands is thus held to indicate 'overgrazing' or 'over-stocking'. Indicators of desertification were considered important from the beginning (Reining 1978). On rangeland, they may include: removal or thinning of grassland communities leaving bare soil surfaces open to erosion; formation of sand dunes or re-activation of stabilized dunes; localized damage around wells and boreholes; changes in the species composition of grasslands in favour of less palatable or preferred species, reducing the supply of fodder per unit area; and the loss of biodiversity (indigenous trees, shrubs or grasses) from savanna.

Such a view was questioned on the ground that insufficient weight was given to hydrological factors, in particular a declining trend in rainfall from the 1960s to the 1990s (Hulme 1992; Hess et al. 1995; Tiffen and Mortimore 2002). If rainfall variability is taken fully into account, it follows that land use was not solely to blame. On the contrary, crop and livestock production systems proved more sustainable than expected: they did not 'collapse' (a word commonly used). Since rainfall cannot be manipulated, but land use systems can be, this divergence of view (which is still debated) has important implications for policy. 'Useful' moisture (that which contributes to the growth of crops or fodder in a given cropping or grazing system) can be and is manipulated by choices of crop landraces, planting times, and weeding regimes (in dryland farming) or of mobility decisions favouring patches of good growth, desired species and access to water (in pastoralist practice). Rainfall measured in millimeters is only a proxy for moisture that contributes to economic production (see Krätli and Schareika 2012).

Linking degradation exclusively with overstocking and mismanagement side-steps the impact of rainfall variability (with a coefficient of variability characteristically over 25 %). Carrying capacity is not stable but violently variable. Pastoralism in arid and semi-arid rangelands is based on herd mobility, and is

[6]The use of an increasing number of different remote sensing products and analytical processing of data is controversial beyond the scope of this discussion.

opportunistic in responding to this variability (Sandford 1983). The condition of livestock depends on fodder successfully accessed, thus indirectly on rainfall, rather than the status of pastures depending on grazing pressure. Animals are lost during and after drought, but herds are rebuilt during wetter years. Debates on this theme (Benhke et al. 1993; Scoones 1994), have given recognition to opportunistic management of non-equilibrial resources, based on systematic seasonal mobility of the herds, as the most gainful strategy for livestock-dependent peoples. Increasing livestock densities in response to good rainfall, and accepting losses in the bad years, does not necessarily lead to degradation. Pastoralism may be understood as a system for conserving, rather than destroying rangeland resources, described as 'living off uncertainty' (Krätli and Schareika 2012).

Governments have been very slow to recognize the ecological and economic rationale of pastoral mobility and its implication—that over-grazing is caused by impeding movements with barriers, boundaries and regulation, rather than by allowing moving herds to redistribute grazing pressure and thereby better conserve the ecosystems (WISP 2007). Mobile grazing, though essential, may cease to be viable when pastures are appropriated by other land users (e.g., government projects, farmers, plantations), especially where riverine 'reserve pastures' are lost. Straight lines on maps (Scott 1998) cannot reflect the complexity, variability and dynamics of range management as practised by small scale herders (Hof et al. 2006). There is urgent need for land use policies to take account of this reality.

3.6 Deforestation: Destruction or Transition?

This term covers four interlinked dimensions which require separate analysis. Each depends on major data sets whose evolution has driven the perceptions and policies in this sector. The first is the expansion of cultivation at the expense of dry woodland. The second is woodland management and in particular the disputed practice of burning. Third is the energy sector's reliance on the fuelwood industry. Fourth is the transition to sustainable woodland management.

3.6.1 Expansion of Cultivation

Air photography (since the 1940s) and earth satellite data (since the 1980s), have provided a basis for quantifying land use conversions up to half a century (Table 3.4, Figs. 3.2 and 3.3).

Maradi (Niger) is representative of the Sahel, and the period covered includes the historical peak in rural migration and settlement. Since the end years shown in the table, semi-arid and dry sub-humid zones have approached 'saturation'—where all cultivable land (unless restrained by forest reserves) is claimed. This, moreover, is accompanied by a significant decline in the average size of holdings. Expanding

Table 3.4 Expansion of cultivated land, mid-late 20th century

Area	Baseline year	Percent	End year	Percent
Djourbel, Bambey Departments, Senegal	1954	82.2	1999	93.3
Maradi Department, Niger	1975	59.0	1996	73.0
Kano State, Nigeria[a]	1950	77.6	1981	88.4
Jigawa State, Nigeria[a]	1950	35.6	1981	54.6
Yobe State, Nigeria[b]	1950	18.9	1990	16.1

[a]One Village Area
[b]Average of two Village Areas
Sources Ba et al. (2001), Mahamane (2001), Mortimore et al. (2001, 2005)

smallholder agriculture, as recorded in these data sets, and extending throughout the Sahel, is sufficiently widespread to play a role not only locally but also in global transformations (climate change, food security and poverty).

The high visibility of agricultural transformation ensured that deforestation figured large in perception and in policy. The colonial forestry departments perceived their mandate as the protection of trees and woodland from reckless indigenous practices, basing their regulatory strategies on 'scientific' knowledge held in opposition to presumed native understanding (Cline-Cole 1997; Ribot 1995). This was to be achieved by reserving forests, banning burning, and requiring permission to fell trees, even on farms, and enforcing its rules with quasi-military forest guards. From natural woodland (or even secondary forest) to open treeless farmland is still commonly equated with reduced biomass and biodiversity, exposure to wind or water erosion, and low crop yields under smallholder 'subsistence' farming. A hostile policy environment—rather than one informed by local knowledge and needs—has not yet completely disappeared from government agendas.

The authoritarian heritage of forest regulatory systems stands in contrast to the policies inherited from colonial governments with regard to access to land itself. Faced with the growth of rural populations, the need to supply export and (later) urban markets, and the functional efficiency of existing institutions, the colonial regimes adopted a dualistic approach to land administration, with formal statutory laws for urban, corporate and government land but customary tenure that recognized local 'custom' administered by locally recognized authorities within their respective territories (e.g., Mortimore 1997).

Customary law is now under pressure from increasing demand (Cotula 2007). Access to land has become a burning issue (Lavigne-Delville et al. 2002). Legislative reform in virtually every West African country since independence has strengthened central control and reduced safeguards against the mis-use of land laws to grant title to commercial or private interests at the expense of smallholders (for example Nigeria's Land Use Act of 1978). Meanwhile resistance against codifying rights to family land, traditionally responsive to changing social realities, has led to a plethora of informal contracts which may actually increase the efficiency of labour to land ratios, both in the forest zone (Berry 1993), and the Sahel

(Chauvaux et al. 2006). These contracts include cultivation of land under tree crops, rights to plant tree crops on village lands, rights to lend land and guardians' rights to valued common land (Cotula et al. 2006). In many places (such as Senegal: Lo and Dione 2000), customary contracts continue in defiance of central legislation.

While adaptive change—in the form of spontaneous, informal contracts—responds to economic and social pressures, a simple market model oversimplifies the complexity at the village level (Benjaminsen and Lund 2003). The remaining forests have come under new pressure from large-scale capital, as, using the myth of unused or unoccupied spaces, central governments have issued leases to foreign corporations or indigenous entrepreneurs for large-scale commercial agriculture. These are justified in terms of national economic priorities (fuels such as Jatropha, export agriculture such as sugar, meat and grain to new markets). A rapidly expanding literature follows the global 'land grabbing' trend especially in Africa. In West Africa, however, the expansion of smallholder agriculture still poses a greater threat to the livelihoods of pastoralists (Hof et al. 2006; Ariyo and Mortimore 2012; Mortimore 2001).

These tensions form the context of agricultural expansion and the management of remaining woodlands in the Sahel. However, negotiated conventions between stakeholders in forest areas offer a solution to persistent conflicts of interest. The Takiéta Joint Forest Management Project helps local people to manage a common resource in an inclusive way. It covers an area of 6,720 ha in SE Niger (Vogt and Vogt 2000). The aim is to create an effective local management structure and strategy through facilitation and genuine participation. Forest boundaries were first redefined. Legal residents were subjected to agreed conditions of access. Dialogue was commenced between user groups on the future of the reserve and costs of management. A Local Management Structure was created with consensual support. Maps (including soils, pastures and vegetation), inventories and knowledge of the resource were put together. Rules for forest use were agreed and user feedback to the draft planning document were taken into account. Following a workshop, management commenced in 2000. The Takiéta experience has since been replicated.

3.6.2 Burning

This practice was the principal method used in clearing forest for agriculture, and almost universally regarded as destructive in ecological and value terms. The history of attempts to control or prevent burning by West African farmers is a vivid example of the clash between indigenous and external knowledge systems that originated in authoritarian colonialism (Cline-Cole 1997).

Burning—which is done to control spontaneous regeneration on farmland or to clear away unwanted vegetation on new fields or after a fallow cycle—has been accused of causing wind and water erosion, reducing soil moisture and organic carbon, and damaging biodiversity. Little attention was paid to the rationale for burning, which was and still is the core technology for transforming woodland into farmland in rotational cultivation systems where labour is limiting and financial capital scarce. Foresters called burning 'indiscriminate'. Colonial forestry departments tried hard to stop burning, and notwithstanding the ineffectiveness of bans, most governments still try to regulate, if not to prevent, the practice.

Such efforts are wasted. In the sub-humid region of south-west Mali, recent research shows that up to 57 % of the landscape may be burned in a year (Laris and Wardell 2006). Draconian regulations to limit burning have been challenged on the basis of work in the subhumid zone of Guinea (Fairhead and Leach 1996). Burning is less damaging if done early in the dry season. It is a management strategy—and not 'indiscriminate'—most common in woodland areas with low population densities, and least common where densities are high and aridity greatest. The differentiation amongst more and less fire-resistant species may contribute to the spatial and temporal diversity of the ecosystem. Burn-scar mapping shows that burning occurs in patches, and it has been argued that the resulting mosaic of burnt and unburnt patches benefits biodiversity, conserving a mix of species. Recognizing the value of indigenous knowledge, a decentralization to local communities of powers and policy on burning practice has been tried with good effect (Laris and Wardell 2006).

3.6.3 Fuelwood Cutting

More than 90 % of energy used for cooking and domestic use in rural areas, and a substantial (though variable) fraction in urban areas, depends on fuelwood or charcoal. Overall densities of trees on farms, the availability of preferred fuelwood species, and the rate of natural germination and regeneration of indigenous trees were all expected to decline to vanishing point on the basis of 'projections' made around mid-century (Trevallion 1966; Moss and Morgan 1981). Beneficial properties of trees in re-fertilizing soil and protection from wind erosion would be lost, while exponential growth of urban consumers would inflate prices and create an ever-expanding 'desert' aureole surrounding growing towns. This scenario was commonly promoted across the African Sahel. Colonial and post-colonial forestry departments used it as a justification for centralizing the control of forest resources (Cline-Cole et al. 1990b).

Statistics of fuelwood cutting and consumption are difficult to compile. Exaggeration has been rife. For example, estimates of the future impact of wood cutting on the dry forests surrounding Bamako were found to be unjustifiably pessimistic (Foley 2001). Revisions changed the expected impact by an order of magnitude. Methods used to estimate natural forest regeneration are underdeveloped though regeneration is robust. A lack of knowledge on standing stock, mean annual increment and inter-specific variability in dry forests impeded the findings of a major study of wood supply and demand in northern Nigeria (Silviconsult 1991).

In northern Nigeria, a dual sector model is best placed for understanding the fuelwood industry (Cline-Cole et al. 1990a, b). In rural areas, reckless deforestation for fuelwood cutting is a myth wherever an individual's land rights are recognized. Studies conducted at the micro-scale have confirmed that farmers (and settled agro-pastoralists) conserve the trees they own. Every tree is selected for the economic value of its non-timber forest products (IUCN 2009), as well as the branchwood, which is used by householders for cooking and heating. This 'rural consumption model' is self-supporting—or aims to be so—but wood may be sold when income is urgently needed.

An 'urban consumption model', on the other hand, is supported by markets and takes no thought for the future. In northern Nigeria, the fuelwood demand for cooking and heating escalated during the twentieth century (Hyman 1993). It follows demographic trends (increasing populations and rapid urbanization), and competes with alternative energy sources (kerosene, bottled gas and electricity) whose uncompetitive prices, and irregular supplies, are major constraints on their uptake. However, rates of household consumption tend to decline with increasing price, especially in urban areas (Cline-Cole et al. 1990a; Hyman 1993). Higher prices should incentivize the purchase of improved stoves, which have received large promotion from World Bank funded projects in Niger and other countries (Noppen et al. 2004).

In closely settled areas with farm trees, the opportunity costs of foregoing the value of non-timber forest products, and the environmental value of standing stock (e.g., shade, soil nutrients) ensure that urban demand for fuelwood is transferred to residual natural forests. In Nigeria, weakly regulated or illegal woodcutting takes place in vulnerable natural forests, including state forest reserves. Rather than a 'fuelwood desert' the patterning of fuelwood landscapes is better characterized as a 'fuelwood perimeter' located in the woody interstices between villages.

Instead of enforcing regulations which are seen as antithetical to local interests, recent reforms in governance institutions (decentralization) have moved forest policy towards local participation. Sustainable futures now rest, in place of privileging access for favoured entrepreneurs and taxing trade movements, in the transfer of ownership, policing and sales quotas to local community organizations (Noppen et al. 2004). Expected benefits are poverty reduction (cash in hand), forest conservation (community monitoring), local development (forest management funds), private sector development, and planning capacity. However, decentralization has failed so far to deliver the expected benefits (Hesse et al. 2013).

3.6.4 Sustainability

The use of the term 'deforestation' commonly implies irreversible destruction or loss of value. A vast literature surrounds the subject at the global scale. However from a dryland smallholder's perspective, the conversion of woodland (most often already degraded secondary woodland) to small-scale farming creates a productive asset for multiple cropping which also includes NTFPs of a wide range of fruit, fodder, food, medicinal and other useful plants, as well as capturing synergies from indigenous agroforestry (e.g., complementary shade regimes, nitrogen fixation, and soil fertilization through animal droppings)(Davies et al. 2012).[7] Such capitalization is achieved largely through labour investments over long periods, enabling small but significant nutritional and economic benefits in small-scale farming livelihoods. Interviews in northern Nigeria and southern Niger bear out the value in which biodiversity is held (Mortimore et al. 2008), and biodiversity conservation is seen as a poverty reduction strategy (Roe et al. 2013), in which burning may yet have a place, and in which tree pollarding preserves wanted species without impeding the growth of crops.

Farmed parkland is not a new feature. It was observed in Kano in the mid-19th century (Barth 1857), in Senegal in the 1950s (Pelissier 1953), in Zaria in the 1970s (Pullan 1974), and in Kano in the 1980s (Cline-Cole et al. 1990a). Its impact on the landscape is visually impressive and apparently permanent. The density of mature trees on farmland (12–15/ha) was maintained throughout two major drought episodes (1973–74 and 1984–85), when pressure of food insecurity threatened to reduce tree stocks through fuelwood cutting for sale in nearby urban markets. It is sustained by protecting natural regeneration of native species, and supplemented by planting exotic shade and fruit trees on private farms and inside compounds. In the Kano region, 75 tree species are preserved under an average annual rainfall of 600 mm. Under an average annual rainfall of 360 mm, 135 useful tree, shrub, grass and herb species were inventoried, including 67 wild food varieties (Mohammed 1994 cited in Harris and Mohammed 2003).

Ecological sustainability in the Sahel must make economic sense to farmers and pastoralists, and a condition for this is secure access and enjoyment rights. Even so, with price inflation, fuelwood cutting could become irresistibly profitable in the Kano markets, overwhelming the value of biodiversity (Maconachie 2007). This is not a new suggestion, and in the absence of longitudinal data, the sustainability of dryland ecosystems may easily be underestimated. However, while overall tree densities may be sustained, the age or species structure of the trees may indicate imbalances significant for regeneration. For example, in the Serer villages of Senegal, the unwanted shrub *Guiera senegalensis* accounts for 57 % of the regenerating individuals, and the loss of more valued species is reported (Sadio et al. 2000; Lericollais 1999).

Recently, eastern Niger has witnessed a dramatic growth in protected regeneration of indigenous trees on farms, especially *Faidherbia albida* whose leaves and pods become available in the dry season while falling in the wet, when growing

[7]This view would be difficult to demonstrate in measured economic terms because a wide range of non-marketed benefits and costs is involved in both space and time. Environmental economics provides methods for turning many costs and benefits into monetary terms. However, combining monetary and non-monetary items in the same equation must overlook the fact that decisions about resource use may not be made either in 'cash' or 'non-cash' terms, but on social, cultural or other qualitative grounds.

crops can make use of their nitrogen-enhancing effect on the soil (Boubacar, this volume; Sendzimir et al. 2011; Reij et al. 2009; WRI 2008). This trend was facilitated by a conjunction of drivers including the reform of land and tree ownership rights under the Rural Code. Large numbers of farmers are claimed to benefit from this Farmer-Managed Natural Regeneration. While an institutional framework supporting tenure security of land and trees is a necessary condition of sustainable tree management, local factors such as market incentives and livelihood opportunities are the sufficient conditions for its realization. The impact of motivated and consistent extension through NGOs and donor programmes in supporting and improving traditional practice may have been critical (Jouet et al. 1996; Reij et al. 2005). With regard to its future, the impact of very high densities of regenerating trees on the farming system, and the beneficial impact on poverty in Niger (Gubbels 2011), need research.

The mature farmed parklands suggest a theory of transition that proceeds from initial clearance of farmland from forest, through toleration of selected volunteer species along farm boundaries, to their active protection on cropland. If markets have driven conservation in highly valued farmed parklands, the same forces put pressure on remaining areas of natural forest and on forest reserves which are surrounded by competing stakeholder interests (e.g., hunting, wood cutting, grazing and farming incursions, often illegal). A process view of deforestation challenges some orthodoxies of forest management. But smallholders' impact, while considerable, is now small, slow and incremental; most of the colonial waves of migration in search of new lands are completed and the exhaustion of the supply of free land (saturation) is driving a new logic of smallholder intensification, in which increased inputs of labour, capital and fertilization conjoin with the protection of tree regeneration.

3.7 Soil Nutrient Depletion—Man-Made Deserts?

The biological productivity of an ecosystem depends on the health of its soils and on soil moisture during the growing period. Although the management of fertility status in soils is often presented in terms of key chemical nutrients (nitrogen, N, phosphorus, P and potassium, K), being easily manipulated in the short term through inorganic fertilizer amendments, other properties are vitally important in the longer term, including some trace elements, salinity, physical attributes (texture, depth and structure), and organic carbon content (C). In Africa, a dominant narrative of soil degradation and erosion has been influential in debates about dryland management, a key theme being nutrient depletion on farmland.

Early surveys claimed that 332 million hectares (25.8 % of the continent's surface) are affected by soil degradation in the arid, semi-arid and dry sub-humid agro-ecological zones.This work, carried out by ISRIC (Oldeman and Hakkeling 1990) and the Winrand Staring Centre (Stoorvogel and Smaling 1990), was used by the First Edition of the World Atlas of Desertification (UNEP 1992). Extended estimates were later published of the annual depletion of chemical nutrients (Stoorvogel and Smaling 1990), which were promoted by the World Bank and other agencies. These put net combined N, P and K losses at 60–100 kg/ha/yr and increasing (Henao and Banaante 1999; World Bank 2001). All but three African countries were said to be losing >30 kg of N, P and K/h/yr.

This narrative continues to guide policy makers, for example at the Abuja Fertilizer Summit, at which governments committed themselves to higher targets for fertilizer procurement and distribution (African Union 2006). According to this narrative, chemical fertilizers assume the role of palliatives for under-productive African agriculture, though their acquisition is highly dependent on international trade, expensive quarrying and processing, or high consumption of non-renewable energy. Fertilizer use has to be subsidized to make it economically viable; consequently the political economy of fertilizer distribution may be an obstruction to its equitable use.

Critiques of the nutrient depletion scenario focused on the conceptualization and methodologies used in measuring, scaling up and projecting from micro-scale site observations to regional and continental scales (Scoones and Toulmin 1998; Faerge and Magid 2004). Key chemical nutrients cannot alone (that is, without essential micro-nutrients) recreate a sustainably fertile soil, or replenish the critically important stock of organic carbon. What solutions can be found for soil degradation?

The Sahel falls behind more humid regions in its agricultural productivity (Breman and de Wit 1983). But a paradox of Sahelian agricultural systems is their persistence over decades under conditions of population growth and unreliable rainfall. The achievements of small-scale farmers challenge the grand theory of failing African agriculture, and provide a platform for current advances in productivity supported by science and policy (Djurfeldt et al. 2011). Nutrient cycling is the key to soil fertility in relatively closed systems supporting a subsistence household economy (Drinkwater et al. 1998). Mixed crop-livestock systems rotate cultivated fields with fallows and transfer nutrients to cultivation through grazing animals (Powell et al. 2004; Thomas et al. 2006). Where land scarcity eliminates field rotation, crop residues, dry composting, inter-cropping and multiple weeding strategies bolster the supply of organic matter (Harris 1999; Mortimore and Harris 2005). In some more intensive systems, with higher rainfall, crop residues can support higher stocking densities even without rangeland (Harris and Yusuf 2001).

Farmers recognize that organic as well as inorganic fertilization is necessary to approach the potentials shown on research stations. But there is never enough, giving rise to sharp differences between plots. These are shown in micro-scale case studies. At four locations in Maradi (Niger), field sampling showed that by investing in a mix of organic manure and dry compost, and very small doses of inorganic fertilizer (when affordable), a farmer can restore organic matter to levels comparable to uncultivated soil (Issaka 2001). This implies that cultivation has not irreversibly degraded the soil.

Research-based soil fertility strategies build on indigenous knowledge and practices. Scarce and costly inorganic fertilizers can be optimized by mixing 'micro-doses' with organic matter (building on local practice) and 'hill placement' on the individual stands (Dimes et al. 2005; Aune et al. 2005). Crop rotations and mixtures (especially of grain and legumes) also mimic indigenous practices. Where conditions are suitable, water harvesting with ridges and basins improves moisture conditions, notably the successful uptake of the *zaï* technique for concentrating scarce surface water and nutrients (Traoré et al. 2005; Mando et al. 2006). Nitrogen-fixing trees are promoted on farms (Thomas et al. 2006).

Conservation agriculture, based on zero or minimum tillage (Aune et al. 2005; Milder et al. 2011) and a range of sustainable land management (SLM) technologies (see WOCAT 2007) is largely unproven under Sahelian conditions of low bio-productivity and integrated crop-livestock systems. Claims that it increases yields, reduces labour requirements, improves soil fertility and reduces erosion have been challenged on the basis of weak or inconsistent evidence (Giller et al. 2009). The need for agrochemical inputs is also unclear in many accounts.

'Ecological intensification'—which includes intercropping systems, integrated pest management, and organic farming—has broadened the scope of extension efforts globally, estimated to have benefited 40 million smallholders (Pretty et al. 2011). 'Sustainable intensification' is an imperative across ecological, genetic and socio-economic spheres (Montpellier Panel 2013). Currently, increasing attention is being paid to biological attributes, soil organisms and the carbon cycle (Uphoff et al. 2006). This new (or 'second') paradigm advocates 'more knowledge of and reliance on biological processes' (Uphoff et al. 2006, pp. 12, 693, 698). It replaces dependence on external inputs—a paradigm characterized as 'Overcome soil constraints through the application of fertilizers and amendments to meet plant requirements'—with 'Rely more on biological processes by adapting germplasm to adverse soil conditions, enhancing soil biological activity, and optimizing nutrient cycling to minimize external inputs and maximize the efficiency of their use'. The most striking feature of the change proposed is a shift from relying on exogenous inputs to an emphasis on endogenous processes within the system—and this no longer the perceived closed system of 'subsistence farming' but a system open to environmental change and the political economy.

A shift from sectoral (commodity-based) research and development to systems approaches translates integrated soil fertility management (ISFM)—no longer perceived mainly in terms of affording chemical amendments—into part of a 'complex adaptive system' (Schiere et al. 2006). It includes a range of interacting drivers and components such as soil and water conservation, ecosystem services, markets, fertilization (organic plus inorganic), integrated pest management, markets, institutions and policy (Bationo et al. 2005; Vanlauwe et al. 2006). Such a change challenges the scientist to integrate research-based knowledge with that of the Sahelian farmer. In place of a policy strategy built on averages and assuming homogeneity in the farmers' response, the system offers space for multiple pathways (Scoones and Wolmer 2000), and variance becomes a virtue, not a nuisance (Schiere et al. 2006).

The management of soil processes—whether sustainable or degrading—is central to desertification, and justifies this excursion into the changing narratives that link people, their knowledge, and environment. Soil bio-productivity is literally a matter of life and death, and the challenges for technology are plain for all to see: invest through participatory research on organic matter management, dry composting, rain-water harvesting, conservation agriculture, crop rotations and mixtures, micro-dose fertilization and other strategies. Nutrient cycling is site-specific.

3.8 Conclusion: Escaping Malthus?

In the preceding sections of this chapter a case is made for revising the conventional wisdom that supports a desertification paradigm for the Sahel. If a new paradigm of adaptive resilience is justified by the science, what should be concluded about poverty, population growth and the future?

3.8.1 Poverty

The six countries of the Sahel (Nigeria must be excluded here because half its area and population lie outside the ecological Sahel) score behind others, even in Africa, on indicators of poverty, vulnerability and well-being (Dobie and Goumandakoye 2005). In these six (The Gambia, Senegal, Mali, Burkina Faso, Niger and Chad), an average 44 % of their populations live below the international poverty line (US $1.25/day). The UNDP's Multi Dimensional Poverty Index gives them an average value of 0.473, which can be compared with South Africa (0.057), Ghana (0.144) and Kenya (0.229) (UNDP 2011).

The Sahel has a combination of increasing rural population densities, erratic rainfall, infertile soils, deforestation and poverty which appears, on the face of it, to pose a classic neo-Malthusian crisis and, given a population doubling-time as low as 22 years (on recent fertility rates), an extremely urgent one. Furthermore, there is evidence that the demographic transition to lower fertility is faltering in some African countries including those of the Sahel (Economist 2014; for Niger, see Boubacar, this volume). For many observers, an intuitive barrier struggles to accommodate a projected 500 million Sahelians. In the Horn of Africa, the case for a neo-Malthusian crisis in pastoralism, and its dependent human populations, is provocatively argued by Sandford (2011). In western Africa, the Sahel has been shedding its pastoral population for generations (Baier 1980). Following the great Sahel famine of 1972–74, many observers were quick to conclude such a crisis, not only for pastoral populations but also for farmers (Mortimore 1998).

Overall, per capita food production in Sahelian countries has stagnated (according to FAO statistics (Mortimore 2003; FAOSTAT 2014). But according to these data, neither did it decline (1961–2001), and a virtual doubling of rural population indicates that total production increased substantially. The answer to this anomaly appears to lie in economic differentiation which condemns a third of the population, lacking enough land, labour or capital, to chronic food insecurity (Gubbels 2011). They depend on food commodity markets even in years of good rainfall. Under conditions of variable rainfall, yields and harvests, weak demand on account of poverty leaves no space for building reserves for the bad year when prices 'sky-rocket'. The most vulnerable of this population are young children. The co-existence of poverty with evidence of adaptive resilience suggests that poverty depresses demand and restricts the investment necessary to achieve sustainable NRM.

Using earth satellite data and population projections, Abdi et al. (2014) modeled increases or decreases in net primary production (NPP) 'supply' and 'demand' during the period 2000–2010 for 22 Sahelian and neighbouring countries. NPP is produced not only on cropland, where production is estimated using FAO crop yield statistics in conjunction with land use classes, but also in woodland, forest (absent in some countries), and savanna grassland. Given a rate of natural increase of 2.8 % per year, the regional population grew from 367 million in 2000 to 471 million in 2010. Assuming that 80 % of crop calories are used for food, estimated NPP 'demand' per capita increased from 19 to 41 % of the 'supply', suggesting a threat to food security. However, only two countries saw negative change in total NPP, which registered significant increases in cropland, woodland and grassland overall, probably due to improved rainfall.

3.8.2 *Adaptation*

Variability characterizes both the 'given' elements of the Sahelian environments and the adaptive responses documented by many observers. In a new and fundamental argument for a shift in paradigm (Krätli 2015), a case is made for 'valuing variability' as an asset of smallholder farmers as well as pastoralists adept in managing incumbent risk through applying local environmental knowledge. Such a model can resolve the contradictions between 'top-down' and 'bottom-up' development strategies and capitalize on the tenacity evident in these social-environmental systems.

Given the variability of a non-equilibrial environment, as shown in the droughts of 1972–74 and 1983–84, the adaptive capacities of Sahelian peoples were underestimated (Mortimore 1989, 2001, 2010). Early projections of mortality in hundreds of thousands—there are no statistics—were not realized (Caldwell 1975), and notwithstanding food security crises in subsequent years, and persistent calls for food aid, the Sahelian farming systems have not collapsed but today support (though inadequately) twice as many people (Tiffen and Mortimore 2002). Adaptation (for farming families) took place at four levels:

- Adjustment of labour on-farm (sowing, weeding, harvesting) to short-term variability of rainfall (Fig. 3.4)
- Use of famine foods from wild plants to supplement or replace cereal grains

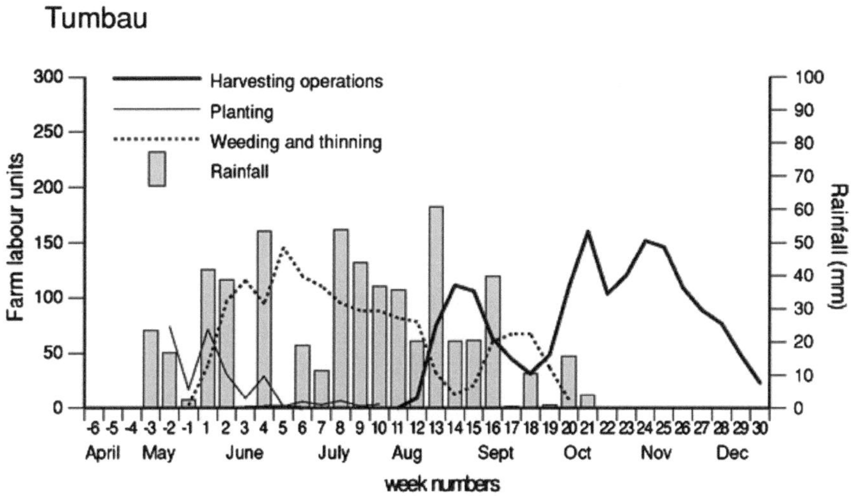

Fig. 3.4 Adapting farm labour inputs to planting, weeding and harvesting under variable rainfall in Kano, Nigeria (units of one week). *Source* Mortimore and Adams 1999.

- Searching for alternative income sources locally
- Migration (short term) seeking employment or trade in towns

Excepting the first, these strategies were also employed by WoDaaBe pastoralists, whose herd's survival depended on the speed and distance of their movements (Bernus 1977a, b). Flexibility, diversity and mobility enabled such adaptation which took most observers by surprise. They are documented over the long term and notwithstanding setbacks and contradictions, they suggest a process of economic and social development through adaptation and diversification as in Senegal and Niger (Faye et al. 2001; Mortimore et al. 2001). It is not possible to consider adaptation to desertification separately from adaptation to drought, since actions taken in the context of short-term food insecurity aggregate over time into a management process.

These capacities, demonstrated at the local level, suggest that a two-step escape route from a Malthusian outcome has been exploited by households in Sahelian social-ecological systems: first a 'smallholder labour-intensification' pathway from rotational subsistence farming to multiple purpose crop-livestock-tree cultures, and second an income diversification pathway that exploits regional income opportunities as well as local. For pastoralists, the first step is to protect the right to do what is already known well. Mobile herding and breeding systems are labour-intensive. Many pastoralists embark on the second step, which may evolve into sedentarization.

For the first step, research and practice based on labour intensification needs to include the stated goal of strengthening capacities to manage non-equilibrial environments, whether those exploited by pastoralists or those cultivated by farmers and agro-pastoralists.[8] This goal must include negotiating rights of access and benefits among contesting stakeholders, and strategies for sustainable use of natural resources. Insurance schemes against risk of loss (of crops or animals) are attracting attention (Hesse et al. 2013), and lessons are being learnt including the need for index-linked insurance to be embedded in a wider portfolio of credit, inputs and risk management. They may also offer social protection for those whose adaptive options have run out (Hazell and Hess 2007). Insurance is becoming more of a priority with contemporary changes such as land 'saturation', market penetration, state interventions in land and water, and conflict.

For the second step, barriers to movement (whether of herds or of migrants) need removal and such regulation as necessary to guarantee mobility when and where needed. This is not always in the forefront of government's concerns. Its necessity arises from increasing regional economic integration, both national and international. Both rural-urban and cross-border movements will increase. Short term movements and circulation are on the increase. Dryland inhabitants have a stake in urbanization as well as their right of cross-border movement under the terms of the Economic Union of West African States (ECOWAS).

Conventional development theory and practice perceives non-equilibrial environments—especially those of the West African Sahel—as a management challenge, subject to increasing demographic, climatic and economic pressures. Public, private and external investment holds the key to reversing poverty and—in time—rebalancing the Sahelian social-environmental systems. Such a view returns to an equilibrial understanding of change. Knowledge systems offer new investment options and scope (UN/EMG 2012).

However, a non-equilibrial model urges that greater attention be paid to local, as well as science-based, knowledge and social capital in engaging realistically with variability and uncertainty in adaptive food production systems, multiple risk management options, mobility, and other parameters at the small or household scale —in essence 'giving small-scale producers a second chance' (Krätli 2015).

This chapter has traced the evolution of development practice and the beginning of a paradigmatic shift towards a better match between environmental change and its management in the West African Sahel. Many questions remain unanswered. Meanwhile the challenge intensifies.

References

Abdi, A. M., Seaquist, J., Tenenbaum, D. E., Eklundh, L., & Ardö, J. (2014). The supply and demand of net primary production in the Sahel. *Environmental Resarch Letters, 9*(9). http://topscience.iop.org/1748-9326/9/9/094003/article

[8]The options for widening the scope of farming and agroforestry systems in drylands are developed in Omanya and Pasternak (2003).

Adams, W. M., & Mortimore, M. (1997). Agricultural intensification and flexibility in the Nigerian Sahel. *The Geographical Journal, 163*, 150–160.

African Union. (2006). *Abuja Declaration on Fertilizer for the African green revolution*. African Union Special Summit of the Heads of State and Government, Abuja, Nigeria, June 13, 2006.

Anyamba, A., & Tucker, C. J. (2005). Analysis of Sahelian vegetation dynamics using NOAA-AVHRR NDVI data from 1981–2003. *Journal of Arid Environments, 63*(3), 596–614.

Ariyo, J. A., & Mortimore, M. (2012). Youth farming and Nigeria's development dilemma: the Shonga experiment. *IDS Bulletin [Institute of Development Studies], 43*(6), 58–66.

Aubréville, A. (1949). *Climats, forêts et désertification de l'Afrique tropicale*. Paris: Société d'Editions Géographiques, Maritimes et Coloniale.

Aune, J. B., Berthe, A., & Doumbia, M. (2005). Conservation agriculture as an approach to agricultural development in dryland areas of Mali. In: G.O. Omanya & D. Pasternak (Eds.), *Sustainable systems for the drylands. Proceedings of the International Symposium for Sustainable Dryland Agriculture Systems* (pp. 239–248), December 2–5, 2003, Niamey, Niger. Niamey: International Crops Research Institute for the Semi-Arid Tropics.

Ba, M., Mbaye, M., Ndao, S., Wade, A., & Ndiaye, L. (2001). *Région Diourbel: cartographie des changements d'occupation-utilisation du sol dans la zone agricole du Sénégal occidentale*. Drylands Research Working Paper 21. Crewkerne, UK: Drylands Research.

Bai, Z. G., Dent, D. L., Olsson, L., & Schaepman, M. E. (2008). *Global assessment of land degradation and improvement: 1, Identification by remote sensing*. World Soil Information Report 2008/01. Rome/Wageningen: FAO/ISRIC.

Baier, S. (1980). *An economic history of central Niger*. Oxford: The Clarendon Press.

Barrow, E. G. C. (1996). *The drylands of Africa. Local participation in tree management*. Nairobi: Initiatives Publishers.

Barth, H. (1857). *Travels and discoveries in North and Central Africa* (Vol. 3). London: Longman.

Bationo, A., Tabo R., Kihara J., Kimeta J., Adamou A., & Koala S. (2005). In G. O. Omanya & D. Pasternak (Eds.), Sustainable systems for the drylands. In *Proceedings of the International Symposium for Sustainable Dryland Agriculture Systems, Niamey, Niger* (pp. 100–128). Niamey: International Crops Research Institute for the Semi-Arid Tropics, December 2–5, 2003.

Benhke, R. H., Scoones, I., & Kerven, C. (Eds.). (1993). *Range ecology at disequilibrium: new models of natural variability and pastoral adaptation in African savannas*. London: Overseas Development Institute.

Benjaminsen, T. A., & Lund, C. (Eds.). (2003). *Securing land rights in Africa*. London: Frank Cass.

Bernus, E. (1977). *Case study on desertification. The Eghazer and Azawak region, Niger*. UN Conference on Desertification. Nairobi: United Nations Environment Programme.

Bernus, E. (1977b). Les tactiques des eleveurs face à la sécheresse: le cas du sud-ouest de l'Aïr, Niger. In J. Gallais (Ed.), *Stratégies pastorales et agricoles des sahéliens durant la sécheresse* (pp. 1969–1974). Bordeaux: CEGET/CNRS.

Berry, S. (1993). *No condition is permanent. The social dynamics of agrarian change in Sub-Saharan Africa*. Madison, Wisconsin: University of Wisconsin Press.

Bovill, E. W. (1921). The encroachment of the Sahara on the Sudan. *Journal of the African Society, 20*(174–85), 259–269.

Breman, H., & de Wit, C. T. (1983). Rangeland productivity and exploitation in the Sahel. *Science, 221*, 1341–1347.

Caldwell, J. C. (1975). *The African drought and its demographic implications*. New York: The Population Council.

Chauvaux, J.-P., Colin, J.-P., Jacob, J.-P., Lavigne Delville, P., & Le Meur, P.-Y. (2006). *Changes in land access and governance in West Africa: Markets, social mediations and public policies. Results of the CLAIMS research project*. International Institute for Environment and Development: London.

Chevalier, A. (1950). La progression de l'aridité, du desséchement et de l'ensablement et la décadence des sols en Afrique Occidentale française. *Comptes Rendu Hebdomadaires des Séances de l'Academie des Sciences, 230*, 2140–2142.
Chevalier, A. (1952). La décadence des sols et de la végétation en Afrique Occidentale française et la protection de la nature. *Bois et Forêts des Tropiques, 16*, 335–353.
Cline-Cole, R. A. (1997). Promoting (anti-) social forestry in northern Nigeria. *Review of African Political Economy, 74*, 515–536.
Cline-Cole, R. A., Falola, J. A., Main, H. A. C., Mortimore, M., Nichol, J. E., & O'Reilly, F. D. (1990a). *Wood fuel in Kano*. Tokyo: United Nations University Press.
Cline-Cole, R. A., Main, H. A. C., & Nichol, J. E. (1990b). On fuelwood consumption, population dynamics and deforestation in Africa. *World Development, 18*(4), 513–527.
Cotula, L., Toulmin, C., & Quan, J. (2006). *Better land access for the rural poor. Lessons from experience and challenges ahead*. London: International Institute for Environment and Development.
Cotula, L. (Ed.). (2007). *Changes in "customary" land tenure systems in Africa*. London: International Institute for Environment and Development.
Cour, J. M., & Snrech, S. (1994). *West Africa long term perspective study: Vision of West Africa in the year 2020*. Paris: OECD.
Davies, J., Poulsen, L., Schulte-Herbrüggen, B., Mackinnon, K., Crawhall, N., Henwood, W. D., et al. (2012). *Conserving dryland biodiversity*. Nairobi: International Union for the Conservation of Nature.
Derrick, J. (1977). The Great West African Drought, 1972–74. *African Affairs, 76*, 537–586.
Dimes, J., Twomlow, S., Rusike, J., Gerard, B., Tabo, R., Freeman, A., & Keatinge, J. D. H. (2005). Increasing research impacts through low-cost fertility management options for Africa's drought-prone areas. In G.O. Omanya & D. Pasternak (Eds.), *Sustainable systems for the drylands. Proceedings of the International Symposium for Sustainable Dryland Agriculture Systems, Niamey, Niger,* (pp. 82–92). Niamey: International Crops Research Institute for the Semi-Arid Tropics, December 2–5, 2003.
Djurfeldt, G., Aretey, E., & Isinika, A. (Eds.). (2011). *African smallholders: Food crops, markets and policy*. Wallingford, UK: CABI International.
Dobie, P., & Goumandakoye, M. (2005). *The Global Drylands Imperative: achieving the Millennium Goals in the drylands of the world*. Nairobi: United Nations Development Programme Drylands Development Centre.
Dregne, H. E. (1983). *Desertification in arid lands. Advances in desert and arid land technology and development 3*. USA: Hardwood Academic Publishers.
Drinkwater, L. E., Wagoner, P., & Sarrantonio, M. (1998). Legume-based cropping systems have reduced carbon and nitrogen losses. *Nature, 396*, 262–265.
Dupire, M. (1962). *Peuls nomade: Etude descriptive des Wodaabe du Sahel Nigérien*. Paris: Institut d'Ethnologie.
Eckholm, E., & Brown, L. (1977). *Spreading deserts: The hand of man*. Worldwatch Paper 13. Washington DC: Worldwatch Institute.
Eklundh, L., & Olsson, L. (2003). Vegetation index trends for the African Sahel 1982–1999. *Geophysical Research Letters, 30*(8), 143.
Faerge, J., & Magid, J. (2004). Evaluating NUTMON nutrient balancing in Sub-Saharan Africa. *Nutrient Cycling in Agro-ecosystems, 69*(2), 101–110.
Fairhead, J., & Leach, M. (1996). *Misreading the African landscape: Society and economy in a forest savanna land*. Cambridge: Cambridge University Press.
Faye, A., Fall, A., Tiffen, M., & Nelson, J., (2001). *Région de Diourbel: Synthesis*. Drylands Research Working Paper 23e. Crewkerne, UK: Drylands Research.
Foley, G. (2001). *Sustainable wood fuel supplies from the dry tropical woodlands*. ESMAP Technical Paper 013. Washington DC: The World Bank.
FAOSTAT. (2014). *UN Food and Agriculture Organisation Statistics*. http://faostat3.fao.org/faostat-gateway

Garcia, R.V. (Ed.) (1981). *Drought and man. The 1972 case history* (Vol. 1). *Nature pleads Not Guilty*. Oxford and New York: Pergamon.

Garcia, R. V., & Escudero, J. C. (Eds.) (1982). *Drought and man. The 1972 case history* (Vol. 2), *The constant catastrophe—malnutrition, famines and drought*. Oxford and New York: Pergamon.

Garcia, R. V., & Spitz, P. (Eds.) (1986). *Drought and man. The 1972 case history (Vol. 3), The roots of catastrophe*. Oxford and New York: Pergamon.

Giller, K. E., Witter, E., Corbeels, M., & Tittonell, P. (2009). Conservation agriculture and smallholder farming in Africa: The heretics' view. *Field Crops Research, 114*, 23–34.

Gubbels, P. (2011). *Escaping the hunger cycle: pathways to resilience in the Sahel*. Sahel Working Group, CAFOD and 8 other agencies.

Harris, F. (1998). Farm-level assessment of the nutrient balance in northern Nigeria. *Agriculture, Ecosystems and Environment, 71*, 201–214.

Harris, F. (1999). Nutrient management strategies of small-holder farmers in a short-fallow farming system in north-east Nigeria. *The Geographical Journal, 165*(3), 275–285.

Harris, F., & Yusuf, M. A. (2001). Manure management by small farmers in the Kano Close-Settled Zone, Nigeria. *Experimental Agriculture, 37*, 319–332.

Harris, F., & Mohammed, S. (2003). Relying on nature: Wild foods in northern Nigeria. *Ambio, 32*(1), 24–29.

Hazell, P. B. R., & Hess, U. (2007). Drought insurance for agricultural development and food security in dryland areas. *Food Security, 2*(4), 395–405.

Henao, J., & Banaante, C. (1999). *Estimating rates of nutrient depletion on soils of agricultural lands of Africa*. Alabama, USA: International Fertilizer Development Centre.

Herrmann, S. M., Anyamba, A., & Tucker, C. J. (2005). Recent trends in vegetation dynamics in the African Sahel and their relationship to climate. *Global Environmental Change, 15*, 394–404.

Herrmann, S. M., & Hutchinson, C. F. (2005). The changing contexts of the desertification debate. *Journal of Arid Environments, 63*(3), 538–555.

Hess, T. M., Stephens, W., & Mariah, U. M. (1995). Rainfall trends in the north-east arid zone of Nigeria 1961–90. *Agricultural and Forest Meteorology, 74*, 87–97.

Hesse, C., Anderson, S., Cotula, L., Skinner, J., & Toulmin, C. (2013). *Building climate resilience in the Sahel*. Paper Presented to the DFID Consultation Workshop on Building Resilience in the Sahel and Planned Building Resilience and Adaptation to Climate Extremes and Disasters (BRACED) Programme—17th July 2013. London: International Institute for Environment and Development.

Hof, A., Malami, B. S., & Rischkowvski, B. (2006). *The use of geoinformation processing to illuminate the dimensions of land cover change in the Zamfara Reserve, northwest Nigeria*. Available online http://ubt.opus.hbz-nrw.de/voltexte/2006/362/

Holling, C. S. (1973). Resilience and stability of ecological systems. *Annual Review of Ecology and Systematics, 4*, 1–23.

Holling, C. S. (2001). Understanding the complexity of economic, ecological and social systems. *Ecosystems, 4*, 390–405.

Hubert, H. (1920). Le desséchement progressif en Afrique Occidentale. *Bulletin du Comité d'Etudes Historique et Scientifiques d'AOF*, 401–67

Hulme, M. (1992). Rainfall changes in Africa 1931–60 and 1961–90. *International Journal of Climatology, 12*, 685–699.

Hyman, E. (1993). Forestry policies and programmes for fuelwood supply in northern Nigeria. *Land Use Policy, 10*(1), 26–43.

Ibrahim, F. (1978). Anthropogenic causes of desertification in western Sudan. *GeoJournal, 2*(3), 243–254.

Issaka, M. (2001). *Evolution à long terme de la fertilité de sol dans la Région de Maradi*. Drylands Research Working Paper 30. Crewkerne, UK: Drylands Research.

IUCN. (2009). *Dryland opportunities: A new paradigm for people, ecosystems and development*. Gland, Switzerland: International Union for the Conservation of Nature.

Jouet, A., Jouve, P., & Banoin, M. (1996). Le défrichement amelioré au Sahel. Une pratique agroforestiére adopté par les paysans. In P. Jouve (Ed.), *Gestion des terroirs et des ressources naturelle au Sahel* (pp. 34–42). Montpellier: CNEARC.

Knauer, K., Gessner, U., Dech, S., & Kuenzer, C. (2014). Remote sensing of vegetation dynamics in West Africa. *International Journal of Remote Sensing, 35*(17), 6357–6396.

Krätli, S. (2008). Cattle breeding, complexity and mobility in a structurally unpredictable environment: the WoDaaBe herders of Niger. *Nomadic Peoples, 12*(1), 11–41.

Krätli, S. (2015). In H. de Jode (Ed.), *Valuing variability: New perspectives on climate resilient drylands development.* London: International Institute for Environment and Development.

Krätli, S., & Schareika, N. (2012). Living off uncertainty: the intelligent animal production of dryland pastoralists. In L. Mol & T. Sternberg (Eds.), *Changing deserts: Integrating people and their environment* (pp. 154–75).

Lamprey, H. S. F. (1975). *Report on the desert encroachment reconnaissance in northern Sudan: 21 October to 10 November, 1975.* Nairobi: United Nations Environment Programme.

Land Degradation & Development. (2011). Special Issue on understanding land degradation trends, 22/2.

Laris, P., & Wardell, D. A. (2006). Good, bad or "necessary evil"? Reinterpreting the colonial burning experiments in the savanna landscapes of West Africa. *The Geographical Journal, 172*(4), 271–290.

Lavigne-Delville, P., Toulmin, C., Colin, J.-P., & Chauveau, J.-P. (2002). *Negotiating access to land in West Africa: a synthesis of findings from research on derived rights to land.* London: International Institute for Environment and Development.

Lericollais, M. (Ed.). (1999). *Paysans Sereer: Dynamiques agraire et mobilités au Sénègal.* Paris: Editions de l'IRD.

Lo, H., & Dione, M. (2000). *Région de Diourbel: évolution des régimes fonciers.* Drylands Research Working Paper 19. Crewkerne, UK: Drylands Research.

Maconachie, R. (2007). *Urban growth and land degradation in developing cities. Change and challenges in Kano, Nigeria.* Aldershot: Ashgate.

Magnan, A., Schipper, E. L. F., Gemenne, F., Bharwani, S., Burkett, M., Burton, I., …& Singh, H. Reframing maladaptation to climate change. In *Proceedings of a Workshop held at the Rockefeller Centre, Bellagio, Italy*, November 6–9, 2012 (in preperation).

Mahamane, A. (2001). *Usages des terres et évolution végétales dans le Département de Maradi.* Drylands Research Working Paper 27. Crewkerne, UK: Drylands Research.

Mando, A., Fatondji, D., Zougmore, R., Brousard, L., Bielders, C. L., & Martius, C. (2006). Restoring soil fertility in semi-arid West Africa: assessment of an indigenous technology. In N. Uphoff, A. S. Ball, E. Fernandes, H. Herren, O. Husson, M. Laing, C. Palm, J. Pretty, P. Sanchez, N. Sanginga, & J. Thies (Eds.), *Biological approaches to sustainable soil systems* (pp. 391–400). CRC/Taylor & Francis: Boca Raton (USA).

Mazzucato, V., & Niemeijer, D. (2000). *Rethinking soil and water conservation in a changing society.* Tropical Resource Management Paper 32. Wageningen: Wageningen University and Research Centre.

Mbow, C., Brandt, M., Ouedraogo, I., de Leeuw, J., & Marshall, M. (2015). What four decades of Earth observations tell us about land degradation in the Sahel? *Remote Sensing, 7*, 4048–4067.

Menshing, H. G. (1980). The Sahelian zone and the problems of desertification. Climatic and anthropogenic causes of desert encroachment. *Paleoecology of Africa and the Surrounding Islands, 12*, 257–266.

Mensching, H., & Ibrahim, F. (1977). The problem of desertification in and around arid lands. *Applied Sciences and Development, 10*, 7–43.

Middleton, N., & Thomas, D. G. (1997). *World atlas of desertification* (2nd ed.). Nairobi/London: Edward Arnold for United Nations Environment Programme.

Milder, J. C., Majanen, T., & Scherr, S. J. (2011). *Performance and potential of Conservation Agriculture for climate change adaptation and mitigation in Sub-Saharan Africa.* Nairobi: Rural Futures Initiative, WWF/CARE.

Montpellier Panel. (2013). *Sustainable intensification: a new paradigm for African agriculture. A 2013 Montpellier Panel Report*. Montpellier Panel: Montpellier/London.
Mortimore, M. (1989). *Adapting to drought: Farmers, famines and desertification in West Africa*. Cambridge: Cambridge University Press.
Mortimore, M. (1993). The intensification of peri-urban agriculture: The Kano Close-Settled Zone, 1964–1986. In B. L. Turner, R. W. Kates, & G. Hyden (Eds.), *Population growth and agricultural change in Africa* (pp. 358–400). Gainesville: University Press of Florida.
Mortimore, M. (1997). *History and evolution of land tenure and administration in West Africa*. Issue Paper 71, Drylands Network. London: International Institute for Environment and Development.
Mortimore, M. (1998). *Roots in the African dust: Sustaining the Sub-Saharan drylands*. Cambridge: Cambridge University Press.
Mortimore, M. (2001). Hard questions for "pastoral development': A northern Nigeria perspective. In E. Tielke, E. Schlecht, & P. Hiernaux (Eds.), *Elevage et gestion de parcours au Sahel, implications pour le développement* (pp. 101–115). Stuttgart: Verlag Grauer.
Mortimore, M. (2003). *The future of family farms in West Africa: what can we learn from long-term data?* Drylands Programme Issue Paper 119. London: International Institute for Environment and Development.
Mortimore, M. (2010). Adapting to drought in the Sahel: Lessons for climate change. *Wiley Interdisciplinary Reviews Climate Change, 1*(1), 134–143.
Mortimore, M., & Adams, W. M. (1999). *Working the Sahel. Environment and society in northern Nigeria*. London: Routledge.
Mortimore, M., Tiffen, M., Boubacar, Y., & Nelson, J. (2001). *Synthesis of long-term change in Maradi Department, Niger, 1960–2000*. Drylands Research Working Paper 39e. Crewkerne, UK: Drylands Research.
Mortimore, M., & Harris, F. (2005). Do small farmers' achievements contradict the nutrient depletion scenarios for Africa? *Land Use Policy, 22*, 43–56.
Mortimore, M., & Turner, B. (2005). Does the smallholder's management of woodland, farm trees, rangeland support the hypothesis of human-induced desertification? *Journal of Arid Environments, 63*, 567–595.
Mortimore, M., Ba, M., Mahamane, A.,Rostom, R.S., Serra del Pozo, P., & Turner, B. (2005). Changing systems and changing landscape: measuring and interpreting land use transformations in African drylands. *Geografisk Tidsskrift (Danish Journal of Geography), 105*(1), 101–118.
Mortimore, M., Ariyo, J., Bouzou, I.M., Mohammed, M., & Yamba, B. (2008). Niger and Nigeria: The Maradi-Kano region. A dryland case study of local natural resource management. In: Shepherd, G. (Ed.), *The ecosystem approach. Learning from experience*. Gland (Switzerland): Commission for Ecosystem Management, World Conservation Union.
Moss, R. P., & Morgan, W. B. (1981). *Fuelwood and rural energy production and supply in the humid tropics*. Tokyo: United Nations University.
Nicholson, S. (2005). On the question of the "recovery' of the rains in the West African Sahel. *Journal of Arid Environments, 63*(3), 615–641.
Nkonya, E., Winslow, M., Reed, M. S., Mortimore, M., & Mirzabaeve, A. (2010). Monitoring and assessing the influence of social, economic and policy factors on sustainable land management in drylands. *Land Degradation and Development, 22*(2), 24–247. doi:10.1002/ldr.v22.2/issuetoc
Noppen, D., Kerkhof, P., & Hesse, C. (2004). *Rural fuelwood markets in Niger. An assessment of Danish support to the Niger Household Energy Strategy 1989–2003*. London/Copenhagen: International Institute for Environment and Development/Danida.
Oldeman, R., & Hakkeling, R. (1990). *World map of the status of human-induced soil degradation: An explanatory note*. Nairobi: United Nations Environment Programme.
Olsson, L., Eklundh, L., & Ardö, J. (2005). A recent greening of the Sahel—trends, patterns and potential causes. *Journal of Arid Environments, 63*(3), 556–566.

Pelissier, P. (1953). *Les paysan de Sénégal: les civilisations agraires du Cayor à la Casamance*. Saint-Yrieix: Imprimerie Fabregue.

Powell, J. M., Pearson, R. A., & Hiernaux, P. H. (2004). Crop-livestock interactions in the West African drylands. *Agronomy Journal, 96*, 469–483.

Pretty, J., Toulmin, C., & Williams, S. (2011). Sustainable intensification in African agriculture. *International Journal of Agricultural Sustainability, 9*(1), 9–24.

Pullan, R. A. (1974). Farmed parkland in West Africa. *Savanna, 3*(2), 119–152.

Raynaut, C., Grégoire, E., Janin, P., Koechlin, J., & Lavigne Delville, P. (1997). *Societies and nature in the Sahel*. London: Routledge.

Reij, C., Tapan, G., & Belemvire, A. (2005). Changing land management practices and vegetation on the central plateau of Burkina Faso (1968–2002). *Journal of Arid Environments, 63*(3), 642–659.

Reij, C., & Thiombiano, T. (2003). *Développement rural et environnement au Burkina Faso: la réhabilitation de la capacité productive des terroirs sur la partie nord du Plateau Central entre 1980 and 2001*. Amsterdam: CIS, Vrije Universiteit.

Reining, P. (1978). *Handbook on desertification indicators*. Washington DC: American Association for the Advancement of Science.

Reynolds, J. F., & Stafford Smith, D. M. (Eds.). (2001). *Global desertification: do humans cause deserts?*. Berlin: Dahlem University Press.

Reynolds, J. F., Stafford Smith, M., Lambin, E. F., Turner, B. L., Mortimore, M., Batterbury, S. P. J., …& Walker, B. (2007) Global desertification: building a science for dryland development. *Science,* 316, 847–851 (with supporting online material. 2007, May 11, 2007).

Ribot, J. C. (1995). From exclusion to participation: Turning Senegal's Forestry Policy around. *World Development, 23*(9), 1587–1599.

Rippstein, G., Peyre de Fabregues, B., et al. (1972). *Modernisation de la zone pastorale du Niger, 2 vols., Etude Agrostologique 33*. Maison-Alfort: Institut d'Elevage et de Médecine Vétérinaire des Pays Tropicaux.

Roe, D., Elliott, J., Sandbrook, C., & Walpole, M. (Eds.). (2013). *Biodiversity conservation and poverty alleviation. Exploring the evidence for a link*. Chichester: Wiley.

Sadio, S., Dione, M., & Ngom, M. S. (2000). *Région de Diourbel: gestion des ressources forestières et de l'arbre*. Drylands Research Working Paper 17. Crewkerne (UK): Drylands Research.

Safriel, Z., Niemeijer, D., & White, R. (2005). *Ecosystems and human well-being. Desertification synthesis: a report of the Millennium Ecosystem Assessment*. Washington, DC: World Resources Institute.

Sandford, S. (1983). *Management of pastoral development in the Third World*. Chichester: Wiley.

Sandford, S. (2011). *Pastoralists and irrigation in the Horn of Africa: Time for a rethink?*. Paper Future Agricultures Consortium: University of Sussex.

Schiere, H. (JB), Baumhardt R. L., Van Keulen, H., Whitbread, H. M., Bruinsma, A. S., Goodchild, T. (AV), Gregorini, P., Slingerland, M. (MA), & Wiedemann-Hartwell, B. (2006). Mixed crop-livestock systems in semi-arid regions. In G. A. Peterson (Ed.), *Dryland Agriculture, 2nd ed. Agronomy Monograph 23*. Madison, USA: American Society of Agronomy, Crop Science Society of America, Soil Science Society of America.

Scoones, I. (Ed.). (1994). *Living with uncertainty: New directions in pastoral development in Africa*. London: Intermediate Technology Publications.

Scoones, I., & Toulmin, C. (1998). Soil nutrient balances. What use for policy? *Agriculture, Ecosystems and Environment, 71*, 255–267.

Scoones, I., & Wolmer, W. (Eds.). (2000). *Pathways of change: crops, livestock and livelihoods in Africa. Lessons from Ethiopia, Mali and Zimbabwe*. Falmer, Brighton: Institute of Development Studies, University of Sussex.

Scott, J. J. (1998). *Seeing like a state: How certain schemes to improve the human condition have failed*. Yale University Press, New Haven.

Sendzimir, J., Reij, C. P., & Magnuszewski, P. (2011). Rebuilding resilience in the Sahel: Regreening in the Maradi and Zinder regions of Niger. *Ecology and Society.*

Silviconsult. (1991). *Household energy demand and supply, Northern Nigeria*. Report to the Nigerian Federal Forestry Evaluation and Coordinating Unit, World Bank Forestry Programme. Thame(UK): Silviconsult Ltd.

Stebbing, E. P. (1935). The encroaching Sahara: The threat to the West African colonies. *The Geographical Journal, 85*, 506–524.

Stebbing, E. P. (1937). *The forests of West Africa and the Sahara. A study of modern conditions*. Edinburgh: Chambers.

Stenning, D. J. (1959). *Pastoral nomads. A study of the Wodaabe pastoral Fulani of western Bornu Province, Northern Region, Nigeria*. Oxford: Oxford University Press.

Stoorvogel, J. J., & Smaling, E. M. A. (1990). *Assessment of soil nutrient depletion in Sub-Saharan Africa: 1983–2000* (Vol. 1). Main Report (2nd edn.). Report 28. Wageningen, The Netherlands: Winand Staring Centre.

Thomas, D., & Middleton, N. (1992). *World Atlas of Desertification*. London: Edward Arnold for UNEP.

Thomas, R. J., El-Dessougi, H., & Tubeileh, A. (2006). Soil system management under arid and semi-arid conditions. In N. Uphoff, A. S. Ball, E. Fernandes, H. Herren, O. Husson, M. Laing, C. Palm, J. Pretty, P. Sanchez, N. Sanginga, & J. Thies (Eds.), *Biological approaches to sustainable soil systems* (pp. 41–58). CRC/Taylor & Francis: Boca Raton (USA).

Tiffen, M., & Mortimore, M. (2002). Questioning desertification in dryland sub-Saharan Africa. *Natural Resources Forum, 26*(3), 218–233.

Timberlake, L. (1986). *The encroaching desert. The consequences of human failure*. A Report for the Independent Commission on International Humanitarian Issues. London: Zed Books.

Tolba, M. K. (1986). Desertification. *World Meteorological Organisation Bulletin, 35*, 17–22.

Traoré, K., Hien, F. G., & Hien, V. (2005). Durabilité du système zaï dans la zone nord du Burkina Faso. Etude des flux de matière organique dans l'exploitation. In G. O. Omanya & D. Pasternak (Eds.), *Sustainable systems for the drylands. Proceedings of the International Symposium for Sustainable Dryland Agriculture Systems, Niamey, Niger* (pp. 50–58). Niamey: International Crops Research Institute for the Semi-Arid Tropics, December 2–5, 2003.

Trevallion, B. W. (1966). *Metropolitan Kano twenty year development plan*. Kano, Nigeria: Greater Kano Development Authority.

Tucker, C. J., Dregne, H. E., & Newcombe, W. W. (1991). Expansion and contraction of the Sahara Desert from 1980 to 1990. *Science, 253*(5017), 299–301.

UNCCD. (1993). *United Nations convention on combatting desertification*. Bonn, Germany: UNCCD.

UNCOD, (UN Conference on Desertification). (1977). Nairobi: UNEP.

UNDP/UNCCD. (2011). *The forgotten billion: MDG achievement in the drylands*. New York: United Nations Development Programme; Bonn, Germany: United Nations Convention to Combat Desertification.

UNEP. (2007). *Sudan post-conflict environmental assessment*. Nairobi: United Nations Environment Programme.

UN/EMG. (2011). *Global drylands: A UN system-wide response*. United Nations Environment Management Group.

Uphoff, N., Ball, A. S., Fernandes, S., Herren, H., Husson, O., Laing, M., et al. (Eds.). (2006). *Biological approaches to sustainable soil systems*. Boca Raton, USA: CRC Press.

USAID. (1972). *Desert encroachment on arable lands: Significance, causes and control*. Washington DC: USAID.

Vanlauwe, B., Ramisch, J. J., & Sanginga, N. (2006). Integrated soil fertility management in Africa: From knowledge to implementation. In N. Uphoff, A. S. Ball, S. Fernandes, H. Herren, O. Husson, M. Laing, C. Palm, J. Pretty, P. Sanchez, N. Sanginga, & J. Thies (Eds.), *Biological approaches to sustainable soil systems* (pp. 257–272). Boca Raton, USA: CRC Press.

Vogt, G., & Vogt, K. (2000). *Hannu biye ke Tchuda Juna—strength in unity. Shared management of common property resources*. A case study from Takieta, Niger. Securing the commons No. 2. London: International Institute for Environment and Development/SOS Sahel.

WISP. (2007). *Pastoralism as conservation in the Horn of Africa*. Nairobi: World Initiative for Sustainable Pastoralism.
WOCAT. (2007). Where the land is greener: case studies and analysis of soil and water conservation initiatives worldwide. In H. Liniger & W. Critchley. Wageningen: CTA.
World Bank. (2001). *World Development Report for 2003*. Washington, DC: The World Bank.
WRI. (2005). *Millennium ecosystem assessment*. Washington, DC: World Resources Institute.
WRI. (2008). *World resources 2008: Roots of resilience—growing the wealth of the poor*. Washington, DC: World Resources Institute.

Chapter 4
Does Climate Change Lead to Conflicts in the Sahel?

Tor A. Benjaminsen

Abstract A dominant narrative in international politics and media presentations holds that there is a close connection between climate change and conflicts, especially conflicts involving pastoralists in the Sahel. The narrative consists of two elements: (1) Global climate change leads to drought and desertification, which in turn lead to resource scarcity; (2) This scarcity leads to migration and the emergence of new conflicts, or it triggers existing, latent conflicts. This chapter is a critical assessment of these claims based on two case studies from Mali and a review of international research. The narrative is attractive to politicians and bureaucrats, in particular, and is championed by some influential scholars. In both case studies, the drought in the 1980s only played a minor role in explaining the conflict, while the root causes were political and historical. In addition, there does not seem to be any clear link between resource scarcity in the Sahel in the 1980s and global climate change. An association between scarcity and increased conflict levels cannot, however, be dismissed, even if empirical results from international research question the validity of such a correlation. The causes of conflicts in the Sahel are in general associated with state policies, which result in the marginalization of pastoralists. In areas where pastoralism and farming overlap as the main forms of land use, there are continuous conflicts of varying scale and intensity. These conflicts are primarily caused by politics, not climate change.

Keywords Climate change · Conflicts · Sahel · Mali · Tuareg rebellion · Farmer-herder conflicts

---

T.A. Benjaminsen (✉)
Department of International Environment and Development Studies, Norwegian University of Life Sciences, PO Box 5003 1432 Ås, Akershus, Norway
e-mail: torbe@nmbu.no

R.H. Behnke and M. Mortimore (eds.), *The End of Desertification?*
Springer Earth System Sciences, DOI 10.1007/978-3-642-16014-1_4

4.1 Introduction

With climate change becoming a leading global political issue, the idea that there is a close link between global warming and violent conflicts has also caught international attention. The Sahel, in particular, is pointed out as the clearest example where there are climate-driven conflicts. Many politicians and international civil servants in particular seem attracted to this idea. For instance, in a newspaper article UN Secretary-General Ban Ki-Moon made a connection between global warming and the Darfur conflict (Ki-Moon 2007). The idea was also at the crux of the decision to award the 2007 Nobel Peace Prize to former US Vice President Al Gore and the Intergovernmental Panel on Climate Change (IPCC). According to the Norwegian Nobel Committee, human-induced climate change is one of the main causes of violent conflict and war in the world today. In his presentation speech at the award ceremony in Oslo, chair of the committee Professor Ole Danbolt Mjøs said:

> The consequences (of climate change) are most obvious, however, among the poorest of the poor, in Darfur and in large sectors of the Sahel belt, where we have already had the first "climate war". The wind that blows the sand off the Sahara sets people and camels moving towards more fertile areas. The outcome is that nomads and peasants, Arabs and Africans, Christians and Muslims from many different tribes clash in a series of conflicts. There are many dimensions to this, but it is growing increasingly obvious that desertification is a central underlying factor. The pattern from Darfur has now spread to Chad and the Central African Republic. Large parts of the Sahel belt, from the Sudan to Senegal, are coming under threat (http://nobelpeaceprize.org/en_GB/laureates/laureates-2007/presentation-2007/).

This is the essence of a narrative about the climate-conflict link in the Sahel that consists of two elements:

1. Global climate change leads to drought and desertification, which in turn lead to resource scarcity.
2. This resource scarcity leads to migration and the emergence of new conflicts, or it triggers existing, latent conflicts.

This chapter takes a critical look at both these claims and assesses them on the basis of available international research. But before assessing these two claims, a brief review the climate security literature on the Sahel will be presented.

4.2 The Climate Security Debate: Theories, Politics and Missing Evidence

The idea that climate change leads to violent conflicts in general can be regarded as a continuation or revised version of the Malthusian concept of scarcity of resources as a cause of environmental degradation, poverty and an escalating struggle for resources. Thomas Homer-Dixon is the best-known proponent of the so-called environmental security school. According to Homer-Dixon, resource scarcity can be caused by population growth, environmental degradation or social inequality (Homer-Dixon 1994, 1999). He also believes that arid regions in Africa are particularly prone to scarcity-induced conflicts. More recently, he has also focused on climate change as a cause of resource scarcity and war (Homer-Dixon 2007).

A team of Swiss researchers associated with the Swiss Peace Foundation have also been prominent representatives of the environmental security school. They have had an even more pronounced focus on the Sahel as a crisis area (Bächler and Spillmann 1996; Bächler 1998). According to Bächler (1998), the Sahel is a typical example of an area where conflicts are caused by environmental degradation. In this region, animal husbandry and farming have led to erosion of the landscape (p. 69), population growth has led to deterioration of the vegetation (pp. 67 and 70), and livestock herding has led to general overgrazing (p. 69). Bächler (1998) presents a list of 11 conflicts that allegedly demonstrates the link between environmental degradation, socio-economic change and violence in the Sahel.

The concept of scarcity as a cause of war and violence has taken root in the media and popular scientific publications too. The American journalist Robert Kaplan has been particularly influential. In a well-publicized article from 1994, he claimed that the conflicts in Liberia, Rwanda and Somalia were unequivocally the result of overpopulation and subsequent environmental crisis in Africa (Kaplan 1994).

Since then, a range of journalists and popular-science writers have played a significant role in spreading a Neo-Malthusian message to politicians and the general public (e.g. Diamond 2005). The scarcity perspective has also been promulgated in a special edition of National Geographic Magazine focusing on Africa in September 2005 (see Moseley 2005 for a critical commentary).

A major criticism of the environmental security school is that the term ‘resource scarcity’ is defined so vaguely and broadly that it loses all meaning (Gleditsch 1998; Fairhead 2001; Richards 2005). Since armed conflicts are almost without exception about control over land, such conflicts will necessarily have a resource dimension. However, it does not follow that this dimension explains the conflicts. It is also misleading when such different processes as environmental degradation, increased population pressure and inequitable access to resources are forced together into a single concept of resource scarcity. In this way, the concept loses its analytical power, critics argue.

Peluso and Watts (2001) also hold that conflicts cannot be understood on the basis of a simple chain of events triggered by resource scarcity, via reduced economic activity and migration, to a violent outcome. Instead, violence is context-specific, and at the same time it is a result of overarching power and production relations.

Elinor Ostrom, who received the Nobel Prize in economics in 2009, also represents a perspective that stands in contrast to the environmental security school. She and her collaborators have shown how scarcity of resources might just as well lead to cooperation and sustainable use instead of conflicts (Ostrom 1990; Poteete et al. 2010).

Some critics claim that population growth can actually serve to increase the resource base and lead to sustainable agricultural intensification in keeping with the Danish agricultural economist Ester Boserup's theory (Boserup 1965). There are in fact examples from Africa of increases in population combined with favourable government policies that have led to increased investments per unit area in the form of work and capital and thus an improved resource base (e.g. Tiffen et al. 1994; Mortimore 1998; Benjaminsen 2001).

While until recently, the focus in the scarcity literature was on 'overpopulation' and the associated 'overuse' of renewable natural resources, climate change has been increasingly in focus as a prime cause of conflicts during the last few years. As already mentioned, the Darfur conflict is then often presented as the best example of the correlation between climate and conflict. For example, Sachs (2007: 24) states that

> Darfur's extreme poverty, rising population, growing water stress and desertification are all important contributors to the Darfur crisis. (…) extreme poverty, falling incomes and failing rains… are the crucial drivers of conflict in less developed countries; much less solid evidence implicates political repression.

This is a good example of how Malthusian factors and climate change are merged into one story to explain conflicts. Homer-Dixon has also claimed—without undertaking any empirical studies of the politics, climate or ecology in Darfur—that climate change is one of the causes of this conflict: 'There is evidence that warming's effect on crops and pastureland is a cause of the Darfur crisis' (Homer-Dixon 2007). In an online discussion forum, however, he refuses to go so far as to say that climate change is the main cause of the conflict in Darfur ('In the case of Darfur, it's pointless to ask about, or to argue over, the relative importance of climate change as a cause of the violence. But based on the evidence available, we can say with considerable confidence that any adequate description or explanation of the crisis must include climate change as a causal factor' http://www.ssrc.org/blogs/darfur/2007/08/02/cause-and-effect/).

A report published by the United Nations Environmental Programme (UNEP) in 2007, which received extensive media coverage and obtained political influence, also claims that there is a close link between climate change, desertification and the conflict in Darfur (UNEP 2007). The report attaches a great deal of importance to the fact that the average rainfall in some parts of Darfur has decreased by 16–34 %, if the periods 1946–1975 and 1976–2005 are compared. However, the report fails to mention that since the mid-1980s, rainfall has increased again. For example, if we look at the 30-year period prior to the conflict breaking out in 2003 there is no decreasing trend (Kevane and Gray 2008). In fact, there is no evidence of a falling or a rising trend in rainfall in Darfur in the period 1972–2002 (Fig. 4.1).

This fact, however, does not impress Mazo (2010) who insists that Darfur is 'the first modern climate change conflict'. He argues that although climate change was not a necessary or sufficient condition for the conflict, it was 'a critical factor

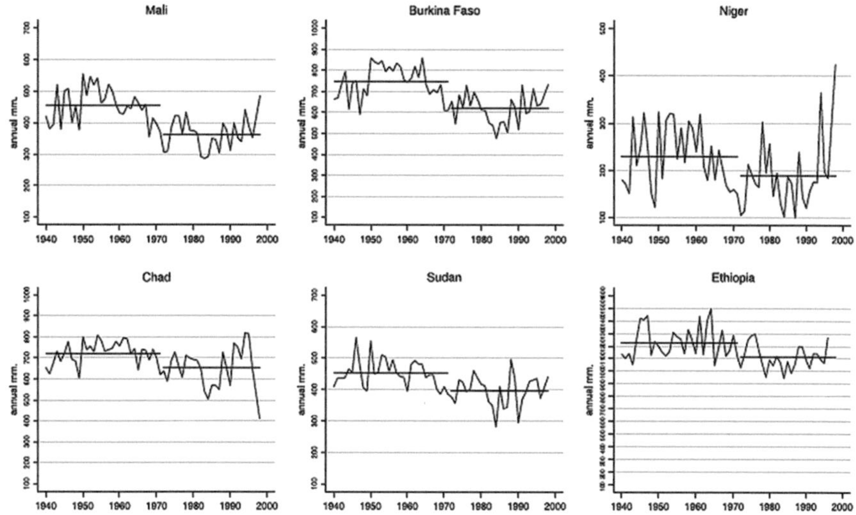

Fig. 4.1 Rainfall in the Sahel. *Source* Climate Research Unit, University of East Anglia, reproduced with permission.

underlying the violence' and that 'to say that other factors were equally, or even more, important politically or morally is not to deny that Darfur was a climate-change conflict' (Mazo 2010: 85–86). He supports this argument primarily by using the above-mentioned UNEP report as well as an article by Burke et al. (2009). The latter study focused on temperature instead of rainfall and reported a strong correlation between annual temperature and the incidence of civil war in Sub-Saharan Africa during 1981–2002. However, according to Buhaug (2010), there are reasons to be sceptical about the results of Burke et al. (2009). The study applies an unconventional definition of civil war, studying only years that generated at least 1000 battle deaths and failing to distinguish between lesser war episodes and peace. This restricted sample implies that many relatively large conflicts are excluded from the analysis. In addition, Buhaug shows that the original findings are easily influenced by small changes in the climate parameters and model specification. Finally, he pointed out that since 2002, the final year of the sample, civil war had decreased in incidence and severity in Africa while warming had persisted.

The idea that climate change leads to more violent conflict had earlier been criticized, for instance by Barnett (2003), Nordås and Gleditsch (2007), Theisen (2008) and Salehyan (2008), who did not find any evidence for scarcity being a driver of conflicts. Nordås and Gleditsch (p. 628) also remark that 'even the IPCC, which prides itself on being a synthesis of the best peer-reviewed science, has fallen prey to relying on second- or third-hand information with little empirical backing when commenting on the implications of climate change for conflict'.

Finally, to look beyond the aggregate level and the general concepts usually used in this debate, Buhaug et al. (2010) operationalized the hypothesized link between climate change and conflicts. They identified three main processes that this link could potentially consist of: intensification of natural disasters, increasing resource scarcity, and sea-level rise. These processes could then cause destruction of infrastructure, increased health risk, and loss of livelihood. It should be stressed, however, that these are hypothesized links that find little support in the empirical literature, whether based on qualitative case studies or quantitative studies.

4.3 Does Climate Change Lead to Desertification?

Claims about desertification in the Sahel are as old as European presence in the region. Already in the early 1900s, there were debates about whether desertification in French-occupied West Africa was a man-made process or caused by desiccation (Benjaminsen and Berge 2004). With time, however, the view that it was created by local overuse of natural resources prevailed and even during the droughts of the 1970s and 1980s this view dominated in research, policy and media presentations.

From the late 1980s, claims of widespread degradation and desertification in the Sahel have been undermined by a number of studies. For instance, scientists at the National Aeronautics and Space Administration (NASA) in the USA have studied satellite images of the southern limits of the Sahara and concluded that the edge of the desert moves back and forth as a direct result of annual rainfall (Herrmann and Sop, this volume; Tucker et al. 1991; Tucker and Nicholson 1999).

Furthermore, a number of studies published from the late 1980s led researchers increasingly to question the idea of desertification in the Sahel. Some of this research was reported in a New Scientist article entitled ‘The myth of the marching desert’ (Forse 1989). This research led to what has been termed a paradigm shift in drylands research (Warren and Khogali 1992; Behnke and Scoones 1993; Benjaminsen 1997). It recognizes the resilience and variability of drylands and stresses the need for flexibility in coping with a highly unstable environment. These ideas have led to the questioning of ecological theory based on notions of equilibrium, carrying capacity, succession and climax as applied to tropical drylands. Instead, non-equilibrial ecological theory states that the vegetation in drylands varies with the annual rainfall and that external factors such as climate, rather than livestock numbers, tend to determine the vegetation composition and cover (Ellis and Swift 1988; Behnke et al. 1993). Moreover, unavailability of forage in bad years may depress livestock populations to the point where the impact of grazing on vegetation is minimal (Sullivan and Rohde 2002). Therefore, in areas of fluctuating climates, rainfall rather than density-dependent factors related to herbivore numbers may ultimately be the most significant variable determining herbivore populations. Wet season pastures such as in the West African Sahel, with its short rainy season, domination of annual grass species, and high resilience, is a good example of a non-equilibrial system (Hiernaux 1993; Turner 1993). The herders’ use of pastures is adapted to the seasonal changes in these drylands. During the rainy season, when the grass grows, herders often move, and therefore exercise little pressure on the vegetation.

Since it is largely rainfall that drives the Sahelian ecosystem, global warming might in the long run lead to desertification—if it reduces rainfall. However, as demonstrated by Buontempo et al. (2010), there is currently considerable uncertainty about current rainfall trends and projections in the Sahel. Not only are there uncertainties about future scenarios, but there are also some disagreements about how to read available climate data. For instance, Hulme (2001) and Chappell and Agnew (2004) disagree on how to interpret rainfall data from the Sahel for the period 1930–1990. While Hulme holds that there was a 20–30 % decline, Chappell and Agnew argue that this decline was largely produced by historical changes in the climate station network.

Climate modellers in general stress that there is uncertainty as to how global warming will affect the climate in the Sahel. This is underlined by the IPCC in its Fourth and Fifth Assessments (Boko et al. 2007: 444; Niang and Ruppel 2014). The former points out that the various models do not concur concerning future climate scenarios for the Sahel. While some models support the theory that this region will become drier, other models suggest that it may rain more in the future (e.g. Haarsma et al. 2005; Odekunle et al. 2008).

Buontempo et al. (2010) also highlight the inability of current generation climate models to capture processes driving Sahelian climate in the 21st century, precipitation in particular. They advise against basing assessments of future climate change in the Sahel on the results from any single model in isolation. Until the processes responsible for the projected changes can be understood and constrained, the long term future will remain uncertain. However, Biasutti (2013) finds that most models conclude that the rainy season will be 'more feeble at its start' and 'more abundant at its core'. Hence, the overall trend seems to be towards wetter conditions, but with rainfall more concentrated in time and with higher average temperatures. Of 20 models only four are outlier models coming to other conclusions. But the Fifth IPCC Assessment accords low to medium confidence in these projected changes of heavier rainfall (Niang and Ruppel 2014).

Throughout the Sahel, there has been a partial recovery of rainfall over the last 20 years. Research on the Sahel is thus no longer discussing desertification, but the fact that the Sahel has become greener. For instance, in November 2005, the Journal of Arid Environments published a special issue on 'The Greening of the Sahel' (Hutchinson et al. 2005; Olsson et al. 2005).

Hence, climate change may lead to drier conditions and desertification in the long term if rainfall declines. But it is problematic to conclude that current rainfall trends are on the decline. Uncertainty remains characteristic of climate scenarios for the Sahel (Giannini, this volume).

4.4 Do Sahelian Droughts Lead to More Conflicts?

There is not a lot of research to build on in order to answer this question. As already mentioned, there is some disagreement in how to interpret the role of drought in explaining the case of the conflict in Darfur. In order to illustrate the potential role of drought in such conflicts, I will, in this section, dwell on two cases from Mali taken from my own research (Benjaminsen and Ba 2009; Benjaminsen 2008).

The first case deals with a conflict between settled farmers and migrating pastoralists in the inland delta of the Niger River (see also Benjaminsen et al. 2012). The other example is the Tuareg rebellion in northern Mali during 1990–1996, which was a major conflict that escalated to civil war proportions.

4.4.1 A Farmer-Herder Conflict in the Inland Delta of the Niger River in Mali

Historically, the delta is one of West Africa's richest regions, in terms of farming, herding and fishing. The Niger and its delta allow farmers to grow crops farther north than anywhere else in the West African Sahel. At the same time, the delta represents an essential resource for pastoralists in the dry season. Herders and their livestock congregate in the delta region in the dry season from December to June, while in the rainy season (July to September) and in the early part of the dry season, they migrate up to several hundred miles north-east and north-west to reach good pastures in the savanna (Fig. 4.2).

In addition to being a source of drinking water for livestock, nutrient-rich pastures called 'burgu' grow in the water here. These various water plants are found in deeper water than rice. During the last few decades, paddy fields have been extended, at the expense of burgu. It is reckoned that roughly a quarter of the burgu areas have been turned into paddy fields since the 1950s (Kouyaté 2006). This is partly the result of reduced water levels in the river during the droughts in the 1970s and 1980s (see Fig. 4.3), when the paddy fields dried out and new ones were established in burgu areas. In addition, the development of a hydroelectric dam in Sélingué in southern Mali, which was completed in 1982, is a major cause of the lower water levels downstream (Turner 1992; Cotula and Cissé 2006).

Over the last few decades, there has been a large number of conflicts about who has control of the land in the delta (Barrière and Barrière 2002; Ba 2008). In order to understand the current resource management regime and the ensuing conflicts, we need to look at the history of the region.

In 1818, Islamic Marabouts mobilized a Jihad and conquered the delta region under the leadership of Cheikou Amadou. This resulted in the establishment of an Islamic theocratic state, the Dina, based in Hamdallahi, south of Mopti. The Dina formalized many of the customary resource management principles and rights in the

Fig. 4.2 Mali with the inland delta of the Niger River. *Source* Benjaminsen et al. (2012).

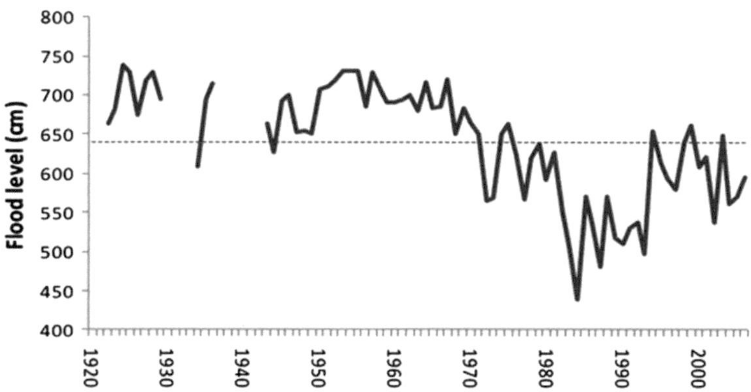

Fig. 4.3 Niger River flow variability in Mopti, 1922–2006. The graph shows maximum annual flood level of the Niger River in Mopti. Several years are missing in the early period. The *dotted line* represents the mean value for all years. Data received from Direction Nationale de l'Hydraulique et de l'Energie in Bamako.

delta region. As part of this formalization, the delta plain was divided up into administrative units called 'leyde' (singular: 'leydy'). Traditional village chiefs, called 'Jowros' were authorized to manage these units, and all users of the burgu pastures had to pay a fee to the local Jowro. This is still the basis of the current system. Today there are 31 leyde in the delta region.

The Jowros, who are noble Fulani pastoralists ('Rimbé'), were to manage the pastures, while responsibility for allocating farmland was delegated to a 'Bessema', who was the chief of the low-caste 'Rimaybé'. Both the Rimbé and Rimaybé are Fulani, but while the former are traditionally pastoralists and thus have high status, the latter are primarily farmers and have low social status.

When this area came under French rule in 1895, the French retained many of the administrative principles of the Dina regime. For example, the Jowro were allowed to continue to operate as 'masters of the pastures' and collect fees from users of these pastures. Then, in 1960, Mali became independent under a socialist government led by Modibo Keita. The new government viewed 'development' in terms of industrialization and modernization of farming. Pastoralism was regarded as an obstacle to this kind of modernization. Nomadic herding was also seen as counter to rational resource management. Modibo Keita said that settlement of all nomads was one of the most important tasks of the new state. Only then could herders become productive citizens (Benjaminsen and Berge 2004). The socialist government also regarded the Jowros as feudal lords and generally tried to undermine their authority.

In 1968, Lieutenant Moussa Traoré led a coup that resulted in a military government, which eventually to some extent reinstated the Jowros. By the next coup d'état in March 1991, which introduced democratic elections, the Jowros had once again become powerful local actors through alliances with the cadres in the only permitted political party.

The sample village Saremala is located in Kounary leydy in the heart of the delta. In the village live the local Jowro and a small number of Rimbé, while the vast majority of the villagers are Rimaybé. The Office Riz Mopti (ORM)—the state organization for the promotion of rice cultivation—is active in this region. Since the state formally owns all the land in Mali, the ORM can confiscate land at will. In particular, much of the burgu pasture controlled by the Jowro has been confiscated and turned into paddy fields. These fields have been divided up into equal-sized parcels of land and leased out to people who have applied to the ORM for land. In addition, there has been widespread random cultivation of burgu pastures by local Rimaybé farmers. The massive loss of burgu pastures, which constitute the power base and main source of income for the local Jowro, is leading to a gradual transfer of local power from the Jowro to the Bessema. A positive aspect of this is that the previously underprivileged Rimaybé now have more power and a higher standard of living. A negative aspect is that important pastures used in the dry season are disappearing and being replaced by paddy fields. These are the pasturelands on which the entire pastoral system in the delta region depends.

In the wake of the transition to democracy in 1991, the state's presence in rural areas was reduced for a period. This was in general a time of great uncertainty about the future direction of the political and administrative system in Mali. State bodies were reorganized, and plans were laid for a new, major decentralization reform. Many local actors took advantage of the power vacuum that arose in the early phase of the decentralization process, by taking possession of land in various ways. This also happened in Saremala. Farmers extended their fields into pasturelands, while the Jowro tried to regain control over the lost burgu areas that had been converted into cultivated farmland. Farmers usually give a small symbolic share of the harvest (usually about 5 kg) in recognition of the person who owns the land according to customary law. However, the political changes and the state's temporary withdrawal since 1991 had whetted the Jowro's appetite, and he decided to try to take control of the cultivated land and introduce a clearer tenant farming system with a larger share of the yield for the owner of the land. This strategy failed, however, because of strong resistance from the Rimaybé.

Frustrated at his loss of power and the loss of the burgu pastures, the Jowro decided to take the Rimaybé to court in 1994. On 25 August 1994, the local court in Mopti ruled in favour of the Jowro, establishing that he had customary rights over all the land in Saremala, not just the pastureland. However, the Rimaybé appealed the case to the Appeal Court in Mopti, which ruled on 31 May 1995 that while the Jowro had customary rights, the Rimaybé had usage rights to the same land. While both parties interpreted this ambiguous ruling in their own favour, the Jowro appealed the case to the Supreme Court. At the same time, the Jowro of Saremala started acting as if his ownership rights had been finally confirmed by the legal system, banning Rimaybé farmers from cultivating the land. At the harvest in December 1995, he announced a general ban on harvesting rice, stating that all harvested rice would be confiscated by force. The Rimaybé (farmers) then paid 18 armed guards to protect them while they harvested their crops. Despite the guards, there was an armed confrontation between the Rimbé (herders) and the Rimaybé on 23 January 1996 resulting in two dead and 16 injured farmers and herders. The village chief, a Rimaybé, claims that the Jowro had bribed the guards to look another way and not intervene when the Rimbé tried to force the Rimaybé to stop harvesting the rice.

On 18 February 1997, the Supreme Court declared that the Appeal Court's decision was invalid and sent the case back to that court, which upheld its earlier decision on 2 July 1997. Indeed, it even increased the ambiguity of its former ruling by ascribing to the Jowro all three aspects of ownership under French law: usus, fructus and abusus. Usus is the right to use, fructus is the right to enjoy the fruits of (harvest, rental income, etc.) and abusus is the right to get rid of the property by giving it away or selling it. This judgment actually gave the Jowro more extensive rights than any reasonable interpretation of customary rights might ascribe to him. It is also self-contradicting in another way, because according to Malian law, only the state has the abusus rights to land without deeds. To top it all, the Appeal Court granted the farmers usage rights to the land they have been cultivating for several decades. This means that the farmers are ascribed the rights of usus and fructus. In practice, this means that the Jowro and the Rimaybé farmers were granted basically identical rights (usus and fructus), which both parties interpreted as a victory.

However we can probably conclude that the Rimaybé have more to celebrate because the court granted them rights they would otherwise have had little chance of establishing under the customary system.

It seems that the legal wrangling back and forth and the court's ambiguous rulings are the result of both parties having paid bribes to the judges and their entourage. Bringing a conflict before a court of law is often a final desperate attempt on the part of one of the parties. Prior to this, the parties have usually already spent vast sums of money on trying to influence the administration.

This case study reviewing the political and economic context of a herder-farmer conflict in Mali shows that the drought in the 1980s was one of several factors that contributed to the loss of burgu pastures and marginalization of pastoralists—and thus indirectly to more conflicts. The drought contributed temporarily to the shrinking of available burgu pastures for pastoralists. With less pastoral space available, herders and livestock will more easily trespass and damage agricultural crops, and conflicts might emerge. However, since the end of the 1980s, there has been more rain and the water levels in the Niger River have increased. Yet, the number of conflicts in the delta has continued to be high due to a continued political and economic marginalization of pastoral space and pastoralists. So, while theoretically general resource scarcity might lead to more conflicts, the state's policy, which led to marginalization of pastoralists—and in turn to increased scarcity of pastoral land—plays a far larger part in explaining the increase in the number of conflicts in the inland delta region of Mali.

Hence, this case study together with the study by Benjaminsen et al. (2012) also carried out in the delta region conclude that land-use conflicts in the area are primarily shaped by political and economic factors rather than climate variability. These factors include first agricultural encroachment on productive key resources for pastoralism and on livestock corridors, obstructing the necessary mobility of herders and animals. This trend is primarily caused by agricultural policies and laws promoting farming at the expense of pastoralism. Second, decentralization from the early 1990s caused a political vacuum that led rural actors to follow opportunistic strategies to claim ownership of land and natural resources. Third, rent-seeking among government officials has undermined rural people's trust in government institutions and the willingness and interest of officials to solve conflicts. This lack of trust may have contributed to some actors taking action on their own, including using violence to lay claim to resources. Climate variability may only play a secondary or tertiary role in increasing or decreasing the conflict level.

4.4.2 The Tuareg Rebellion in Mali

The second Tuareg rebellion in northern Mali after independence took place between 1990 and 1996, and resulted in several thousand deaths and a quarter of a million refugees fleeing to neighbouring countries. This is a conflict that advocates of the environmental security school explain through desertification and an ensuing scarcity of natural resources. For example, Kahl (2006: 234) claims that in northern Mali, the combination of 'population pressures, poor land use practices, and a fragile ecology… made soil erosion, desertification, and freshwater scarcity serious problems.' He also claims that what he calls 'demographic and environmental stress' were important causes of the Tuareg rebellion, without presenting any documentation to back up these claims.

It is, however, difficult to demonstrate any 'desertification' in northern Mali. Since rainfall has increased over the last 20 years, the forests have grown back and there has therefore been a reversed desertification process or 'greening' in this area as in the Sahel in general (see Hiernaux, this volume, for a more detailed description of recent environmental change in northern Mali; Hiernaux et al. 2009; Mougin et al. 2009).

A brief look at the political history of this part of the Sahel allows us to assess the variables that have played a role in the conflict dynamics. After Mali's independence in 1960, the new government initiated, as mentioned earlier, a policy to modernize agriculture. This policy was associated with the nomadic lifestyle being regarded as old-fashioned and unproductive. The new President, Modibo Keita, argued, for instance, that sedentarization of nomads was important in order to develop the new nation and to convert nomads into 'productive' citizens by having them take up farming (Benjaminsen and Berge 2004). Hence, the nomadic way of life was considered backward, unproductive and undesirable, and the enormous grasslands in northern Mali were referred to as the 'useless' part of the country (Benjaminsen and Berge 2004). The main implication of this policy for pastoralists was that many important dry season wetland pastures in flooded areas close to the Niger River would be converted to rice fields through state intervention.

Many Tuareg regarded the anti-pastoral policy of the Malian government as a form of neocolonialism—only this time by the authorities in southern Mali, instead of Europe. The policy led to further marginalization of nomads such as the Tuareg. According to Ag Baye (1993), through frequent confiscations, humiliations and violence the new Malian administration was even more hostile to the Tuareg than the French administration had been. One of the results of this policy was that many Tuareg did not develop a feeling of being Malian (Poulton and Ag Youssouf 1998). This marginalization, including the taxes nomads had to pay without receiving any benefits, precipitated the revolt of 1963 (Ag Baye 1993; Lecocq 2004), which took place during a wet period in the Sahel. The uprising was suppressed by the Malian army using fighter planes and public executions.

After the coup d'état in 1968, the new government continued the agricultural policies of the previous government. Nomadic groups in northern Mali had little influence on national or local policy, and in practice, the region was governed by a military governor throughout the whole period until the rebellion started in 1990. Many Tuareg experienced this as a form of military occupation (Poulton and Ag Youssouf 1998).

The drought in the 1970s and 1980s also played a role in the uprising, but not in the way the argument about the correlation between climate and conflict assumes. Firstly, the droughts led many Tuareg to move to Algeria and Libya. Many of them became politically radicalized by Gaddafi's ideology—a melange of Islam and socialism. A large number of the Tuareg who moved to Libya also ended up as professional soldiers in Gaddafi's army and gained practical experience in warfare in Palestine, Lebanon and Chad. It was these soldiers that started planning a new uprising in Mali in the 1970s (Lecocq 2004).

Secondly, the droughts led to Mali receiving large amounts of emergency aid. Much of this relief aid intended for northern Mali was allegedly embezzled by government officials. The rumours of corruption further stoked the anger of many Tuareg at the Malian authorities (Klute 1995).

There is also increasing competition between the Tuareg and the settled Songhay over land along the Niger River in northern Mali. The Tuareg, who are primarily pastoralists, are dependent on access to the burgu pastures along the river as an important resource in the dry season, while the Songhay want to cultivate as much land as they can to grow rice. The situation is parallel to the one described earlier in the delta further south. The ensuing competition for land is a constant source of minor conflicts, some of which are violent. However, these types of conflicts had nothing to do with the uprising, which was started by people with roots in Kidal deep in the Sahara a long way from the river. In fact the uprising came as much of a surprise to the Tuareg further south along the river as it did to everyone else (Poulton and Ag Youssouf 1998).

Thus the main cause of the Tuareg rebellion that started in 1990 was a modernization policy that led to the marginalization of nomads, combined with anger at what was perceived as a predatory state. The first Tuareg uprising took place in 1963 (in an unusually wet period) against what was seen as a new form of colonization—this time not from Europe but from the south. The uprising was severely suppressed by the Malian army and increased the Tuaregs' bitterness and animosity towards the state. The droughts of the 1970s and 80s led many young Tuareg to move to Algeria and Libya where they became further radicalized and many also gained practical training and experience in warfare in the Libyan army. It was these professional soldiers that started the rebellion in 1990. Hence, this case also shows how political factors are the root causes of the conflict, while climate factors only play a minor role.

4.5 Conclusions

A dominant narrative in international politics and media presentations says that there is a close connection between climate change and conflicts, especially in the Sahelian zone in Africa. In this chapter, I have presented a critical review of this narrative. It consists of two elements:

1. Global climate change leads to drought and desertification, which in turn lead to resource scarcity.
2. This resource scarcity leads to migration of ethnic groups and new conflicts, or it triggers existing, latent conflicts.

Contrary to the assumptions behind the first narrative element, rainfall in the Sahel has increased since the drought in the 1980s. Most climate models also predict a wetter, although more concentrated, rainfall pattern in the future. The Sahel has actually become greener and richer in renewable resources over the last 25–30 years. In both case studies presented from Mali, the drought in the 1980s only played a minor role in explaining the conflict level. In both cases, however, the root causes of the conflicts are political and historical. In addition, there is no clear link between resource scarcity in the Sahel in the 1980s and global climate change. An association between resource scarcity and increased conflict levels cannot, however, be dismissed, even if empirical results from international research question the validity of such a correlation. Quantitative studies undermine the validity of a general link between climate and conflict, while case studies in central parts of the Sahel—such as the two cases I have presented here—indicate that the conflicts have other causes.

The main cause of the two conflicts in Mali is therefore not related to climate change; it has much more to do with the state's policies and legislation, which result in the marginalization of pastoralists. In the dry parts of Africa where pastoralism and farming overlap as the main forms of land use, there are continuous conflicts of varying scale and intensity. These conflicts are primarily caused by politics, not climate change.

But if climate change in the long term leads to drier conditions in the Sahel contrary to what most climate models predict, more scarcity of resources in some areas will follow. If water levels in the Niger River decrease, this will lead to continued loss of burgu pastures. In addition, drier conditions in the wet season pastures might lead to further increased dependency on the burgu areas. This might again increase conflict levels depending on the policies of the state. But again, such a scenario goes against most current predictions for climate change in the Sahel.

References

Ag Baye, C. (1993). The process of a peace agreement. Between the movements and the United Fronts of Azawad and the government of Mali. In H. Veber, J. Dahl, F. Wilson & E. Wæhle (Eds.), "*...Never drink from the same cup*". *Proceedings of the conference on indigenous peoples in Africa* (pp. 247–256). Tune, Denmark. IWGIA Document no. 74.

Ba, B. (2008). *Pouvoirs, ressources et développement dans le delta central du Niger*. Paris: L'Harmattan.

Bächler, G. (1998). Violence through environmental discrimination. *Causes, Rwanda arena, and conflict model*. Dordrecht/Boston/London: Kluwer Academic Publishers.

Bächler, G., & Spillmann, K. R. (Eds.). (1996). *Environmental degradation as a cause of war. Country studies of external experts*. Chur-Zürich: Rüegger.

Barnett, J. (2003). Security and climate change. *Global Environmental Change, 13*, 717.

Barrière, O., & Barrière, C. (2002). Un droit à inventer. Foncier et environnement dans le delta intérieur du Niger. Editions IRD, collection "A travers champs". Paris: Institut de Recherche pour le Développement (IRD).

Behnke, R., & Scoones, I. (1993). Rethinking range ecology: Implications for rangeland management in Africa. In R. Behnke, I. Scoones & C. Kerven (Eds.), *Range ecology at disequilibrium. New models of natural variability and pastoral adaptation in African savannas*. London: Overseas Development Institute and International Institute for Environment and Development.
Behnke, R., Scoones, I., & Kerven, C. (Eds.). (1993). *Range ecology at disequilibrium. New models of natural variability and pastoral adaptation in African savannas*. London: Overseas Development Institute and International Institute for Environment and Development.
Benjaminsen, T. (1997). Natural resource management, paradigm shifts and the decentralization reform in Mali. *Human Ecology, 25*(1), 121–143.
Benjaminsen, T. A. (2001). The population-agriculture-environment nexus in the Malian cotton zone. *Global Environmental Change, 11*(4), 27–39.
Benjaminsen, T. A. (2008). Does supply-induced scarcity drive violent conflicts in the African Sahel? The case of the Tuareg rebellion in northern Mali. *Journal of Peace Research, 45*(6), 831–848.
Benjaminsen, T. A., Alinon, K., Buhaug, H., & Buseth, J. T. (2012). Does climate change drive land-use conflicts in the Sahel? *Journal of Peace Research, 49*(1), 97–111.
Benjaminsen, T. A., & Ba, B. (2009). Farmer-herder conflicts, pastoral marginalisation and corruption: a case study from the inland Niger delta of Mali. *The Geographical Journal, 174* (1), 71–81.
Benjaminsen, T. A., & Berge, G. (2004). *Une histoire de Tombouctou*. Arles: Actes-Sud.
Biasutti, M. (2013). Forced Sahel rainfall trends in the CMIP5 archive. *Journal of Geophysical Research: Atmospheres, 118*, 1613–1623.
Boko, M., Niang, I., Nyong, A., Vogel, C., Githeko, A., Medany, M., et al. (2007) Africa. Climate change 2007: Impacts, adaptation and vulnerability. In M. L. Parry, O.F. Canziani, J. P. Palutikof, P. J. van der Linden & C. E. Hanson (Eds.), *Contribution of working group II to the fourth assessment report of the intergovernmental panel on climate change* (pp. 433–467). Cambridge: Cambridge University Press.
Boserup, E. (1965). *The conditions of agricultural growth*. London: Earthscan.
Buhaug, H. (2010). *Climate not to blame for African civil wars*. Paper presented at the 'Climate Change and Security' conference, Trondheim, June 21–24, 2010. http://climsec.prio.no/papers/Climate%20wars%20DKNVS2010.pdf
Buhaug, H., Gleditsch, N. P., & Theisen, O. M. (2010). Implications of climate change for armed conflict. In R. Mearns & A. Norton (Eds.), *Social dimensions of climate change: Equity and vulnerability in a warming world* (pp. 75–101). World Bank, Washington, DC: New Frontiers of Social Policy.
Buontempo, C., Booth, B., & Moufouma-Okia, W. (2010). The climate of the Sahel. In P. Heinrigs & M. Trémolières (Eds.), *Global security risks and West Africa. Development challenges* (pp. 58–71). Paris: OECD.
Burke, M. B., Miguel, E., Satyanath, S., Dykema, J. A., & Lobell, D. B. (2009). Warming increases the risk of civil war in Africa. *Proceedings of the National Academy of Sciences, 106*, 20670–20674.
Chappell, A., & Agnew, C. (2004). Modelling climate change in West African Sahel rainfall (1931–90) as an artifact of changing station locations. *International Journal of Climatology, 24*, 547–554.
Cotula, L., & Cissé, S. (2006). Changes in 'customary' resource tenure systems in the inner Niger delta, Mali. *Journal of Legal Pluralism and Unofficial Law, 52*, 129.
Diamond, J. (2005). *Collapse. How societies choose to fail or survive*. London: Allen Lane.
Ellis, J. E., & Swift, D. M. (1988). Stability of African pastoral ecosystems: Alternate paradigms and implications for development. *Journal of Range Management, 41*(6), 450–459.
Fairhead, J. (2001). International dimensions of conflict over natural and environmental resources. In N. L. Peluso & M. Watts (Eds.), *Violent environments* (pp. 213–236). Ithaca and London: Cornell University Press.
Forse, B. (1989). The myth of the marching desert. *New Scientist, 4*, 31–32.

Gleditsch, N. P. (1998). Armed conflict and the environment: A critique of the literature. *Journal of Peace Research, 35*(3), 363–380.
Haarsma, R. J., Selten, F., Weber, N., & Kliphuis, M. (2005). Sahel rainfall variability and responses to greenhouse warming. *Geophysical Research Letters, 32*, L17702. doi:10.1029/2005GL023232.
Hiernaux, P. (1993). *The crisis of Sahelian pastoralism: Ecological or economic?* Addis Ababa: International Livestock Centre for Africa.
Hiernaux, P., Mougin, E., Diarra, L., Soumaguel, N., Lavenu, F., Tracol, Y., et al. (2009). Sahelian rangeland response to changes in rainfall over two decades in the Gourma region, Mali. *Journal of Hydrology, 375*(1–2), 114–127.
Homer-Dixon, T. (1994). Environmental scarcities and violent conflict. Evidence from cases. *International Security, 19*(1), 5–40.
Homer-Dixon, T. (1999). *Environment, scarcity, and violence*. Princeton: Princeton University Press.
Homer-Dixon, T. (2007). *Terror in the weather forecast*. The New York Times, April 24.
Hulme, M. (2001). Climatic perspectives on Sahelian desiccation: 1973–1998. *Global Environmental Change, 11*(1), 19–29.
Hutchinson, C. F., Herrmann, S. M., Maukonen, T., & Weber, J. (2005). Introduction: The 'Greening' of the Sahel. *Journal of Arid Environments, 63*, 535–537.
Kahl, C. H. (2006). *States, scarcity, and civil strife in the developing world*. Princeton: Princeton University Press.
Kaplan, R. D. (1994). *The coming anarchy: How scarcity, crime, overpopulation, and disease are rapidly destroying the social fabric of our planet* (pp. 44–76). Atlantic Monthly, February.
Kevane, M., & Gray, L. (2008). Darfur: Rainfall and conflict. *Environmental Research Letters, 3*, 034006. doi:10.1088/1748-9326/3/3/034006.
Ki-Moon, B. (2007). A climate culprit in Darfur. Washington Post 16 June.
Klute, G. (1995). Hostilités et alliances. Archéologie de la dissidence des touaregs au Mali. *Cahiers d'Études Africaines, 137*(XXXV-1), 55–71.
Kouyaté, S. (2006). Etude des enjeux nationaux de protection du basin du fleuve Niger. Report to Groupe de Coordination des Zones Arides (GCOZA). Bamako: GCOZA.
Lecocq, B. (2004). Unemployed intellectuals in the Sahara: The teshumara nationalist movement and the revolutions in Tuareg society. *IRSH, 49*, 87–109.
Mazo, J. (2010). *Climate conflict. How global warming threatens security and what to do about it*. London: Routledge.
Mortimore, M. (1998). *Roots in the African dust: Sustaining the sub-Saharan drylands*. Cambridge: Cambridge University Press.
Moseley, W. G. (2005). Reflecting on National Geographic Magazine and Academic Geography: The September 2005 special issue on Africa. *African Geographical Revie, 24*, 93–100.
Mougin, E., Hiernaux, P., Kergoat, L., Grippa, M., De Rosnay, P., Timouk, F., et al. (2009). The AMMA-CATCH Gourma observatory site in Mali: Relating climatic variations to changes in vegetation, surface hydrology, fluxes and natural resources. *Journal of Hydrology, 375*(1–2), 34–51.
Niang, I. & Ruppel, O. C. (Coordinating Lead Authors). (2014). Africa, in IPCC 5th assessment report (Chapter 22).
Nordås, R., & Gleditsch, N. P. (2007). Climate change and conflict. *Political Geography, 26*, 627–638.
Odekunle, T. O., Andrew, O., & Aremu, S. O. (2008). Towards a wetter Sudano-Sahelian ecological zone in twenty-first century Nigeria. *Weather, 63*(3), 66–70.
Olsson, L., Eklundh, L., & Ardö, J. (2005). A recent greening of the Sahel—Trends, patterns and potential causes. *Journal of Arid Environments, 63*, 556–566.
Ostrom, E. (1990). *Governing the commons: The evolution of institutions for collective action*. New York: Cambridge University Press.
Peluso, N. L., & Watts, M. (Eds.). (2001). *Violent environments*. Ithaca & London: Cornell University Press.

Poteete, A. R., Janssen, M. A., & Ostrom, E. (2010). *Working together. Collective action, the commons, and multiple methods in practice*. Princeton: Princeton University Press.

Poulton, R. E., & Ag Youssouf, I. (1998). *A peace of Timbuktu. Democratic governance, development and African peacemaking*. New York and Geneva: UNIDIR.

Richards, P. (Ed.). (2005). *No peace, no war: An anthropology of contemporary armed conflicts*. Oxford: James Currey.

Sachs, J. (2007). Poverty and environmental stress fuel Darfur crisis. *Nature, 449*, 24. September 6, 2007.

Salehyan, I. (2008). From climate change to conflict? No consensus yet. *Journal of Peace Research, 45*(3), 315–326.

Sullivan, S., & Rohde, R. F. (2002). On non-equilibrium in arid and semi-arid grazing systems. *Journal of Biogeography, 29*, 1–26.

Theisen, O. M. (2008). Blood and soil? Resource scarcity and internal armed conflict revisted. *Journal of Peace Research, 45*(6), 813–830.

Tiffen, M., Mortimore, M., & Gichuki, F. (1994). More people, less erosion. Environmental recovery in Kenya. Chichester: Wiley.

Tucker, C., Dregne, H. E., & Newcomb, W. W. (1991). Expansion and contraction of the Sahara desert from 1980 to 1990. *Science, 253*, 299–301.

Tucker, C., & Nicholson, S. (1999). Variations in the size of the Sahara Desert from 1980 to 1997. *Ambio, 28*(7), 587–591.

Turner, M. (1992). Living on the edge: Fulbe herding practices and the relationship between economy and ecology in the inland Niger Delta of Mali. Ph.D. dissertation, University of California, Berkeley.

Turner, M. (1993). Overstocking the range: A critical analysis of the environmental science of Sahelian Pastoralism. *Economic Geography, 69*(4), 402–421.

UNEP. (2007). *Sudan: Post-conflict environmental assessment*. Nairobi: UNEP.

Warren, A., & Khogali, M. (1992). *Assessment of desertification and drought in the Sudano-Sahelian region 1985–1991*. New York: UNSO.

Chapter 5
The Map Is not the Territory: How Satellite Remote Sensing and Ground Evidence Have Re-shaped the Image of Sahelian Desertification

Stefanie M. Herrmann and Tene Kwetche Sop

Abstract Satellite remote sensing, in particular the analysis of coarse resolution time series of vegetation indices, has played an important role in challenging earlier assumptions of widespread desertification in the Sahel. Findings of such analyses show a greening trend in much of the region since the early 1980s, which seems to suggest a positive development. On the other hand, a growing number of field studies of vegetation dynamics across the Sahel offer a more fine-scaled and nuanced picture of changes. Of particular interest with respect to degradation and rehabilitation is the woody component of the vegetation cover, which is less affected by short-term fluctuations in precipitation than the herbaceous component. We synthesized findings from published field studies on changes in the abundance and diversity of woody vegetation across the Sahel and spatially compared them with the remotely sensed greenness trends. Many field sites reported a decline in the abundance of woody vegetation since before the great droughts, in particular of large trees. In addition, the woody vegetation shifted from a diverse species composition towards fewer and more drought tolerant species in the majority of sites. However, some success stories of agroforestry management stood out as well, where formerly degraded farmlands were rehabilitated and in some cases have reached even higher tree densities than in the 1960s. The discrepancy between satellite-observed greening trends and changes in woody vegetation on the ground

"The Map is not the Territory" is an aphorism that goes back to the Polish-American scientist Alfred Korzybski and emphasizes that a representation of reality (=map) must not be confused with reality itself (=territory). Thus, maps and graphical data convey images that risk developing a life of their own, with the map preceding and even becoming the territory.

S.M. Herrmann (✉)
The University of Arizona, 1955 E. Sixth St., Tucson, AZ 85719, USA
e-mail: stef@email.arizona.edu

T.K. Sop
Biocentre Klein Flottbek and Botanical Garden, Research Unit "Biodiversity Evolution and Ecology of Plants", University of Hamburg, Ohnhorststr. 18, 22609 Hamburg, Germany

R.H. Behnke and M. Mortimore (eds.), *The End of Desertification?*
Springer Earth System Sciences, DOI 10.1007/978-3-642-16014-1_5

—in both directions—emphasizes the need of integrating multiple perspectives and scales in the interpretation of greening trends with respect to desertification.

Keywords Remote sensing · Time series · (Re-)greening · Tree cover · Ground truthing · Longitudinal studies · Size class distributions · Local knowledge

5.1 Introduction: Putting Desertification on the Map

The West African Sahel region has been characterized as a 'hotspot' of desertification since it was afflicted by a series of drought years from the late 1960s through the 1980s, following a decade of above-average rainfall years (Raynaut et al. 1996; Reynolds and Stafford Smith 2002; Middleton and Thomas 1997). The amount of human suffering that resulted from the droughts, as well as the perceived degradation of the environment, triggered much political and scientific interest in the phenomenon of desertification. In the late 1970s, a study commissioned by the United Nations to assess the spatial extent of the problem had estimated that the Sahara desert was expanding southward at the startling pace of 5–6 km annually (Lamprey 1988) (Fig. 5.1). A decade later, the expert opinion-based Global Assessment of Soil Degradation (GLASOD) provided a new official perspective on the extent of degradation (Oldeman et al. 1991), in which the entire Sahel region was classified as severely degraded (Fig. 5.2). GLASOD also formed the basis of two editions of UNEP's World Atlas of Desertification (Middleton and Thomas 1992, 1997).

The lingering imprecision of the desertification concept and the lack of measurable criteria, however, fueled debate and controversy among scientists (Pearce 1992; Thomas and Middleton 1994; Herrmann and Hutchinson 2005). Subjects of the debate have been the spatial extent of desertification, its reversibility, as well as the relative contributions of climatic (i.e., drought) and anthropogenic (i.e., overgrazing, deforestation) driving forces of desertification.

Despite the controversy, the notion of regional desertification in the Sahel was widely accepted without question until several remote sensing-based studies challenged the mainstream paradigm of irreversible degradation by documenting a greening trend across much of the Sahel over the past 30 years (Hutchinson et al. 2005). This greening trend, which suggests a possible recovery of the vegetation, was derived from time series of the Normalized Difference Vegetation Index (NDVI) computed from reflectance measurements by a series of Advanced Very High Resolution Radiometer (AVHRR) sensors since 1981. The Land Degradation Assessment in Drylands (LADA) global project, which superseded GLASOD in 2006, adopted the analysis of time series of NDVI as part of their strategy to determine status and trends of land degradation and rehabilitation (Bai et al. 2008). A third edition of the World Atlas for Desertification is currently being compiled,

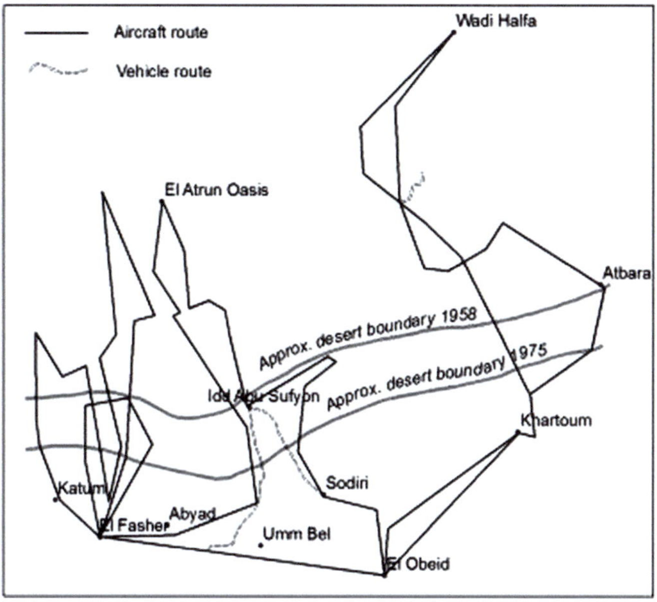

Fig. 5.1 Lamprey's report on desert encroachment in the northern Sudan, based on a reconnaissance survey to map the desert boundary and its comparison with a previously published vegetation map (Harrison and Jackson 1958), became one of the most widely cited references for the rate of southward expansion of the Sahara desert. Different interpretations of the desert boundary, owing in part to differences in rainfall conditions in the years of assessment, led to the misnomer that the desert was advancing. Figure adapted from Lamprey (1988).

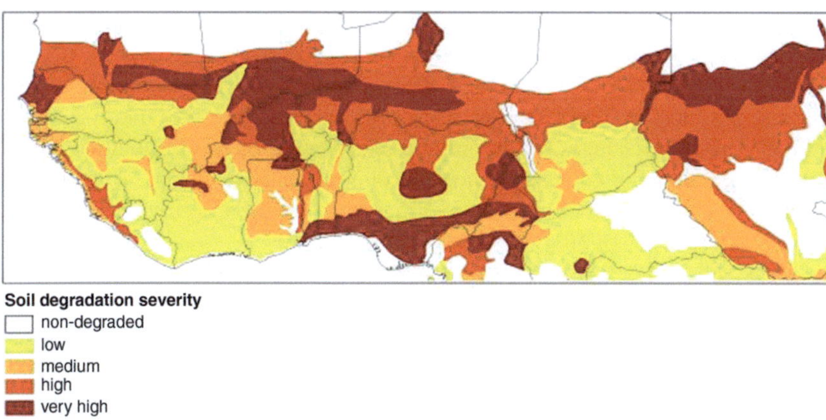

Fig. 5.2 Map of the severity of soil degradation as an indicator of desertification, compiled by the Global Assessment of Soil Degradation (GLASOD) (Oldeman et al. 1991). GLASOD has been one of the most influential global appraisals of land degradation and has majorly contributed to the perception of widespread degradation in the Sahel. For a more in-depth discussion of the GLASOD suite of maps, see Prince (this volume).

the mapping of which is also built on this satellite-based vegetation greenness data set (Joint Research Centre of the European Commission 2015).

While the ever lengthening time series of NDVI data constitute an invaluable and irreplaceable archive of land cover information (Tucker et al. 2005; Dardel et al. 2014a), their interpretation with respect to the nature of vegetation changes is not straightforward. In order to be useful for assessing and monitoring land degradation and rehabilitation, the synoptic overview provided by satellites needs to be related to the physical realities on the ground. The process of "ground truthing"[1]—gathering field data to complement remote sensing data and to establish a link between the radiometric characteristics of a surface and the information desired by a particular application or study—is an integral part of every remote sensing project (Steven 1987). Although the need for field data appears self-evident, there are a number of practical and conceptual challenges, ranging from the timing of the field data collection to coincide with the satellite overpass to the decision of what variables to collect and at what spatial scale. Relating remote sensing data to ground observations is further complicated when working with historical or time series of remote sensing data, as field visits allow direct assessment only of the current state of the environment.

In the case of the satellite-observed NDVI trends in the Sahel, simply visiting a number of locations on the ground and taking stock of the current state of the land cover alone cannot confirm or dispute that greening has taken place. Nor does it inform about the nature of the greening in the sense of a rehabilitation or recovery of the vegetation cover. Inter-comparison between NDVI trends derived from different satellite sensors has been used to verify the direction of trends (e.g., Fensholt et al. 2009; Fensholt and Proud 2012). For some sites, historical field data on woody and herbaceous vegetation are available from prior studies, which give insights into the land cover and vegetation conditions in the past (e.g., Herrmann and Tappan 2013; Dardel et al. 2014a). Although those earlier studies had not necessarily been designed as baselines for longitudinal studies and might not have collected the most desired variables, they present a valuable resource for characterizing vegetation changes on the ground and relating them to satellite-observed trends. In addition, the knowledge of the local inhabitants on changes in vegetation cover over time has been used to support findings from ground surveys and remote sensing data (Herrmann et al. 2014).

In the following, we will outline how regional-scale remote sensing-based findings have shaped the image of Sahelian desertification over past two decades (Sect. 5.2). We will then review different approaches using field data to reconstruct vegetation changes, focusing on the woody component of the vegetation (Sect. 5.3), and synthesize findings from published field studies across the Sahel region

[1]Although the term "ground truth" is widely used in the remote sensing community to describe field observations that help interpret remote sensing imagery, it is disliked by many remote sensing scientists, because it might imply that satellite data are erroneous in some way and because reference data can come from sources not necessarily involving ground investigations. Despite its shortcomings, we use this term here for lack of a better and equally short alternative.

(Sect. 5.4). Finally, we will discuss agreement and discrepancies between coarse-resolution regional scale remote sensing-based observations of trends in vegetation greenness (NDVI) and local scale findings from field studies of woody vegetation dynamics (Sect. 5.5).

5.2 Remote Sensing Perspectives on Regional Greening

Remote sensing products, such as air photos and early Landsat images, were used in early mapping of the formation of desert patches around villages (Ibrahim 1978) and the southward expansion of the Sahara desert (Lamprey 1988) (Fig. 5.1). However, this early use of remote sensing was limited to local scales and did not take into account the seasonality and interannual variability of rainfall and vegetation cover. Based on spatial extrapolation of such studies, often without disclosure of methodologies employed or presentation of primary data, widespread degradation in the Sahel was largely treated as an established fact in the 1970s. A few case studies on the ground (Warren and Agnew 1988; Mortimore 1989) and using remote sensing (e.g., Hellden 1991) had begun questioning the extent of desertification as early as the 1980s. At a regional scale, however, it was not until the 1990s that sufficiently long time series of high frequency, coarse resolution remote sensing observations became available, the analyses of which shed new light on vegetation dynamics and desertification (National Research Council 2008).

These analyses have been based on the Normalized Difference Vegetation Index (NDVI), a greenness index derived from the red and near infrared spectral bands, which is strongly correlated with photosynthetically active biomass (Rouse et al. 1973; Tucker 1979). Time series of bi-weekly NDVI at a spatial resolution of 8 km, derived from reflectance measurements by the Advanced Very High Resolution Radiometer (AVHRR) sensors (James and Kalluri 1994; Tucker et al. 2005), begin in 1981. This series is the longest continuous data record of its kind, and has formed the basis of several influential research studies on vegetation trends in West Africa:

Tucker et al. (1991) were the first to show the high interannual variability of NDVI in the Sahel zone, which directly responds to precipitation. Their findings helped refute the hypothesis of a continuous expansion of the Sahara desert to the south. In a follow-up study, Tucker and Nicholson (1999) confirmed the variability of NDVI over an extended time period as well as the interannual movements of the critical 200 mm precipitation isohyet. Both studies show the close link between precipitation and bio-productivity, contesting the existence of man-made desertification.

Prince et al. (1998) used rain-use efficiencies—the ratio of satellite-derived net primary production (NPP) to precipitation—as an indicator of vegetation degradation, postulating that degrading lands would be marked by declining rain-use efficiencies. However, their analysis showed locally variable rain-use efficiencies with overall upward trends over the period studied, and thus did not demonstrate widespread Sahelian desertification.

Eklundh and Olsson (2003) found strong increases in seasonal NDVI over large areas of the Sahel and Sudan ecological zones from 1982 to 1999, which they interpreted as a vegetation recovery after the drought years of the 1980s. This positive trend was confirmed by Anyamba and Tucker (2005), who described two distinct periods of NDVI patterns, with below average NDVI in the 1980s and above average NDVI in the 1990s and 2000s, in agreement with temporal patterns of rainfall anomalies. Since the AVHRR NDVI time series happens to begin in a very dry period in the Sahel, when the vegetation was arguable in its most desiccated state, it is not too surprising that the positive trend in precipitation since then (which has not canceled out the secular negative trend in precipitation) is accompanied by a greening trend over the same period (Fig. 5.4; Giannini, in this volume).

Herrmann et al. (2005) analysed linear trends in the residual NDVI after removal of the precipitation signal and concluded that precipitation is indeed an important causative factor for the positive NDVI trend, but not the only one, leaving room for interpretations of a potential anthropogenic factor. Seaquist et al. (2009) used a modeling approach to test the hypothesis that humans have had a measurable impact on vegetation dynamics in the Sahel. Their model, which included coarse resolution demographic, pasture and cropping data, suggested that land use pressures do not have a significant influence on the observed NDVI dynamics in the Sahel.

Investigating vegetation phenology—the characteristics of the seasonal vegetation cycle—Heumann et al. (2007) differentiated between two types of greening trends. In the Sudanian zone, where perennial grasses and woody cover are common, greening was associated with an increase in the length of the growing period; in the Sahel zone, where annual grasses predominate, greening was associated with an increase in NDVI at the peak of the growing season.

More recent studies using a new and extended AVHRR NDVI time series have corroborated earlier findings of the close link between vegetation dynamics and precipitation variability. They show positive trends in rain use efficiencies in most of the Sahel, confirming once more that land degradation unrelated to precipitation is not widespread (Fensholt et al. 2013). In addition to the increased precipitation during the period of observations, CO_2 fertilization (the enhancement of photosynthesis due to rising CO_2 levels) may have played an important role in the greening across the warm drylands globally, including the Sahel (Donohue et al. 2013).

In one of the few studies that compare NDVI trends with long-term field data collection of herbaceous biomass, Dardel et al. (2014a) attributed the greening trend to a large-scale increase in herbaceous production over the time period 1981–2011, spurred by increasing rainfall. While greening dominates across the Sahel, their study also found a smaller occurrence of decreasing NDVI in southwestern Niger, taking only the months of August and September into consideration. Although they found bio-productivity had increased and rain-use efficiencies were stable—both indicators of recovery or resilience rather than degradation—Dardel et al. (2014b) also noted an increase of runoff coefficients over the same period, which can be interpreted as a sign of degradation. They concluded that degradation in a small part of the landscape, in particular on shallow soils, is counteracted by resilience on the more productive sandy soils.

Some differences between studies in the reported greening trends may be due to slight differences in the calibrated time series developed from the AVHRR data (McCloy et al. 2005), in the length of the time series (Dardel et al. 2014a), in the inclusion or exclusion of particular months and in trend analysis procedures employed (Fensholt et al. 2013).[2] Despite disagreements on best methodological approaches, underlying assumptions, and interpretation of results (Hein and De Ridder 2006; Prince et al. 2007), two findings have emerged as rather robust across all studies: (1) It seems indisputable that parts of the Sahel and Sudan zones have become greener since the early 1980s, at least from the perspective of coarse-resolution satellite-derived NDVI data (Fig. 5.3). (2) This greening trend has been paralleled by and is closely related to an increase in rainfall amounts over the same time period (Fig. 5.4).

What these findings imply in terms of vegetation changes on the ground, however, remains mostly unaddressed in these satellite-based studies. The land cover of this semi-arid environment is characterized by a fine mosaic of woody vegetation, annual and perennial herbaceous cover, agricultural crops, and bare ground. Virtually all pixels at the 8-km AVHRR resolution are mixed pixels, made up of those components in varying proportions. The proportions of woody and herbaceous vegetation are difficult to extract from the integrated satellite signal, as their seasonal greening cycles are very similar in this region (Akpo 1997). Global-scale efforts to estimate proportions of woody, herbaceous, and bare soil cover per pixel, while adequate for regions with medium to high tree cover, have proven notoriously unreliable for semi-arid regions where tree cover is low and interannual variability of the underlying grass cover high (Hansen and DeFries 2004; Schwarz et al. 2004). Hence, the relative contributions of woody and herbaceous vegetation, let alone different species compositions, to the observed greening trend cannot be assessed by means of remote sensing alone. Yet, such changes in the vegetation composition can be as important indicators of degradation or rehabilitation as changes in bio-productivity. Regional scale remote sensing observations have contributed to uncovering the interannual variability of

[2]Since the NOAA AVHRR NDVI time series from 1981 to present combines observations from two different sensors flown on a series of fourteen satellites, inter-sensor calibration and bias correction for orbital drifts are prerequisites for the creation of long term stable NDVI time series (Pinzon and Tucker 2014). Much of the time series development has been done by the Global Inventory Modeling and Mapping Studies (GIMMS) group. As calibration and correction algorithms have evolved over time, each generation of the GIMMS NDVI time series shows slight differences from the previous one. One of the authors of this chapter compared trends of the most recent 3rd generation GIMMS NDVI with trends of the previous 2nd generation and found considerable differences in the spatial patterns of those trends for West Africa over an identical time period. Such differences, which are explained by data preprocessing alone, call into question the interpretability of NDVI trends from the NOAA AVHRR time series with respect to land degradation and desertification.

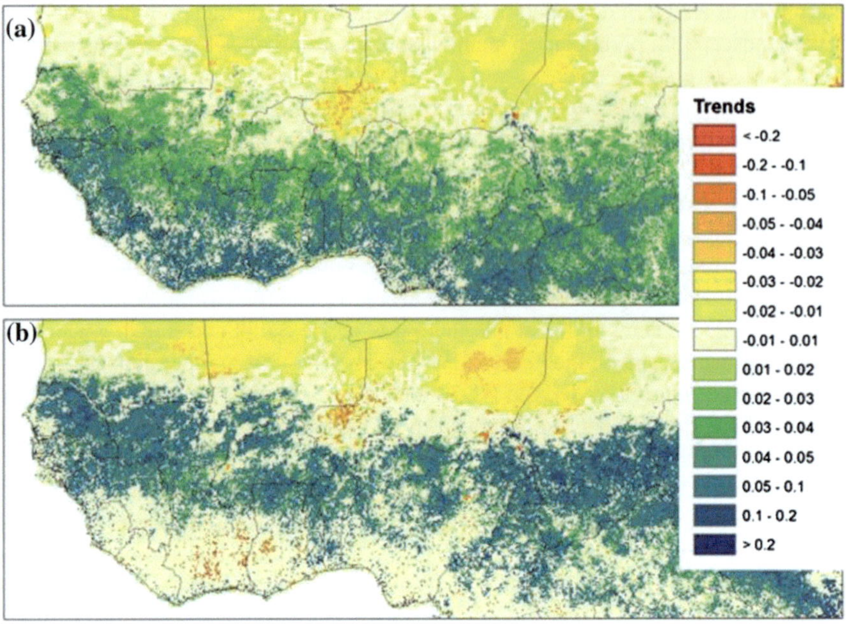

Fig. 5.3 Regional trends in **a** mean annual NDVI and **b** mean growing season (July–October) NDVI, derived from the NOAA AVHRR time series 1982–2011 (GIMMS 3G dataset) (Pinzon and Tucker 2014), show predominantly positive trends in the Sahel and Sudan zones. Note the differences in spatial patterns of greening depending on inclusion of annual or seasonal NDVI.

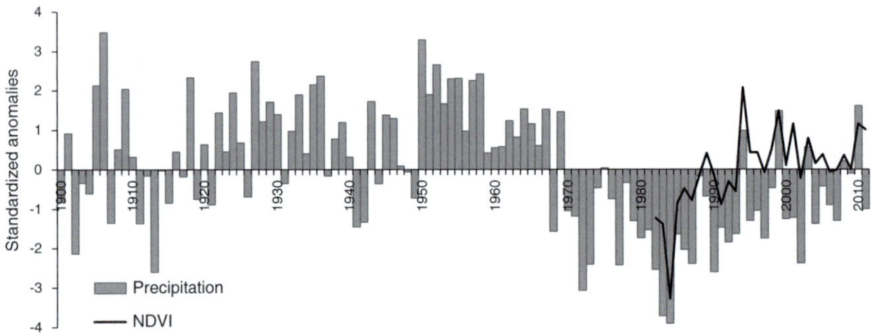

Fig. 5.4 Standardized June–October Sahel precipitation anomalies (10°–20°N and 20°W–10°E) from 1900 to 2011 (doi: 10.6069/H5MW2F2Q) and standardized NOAA AVHRR NDVI anomalies for the same region from 1982 to 2011. Both indices show a similar temporal pattern (Pearson's linear correlation coefficient: 0.82) for the three decades of overlap, illustrating the strong link between rainfall and vegetation greenness. Temporal trends in both variables are positive since 1982; at a secular timescale, however, the rainfall trend is negative.

vegetation cover in the Sahel and to refuting exaggerated claims of the extent of desertification. But longitudinal ground studies remain indispensable for providing a more detailed understanding of the nature of vegetation changes, and their ecological and livelihood implications. Of particular importance in this respect are assessments of changes in the woody vegetation cover, which as a slow variable is a more robust indicator of ecosystem health than the highly dynamic herbaceous cover.

5.3 Approaches and Challenges of "Ground Truthing"

When the AVHRR NDVI data first became available, there were no adequate ground observations of vegetation productivity against which the new index could be directly compared, prompting the establishment of a new field data collection programme in the Sahelian grasslands of Senegal (Prince 1991a). The studies emerging from this programme all indicated a positive correlation between integrated growing season NDVI and total dry biomass collected at the end of each growing season for a range of dry and wet years of the 1980s. This positive correlation suggested that the NDVI could be used for monitoring the spatial and temporal variability of biomass production over large areas (Tucker et al. 1983; Tucker et al.1985; Prince 1991a)—a promising technical advance of great interest for the desertification and global change communities.

Analogous studies in the dry grasslands of the US-Mexican border region, however, found that the NDVI failed to capture significant differences in biomass between sites (Beck et al. 1990; Huete and Jackson 1987). They warned of the indiscriminate use of the NDVI in every grassland and called for more research into the appropriate interpretation of NDVI for biomass parameters. A study by Prince (1991b) established that the relationship between biomass production and NDVI is not static but is also influenced by variable factors such as incident radiation, temperature and water stress of plants. Despite the obvious inconsistencies and documented need for more research on the interpretation of the NDVI and its trends over time, longitudinal studies linking NDVI time series data and ground observations of vegetation cover, let alone degradation status, are few and far between (e.g., Milich and Weiss 2000; Dardel et al. 2014a, b).

Even if the relationship between NDVI and biomass were assumed perfect, desertification remains difficult to assess, as the relationship between biomass and ecosystem health is complex: increasing biomass in the Sahel does not necessarily indicate continued improvement (Warren 2002; Warren and Olsson 2003). In the following sections, we will review three ways in which vegetation changes and trends in the Sahel, assessed in the field, could be used for "ground-truthing" NDVI trends with respect to land degradation: comparison of current with historical vegetation data (Sect. 5.3.1); assessment of regeneration potential of woody vegetation from current size class distributions (Sect. 5.3.2); and ethnobotanical knowledge (changes in vegetation perceived by local inhabitants, Sect. 5.3.3) (Fig. 5.5).

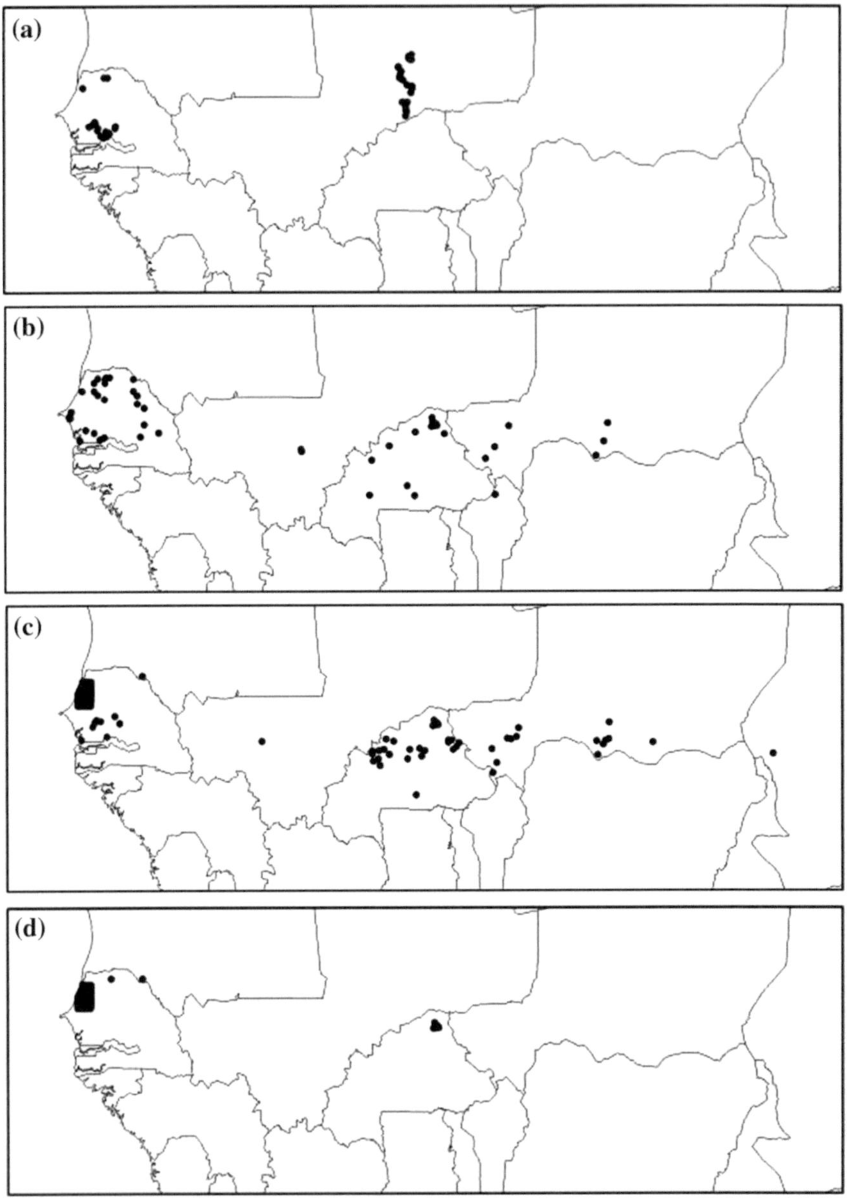

Fig. 5.5 Location of field sites of published studies (see Sects. 5.3.1–5.3.3) of woody vegetation change in the Sahel and Sudanian zones: **a** sites of studies using historical data, **b** sites of studies using size class distributions, **c** sites of studies using ethnobotanical knowledge, **d** sites of studies using high resolution remote sensing imagery.

5.3.1 Comparison of Current with Historical Vegetation Data

The availability of repeatedly collected historical field data of vegetation cover obviously presents the ideal case for "ground-truthing" remote sensing-derived trends. However, only for few locations throughout the area of interest are such historical datasets available. While some observations may have been made by environmental, forestry or livestock agencies during past decades, not all records were systematically archived or made accessible to researchers. Moreover, the available and accessible vegetation data were not necessarily collected with the goal of longitudinal studies of land degradation in mind, and may not have focused on the most pertinent variables for this type of study, nor did sampling designs and data collection always follow rigorous quantitative research protocols.

A few notable field studies in the semi-arid Sahel and Sudanian zones have nonetheless contributed to land cover change and desertification research. Long-term controlled grazing experiments were carried out at a field site in the Senegalese Ferlo region (Wendou Thiengoly), where quantitative field data on herbaceous biomass, composition of the herb layer and rainfall were recorded annually over a period of 27 years starting in 1981 (Miehe et al. 2010). Herbaceous biomass data were also collected annually in sites along an aridity gradient in the Malian Gourma region over 28 years. These data collections were carried out first by the International Livestock Research Institute (Hiernaux and Turner 1996), to assess the impact of droughts on fodder resources, and then as part of the African Monsoon Multidisciplinary Analysis (AMMA) project (Mougin et al. 2009; Hiernaux et al. 2009a, b) to better understand ecosystem functioning and dynamics. Another well studied field site is located in southwestern Niger (Fakara), where field observations of vegetation in a set of 24 sites began in 1994 (Hiernaux and Ayantunde 2004) and were carried on until 2010 (Cappelaere et al. 2009). The temporal trends in herbaceous biomass data from the field sites in Mali and Niger were later re-analysed by Dardel et al. (2014a) and compared with the satellite-observed NDVI trends.

Woody plant population dynamics were assessed in the Malian Gourma region (Hiernaux et al. 2009a, b; this volume). In this study, patterns of changes in woody plant population density, size, foliage and woody biomass and species composition were documented between 1984 and 2006. Vincke et al. (2010) analysed the temporal dynamics of the woody vegetation in the Senegalese Ferlo and explored spatial differences in regeneration. Herrmann and Tappan (2013) repeated an inventory of woody vegetation in 18 field sites in central Senegal after 25 years, albeit with the challenge of reliably replicating the expert-driven, semi-quantitative method used in the baseline assessment. In addition to estimating relative abundance by woody species and overall woody cover densities, the authors used repeat photography to document vegetation cover changes. In a similar fashion, Ganaba and Guinko (1995) compared the abundance and diversity of woody species to historical data in the Mare d'Oursi region of Burkina Faso.

A number of studies used historical air or satellite photography as a substitute for field assessments of woody vegetation, because historical field assessments are only scarcely available (e.g., Lykke et al. 1999; Gonzalez 2001; Gonzalez et al. 2012). Air photos and very high resolution satellite imagery resolve enough spatial detail for reliably estimating woody vegetation densities, but like all remotely-sensed observations, cannot replace field data when it comes to determining species composition.

5.3.2 Assessment of Regeneration Potential from Current Size Class Distributions of Woody Vegetation

In the absence of historical ground data, potential changes in woody vegetation have been inferred from an analysis of size class distributions of current tree stands. Size class distributions refer to the frequencies of tree species in a number of defined size classes, with size commonly denoted by the trunk diameter at breast height. Originating from forest ecology, size class distributions have commonly been used to describe the population structures of forests in humid West Africa (e.g., Aubréville 1938; Poorter et al. 1996) and other tropical forests (e.g., Condit et al. 1998). However, they were also found to offer potential for identifying declining and increasing species in dry ecosystems characterized by slow growing plants (Lykke 1998).

These ecosystems are particularly vulnerable to both anthropogenic and natural disturbances, which directly affect population structures (Feely et al. 2007; Venter and Witkowski 2010). Based on the sampling of size class distributions, variables such as health, vitality, regeneration potential and population structures of tree species can be revealed (Sop et al. 2011). Figure 5.6 illustrates the size class distributions of three fictitious species: The size class distribution of Species A drops abruptly with increasing diameter and is described as a reverse J shape, which is characteristic of species with good regeneration and recruitment. Species B shows a flat size class distribution, which is indicative of lack of recruitment and a possible change in species composition (Hall and Bawa 1993; Lykke 1998). The positive slope of Species C denotes poor regeneration (Shackleton 1993).

Several studies have analysed size class distributions to infer past changes and predict future variations in species populations in the Sahel. Lawesson (1990) sampled woody vegetation along strip transects in 15 sites located in reserves or protected forests across northern Senegal, with the goal of assessing the structure and composition of woody vegetation along soil and precipitation gradients. Lykke (1998) used size class distributions to analyse trends in 22 common woody species in a fire-disturbed savanna ecosystem in the Saloum Delta on the west coast of Senegal.

Analysis of size class distributions was also a part of the methodology employed by Lykke et al. (1999) to document vegetation changes and their consequences for local resource users in Sahelian Burkina Faso, in conjunction with air photo interpretation and ethnobotanical surveys. Sop et al. (2011) focused their analysis on three widely used multipurpose species in sub-Sahelian Burkina Faso—Acacia seyal, Balanites aegyptiaca and Pterocarpus lucens—and compared their population structures and

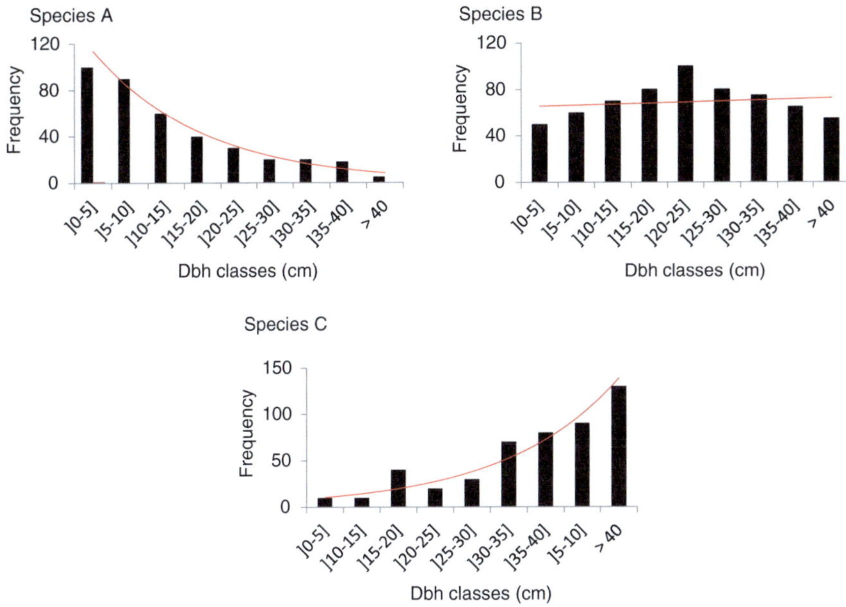

Fig. 5.6 Size class distributions (Dbh = diameter at breast height) illustrate high regeneration potential (*Species A*), lack of recruitment with possible decline (*Species B*) and no regeneration leading to extinction (*Species C*).

regeneration patterns. Traoré et al. (2013) carried out a similar study in the Sudanian phytogeographical zone of western Burkina Faso, focusing on the Sudanian species *Diospyros mespiliformis*, Prosopis africana and *Sterculia setigera*, with the goal of evaluating the effectiveness of protected areas in conserving these species.

The analysis of size class distributions is not without limitations for estimating potential trends in past and future vegetation compositions: (1) Size class distributions of trees in a stand are static representations of the population composition at a certain time and inferring rates of change from them can be problematic (Sokpon and Biaou 2002). (2) Size class distribution curves depend not only on population health but also differ between fast growing species with higher survival rates and slow growing species with lower survival rates (Condit et al. 1998). Notwithstanding these shortcomings, the analysis of size class distributions has made important contributions to our understanding of changes in vegetation structure and species compositions in the Sahel, in particular in combination with ethnobotanical knowledge.

5.3.3 *Ethnobotanical Knowledge*

In contrast to researchers, who might visit field sites only infrequently and over a limited time period, local populations are permanently present in specific sites or areas and as plant resource users have a strong interest in the fate of the vegetation. Thus, while they do not make systematic vegetation assessments as researchers would do, their observations and recollections provide invaluable evidence of stability and changes in vegetation. Ethnobotanical knowledge has been shown to be a relatively accessible and reliable source of information on historical distribution ranges of species, especially rare or endangered species which are generally difficult to assess using classical ecological methods (Lykke et al. 2004). In the past two decades, scientists have increasingly taken account of the local knowledge of rural farmers to reconstruct vegetation dynamics in the Sahel, particularly in combination with other botanical assessment and analysis methods (e.g., Lindskog and Tengberg 1994; Lykke 1998; Lykke et al. 1999, 2004; Wezel and Haigis 2000; Wezel and Lykke 2006; Ayantunde et al. 2008; Sop et al. 2011, 2013; Sop et al. 2012; Herrmann and Tappan 2013).

A number of qualitative and quantitative social science methods have been used to elicit information on the status and the cultural importance of plant species in the Sahel from informants, including structured and semi-structured interviews and focus group discussions. Typically, informants are asked to list woody plants occurring in their communities, rank them by their use value, and identify those perceived as increasing or declining in abundance or that need priority for conservation initiatives (Fig. 5.7).

Fig. 5.7 Local informants—pastoralists of the Peulh ethnic group—use illustrated cards of woody species as visual aids in a discussion of changes in vegetation composition in the Senegalese Sahel (Herrmann et al. 2014).

Lindskog and Tengberg (1994) used local knowledge as part of their multidisciplinary case study of land degradation in Sahelian Burkina Faso. They solicited villagers' perceptions on environmental deterioration by means of in-depth household interviews and used this qualitative information in support of assessments of degradation of the vegetation cover from air photos and a satellite image. Also in Sahelian Burkina Faso, Lykke (1998) and Lykke et al. (2004) assessed the use preferences and dynamics of woody species using a quantitative ethnobotanical method based on structured interviews. In the context of the much reported satellite-observed greening trend in the Sahel, Sop and Oldeland (2013) conducted semi-structured interviews with 87 groups of informants from twenty villages belonging to the Mossi, Fulani and Samo ethnic groups in order to sample the perceptions of local people on vegetation dynamics in the sub-Sahel of Burkina Faso. They compared the information obtained on usage, abundance, conservation status and priority as well as ranking of socio-economic importance of all the woody plants listed by the informants with an analysis of size class distributions of key species (Sop et al. 2011, 2012).

In semi-arid Niger, Wezel and Haigis (2000) solicited observations of past and present species occurrence and abundance from men and women and explored preferences for leaving certain species on cultivated fields. Wezel (2005) re-analysed local knowledge on changes in woody vegetation species from five cast studies in Burkina Faso, Senegal and Niger, which had been obtained in household interviews and focus group discussions. Wezel and Lykke (2006) extended the scope of the re-analysis to seven case studies.

Gonzalez (2001) conducted an extensive study in northwestern Senegal, in which he systematically sampled villages in an area of 7600 km^2 and recorded elders' recollections on the presence and absence of each of a predefined set of 126 woody species around 1945 and 1993. Biomass estimates made by the author as well as quantification of tree densities from air photos complemented the perceptions of the local inhabitants. In a study spanning 15 research sites in Senegal, Mali, Burkina Faso, Niger and Chad, a similar methodology was employed to obtain perceptions of species abundance in 1960 and 2000, which were analysed in conjunction with changes in tree densities estimated from air photos and very high resolution satellite imagery (Gonzalez et al. 2012). Herrmann and Tappan (2013) included local perceptions of occurrence, abundance and trends of woody species in a study which compares ground evidence of woody vegetation change in sites that appear to have greened up according to coarse resolution satellite observations with change in sites that appear not to have greened up.

Although a good correlation between local knowledge and the physical reality of the environment has been documented (e.g., Lindskog and Tengberg 1994; Reed et al. 2007), some challenges associated with the use of such knowledge in ethnobotanical assessments should not be ignored: People's perceptions reflect the informants' subjective experience of environmental change, which may be subject to biases and exaggerate or mask trends (Daw 2010). Informants tend to focus on the species most useful to them, while overlooking rare and less used species. Environmental knowledge also depends on the age of the informants (Hanazaki et al. 2013). The absence of standardized data collection protocols to assess local perceptions and knowledge can make data integration problematic. Despite these shortcomings when judging ethnobotanical knowledge by scientific standards, the first-hand experience of local land users is an invaluable source of information, and large sample sizes can at least in part make up for these shortcomings.

5.4 Synthesis of Ground-Based Vegetation Studies Across the Sahel

The vegetation studies reviewed in this chapter followed various methodological approaches and data collection protocols and they do not cover identical time periods, which precludes quantitative, statistical meta-analysis. On the other hand, the qualitative synthesis of these studies offers a more reliable picture of vegetation changes across the Sahel than any single study could. The consistency of the findings, despite the different approaches—the comparison of current with historical data, the analysis of size class distributions, and people's perceptions—adds validity to the changes we document. A total of 301 field sites from 20 of the studies reviewed could be located precisely in space (Fig. 5.5) and interpreted with respect to changes in abundance and/or diversity of woody vegetation. By abundance, we mean any indication of the number of trees, density or percent cover; by diversity, we mean any indication of the number of different species present at a site. We summarized the changes in woody vegetation at each site into the three simple categories of "increasing", "stable" or "decreasing" over time, in order to obtain a general representation of change and to allow for comparison with remote sensing-based trends over time (Figs. 5.8, 5.9 and 5.10).

This synthesis documents findings on the changes in woody vegetation at a point in time (May 2014), which will continue to be revised as new studies are carried out in this very dynamic field.

5.4.1 *Long Term Changes in the Abundance of Woody Vegetation*

Most of the studies across the region comparing historical to current vegetation data point to an overall decline in the abundance of woody vegetation between the pre-drought period and the late 1990s or early 2000s (e.g., Gonzalez 1997, 2001; Lykke et al. 1999; Vincke et al. 2010; Gonzalez et al. 2012; Herrmann and Tappan 2013; Herrmann et al. 2013) (Fig. 5.8a). Local inhabitants also observed a substantial reduction of woody vegetation with the disappearance or decreasing abundance of many species (e.g., Lykke et al. 1999; Wezel and Haigis 2000; Gonzalez 2001; Sop et al. 2011) (Fig. 5.8b). Analyses of size class

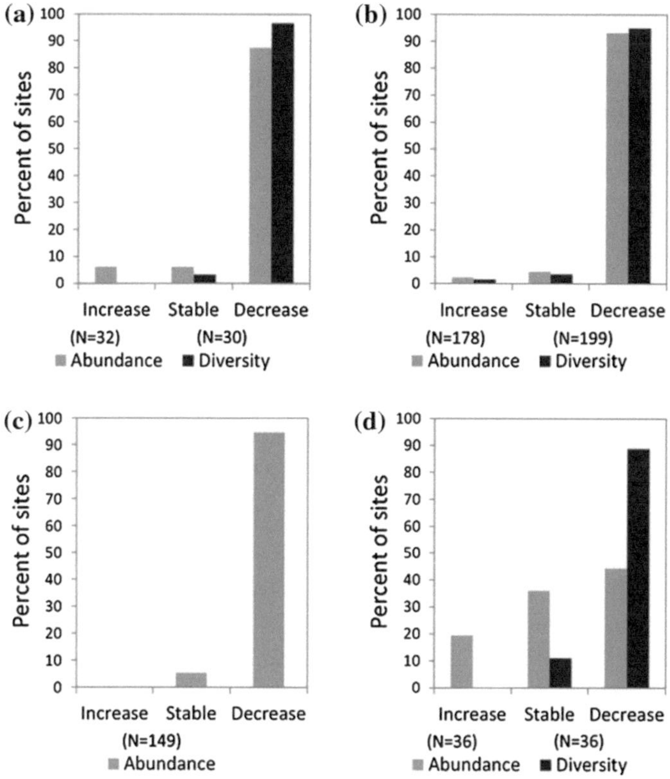

Fig. 5.8 Proportion of field sites showing increasing, stable and decreasing abundance and diversity of woody vegetation as assessed by **a** studies based on historical ground data, **b** studies based on people's perceptions of vegetation change, **c** studies based on very high resolution imagery, **d** studies based on size class distributions. N is the number of sites included in the count for each variable. The number of sites per study ranged from only one to 135, meaning that findings from some studies are more heavily represented in this assessment than from others.

distributions generally confirmed the lack of rejuvenation of many, but not all woody species (e.g., Lykke 1998; Lykke et al. 1999; Bognounou et al. 2009; Sop et al. 2013) (Fig. 5.8c).

The observed changes are not uniform. At finer scales, the magnitude of the decline appears to vary depending on the location of the site within the landscape. Lykke et al. (1999) found that valley sites in Sahelian Burkina Faso were most severely affected by the loss of trees. In contrast, Ganaba and Guinko (1995) and Vincke et al. (2010) found that regeneration of the vegetation was better in depressions and along valleys, where runoff induces greater accumulation of water, fine soil particles and organic matter, than on the plateaus.

Exceptions to the general reduction in the abundance of the woody vegetation cover were found in the shrub layer, which, as opposed to the tree layer, has increased in some sites studied in central Senegal (Herrmann and Tappan 2013) and in the Gourma region of Mali (Hiernaux et al. 2009a, b). Expansion of fruit tree plantations have also accounted for localized increases in tree densities over the past four decades, particularly in the vicinity of the urban centers of Dakar and Thies (Herrmann et al. 2013). The most noteworthy exceptions to the general decline in tree cover, however, are the success stories of farmer-managed natural regeneration (Reij et al. 2009; see Sect. 5.4.3).

5.4.2 Long Term Changes in Woody Vegetation Composition

With the exception of farmer-managed natural regeneration, virtually all ground studies point to a progressive impoverishment of the woody vegetation cover in several respects: overall species diversity has been decreasing; a shift towards more xeric species has taken place; and the species most valued by the local inhabitants have declined most dramatically.

For example, studies comparing current and historical woody vegetation documented reductions in or disappearance of a number of woody species at sites in the Senegalese Ferlo (Vincke et al. 2010), central Senegal (Herrmann and Tappan 2013), and the Mare d'Oursi in northern Burkina Faso (Ganaba and Guinko 1995), leading to a less diverse woody cover. These findings are corroborated by ethnobotanical studies, in which informants listed more decreasing than increasing species at a majority of sites. Even parkland species such as *Parkia biglobosa* and *Vitellaria paradoxa*, which are traditionally protected on farmers' fields for soil fertility management and fruit production, were decreasing in some villages in Niger (Wezel and Haigis 2000). The only exceptions were villages where tree planting had been promoted by development projects, mostly of fast growing exotic species, including *Azadirachta indica* (neem), *Mangifera indica* (mango) and *Eucalyptus camaldulensis* (Wezel and Haigis 2000; Wezel and Lykke 2006). While those newly planted species provided some wood and other tree products for rural households, they were not able to compensate for the cutting of the natural vegetation (Wezel and Haigis 2000).

Among the most dramatically declining tree species were the more mesic and less drought-resistant Sudanian and Guinean species (Ganaba and Guinko 1995; Gonzalez 2001; Vincke et al. 2010; Sop et al. 2011; Herrmann and Tappan 2013). The more drought-resistant Sahelian woody species were left to dominate not only in the Sahel but increasingly also in the Sudanian zone, indicating a possible shift of the vegetation belts southward (Gonzalez 2001). The mesic species being lost include large canopy trees, characterized by hard wood and edible fruits, which makes them highly valued. The most severely declining species listed in a study of 25 villages in Burkina Faso, Niger and Senegal are *Acacia ataxacantha*, *A. senegal*, *A. seyal*, *A. ehrenbergiana*, *Adansonia digitata*, *Boscia senegalensis*, *Ceiba pentandra*, *Dalbergia melanoxylon*, *Ficus gnaphalocarpa*, *Grewia bicolor*,

Khaya senegalensis, *Maerua crassifolia*, *Pterocarpus lucens*, *Sclerocarya birrea* and *Tamarindus indica* (Wezel and Lykke 2006). This list includes highly valued species that are relied upon for food, fodder, medicines, construction and craft materials, and as energy sources (Sop et al. 2012).

Size class distribution analyses show a mixed picture, with good regeneration potential, i.e., higher frequency of young individuals, in some sites (Bognounou et al. 2009; Sop et al. 2012, 2013). Rejuvenation was however limited to a single or very few hardy species per site, indicating that the reduction in species diversity seen in the past is likely to continue into the future. Some examples include increasing dominance of Acacia raddiana in the Mare d'Oursi (Ganaba and Guinko 1995), of *Calotropis procera*, *Boscia senegalensis* and *Balanites aegyptiaca* in the *Senegalese Ferlo* (Lawesson 1990; Vincke et al. 2010), of *Acacia tortilis* and *Leptadania pyrotechnica* around Gorom-Gorom in Sahelian Burkina Faso (Lykke et al. 2004), and of *Combretum glutinosum* and *Guiera senegalensis* in the Sudanian domain of central Senegal (Herrmann and Tappan 2013).

As shown by the good agreement between historical vegetation surveys, ethnobotanical studies and analyses of size class distributions, the degradation of the woody vegetation cover does not appear to be a localized problem, but has been documented at sites throughout the Sahel and Sudanian zones.

5.4.3 The Case of Farmer-Managed Natural Regeneration

In contrast to the widespread negative changes in the woody vegetation cover, two "success stories" stand out: (1) farmer-managed soil and water conservation in the central plateau of Burkina Faso, where vast stretches of degraded and barren farmland have been rehabilitated and (2) farmer-managed natural regeneration of agroforestry parklands in southern Niger (Reij et al. 2009). Both practices began to be implemented at the height of the environmental crisis in the early 1980s, when innovative farmers and non-governmental organizations experimented with traditional planting pits and stone bunds to reclaim severely degraded farmland (Reij et al. 2005) and protected on-farm trees to reduce wind speeds, improve soil fertility and provide fodder for livestock (Tougiani et al. 2009).

Reportedly large in scale—an estimated 200,000–300,000 ha rehabilitated in Burkina Faso (Botoni and Reij 2009) and an area of nearly 5 million ha under farmer-managed natural regeneration in Niger (Reij et al. 2009)—these developments have not been rigorously quantified for such large areas. However, evidence from air photography and high resolution satellite imagery in 30 sample locations in the Mirriah-Magaria-Matameye triangle of Niger show that average field tree cover, after having dropped from 2.8 % in 1957 to 1.5 % in 1975, had almost tripled to 4.4 % in 2005, with local concentrations of up to 16 % (Reij et al. 2009). Thus, the practice of farmer-managed natural regeneration seems to have allowed tree densities in the parkland to reach higher levels than in 1957. With a high percentage of young trees among the current field tree population, the percent cover is expected to continue to increase significantly over the next decade. Mahamane et al. (2012) confirm the regenerative potential of field tree cover in three village territories in the Maradi region.

In the Central Plateau of Burkina Faso, Belemviré (2003) counted an average of 126 trees per ha on rehabilitated fields, compared with 103 trees per ha in control fields. Trees in rehabilitated fields were also larger and of higher species diversity than those in the control fields. Species that persisted through the environmental degradation crisis include *Combretum glutinosum*, *Guiera senegalensis* and *Piliostigma reticulatum*. These species continue to be dominant, while species that had been severely decimated or lost—such as *Diospyros mespiliformis*, *Anogeissus leiocarpus*, *Sclerocarya birrea*, *Butyrospermum parkii* and *Lannea sp.*—have re-appeared in rehabilitated fields (Reij et al. 2005).

As farmer-managed natural regeneration has shown, human population growth and agricultural intensification do not have to entail losses in vegetation productivity and environmental degradation (Mortimore et al. 1999). On the contrary, where sound management is practised, farmers' fields stand out by improved vegetation cover and diversity, while unprotected woodlands and unmanaged fields continue to degrade.

5.5 Linking Remote Sensing Observations to Ground Evidence

With few exceptions, regional-scale trends of vegetation productivity in the Sahel, observed from satellite data, and local-scale changes in the vegetation cover on the ground have so far mostly been addressed separately and by different researchers. The findings from those two perspectives have led to differing and sometimes conflicting conclusions on the nature and extent of degradation/desertification in the Sahel.

The simplified three-class summarization of changes in the abundance and diversity of woody vegetation as "increasing", "stable" or "decreasing" (see Sect. 5.4) was spatially overlaid with an equally simplified interpretation of the satellite observed greening trends, (binary data layer of "greening" versus "not greening"). "Greening" denotes a significantly ($p < 0.05$) positive slope of the linear fit of a trend line to the time series of AVHRR NDVI from 1982 to 2011 for each 8 km pixel, disregarding the magnitude of the trend; "not greening" is the absence of such a trend. No significantly negative trends in vegetation greenness were present at the field sites. While trends in both integrated annual and growing season (July-October) NDVI were considered, only trends in growing season averages are shown here, as their spatial patterns appear to be more consistent across different versions of the NDVI dataset and across different studies. Figure 5.9 illustrates the differences between the two types of information in just one 100 × 100 km subset in

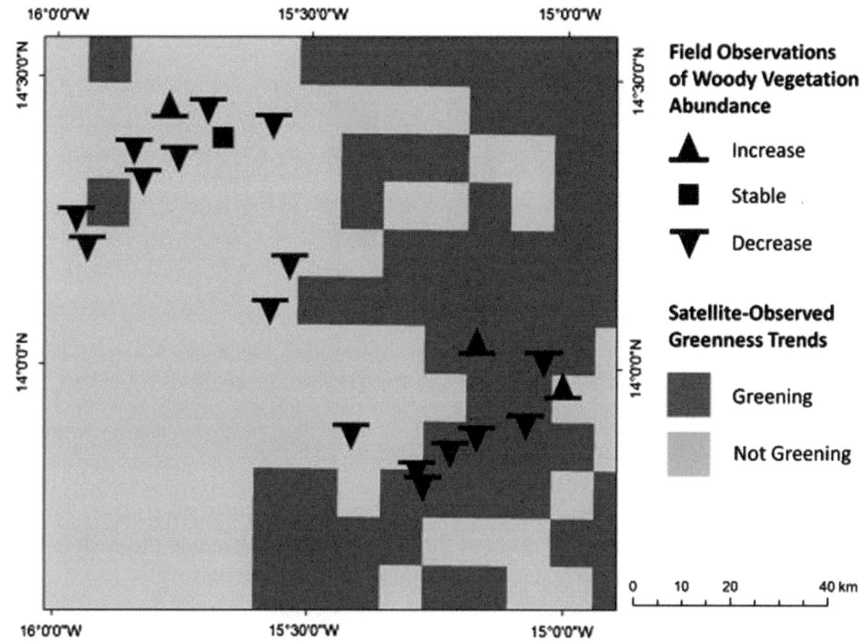

Fig. 5.9 Spatial overlay of the direction of changes in woody vegetation abundance as observed on the ground and of a binary map of satellite-observed greenness trends (greening vs. not greening) for a subset of study sites in central Senegal. Ground observations of decreasing woody vegetation cover appear to dominate throughout the area, whether coinciding with the satellite-observed greening trend or not.

central Senegal, as the clustered distribution of field sites (Fig. 5.5) precludes a legible display of changes at a small scale over the entire Sahel and Sudan zones.

Whereas positive trends in vegetation greenness prevail—65 % of the full set of field sites are associated with significant greening according to growing season NDVI trends—ground evidence shows that a perceived degradation of the woody vegetation cover is predominant. In 85 % of the field sites, a majority of which was associated with positive greening trends, a decrease in woody vegetation density was noted (Fig. 5.10). In 10.5 and 4.5 % of the field sites, woody vegetation cover was stable and increasing respectively. Paradoxically, more than half of the sites with increasing woody vegetation cover did not show significant greening in satellite observations. This is in line with earlier observations that the areas of farmer-managed natural regeneration in southern Niger, where field tree cover is said to have improved, do not stand out in the satellite-derived greenness trend maps (Reij et al. 2009). It has to be noted that in our comparison of remote sensing observations and ground evidence, farmer-managed natural regeneration is not well represented, as very little data on increases in tree cover have been reported in the spatially explicit way necessary for a direct comparison with NDVI trends. Another limitation and potential source of bias in the comparison is introduced by the

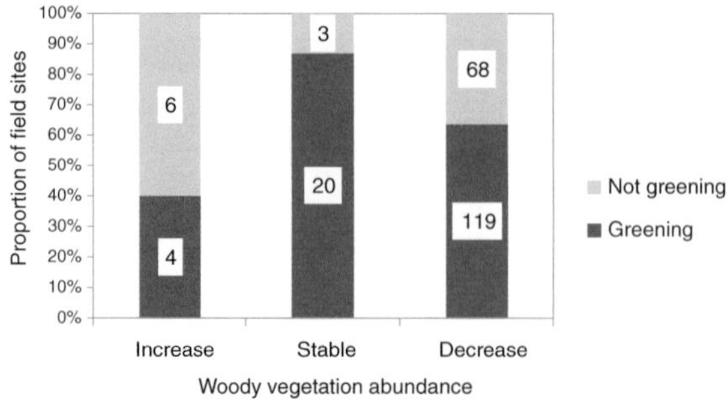

Fig. 5.10 Proportion and number of field sites with increasing, stable and decreasing woody vegetation abundance that spatially coincide with satellite-observed greening and not greening respectively. Satellite-observed greening—represented by growing season (July–October) trends in NDVI from 1982 to 2011—dominates by 65–35 %, whereas on the ground increase in woody cover was observed in only 10 sites compared to decrease in 187 sites.

constraint that the start and end dates of the various field studies and the period over which NDVI trends were computed correspond only approximately.

Diversity in the woody vegetation cover was reported to have decreased even more dramatically than its abundance. A majority of the 95 % of field sites that reported decreasing species diversity are associated with significant greening trends according to satellite observations. There is no reason to expect a causal link between the two variables; rather, they are independent indicators of degradation or desertification. However, what these findings tell is that any increase in bio-productivity was generally accomplished by very few species at the expense of species diversity.

Discrepancy between the two indicators (woody vegetation changes observed in the field and satellite observations of bioproductivity) can be explained by the fact that much of the biomass production in the Sahel and Sudan zones is made up of herbaceous vegetation, which responds fast to precipitation variability (Anyamba and Tucker 2005; Hiernaux, this volume). Precipitation has increased over the past two decades (Nicholson 2013), triggering a growth response of the herbaceous vegetation and consequently an increase in biomass production (Anyamba et al. 2014). This increase, however, is not to be mistaken for a recovery from degradation and a return to pre-drought vegetation conditions, as shown by the simultaneous impoverishment of the woody vegetation cover. In an environment where trees form a small, yet ecologically and economically crucial, fraction of the overall green vegetation cover, the coarse spatial resolution of long time series of NDVI has proven insufficient for capturing the woody component of vegetation dynamics. Thus, in the absence of ancillary data that allow for more intricate analyses but are rarely available at regional to global scales, NDVI time series alone are of limited use for identifying or monitoring desertification or land degradation (Prince, this volume).

5.6 Conclusions

Satellite remote sensing has given us an unprecedented synoptic perspective on global vegetation dynamics by providing consistent, repeatable observations. As the only source of long term vegetation greenness data, the NDVI time series derived from the NOAA AVHRR satellite sensors (Tucker et al. 2005; Pinzon and Tucker 2014) have played an important role in uncovering the interannual variability of vegetation cover in the Sahel and have helped refute exaggerated claims of the extent of desertification (e.g., Prince et al. 1998). Based on NOAA AVHRR NDVI data, a growing number of satellite-based studies has documented a greening trend in the Sahel since the early 1980s.

While these findings indicate that biomass production has likely increased in response to a moderate recovery in rainfall since the great droughts, it can be argued that bio-productivity is not the only indicator of the state of desertification or rehabilitation. Rain use efficiencies and trends in residual NDVI have been used to better distinguish between the effects of rainfall variability and degradation on biomass production, neither of them suggesting regional-scale degradation.

A number of field studies, however, have painted a less positive picture of vegetation changes at local scales. Focusing on the woody vegetation, a key component of the savanna ecosystem and of the livelihood system of rural populations, the prevailing story is one of degradation. The abundance of woody vegetation has declined in many sites. Most importantly, the loss of large trees is ubiquitous, while shrub cover was observed to have increased in some locations. The remaining woody cover has shifted towards fewer and more drought tolerant species, with a few dominating and replacing a more diverse vegetation. This trend, which is not incompatible with increases in bio-productivity as measured by the NDVI, has been observed in longitudinal botanical field studies and in ethnobotanical studies alike. An important exception has emerged where field trees are explicitly protected and diversity promoted in farmer-managed natural regeneration initiatives. Such successes have been documented mainly in the Central Plateau of Burkina Faso and in southern Niger, with positive impacts on soil fertility, crop productivity, and food security. However, these positive developments on the ground, able to halt or reverse degradation, do not stand out above other areas by their NDVI trend.

While the satellite-observed, region-wide greening trend is encouraging in the face of earlier claims of widespread desertification in the Sahel, it is based on rather coarse resolution data and masks developments at finer scales, both negative and positive. The contrasting interpretations of findings from the field and of remote sensing observations with respect to desertification illustrate that the availability of remote sensing data does not ultimately remove the need for ground-based data. The perspective from the field remains indispensable in order to provide "ground truth" for the NDVI data and to put desertification and land rehabilitation on the map. The divergent findings also highlight the need for more research that bridges scale differences and takes account of both physical and social science perspectives.

In particular, field studies explicitly targeting greening and control areas should be encouraged. Individual field studies are spatially confined, so sharing regional archives, field data and collection protocols from different studies across the region would benefit our understanding of NDVI trends and derivative metrics in the context of land degradation and rehabilitation.

References

Akpo, L.-E. (1997). Phenological interactions between tree and understory herbaceous vegetation of a Sahelian semi-arid savanna. *Plant Ecology, 131*, 241–248.

Anyamba, A., Small, J. L., Tucker, C. J., & Pak, E. W. (2014). Thirty-two years of Sahelian Zone growing season non-stationary NDVI3 g patterns and trends. *Remote Sensing, 6*, 3101–3122.

Anyamba, A., & Tucker, C. J. (2005). Analysis of Sahelian vegetation dynamics using NOAA-AVHRR NDVI data from 1981–2003. *Journal of Arid Environments, 63*, 596–614.

Aubréville, A. (1938). La forêt coloniale. *Annals of the Academy Science Colon, 9*, 1–245.

Ayantunde, A. A., Briejer, M., Hiernaux, H., Henk, M., Udo, J., & Tabo, R. (2008). Botanical knowledge and its differentiation by age, gender and ethnicity in south western Niger. *Human Ecology, 36*, 881–889.

Bai, Z. G., Dent, D. L., Olsson, L., & Schaepman, M. E. (2008). *Global assessment of land degradation and improvement 1: Identification by remote sensing*. Report 2008/01, FAO/ISRIC—Rome/Wageningen.

Beck, L. R., Hutchinson, C. F., & Zauderer, J. (1990). A comparison of greenness measures in two semi-arid grasslands. *Climatic Change, 17*, 287–303.

Belemvire, A. (2003). Impact de la conservation de l'eau et des sols sur la régénération naturelle assistée. Développement rural et environnement au Burkina Faso: La réhabilitation de la capacité des terroirs sur la partie Nord du Plateau central entre 1980 et 2000. Rapport de travail no. 1. .Ouagadougou, Burkina Faso: Conseil National pour la Gestion de l'Environnement.

Bognounou, F., Thiombiano, A., Savadogo, P., Boussim, J. I., Oden, P. C., & Guinko, S. (2009). Woody vegetation structure and composition at four sites along a latitudinal gradient in Western Burkina Faso. *Bois et Forêts des Tropiques, 300*, 29–44.

Botoni, E., & Reij, C. (2009). La transformation silencieuse de l'environnement et des systèmes de production au Sahel: L'impacts des investissements publics et privés dans la gestion des ressources naturelles. Amsterdam, the Netherlands: Comité Permanent Inter-Etats de Lutte Contre la Secheresse dans le Sahel (CILSS) and Vrije University Amsterdam.

Cappelaere, B., Descroix, L., Lebel, T., Boulain, N., Ramier, D., Laurent, J. P., et al. (2009). The AMMA-CATCH experiment in the cultivated Sahelian area of south-west Niger: Investigating water cycle response to a fluctuating climate and changing environment. *Journal of Hydrology, 375*, 34–51.

Cherlet et al. (2015). *World atlas of desertification* (3rd edn.). http://wad.jrc.ec.europa.eu/. Last accessed April 28, 2014.

Condit, R., Sukumar, R., Hubbell, S. P., & Foster, R. B. (1998). Predicting population trends from size distributions: A direct test in a tropical tree community. *American Naturalist, 152*, 495–509.

Dardel, C., Kergoat, L., Hiernaux, P., Grippa, M., Mougin, E., Ciais, P., & Nguyen, C.-C. (2014a). Rain-use-efficiency: What it tells us about the conflicting Sahel greening and Sahelian paradox. *Remote Sensing, 6*, 3446–3474.

Dardel, C., Kergoat, L., Hiernaux, P., Mougin, E., Grippa, M., & Tucker, C. J. (2014b). Re-greening Sahel: 30 years of remote sensing data and field observations (Mali, Niger). *Remote Sensing of Environment, 140*, 350–364.

Daw, T. M. (2010). Shifting baselines and memory illusions: What should we worry about when inferring trends from resource user interviews? *Animal Conservation, 13*, 534–535.

Donohue, R. J., Roderick, M. L., McVicar, T. R., & Farquhar, G. D. (2013). Impact of CO2 fertilization on maximum foliage cover across the globe's warm, arid environments. *Geophysical Research Letters, 40*, 3031–3035.

Eklundh, L., & Olsson, L. (2003). Vegetation index trends for the African Sahel 1982–1999. *Geophysical Research Letters, 30*: 13-11–13-14.

Feely, K. J., Davies, S. J., Noor, N. S., Kassim, A. R., & Tan, S. (2007). Do current stem size distributions predict future population changes? An empirical test of intraspecific patterns in tropical trees at two spatial scales. *Journal of Tropical Ecology, 23*, 191–198.

Fensholt, R., & Proud, S. R. (2012). Evaluation of earth observation based global long term vegetation trends—Comparing GIMMS and MODIS global NDVI time series. *Remote Sensing of Environment, 119*, 131–147.

Fensholt, R., Rasmussen, K., Kaspersen, P., Huber, S., Horion, S., & Swinnen, E. (2013). Assessing land degradation/recovery in the African Sahel from long-term earth observation based primary productivity and precipitation relationships. *Remote Sensing, 5*, 664–686.

Fensholt, R., Rasmussen, K., Nielsen, T. T., & Mbow, C. (2009). Evaluation of earth observation based long term vegetation trends—Intercomparing NDVI time series trend analysis consistency of Sahel from AVHRR GIMMS, Terra MODIS and SPOT VGT data. *Remote Sensing of Environment, 113*, 1886–1898.

Ganaba, S., & Guinko, S. (1995). Etat actuel et dynamique du peuplement ligneux de la région de la Mare d'Oursi (Burkina Faso). *Etudes sur la Flore et la Végétation du Burkina Faso et des Pays Avoisinats, 2*, 3–14.

Gonzalez, P. (1997). *Dynamics of biodiversity and human carrying capacity in the Senegal Sahel.* PhD dissertation, University of California, Berkeley.

Gonzalez, P. (2001). Desertification and a shift of forest species in the West African Sahel. *Climate Research, 17*, 217–228.

Gonzalez, P., Tucker, C. J., & Sy, H. (2012). Tree density and species decline in the African Sahel attributable to climate. *Journal of Arid Environments, 78*, 55–64.

Hall, P., & Bawa, K. (1993). Methods to assess the impact of extraction of nontimber tropical forest products on plant populations. *Economical Botany, 47*, 234–247.

Hanazaki, N., Herbst, D.F., Marques, M.S. & Vandebroek, I. (2013). Evidence of the shifting baseline syndrome in ethnobotanical research. *Journal of Ethnobiology and Ethnomedicine, 9*.

Hansen, M. C., & DeFries, R. S. (2004). Detecting long-term global forest change using continuous fields of tree-cover maps from 8-km advanced very high resolution radiometer (AVHRR) data for the years 1982–99. *Ecosystems, 7*, 695–716.

Harrison, M. N., & Jackson, J. K. (1958). Ecological classification of the vegetation of Sudan. Forests Bulletin 2 and Vegetation Map of Sudan printed by Sudan Survey Department, Khartoum, Topo No. S. 625–40.

Hein, L., & De Ridder, N. (2006). Desertification in the Sahel: a reinterpretation. *Global Change Biology, 12*, 751–758.

Hellden, U. (1991). Desertification—time for an assessment? *Ambio, 20*, 372–383.

Herrmann, S. M., Anyamba, A., & Tucker, C. J. (2005). Recent trends in vegetation dynamics in the African Sahel and their relationship to climate. *Global Environmental Change, 15*, 394–404.

Herrmann, S. M., & Hutchinson, C. F. (2005). The changing contexts of the desertification debate. *Journal of Arid Environments, 63*, 538–555.

Herrmann, S. M., Sall, I., & Sy, O. (2014). People and pixels in the Sahel—A study linking coarse-resolution remote sensing observations to land users' perceptions of their changing environment in Senegal. *Ecology and Society* (under revision).

Herrmann, S. M., & Tappan, G. G. (2013). Vegetation impoverishment despite greening: A case study from central Senegal. *Journal of Arid Environments, 90*, 55–66.

Herrmann, S. M., Wickhorst, A. J., & Marsh, S. E. (2013). Estimation of tree cover in an agricultural parkland of Senegal using rule-based regression tree modeling. *Remote Sensing, 5*, 4900–4918.
Heumann, B. W., Seaquist, J. W., Eklundh, L., & Joensson, P. (2007). AVHRR derived phenological change in the Sahel and Soudan, Africa, 1982–2005. *Remote Sensing of Environment, 108*, 385–392.
Hiernaux, P., & Ayantunde, A. A. (2004). *The Fakara: a semi-arid agro-ecosystems under stress*. Report of research activities of International Livestock Research Institute (ILRI) in Fakara, South-western Niger, between 1994 and 2002, Desert Margins Program. ICRISAT, Niamey.
Hiernaux, P., Diarra, L., Trichon, V., Mougin, E., Soumaguel, N., & Baup, F. (2009a). Woody plant population dynamics in response to climate changes from 1984 to 2006 in Sahel (Gourma, Mali). *Journal of Hydrology, 375*, 103–113.
Hiernaux, P., Mougin, E., Diarra, L., Soumaguel, N., Lavenu, F., Tracol, Y., & Diawara, M. (2009b). Sahelian rangeland response to changes in rainfall over two decades in the Gourma region, Mali. *Journal of Hydrology, 375*, 114–127.
Hiernaux, P., & Turner, M. D. (1996). The effect of clipping on growth and nutrient uptake of Sahelian annual rangelands. *Journal of Applied Ecology, 33*, 387–399.
Huete, A. R., & Jackson, R. D. (1987). Suitability of spectral indices for evaluating vegetation characteristics on arid rangelands. *Remote Sensing of Environment, 23*, 213–232.
Hutchinson, C. F., Herrmann, S. M., Maukonen, T., & Weber, J. (2005). Editorial: Introduction: The "Greening" of the Sahel. *Journal of Arid Environments, 63*, 535–537.
Ibrahim, F. (1978). The problem of desertification in the Republic of the Sudan with spacial reference to Northern Darfur Province. SRC Monograph Series 8.
James, M. E., & Kalluri, S. N. V. (1994). The Pathfinder Avhrr land data set—An improved coarse resolution data set for terrestrial monitoring. *International Journal of Remote Sensing, 15*, 3347–3363.
Joint Research Centre of the European Commission. (2015). *World atlas of desertification* http://wad.jrc.ec.europa.eu/. Last accessed February 7, 2015.
Lamprey, H. F. (1988). Report on desert encroachment reconnaissance in Northern Sudan: 21 October to 10 November 1975. *Desertification Control Bulletin, 17*, 1–7.
Lawesson, J. E. (1990). Sahelian woody vegetation in Senegal. *Vegetatio, 86*, 161–174.
Lindskog, P., & Tengberg, A. (1994). Land degradation, natural resources and local knowledge in the Sahel Zone of Burkina Faso. *GeoJournal, 33*, 365–375.
Lykke, A. M. (1998). Assessment of species composition change in savanna vegetation by means of woody plants' size class distributions and local information. *Biodiversity and Conservation, 7*, 1261–1275.
Lykke, A. M., Fog, B., & Madsen, J. E. (1999). Woody vegetation changes in the Sahel of Burkina Faso assessed by means of local knowledge, aerial photos, and botanical investigations. *Geografisk Tidsskrift (Danish Journal of Geography)*, 57–68.
Lykke, A. M., Kristensen, M. K., & Ganaba, S. (2004). Valuation of local use and dynamics of 56 woody species in the Sahel. *Biodiversity and Conservation, 13*, 1961–1990.
Mahamane, L., Iro, D. G., Matig, O. E., & Idrissa, I. A. (2012). Farmer managed tree natural regeneration and diversity in a Sahelian environment: Case study of Maradi region, Niger. *Continental Journal of Agricultural Science, 6*, 38–49.
McCloy, K. R., Los, S., Lucht, W., & Højsgaard, S. A. (2005). A comparative analysis of three long-term NDVI datasets derived from AVHRR satellite data. *EARSeL Proceedings, 4*, 52–69.
Middleton, N., Thomas, D., & United Nations Environmental Programme. (1992). *World atlas of desertification* (1st edn). London: Arnold.
Middleton, N., Thomas, D., & United Nations Environmental Programme. (1997). *World atlas of desertification* (2nd edn). London: Arnold.
Miehe, S., Kluge, J., von Wehrden, H., & Retzer, V. (2010). Long-term degradation of Sahelian rangeland detected by 27 years of field study in Senegal. *Journal of Applied Ecology, 47*, 692–700.

Milich, L., & Weiss, E. (2000). GAC NDVI interannual coefficient of variation (CoV) images: Ground truth sampling of the Sahel along north-south transects. *International Journal of Remote Sensing, 21*, 235–260.

Mortimore, M. (1989). *Adapting to drought. Farmers, famine and desertification in West Africa*. Cambridge: Cambridge University Press.

Mortimore, M., Harris, F. M. A., & Turner, B. (1999). Implications of land use change for the production of plant biomass in densely populated Sahelo-Sudanian shrub-grasslands in north-east Nigeria. *Global Ecology and Biogeography, 8*, 243–256.

Mougin, E., Hiernaux, P., Kergoat, L., Grippa, M., de Rosnay, P., Timouk, F., et al. (2009). The AMMA-CATCH Gourma observatory site in Mali: Relating climatic variations to changes in vegetation, surface hydrology, fluxes and natural resources. *Journal of Hydrology, 375*, 14–33.

National Research Council. (2008). *Earth observations from space: The first 50 years of scientific achievements*. Washington, D.C.: National Academy Press.

Nicholson, S. E. (2013). The West African Sahel: A review of recent studies on the rainfall regime and its interannual variability. *ISRN Meteorology*, 1–32.

Oldeman, L. R., Hakkeling, R. T. A., & Sombroek, W. G. (1991). World map of the status of human-induced soil degradation: An explanatory note, second revised edition. ISRIC/UNEP.

Pearce, F. (1992). Mirage of the shifting sands. *New Scientist, 136*, 38–42.

Pinzon, J. E., & Tucker, C. J. (2014). A non-stationary 1981–2012 AVHRR NDVI3 g time series. *Remote Sensing, 6*, 6929–6960.

Poorter, L., Bongers, F., van Rompaey, R. S. A. R., & de Klerk, M. (1996). Regeneration of canopy tree species at five sites in west African moist forest. *Forest Ecology and Management, 84*, 61–69.

Prince, S. (1991a). Satellite remote sensing of primary production: Comparison of results for Sahelian grasslands 1981–1988. *International Journal of Remote Sensing, 12*, 1301–1311.

Prince, S. D. (1991b). A model of regional primary production for use with coarse resolution satellite data. *International Journal of Remote Sensing, 12*, 1313–1330.

Prince, S. D., Brown de Colstoun, E., & Kravitz, L. L. (1998). Evidence from rain-use efficiencies does not indicate extensive Sahelian desertification. *Global Change Biology, 4*, 359–374.

Prince, S. D., Wessels, K. J., Tucker, C. J., & Nicholson, S. E. (2007). Desertification in the Sahel: A reinterpretation of a reinterpretation. *Global Change Biology, 13*, 1308–1313.

Raynaut, C., Gregoire, E., Janin, P., Koechlin, J., & Lavigne Delville, P. (1996). *Societies and nature in the Sahel*. London: Routledge.

Reed, M. S., Dougill, A. J., & Taylor, M. J. (2007). Integrating local and scientific knowledge for adaptation to land degradation: Kalahari rangeland management options. *Land Degradation & Development, 18*, 249–268.

Reij, C., Tappan, G., & Belemvire, A. (2005). Changing land management practices and vegetation on the Central Plateau of Burkina Faso (1968–2002). *Journal of Arid Environments, 63*, 642–659.

Reij, C., Tappan, G., & Smale, M. (2009) Agroenvironmental transformation in the Sahel: Another kind of "Green Revolution". In: IFPRI Discussion Paper 00914: International Food Policy Research Institute.

Reynolds, J. F., & Stafford Smith, D. M. (2002) Do humans cause deserts? Global desertification: Do humans cause deserts? In J. F. Reynolds & D. M. Stafford Smith (Eds.) Berlin: Dahlem University Press.

Rouse, J. W., Haas, R. H., Schell, J. A., & Deering, D. W. (1973) Monitoring vegetation systems in the great plains with ERTS. In: *Third Earth Resources Technology Satellite-1 Symposium, (Greenbelt: NASA)* (pp. 309–317).

Schwarz, M., Zimmermann, N. E., & Waser, L. T. (2004). MODIS based continuous fields of tree cover using generalized linear models. *IEEE Transactions on Geoscience and Remote Sensing, 4*, 2377–2380.

Seaquist, J. W., Hickler, T., Eklundh, L., Ardo, J., & Heumann, B. W. (2009). Disentangling the effects of climate and people on Sahel vegetation dynamics. *Biogeosciences, 6*, 469–477.

Shackleton, C. M. (1993). Demography and dynamics of the dominant woodyspecies in a communal and protected area of the eastern Transvaal Lowveld. *South African Journal of Botany, 59*, 569–574.

Sokpon, N., & Biaou, S. H. (2002). The use of diameter distributions in sustained-use management of remnant forests in Benin: case of Bassila forest reserve in North Benin. *Forest Ecology and management, 161*, 13–25.

Sop, T. K., & Oldeland, J. (2013). Local perceptions of woody vegetation dynamics in the context of a 'greening Sahel': A case study from Burkina Faso. *Land Degradation & Development, 24*, 511–527. doi:10.1002/ldr.1144

Sop, T. K., Oldeland, J., Bognounou, F., Schmiedel, U., & Thiombiano, A. (2012). Ethnobotanical knowledge and valuation of woody plants species: A comparative analysis of three ethnic groups from the sub-Sahel of Burkina Faso. *Environment, Development and Sustainability, 14*, 627–649.

Sop, T. K., Oldeland, J., Schmiedel, U., Ouedraogo, I., & Thiombiano, A. (2011). Population structure of three woody species in four ethnic domains of the sub-sahel of Burkina Faso. *Land Degradation & Development, 22*, 519–529.

Steven, M. D. (1987). Ground truth—An underview. *International Journal of Remote Sensing, 8*, 1033–1038.

Thomas, D. S. G., & Middleton, N. J. (1994). *Desertification. Exploding the myth*. Chichester: Wiley.

Tougiani, A., Guero, C., & Rinaudo, T. (2009). Success in improving livelihoods through tree crop management and use in Niger. *GeoJournal, 74*(5), 377–389.

Traore, L., Sop, T. K., Dayamba, S. D., Traore, S., Hahn, K., & Thiombiano, A. (2013). Do protected areas really work to conserve species? A case study of three vulnerable woody species in the Sudanian zone of Burkina Faso. *Environment, Development and Sustainability, 15*, 663–686.

Tucker, C. J. (1979). Red and photographic infrared linear combinations for monitoring vegetation. *Remote Sensing of Environment, 8*, 127–150.

Tucker, C.J., Vanpraet, C.L., Sharman, M.J. and van Ittersum, G. (1985). Satellite Remote Sensing of Total Herbaceous Biomass Production in the Senegalese Sahel: 1980–1984. *Remote Sensing of Environment, 17*, 233–249.

Tucker, C. J., Dregne, H. E., & Newcomb, W. W. (1991). Expansion and contraction of the Sahara Desert from 1980 to 1990. *Science, 253*, 299–301.

Tucker, C. J., & Nicholson, S. E. (1999). Variations in the size of the Sahara Desert from 1980 to 1997. *Ambio, 28*, 587–591.

Tucker, C. J., Pinzon, J. E., Brown, M. E., Slayback, D. A., Pak, E. W., Mahoney, R., et al. (2005). An extended AVHRR 8-km NDVI dataset compatible with MODIS and SPOT vegetation NDVI data. *International Journal of Remote Sensing, 26*, 4485–4498.

Tucker, C. J., Vanpraet, C. L., Boerwinkel, E., & Gaston, A. (1983). Satellite remote sensing of total dry matter production in the Senegalese Sahel. *Remote Sensing of Environment, 13*, 461–474.

Venter, S. M., & Witkowski, E. T. F. (2010). Baobab (Adansonia digitata L.) density, size-class distribution and population trends between four land-use types in northern Venda, South Africa. *Forest Ecology and Management, 259*, 294–300.

Warren, A. (2002). Land Degradation is Contextual. *Land Degradation and Development, 13*, 449–459.

Vincke, C., Diedhiou, I., & Grouzis, M. (2010). Long term dynamics and structure of woody vegetation in the Ferlo (Senegal). *Journal of Arid Environments, 74*, 268–276.

Warren, A., & Agnew, C. (1988). An assessment of desertification and land degradation in arid and semi-arid areas. In: International Institute for Environment and Development paper no. 2, (London: Ecology and Conservation Unit, University College).

Warren, A., & Olsson, L. (2003). Desertification: Loss of credibility despite the evidence. *Annals of Arid Zone, 42*, 271–287.

Wezel, A. (2005). Decline of woody species in the Sahel. In Huber, B.A., Sinclair, B.J. & Lampe, K.-H. (eds.) African Biodiversity, New York: Springer.

Wezel, A., & Haigis, J. (2000). Farmers' perception of vegetation changes in semi-arid Niger. *Land Degradation and Development, 11*, 523–534.

Wezel, A., & Lykke, A. M. (2006). Woody vegetation change in Sahelian West Africa: Evidence from local knowledge. *Environment, Development and Sustainability, 8*, 553–567.

Chapter 6
Desertification, Adaptation and Resilience in the Sahel: Lessons from Long Term Monitoring of Agro-ecosystems

Pierre Hiernaux, Cecile Dardel, Laurent Kergoat and Eric Mougin

Abstract The desertification paradigm has a long history in the Sahel, from colonial to modern times. Despite scientific challenge, it continued to be influential after independence, revived by the dramatic droughts of the 1970s and 1980s, and was institutionalized at local, national and international levels. Collaborative efforts were made to improve scientific knowledge on the functioning, environmental impact and monitoring of selected agricultural systems over the long term, and to assess trends in the ecosystems, beyond their short term variability. Two case studies are developed here: the pastoral system of the arid to semi-arid Gourma in Mali, and the mixed farming system of the semi-arid Fakara in Niger. The pastoral landscapes are resilient to droughts, except on shallow soils, and to grazing, following a non-equilibrium model. The impact of cropping on the landscape is larger and longer lasting. It also induces locally high grazing pressure that pushes rangeland resilience to its limits. By spatial transfer of organic matter and mineral, farmers' livestock create patches of higher fertility that locally enhance the system's resilience. The agro-pastoral ecosystem remains non-equilibrial provided that inputs do not increase stocking rates disproportionately. Remote sensing confirms the overall re-greening of the Sahel after the drought of the 1980s, contrary to the paradigm of desertification. Ways forward are proposed to adapt the pastoral and mixed farming economies and their regional integration to the context of human and livestock population growth and expanding croplands.

Keywords Sahel · Desertification, ecosystem resilience · Pastoral management · Non-equilibrium dynamics

---

P. Hiernaux (✉) · C. Dardel · L. Kergoat · E. Mougin
Geoscience Environment Toulouse (GET), 14 Av. E. Belin, 31400 Toulouse, France
e-mail: pierre.hiernaux@wanadoo.fr

R.H. Behnke and M. Mortimore (eds.), *The End of Desertification?*
Springer Earth System Sciences, DOI 10.1007/978-3-642-16014-1_6

6.1 Introduction

The desertification paradigm took shape in the Sahel, a 500 km wide strip of land stretching 6000 km along the southern edge of the Sahara desert (Fig. 6.1). It contains an estimated 80 million people in 10 states. It is a bioclimatic zone under the reach of the West African Monsoon with a short rainy season peaking in August (Nicholson 2013). Rainfall is variable and occurs from May to October in the south, and in the north for only a few weeks between July and September (Le Houérou 1989).

The first objective of this chapter is to bring some results of long term monitoring of agro-ecosystems to the debate about desertification in the Sahel. The Sahel belt is very wide and diverse with different geologies, people, histories, economies and political institutions. However, the same bioclimatic gradient is found all along the belt with pastoral systems towards the more arid edge and mixed farming towards the semi-arid. After justifying a long term monitoring approach, results are presented for two case studies: one from the arid pastoral system (Gourma, in Mali) and one for a mixed farming system (Fakara, in Niger). Up-scaling these case studies to a regional level is then attempted, using remote sensing data. The regional context in human demography, land use and livestock population changes is then reviewed, and the chapter concludes with a discussion of ways to adapt Sahel pastoral and mixed farmers economies.

The desertification paradigm has a long history in the Sahel (Davis, Mortimore, Toulmin and Brock, this volume). It derived from the desiccation theory in vogue in the 19th century (Grove 1995) but widely developed as part of the colonial ideology in the first half of the 20th century (Hubert 1920; Stebbing 1935; Aubréville 1949). Scientific basis for the paradigm was provided in the then widely accepted vegetation climax theory (Clements 1916; Trochain 1940). Although questioned by a few scientists (De Gironcourt 1912; Chudeau 1921; Jones 1938), the desertification paradigm reinforced a colonial deprecation of local farming and pastoral practices as backward, inefficient, resource wasting and culpable for environment degradation

Fig. 6.1 The Sahel. Location of the two case studies, the Gourma region in Mali and the Dantiandou district (Fakara) in Niger, in the Sahel delineated by the 600 and 100 mm rainfall isohyets, south of the Sahara. Google Earth image, isohyets after Morel (1991).

(Fairhead and Leach 1990). In pastoral zones, 'over-grazing' became the ordinary culprit for rangeland degradation attributed to the 'tragedy of the commons' (Dodd 1994) and the 'livestock accumulation syndrome' (Warren 1995). The paradigm justified land set aside as forest and game reserves, the banning of fire and restriction of pastoralist mobility (Hubert 1920). Initially developed in the context of low rural population densities, the paradigm gained strength from the rapid increase in population, and its impact on natural resources, with the growing concern about saturation of the carrying capacity of the land (Harroy1944).

Later, following independence, the aspiration for agricultural modernity sustained the paradigm in the 1960s and 1970s. The overall failure of agricultural modernization (Dumont 1962) and the extent of the humanitarian and environmental catastrophe that hit the Sahel during the 1968–73 drought (followed by the 1983–84 drought), re-energized the paradigm. 'Desertification'became a part of all narratives and was institutionalized, from local NGOs to international agencies, with the creation of UNEP and UNCCD. More recently still, prospects of global warming and desiccation have been linked to desertification in spite of growing doubts about the evidence. Indeed, research efforts devoted to understanding the biology and ecology of Sahelian ecosystems (Penning de Vries and Djiteye 1982), to analyzing the environmental impact of evolving farming systems (Osbahr 2001; Raynaut 2001) and the long term monitoring of ecosystem dynamics both in the field and by remote sensing (Herrmann, this volume; Hiernaux, this Chapter) all question the pertinence of the desertification paradigm.

6.1.1 Why Are Long Term Monitoring and Multi-Scale Sampling Needed to Assess Ecosystem Dynamics in the Sahel?

The large inter-annual variability in rainfall amounts (Fig. 6.2), in rainfall distribution within rainy seasons (Frappart et al. 2009; Lebel and Ali 2009; Nicholson 2013), and in the patchiness of the rainfall events (Ali et al. 2003) justify the use of long term monitoring to identify trends beyond large temporal variations and spatial heterogeneities in vegetation attributes (Hiernaux et al. 2009a, b). The long-term perspective is also justified by the 'slow' pace of change of some ecosystem components such as woody populations (Hiernaux et al. 2009b; Herrmann and Sop, this volume) and soil fertility parameters (Pieri 1989). Even the increasing pressure on land resources triggered by the human demography (Tabutin and Schoumaker 2004; Guengant et al. 2002) is a slow driver (Walker et al. 2012). Moreover, the agrarian production systems in the Sahel being highly adaptive (Raynaut 2001; Thébaud and Batterbury 2001), long term monitoring is required to separate conjunctive variations from long term evolution of the farming systems. Finally, the trends in global climate change, with the progressive rise of air temperature and

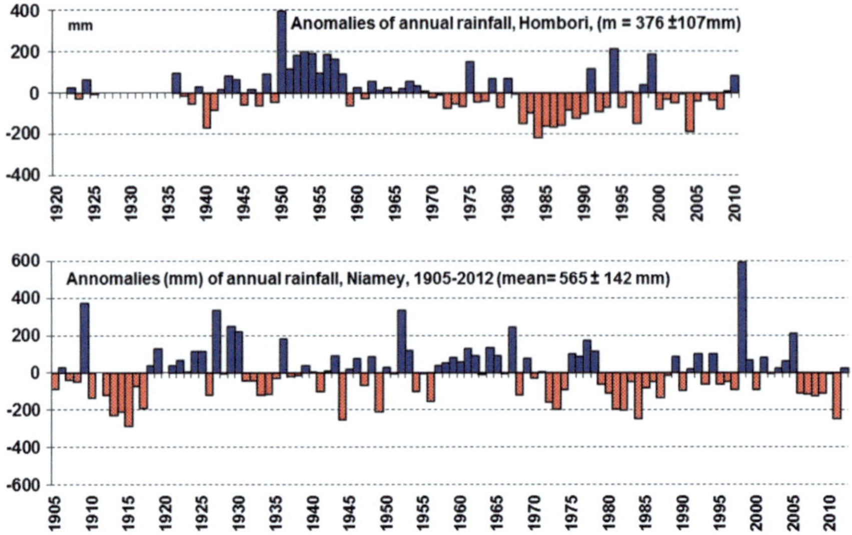

Fig. 6.2 Annual rainfall anomalies in Hombori (Mali) and Niamey (Niger). Histogram of annual rainfall expressed in anomalies (differences from the mean) in mm. Positive anomalies, rainfall superior to the mean are *filled bars*, negative anomalies, rainfall inferior to the mean are in *dotted bars*. Rainfall data from the National Meteorological Services DNM in Mali, DMN In Niger.

concentration in CO_2 (Giannini et al. 2003; Giannini this volume; IPCC 2013) support an even longer term perspective in ecosystem dynamics.

More difficult still, spatial heterogeneity is particularly high in Sahel ecosystems driven by the redistribution of rainfall through run-off (Breman and de Ridder 1991). The rain water redistribution interacts with the pattern of soil texture and soil fertility (Brouwer et al. 1993; Voortman et al. 2004). Soil texture itself reflects co-evolution with past climatic fluctuations during Tertiary and Quaternary periods (Brooks 2004; Grimaud et al. 2011). Soil fertility also relates to land use history (Turner 1998; Le Drézen et al. 2010). The patterned heterogeneity of the ecosystem thus requires an integrated multi-scale approach, from local site to topographic sequences, watersheds, and natural regions (Ludwig and Tonway 1995).

Among the adaptations to both temporal variability and spatial heterogeneity, agrarian systems have adopted a range of diversification strategies, with cropland dispersion in the landscape (Akponikpe et al. 2010), crop associations within a field (Ntaré and Williams 1992), local and regional mobility of grazing livestock (Turner et al. 2014) and crop-livestock association strategies either between specialized communities or increasingly within farms (Powell et al. 1996). This again justifies both long term and multi-scale approaches.

6.2 Two Case Studies: Pastoral in the Gourma (Mali) and Agro-pastoral in the Fakara (Niger)

The Gourma region in Mali extends over 80,000 km^2 to the south of the large loop made by the course of the Niger River between Mopti and Gao down to the border with Burkina Faso (Fig. 6.1), cutting across most of the Sahelian bioclimatic gradient from 100 mm in the north to 550 mm in the south (Frappart et al. 2009). The Gourma is lightly populated (4.9 persons km^{-2}), with an economy that is mostly pastoral, though associated with some staple millet cropping in the south and rice in the narrow valley of the Niger River (Gallais 1984; Ag Mahmoud 1992). A set of 40 field sites were sampled along the bioclimatic gradient on the main soil types and with different levels of grazing pressure, first to assess available fodder resources under drought during the 1983 dry season (details on site sampling are found in Hiernaux et al. 2009b). These have since then been monitored under different research projects documenting the long term dynamics of pastoral sites. Some of these sites were also selected during and after the previous major drought of 1972–1973 (Boudet 1972, 1979) adding historical depth to the study of ecosystem dynamics.

The district of Dandiandou extends to 846 km^2, 80 km to the east of the capital town, Niamey (Fig. 6.1). The district territory stands on the Fakara, a low sandstone plateau dissected by a web of fossil valleys infilled with sands, extending over 8000 km^2 in the interfluve between the Niger River and the Dallol Bosso valleys. The district is relatively densely populated (42.5 persons km^{-2}) with an economy relying on staple crops (millet, sorghum, cowpea, sorrel) and livestock husbandry (zebu cattle, sheep and goats, a few donkeys, horses and camels). The district is representative of the southern Sahel environments with a mean annual rainfall of 492 ± 89 mm (at Banizoumbou, 1990–2013), and dominant infertile sandy soils (Hiernaux and Ayantunde 2004).

The site was selected because of the large available data base from previous studies including Hapex-Sahel (Goutorbe et al. 1994), and the ongoing network of meteorological stations EPSAT (Le Barbé and Lebel 1997; Lebel et al. 2009). Further research was carried out on atmosphere-surface exchanges, hydrology and hydrogeology, wind erosion, vegetation physiology and phenology within the district under the AMMA (African Monsoon Multi-disciplinary Analysis)[1] project

[1]In collaboration with IER (Mali Institute for Rural Economy) ILCA (International Livestock Centre for Africa, Addis Ababa, Ethiopia) first launched a multi-disciplinary assessment of the short term impact of the 1982-83 drought on pastoral systems in the Gourma region, then funded a ten years monitoring of the sites to assess the drought impact over the longer term. After 5 years with few observations, the site monitoring was resumed in 1999 under the African Monsoon Multi-disciplinary Analysis, AMMA project till 2009 (Redelsperger et al. 2006), it continued then under the AMMA-CATCH observatory (Mougin et al. 2009) but was scaled down since 2011 because of civil insecurity. From 2009 on, the monitoring also benefited from the research context provided by a series of ANR (French Research National Agency) funded research project: ECliS ('Livestock Climate and Society' between 2009 and 2012; http://eclis.get.obs-mip.fr/), ESCAPE

untill 2009, afterwards prolonged by the monitoring of some of the studied processes under the AMMA-CATCH observatory (Cappelaere et al. 2009). In the framework of the crop-livestock farming system study initiated in 1994 by ILRI, (International Livestock Research Institute), 54 sites (1 or 2 ha each, half in croplands and half in fallows or rangelands) were sampled from the main topographic and soil types, in order to assess seasonal fodder availability and use at a district scale. Except for an interruption in 2007–2008, the sites have been monitored continuously under successive research projects. The crop-livestock farming system study led by ILRI ended in 1998, and was followed by complementary experiments and surveys till 2006. Meanwhile experimental research on options to intensify millet cropping (fertilizer micro-dose, selected breeds of millet and cowpea, animal traction) as well as surveys of societal and economic issues were developed under the leadership of ICRISAT (International Council for Research in the Semi-Arid Tropics).[2]

6.3 The Pastoral Ecosystem Dynamics

6.3.1 The Large Sensitivity of the Vegetation to Rainfall

The Sahel vegetation, called either steppe or savanna (Le Houérou 1989, 2009) is composed of a herbaceous layer of annuals dominated by C_4 grasses, and a scattered population of C_3 woody plants in which leaf phenology varies from ephemeral to evergreen (Hiernaux et al. 1994). The seasonality of the vegetation phenology and growth is controlled by the West African Monsoon. Seasonality is thus regular with herbaceous annuals germinating with the first rains from May to July, growing in a few weeks, setting flowers in September and wilting just after maturation in

(Footnote 1 continued)

('Environmental and social changes in Africa: past, present and future' from 2012 onwards; http://www.locean-ipsl.upmc.fr/~ESCAPE/) and CAVIARS ('Climate, agriculture and vegetation: impacts on aeolian erosion in the Sahel' from 2013 onwards).

[2]In collaboration with INRAN (Niger National Institute for Agronomy Research), ILRI (International Livestock Research Institute, Nairobi, Kenya) started in June 1994 a multi-disciplinary diagnostic of mix crop-livestock production systems in Southern Sahel. The seasonal monitoring of 54 field sites was part of the farming system study and continued under different ILRI research projects till 2006. After two years' interruption the site monitoring was resumed under the ECLiS (http://eclis.get.obs-mip.fr/) project and the AMMA-CATCH observatory. The Dantiandou district provided one of three sites retained by the international research project on the West African monsoon, AMMA (Cappelaere et al. 2009). In the meantime, ICRISAT (International Crops Research Institute for the Semi-Arid-Tropics, Hyderabad, India) developed in Dantiandou, in collaboration with INRAN and the University of Louvain-la-Neuve (Belgium), experimental research on options to intensify millet crop (fertilizer micro-dose, selected breeds of millet and cowpea, animal traction) as well as surveys on societal and economic issues (Saqali et al. 2010).

September or October (Penning de Vries and Djiteye 1982). On an average of 287 site-years observed between 1984 and 2011, across 25 rangeland sites in the Gourma, 84 ± 21 % (mean ± standard deviation) of the herbaceous production is achieved during the rapid growth phase that lasts 38 ± 16 days, at a mean rate of 2.6 ± 2.1 g $m^{-2}d^{-1}$, 3.7 ± 2.0 when bare soil patches are excluded (Hiernaux et al. 2013).

The large inter-annual variations in herbaceous production are largely explained by the volume and distribution of rainfall, at least on the dominant sandy soils (Fig. 6.3). However, primary production models, such as STEP (Mougin et al. 1995), run with daily rainfall data are quite successful in predicting the growth pattern, but poor in predicting the yield unless they are calibrated for the particular site and flora composition (Tracol et al. 2006). Indeed soil fertility is not accounted for by the model, and annual species composition may change drastically from one year to the next on the same site in relation to seed dispersion and germination ecology (Cissé 1986; Carrière 1989; Seghieri 1996; Hérault and Hiernaux 2004). On upland shallow soils, and lowland fine textured soils, the vegetation structure and production only relate indirectly to rainfall, as the balance between run-off and run-on modifies the soil moisture regime. On shallow soil, the herbaceous yields also depend on the extent of bare soil patches that may vary from year to year. However, the rate of growth on the sole vegetated patches, during the rapid growth phase when soil moisture is not limiting (180 site-years in the Gourma monitoring), reaches 5.2 ± 2.3 g $m^{-2}d^{-1}$ on shallow soils, above the loamy soils at 4.5 ± 1.7 g $m^{-2}d^{-1}$ and the sandy soils at 3.8 ± 1.4 g $m^{-2}d^{-1}$, and below the clay soils at 6.6 ± 2.6 g $m^{-2}d^{-1}$. This was expected from their respective biochemical fertility (Hiernaux et al. 2013).

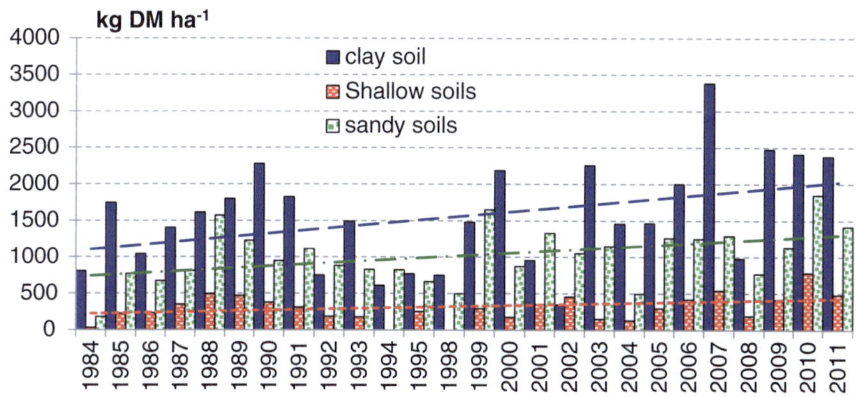

Fig. 6.3 Herbaceous yield dynamics in Gourma rangeland (1984–2011). Histogram of mean maximum standing mass of the herbaceous layer of Gourma rangelands grouped by soil types (deep sandy soils, deep clay soils, shallow sandy loam soils). Overall linear regressions of the mean maximum standing mass (M kg Dry Matter ha^{-1}) are poor: M = 22.5*Yr + 717 (r^2 = 0.20) on sandy soils; M = 36.7*Yr + 1070 (r^2 = 0.15) on clayed soils; M = 8.2* Yr + 215 (r^2 = 0.15) on shallow soils. *Data source* AMMA CATCH observatory.

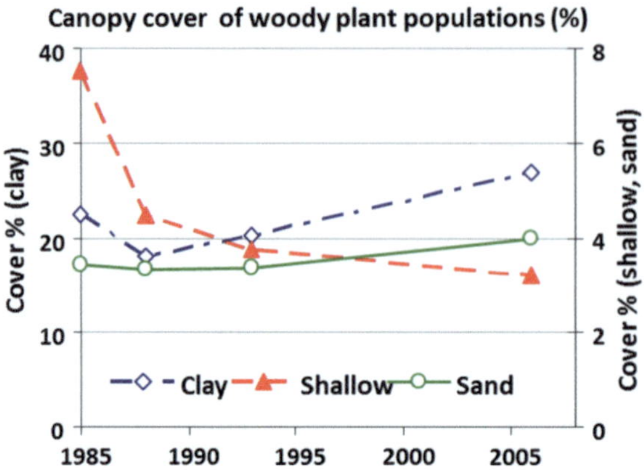

Fig. 6.4 Dynamics of woody canopies in Gourma rangelands (1985–2005). Change over time of the mean canopy cover (%) measured in Gourma rangeland sites, grouped by type of soil: deep sandy soils, deep clay soils, shallow sandy loam soils (Hiernaux et al. 2009b).

The inter-annual dynamic of the vegetation interacts in contrasting ways with the main edaphic components of the landscape as follows:

- The strong resilience of the vegetation on sandy soils. Even when the consecutive drought years in 1972–73 or 1983–84 had wrecked the herbaceous layer and decimated the soil seed stock, it took only a couple of years for the herbaceous layer to recover on most sandy soils (Fig. 6.5 left), though with successive steps in species composition (Boudet 1990; Hiernaux and Le Houérou 2006). The woody population was also decimated by the droughts (with some lag in time) and it took much longer for the populations to rebuild, often starting with pioneer fast growing shrubs such as *Leptadenia pyrotechnica*,[3] *Calotropis procera* and *Acacia ehrenbergiana* (Hiernaux et al. 2009a). The density and aggregated canopy cover of the woody plants measured in 2009 are larger than in 1984, though woody plants which died that year were included (Fig. 6.4).
- The collapse and profound mutation of the vegetation on shallow soils. By contrast with the sandy soils, the patchy herbaceous layer and patterned woody vegetation did not recover from the major droughts between 1985 and 2007 (Fig. 6.6). The main reason is a co-evolution of the vegetation with the run-off system that becomes progressively concentrated with the opening of the vegetation cover. The sheet run-off prevailing on the long gentle slopes of the erosion surfaces (rocky peneplains, ferralitic hardpans), feeding the 'tiger bush' patterned vegetation with dense and narrow woody thickets set perpendicular to the slope (d'Herbès et al. 2001), is progressively replaced by a concentrated run-off

[3]All plant species are named after the flora of West Tropical Africa by Hutchinson and Dalziel (1954–1972).

Fig. 6.5 Year to year changes of rangeland vegetation in the Gourma: sandy and clay soils. Series of photos taken from same point and direction in two monitored rangeland in the Gourma: *Left* Aouli, open savanna on fixed sand dune (N 16.01933°–W 1.50327°), *Right* Kelma Acacia seyal stand in a low land clayed soil, seasonally flooded (N 15.21875°–W 1.58810°). *Photos* Hiernaux and Soumaguel (2011).

in a web of gullies, depriving the remaining patches of vegetation of water and nutrient resources, and thus aggravating the degradation of the vegetation cover (Hiernaux and Gérard 1999).

- The erratic dynamics of the vegetation in fine textured soils in low- lands, which followed the changes in the water run-on regime on which their growth depends.

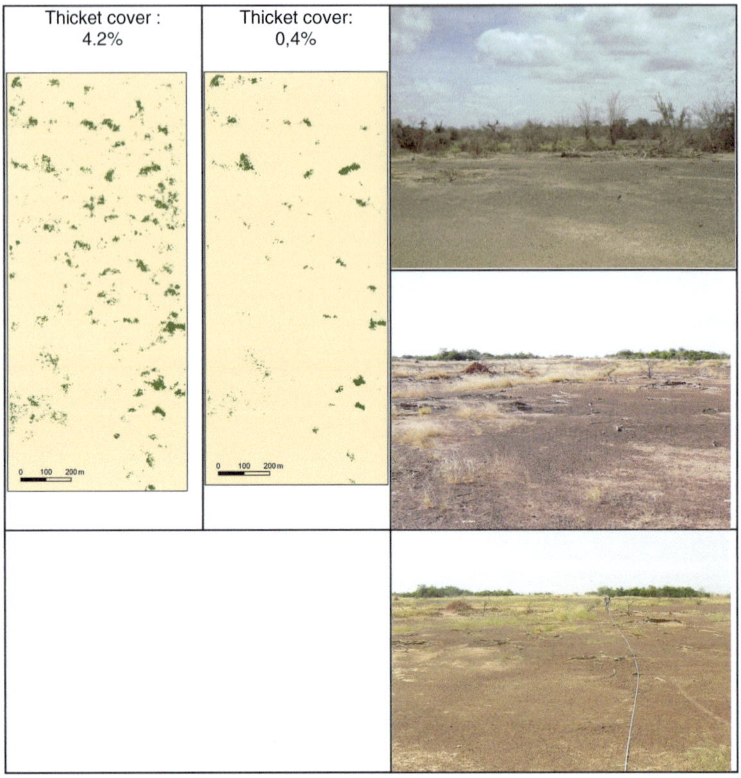

Fig. 6.6 Year to year changes of rangeland vegetation in the Gourma: shallow soils. Series of three photos taken from same point and direction at Ortondé (N 15.15503°–W 1.56292°) showing very open 'tiger bush' on ferralitic hard pan (*top* 25 Sept 1986; *centre* 30 Sept 2007; *bottom* 12 Sept 2011). The woody plant thickets reduce in number and cover over time as shown on the two 1 km long transect maps established by supervised classification of landsat scenes in 1985 (*left*) and 2007 (*right*) (Trichon et al. 2009). *Photos* Hiernaux and Soumaguel (2011).

Successive drought years without flooding lead to the collapse of the herbaceous layer which is dominated by grasses adapted to seasonal flooding, such as *Panicum laetum* and *Echinochloa colona*. After one year's lag the dense stands of woody plants were in turn decimated, especially those of *Acacia seyal*, *A. nilotica* and *Anogeissus leiocarpus*. Then, depending on the restored flood regime, resulting from the increase and concentration of the run-off on the shallow upland soils, the vegetation in the lowlands either re-established (Fig. 6.5 right) or failed to re-establish. The increased volume of run-off from the shallow soils largely modified the web of gullies and wadis. Pond outlets enlarged markedly and new ones became established (Gardelle et al. 2010), either killing previous vegetation by asphyxia or promoting the development of new vegetation at the pond edges, especially pioneer shrubs such as *Acacia ehrenbergiana*.

6.3.2 The Impact of Grazing Livestock

Comparing time series of vegetation data collected in sites under contrasted grazing pressure (Fig. 6.7) helps in assessing the impact of grazing livestock on the dynamics of the Sahel biotopes. On sandy soils and lowland clay soils there is no evidence of losses in herbaceous productivity with increasing grazing pressure (Hiernaux et al. 2009a). While a repeated cutting regime during the growing season may seriously affect the production of herbaceous annuals—depending on the timing and frequency of the cuts (Hiernaux and Turner 1996)—the impact of heavy grazing around livestock concentration points does not systematically translate into a loss of productivity (Hanan et al. 1991). Indeed, depending on the species promoted by heavy grazing (Ayantunde et al. 1999), accompanied by higher trampling and excretion deposition, the productivity is usually enhanced, not only because unpalatable species such as *Sida cordifolia*, *Acanthospermum hispidum*, or *Senna obtusifolia* get promoted but also, with good forage species such as *Cenchrus biflorus* and *Dactyloctenium aegyptium* (Hiernaux 1998). In other sites, the species promoted by heavy grazing are short cycle types, and thus less productive, but may be high quality livestock fodder such as *Zornia glochidiata*, *Tribulus terrestris* and *Tragus berteronianus* (Valenza 1984).

On 14 sandy soil sites (180 site-years) monitored between 1984 and 2011 in the Gourma, the average growth rates during rapid growth slightly increased with grazing pressure from 3.0 ± 0.8 g $m^{-2}d^{-1}$under moderate grazing pressure to 3.9 ± 1.5 g $m^{-2}d^{-1}$under high and 4.0 ± 1.3 g $m^{-2}d^{-1}$under intense grazing (Hiernaux et al. 2013). Moreover, the density of herbaceous perennials and woody plants on sandy soils does not necessarily decrease with increasing grazing pressure. Large tussock perennials such as *Aristida sieberiana* and *Panicum turgidum* are often more dense near livestock concentration points (camps, pools, livestock

Fig. 6.7 Wet and dry season grazing in Gourma rangelands. Zebu cattle grazing during the wet season (lowland with Acacia seyal, Panicum laetum, Echinochloa colona; Kelma), and grazing standing straw and litter during the late dry season (*Acacia raddiana*, *Balanites aegyptiaca*, *Cenchrus biflorus*, *Aristida mutabilis*; Aouéli). *Photos* P. Hiernaux.

paths), in the northern edge of the Sahel, but less often in the southern Sahel with *Andropogon gayanus* and *Cymbopogon shoenanthus*. The same observation applies to woody plants, with dense stands of *Acacia* spp., *Balanites aegyptiaca*, or *Boscia senegalensis* in the vicinity of wells. However, competition between perennial herbaceous and woody plants has been observed in the northern Sahel (Anthelme et al. 2006).

On lowland fine textured soils, extremely high browsing pressure close to water points and settlements does not impede woody plant populations from recovering through successive cohorts of seedlings, as observed in lowland clay soils near Wami, or surrounding Gossi lake where a strong regeneration of *Acacia ehrenbergiana*, *A. seyal* and *A. nilotica* followed the 1983–84 drought.

On the other hand, heavy trampling on loamy and sandy loam shallow soils may contribute to the concentration of run-off, and thus accelerate vegetation degradation (De Wispelaere 1980; Leprun 1992). However, the collapse of herbaceous patches and woody populations observed on shallow soils also occurs under very light grazing pressure where there is a lack of permanent water resources for livestock.

6.3.3 What Would Explain the Minor Impact of Grazing Livestock on Pastoral Ecosystem Dynamics?

There are a number of ecosystem peculiarities that moderate the impact of livestock grazing on the pastoral ecosystem.

- First, the effect of grazing on the herbaceous production is limited by the short growth duration of annuals. A large fraction of herbaceous growth (84 %) takes places in three weeks or less, and synchronically over the region, except in the large alluvial plains whose flood regime is also determined by rainfall but far away upstream, such as in the Inland Niger Delta (Gallais 1967).
- Heavy grazing pressure during the few weeks of rapid growth is transient and can only occur locally. The same animals have to survive the remaining 40 or more weeks on the forage produced during the short growing season. On the other hand, the impact of grazing livestock on the herbaceous layer during the dry season is limited by the mediocre capacity of livestock to pick up straw and litter. Indeed, controlled experiments and surveys (Hiernaux et al. 2012) demonstrated that, at most, one third of the standing straw vegetation at the end of the growing season is potentially ingested by livestock because of the concomitant losses by trampling, insect and rodent herbivory and organic decomposition.
- Moreover, livestock recycle about half of the organic matter and more than 80 % of the main nutrients through urine and faeces excretions (Schlecht et al. 2004).
- With the most palatable species, the ability of livestock to browse leaves, twigs and fruits of woody plants is limited by access, even for specialized browsers such as goats and camels. Livestock browsing is only detrimental to a woody

plant population at an early stage of development (germination, seedlings), or to low shrubs.

- Continuous high grazing pressure in the vicinity of water points and settlements has not impeded the regeneration of *Acacia seyal* stands in the Wami area near Hombori, nor of *Acacia erhenbergiana* and *Balanites aegyptiaca* near Gossi (Hiernaux et al. 2009a).

At broader spatial scales and between years, the impact of livestock on the Sahel pastoral ecosystem is limited by the overall control of the fodder resources on livestock numbers. This control either manifests itself in catastrophic losses of animals during droughts (Toulmin 1987), floods, or other extreme events on top of losses from epidemics, war and cold rainfall in winter. It also manifests itself by affecting the reproduction parameters of animals living in conditions of chronic under-nutrition (Fernandez-Rivera et al. 2005). Statistics of livestock losses during the major droughts of 1972–73 and 1983–84 are poor, but animal losses were large, as shown in case studies documented in the Gourma just after the drought for sheep and goats (Peacock 1983) and retrospectively for cattle (Dawalak 2009). This last study also indicates that recovery of animal numbers takes several decades, and may never occur for pastoral families who are pushed out of the pastoral economy. Indeed, the rates of herd growth (taking account of offtake) rapidly decline to very low figures, even negative, when parameters such as the age of first calving, or the mean time between consecutive calving, increase to the point where the reconstitution of a herd may not be feasible in a herder's life time (Lesnoff et al. 2012).

This overall limitation of the livestock population by the low points of Sahelian rangeland resources (a late dry season in a poor rainfall year) reduces the risks of ecosystem degradation through overgrazing, unless livestock are confined to concentration points on the most fragile shallow soils. Provided that the stocking rate is not artificially maintained by imported feed, the large seasonal impact of livestock grazing on vegetation cover has very limited impact on the longer term dynamics of the Sahel ecosystem, in that regard behaving as a non-equilibrial system (Ellis and Swift 1988; Vetter 2005).

6.4 Agro-pastoral Ecosystem Dynamics

6.4.1 *The Land Cover Diversified by Clearing and Cropping Practices*

Clearing land for cropping affects the woody population in the longer term (Fig. 6.8). In extensive cropping systems, fallows alternate with a few years of cropping, allowing for the sprouting of coppices (mostly *Guiera senegalensis*, *Combretum glutinosum*, *Piliostigma reticulata* at Dantiandou) and the regeneration from seed of some woody plants. The flora of the herbaceous layer evolves progressively as the fallow ages from weed dominance to savanna species (Achard et al. 2001).

Fig. 6.8 Fallow and millet crop field in Dantiandou (Fakara). *Left* Cattle grazing a heavily grazed enclave fallow during the wet season (*Guiera senegalensis*, *Zornia glochidiata*, *Sida cordifolia*). *Right* Non-manured farmer's field in Tigo Tegi with associated crops of millet (*Pennisetum glaucum*) and cowpea (*vigna unguinculata*) prior to harvest in September, 2009 (trees in the field edge are *Combretum glutinosum*). *Photo* P. Hiernaux.

Successive crop and fallow cycles progressively shape the woody population in an agro-forestry parkland, where tree composition depends on the climate, soil texture, fertility and moisture regime. In the sandy soils of Dantiandou, it is dominated by *Combretum glutinosum*, *Prosopis africana*, *Faidherbia albida*, *Detarium microcarpum*, and *Parinari macrophylla*. In addition individual plots are often delineated by living hedges made up of trees, shrubs and the perennial grasses *Andropogon gayanus* and *Aristida sieberiana*. None of the hedge components are planted; they are spontaneous but protected to delineate the field.

Cropping practices determine the sensitivity of the cropland to wind or run-off erosion, including: soil tillage (mostly with hand tools—animal traction is rare); sowing times and densities, organic fertilization (animal corralling, manuring, waste application); mineral fertilization; weeding times and frequency; and harvesting methods (sole panicles or whole plants (Osbahr 2001; Bielders et al. 2004). The management of cereal stalks following grain harvest is particularly important for wind erosion. Leaving only 500 kg ha^{-1} of millet or sorghum stalks laid on the soil during the dry season controls most of the erosion (Michels et al. 1998). However, cereal stalks are increasingly harvested, which is justified by their private use as livestock feed and construction material, increasing the risks of wind erosion on crop fields (Bielders et al. 2004).

6.4.2 The Impact of Livestock on Semi-arid Agro-ecosystems

The spatial distribution of grazing pressure is organized seasonally to avoid livestock presence in croplands during the growing season: this implies close herding and seasonal migration out of the cropland areas (Turner et al. 2005). It also structures the distribution of livestock in the landscape, with extremely high concentration along the livestock paths, around water points, resting camps and in rangeland enclaves (Turner and Hiernaux 2002). This increasing grazing pressure results in a decreasing trend of fallows and rangeland herbage yield (Fig. 6.9) that is not explained by the trend in rainfall (Fig. 6.2). These local high stocking rates, associated with soil trampling, favor short cycle species resistant to repeated grazing, such as *Zornia glochidiata* and *Dactyloctenium aegyptiacum* on sandy soils, or *Microchloa indica* and *Eragrostis pilosa* on sandy loams. Heavy trampling during the wet season may also open bare soil patches. The two processes (high stocking rates and trampling) contribute to lower herbaceous production and increase the risks of local wind and water erosion. In other places the high concentration in livestock excretion, together with intense grazing and trampling, favor the local proliferation of nitrophilous unpalatable species such as *Ipomoea asarifolia*, *Sida cordifolia* or *Acanthospermum hispidum*. Herbaceous production in these patches remains high in spite of the high grazing pressure.

The transfer of organic matter and nutrients by livestock within the agro-ecosystem prevails as a consequence of the spatio-temporal distribution of animal grazing time on one hand, and walking-resting-ruminating time on the other hand (Schlecht et al. 2004). The transfer is effective both to uncropped areas along

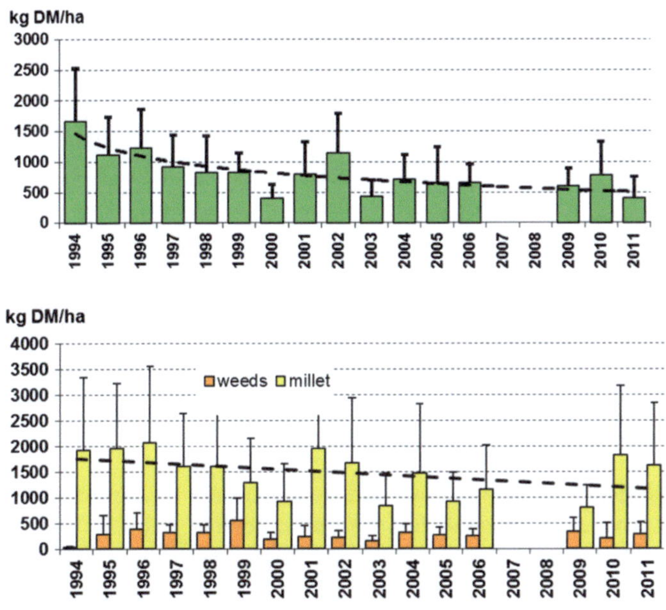

Fig. 6.9 Yield trends in fallows, rangelands and crop fields at Dantiandou, Fakara (1994–2011). Histograms of mean annual yields (M) in kg ha^{-1} in: *top* herbaceous layer from 24 fallows and rangelands sites; *bottom* 24 millet crop fields and associated weeds, monitored from 1994 to 2011 in the Dantiandou district. Yield trends over years (Yr): M = −333 Ln(Yr) +1467 (r^2 = 0.65) in fallows and rangelands; M = −34.2* Yr +1780 (r^2 = 0.18) for millet crop (from Hiernaux et al. 2009c).

livestock paths and around water points, and to cropland, either by corralling livestock in the fields during the dry season and on fallows in the wet season, or else by transporting and applying manure collected in paddocks or yards (Fig. 6.11). Whatever the method, manure application improves or maintains soil fertility on a small fraction of the land cropped in the Dantiandou district from 3 to 8 % depending on the density of the population (Hiernaux and Turner 2002). However, the high level of manure application practised in corralling expands the effect of manuring to about a fifth of the land cropped, considering that the residual benefit of manure application on crop yield is significant for at least 4 years (Gandah et al. 2003), moderating yield decline (Fig. 6.9). Moreover, cattle, donkeys and camels also contribute to the intensification of the cropping system, via animal traction, rarely used for soil tillage (ploughing, weeding, sowing, fertilizer application) but commonly used for transportation (manure, yields, stalks, haulms… and people).

6.4.3 Do Cropping Practices Attenuate the Ecosystem Sensitivity to Rainfall Variations?

Crop phenology differs from that of rangeland and fallow plants. There is an initial delay in growth due to the farmer's choice of a sowing date, most often later than the first rainy day. The phenology of field crops also differs from that of spontaneous vegetation in its lower density and regular distribution of the planted seeds, and by the weeding done to reduce competition from weeds. However, crop landraces have been selected for their capacity to grow fast, to tiller or branch profusely (provided that soil moisture and nutrients are not limiting), so that at the end of the season, primary production is often higher in fields than in surrounding fallows or rangelands, even when these have been protected from grazing (Hiernaux et al. 2009c). This higher yield is also explained by the longer cycle of many crop breeds, remaining green after September while the rangeland and fallow annuals have wilted soon after seeding, due to sensitivity to the photoperiod. In spite of the high adaptation of millet and cowpea landraces, crop yield remains more sensitive to dry spells in rain distribution than fallow and rangeland herbaceous production although less so on manured fields.

The crop harvest subtracts a large export of organic matter and nutrients in consumable grains, beans or seeds. This is aggravated by the export of all of the haulms of cowpea, groundnut, Bambara nuts, sesame and sorrel, and that of an increasing fraction of the cereal stalks (millet, sorghum) and straw (*Andropogon gayanus*, *Ctenium elegans*). When that export is not compensated by mineral fertilization and organic amendments, the biochemical fertility of cropland soils slowly degrades, with losses in organic matter content and cation exchange capacity, and loss in the concentration of nutrients (mostly nitrogen and phosphorus), often aggravated by acidification with the risk of triggering aluminum toxicity (Bationo et al. 2012).

The loss in soil fertility weakens vegetation growth rates, leading to lower yields and also higher sensitivity to dry spells and irregular rainfall patterns (Rockström and de Rouw 1997). Fallowing a field cropped for a series of years is a passive way to restore the soil fertility through the rebuilding of biological activity triggered by the recycling of more organic matter to the soil, also by reducing soil losses through erosion, and eventually by trapping dust and sands blown from the neighboring crop fields (Bielders et al. 2002). In this process, a major role is played by the rapid regrowth of shrubs kept coppiced during the cropping period (Dick et al. 2010). The degradation of soil fertility is unequal in the landscape, depending on the land-use history of each field and its fertilization (Warren et al. 2001). Within fields, the nutrient balance is mediated by the run-off/run-on balance, which in turn is influenced by the topography, and the presence and density of woody plants, resulting in micro-heterogeneity in crop yields (Brouwer et al. 1993; Rockström et al. 1999).

Finally, in extensive cropping with little or no fertilizer or manure application, crop yields are low and more sensitive to rainfall variability than herbaceous production in fallows and rangeland, while more intensive cropping systems with systematic application of manure and fertilizer, and anti-erosive designs, have higher yields and tend to better resist poor rainfall and dry spells. A reason for better adaptation is the faster and larger root growth, allowing the plant to exploit a larger volume of soil for water and nutrients (Akponikpe 2008). However, better adaptation of crops to dry spells in more fertile soil is debated, and contradicted by some experimental results elsewhere in the Sahel (Affholder 1995).

At the landscape scale, the association of staple crops and pastoral livestock in the semi-arid Sahel increases spatial heterogeneity, first because of the dynamic mosaic of crop fields, fallows and rangelands, and second, because of the organic matter and nutrient transfers within the landscape, resulting in an overall reduction in soil fertility except in small high fertility spots. As in the arid pastoral Sahel, the ecosystem dynamic is of a non-equilibrial nature, but the capacity to respond to rainfall vagaries is affected by the impact of very high stocking rates, locally and seasonally, only sustainable because livestock partially feed on inputs from transhumance and purchased agro-industrial livestock feeds.

6.4.4 Up-Scaling with Satellite Remote Sensing: Desertification and 'Greening' Trends in the Sahel?

Assessing the expanse of cropland with high resolution satellite data looks like a straightforward task, as most crop fields are visible on high resolution panchromatic or infra-red aerial photographs, as well as high resolution multispectral satellite imagery, given their sensitivity to seasonal contrasts between the phenology of cropland and neighboring fallows or rangelands. However, attempts to systematically map land use, for example by supervised classification of multispectral satellite images, is fraught with difficulties (Begue et al. 2011). In Dantiandou, part of the challenge comes from the small size of the cropped fields, the presence of trees and shrubs within and around crop fields, the no-tillage practice, the diversity between fields in crop phenology due to the array of breeds and cropping practices, and to the heterogeneity of organic matter distribution between and within fields.

Nevertheless, photo-interpretation of 1950, 1975 and 1992, and carefully supervised classification of a series of SPOT scenes over the Dantiandou district, confirms the rapid increase in areas cropped. Cropland expanded at a decreasing rate, but still averages 3 % annually, close to the rate of increase of the rural population (Hiernaux and Ayantunde 2004). Savanna had already disappeared from deep sandy soils by 1975, owing to the expansion of both cropland and fallows. Since then, areas in fallow have also been shrinking to balance the continuing expansion of croplands (Table 6.1).

Mapping land use over larger regions is less successful and the few global products available compare poorly amongst themselves and with maps established locally (Vintrou et al. 2012). As a consequence their precision is not sufficient to assess trends in land use over time. The same restriction applies to other categorized mapping of soil degradation and desertification for which definition and indicators often differ (UNEP 1992).

'Sahel greening' is a term introduced by remote sensing scientists to name an overall increasing trend in the Normalized Difference Vegetation Index (NDVI) (Eklundh and Olsson 2003; Anyamba and Tucker 2005; Hutchinson et al. 2005; Herrmann et al. 2005; Olsson et al. 2005; Hickler et al. 2005). First proposed by Tucker (1979), it was observed over a series of growing seasons by 1991 (Tucker et al. 1991). The series starts in 1981 for the more widely used satellite data set over Africa, GIMMS (Global Inventory Modeling and Mapping Studies, Tucker et al. 2005) produced by NASA from the AVHRR (Advanced Very High Resolution Radiometer) recordings. Similar and consistent trends have been established with the MODIS NDVI product from 2002 onwards and SPOT VGT data (Fensholt et al. 2006; Dardel et al. 2014b). The general increase—and a few local decreases—in NDVI statistics (Fig. 6.10) have been interpreted as increases of the vegetation

Table 6.1 Land use changes in Dantiandou (1950–2011). Mean annual rates of changes in land use (Crop field/fallow/rangeland) assessed by remote sensing from 1950 to 2011 over the district of Dantiandou (Niger)

Periods	Annual rates of changes in area over the periods %		
	Crop fields	Fallows	Rangelands
1950–1975	+7.7	+5.8	−3.6
1975–1994	+3.3	+0.0	−4.5
1986–2008	+3.0	−2.0	+0.2
2008–2011	+3.3	−3.5	+0.6

Land use maps in 1950, 1975 and 1994 are established by stereoscopic photointerpretation of existing aerial photo coverages, land use maps in 1986, 2008 and 2011 are established by supervised classification of multispectral SPOT images (Hiernaux, unpublished data)

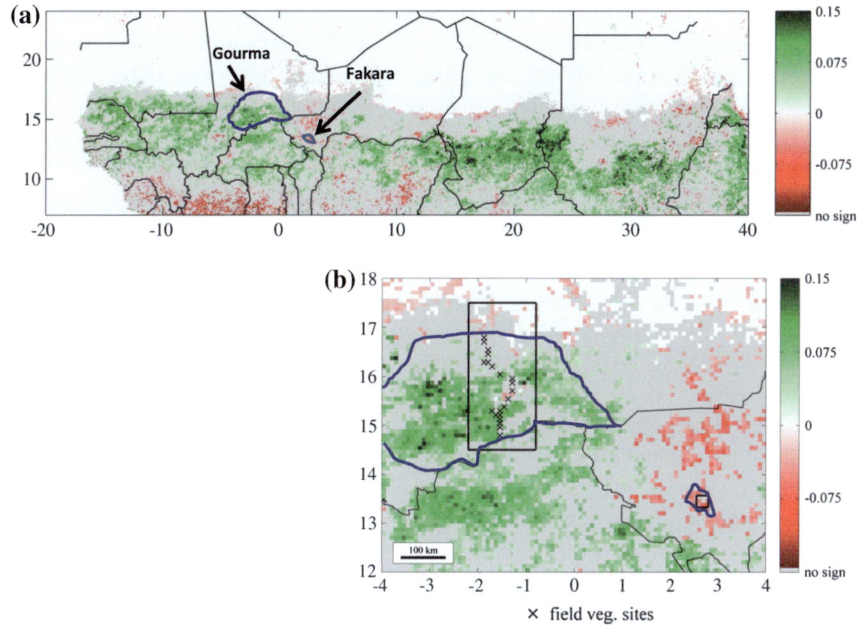

Fig. 6.10 Sahel vegetation greening or degrading. Maps of mean GIMMS-3g NDVI trends (linear increment) from 1981 to 2011 over: **a** the Sahel belt, **b** the Gourma (*large left box*) and the Dantiandou (*small right box*) areas. The more *dark green* is a pixel the more positive is the significant trend (at the 95 % level), the more *dark red* the more negative is the significant trend; the scale is on the left. Trends in *dark grey* areas are not significant, while *light grey* are areas where NDVI standard deviation is lower than 0.015 NDVI units (desert). From Dardel et al. (2014b).

cover and production over the decades that followed the last major drought in the Sahel (1983–84). This contradicts the dominant narrative on desertification, and has been much discussed in relation to the dynamics of rainfall, population density and land use (Fensholt et al. 2006; Heumann et al. 2007; Hellden and Tottrup 2008; Fensholt and Rasmussen 2010; Herrmann and Tappan 2013).

The relationship between the mean (or the NDVI integral) over the vegetational growing season and vegetation production (or the seasonal maximum standing mass, as proxy) have long been studied on individual site-year data (Tucker et al. 1985; Hiernaux 1988; Myneni et al. 1995; Fensholt et al. 2006). It is only recently that NDVI time series have been compared to long term field monitoring data (Dardel et al. 2014b; Herrmann, this volume). Good agreement was found between trends in maximum herbaceous standing mass, measured in the field, and trends of mean NDVI over the growing season, both in the Gourma region (where trends from 1981 to 2011 are positive) and in the Dantiandou district where the trends are negative (Dardel et al. 2014b). Moreover, the positive NDVI trends observed in the Gourma are significant over the whole period, and are mostly due to the positive trends prevailing on sandy soils, while trends on shallow soils are not significant (no trends could be assessed on clay soil because of the coarse resolution of GIMMS data). The overall negative NDVI trends in Dantiandou includes however a positive trend from the 1980s to the 1990s, followed by a negative trend that is not consistent with the improving trend in rainfall since the mid 1990s.

At both sites, the positive and negative trends are not due to changes in the seasonality of growth, but to its intensity, reflected in the seasonal amplitude of NDVI. The dominant positive trend observed on deep soil in the Gourma region is largely explained by a positive trend in the rainfall, while the negative trends observed on the shallow soils and croplands in the Fakara (including Dantiandou) are not explained by a decreasing trend in rainfall (Dardel et al. 2014b).

Trends in rain use efficiency (RUE)—defined as the ratio of above ground net primary production (ANPP) over annual rainfall—have been controversially interpreted (Prince et al. 1998, Hein and de Ridder 2006; Prince et al. 2007; Hein et al. 2011). When RUEs are assessed from field data (herbaceous ANPP and rain gauge data) aggregated over large areas, or from satellite data (NDVI curve integral over the wet season, rainfall estimates) averaged over the same large areas, the trends are not significant (Dardel et al. 2014a).

The integral value of NDVI over the growing season only (unlike the mean over the year cycle as used by Brandt et al. (2014)) varies in good agreement with the herbaceous layer yield. Its overall and robust increase since 1981 thus indicates an overall increasing trend in herbaceous vegetation production, largely explained by the concomitant increase in rainfall. This trend confirms the overall resilience of the Sahel ecosystem both to the drought of the early 1980s and the increased pressure on resources. It denies general desertification. Yet the signal does not tell us much about the trend in woody plant populations (Herrmann, this volume), and the resolution of the satellite data does not exclude local decreasing trends in herbaceous vegetation yields, especially in darker soil enclaves (either shallow soils on hard pans or rock outcrops, or lowland clay soils). In the mosaic of cropland and rangelands, the differential NDVI signal for crops (lower NDVI values at equivalent vegetation mass) and for rangelands, together with expanding croplands could also affect the trend towards lower NDVI values without evidencing real decrease in the herbaceous yield. Caution should thus be taken in interpreting the negative NDVI trends observed locally, as in the Dantiandou district.

6.5 Adapting Crop-Livestock Production Systems to a Fast Evolving Context

The Sahel ecosystems have so far demonstrated their high resilience to droughts and growing pressure on resources, denying a general process of desertification. However, agricultural productivity is low and rural people are poor, encouraging emigration to fast growing cities. In a context of high rural population growth, expanding cropland and increasing livestock populations, the challenge for the coming decades is to find adaptations to enhance agricultural productivity and rural welfare without disrupting this resilience.

In a context of active demographic growth, population density tends to decrease with aridity along the Sahel bioclimatic gradient, with a mean of 42.5 persons/km^2 in the Dantiandou district to 5.9 in the Hombori district in central Gourma. In spite of the gaps and weaknesses of the national population censuses, the rural population has grown at both sites at sustained rates over recent decades—3.2 % annually in Dantiandou (1988–2001) and 1.9 % in Hombori (1950–2009)—in spite of accelerating out-migration from rural areas to towns and coastal countries in the region (Hampshire 2002). These sustained growth rates result from high fertility rates and decreasing infant mortality rates (Guengant et al. 2002). As a result, rural populations are very young with high child dependency ratios, enhanced by the emigration of young men. Detailed analysis of recent population dynamics between surveys at Dantiandou in 1994 and 2008 reveals contrasted family changes, with small net increases in population reduced by migration, and a trend toward smaller families caused by splitting of the patriarchal family units (Saqalli et al. 2013). The pastoral population also increased, but more slowly because of locally negative net migration.

Seasonally mobile pastoral populations have settled widely, but especially in the years following the main drought events. They settled in the vicinity of permanent water points, often adjacent to existing villages and small towns, or in new settlements set along the main roads, claiming community rights to land. This settlement of pastoral populations does not impede the seasonal mobility of the herds which are conducted by herdsmen or shepherds, and facilitates income diversification within pastoral families, especially in trade (Tuner et al. 2014).

6.5.1 A Context of Large and Rapid Changes in Land Use

In the absence of major technological intensification, Sahel agriculture responds to the large and rapid increase in market demand for food, triggered by the demographic upsurge and fast urbanization, by expanding the area cropped in parallel with population increase. Cash crops such as groundnut and cotton have developed in the southern Sahel. An initial phase of rapid cropland expansion was followed by intensification through the use of fertilizers and pesticides, or by the adoption of animal traction for land preparation and weeding. This provides the cash needed to invest in these technologies in an increasingly supportive environment: more effective access to technical and extension services, loans and insurance (Dufumier 2005).

The expansion of cropland is first made at the expenses of the remaining pristine savannas and forests, away from settlements in areas difficult of access (rocky hills, swamps). Then, when all the possible arable lands have once been claimed and cleared (described as saturation), the expansion is continued at the expense of fallows, now shrinking to extinction. Finally all arable land becomes permanently cropped, which is only sustainable by depending on organic and mineral amendments. Crop expansion is not confined to the local scale. Even before reaching saturation, the need for new land to clear has been the motivation for large scale rural migration from densely populated areas toward ‘pioneer fronts’, such as the westward migration from the Mossi plateau in central Burkina Faso (Chauveau et al. 2006), and that from the old groundnut belt in Senegal (‘bassin arachidier’) to the new groundnut belt to the south-east, and northward movement from the Hausa states on the Niger-Nigerian border.

6.5.2 A Context of Increasing Livestock Populations

Livestock numbers have increased as a result of the enhanced investment capacities of the farmers involved in cash crops, and some urban traders and civil servants. Statistics on livestock numbers are unreliable (Lesnoff et al. 2012). However, national statistics often indicate a larger increase of sheep and goats than of cattle, and (locally in the northern Sahel), a rapid increase in the camel population (Touré et al. 2012).

These trends have been interpreted as strategies of adaptation to the reduction in grazing resources. This is debated however. Increasing numbers of small ruminants could also reflect the first step in a strategy of livestock reconstitution after the large losses due to droughts, or reflect the changes in livestock ownership (Turner 1999), or else adaptation to changes in the market demand (Wane et al. 2009). Moreover, from regional and national policies to local arrangements, very little consideration was given to the implications of this increase for access to pastoral resources. At best, the main routes used by herds in their local and regional migrations were protected from clearing for cropping, and the herds’ access to the main water points was kept free. However, access to rangeland results mainly from the sum of individual farmers’ decisions to crop, and from public policies on infrastructure or conservation. Contrary to Eastern and Southern Africa, attempts to privatize rangelands are unusual and have generally failed in the Sahel (Boutrais 1990). External funding and the efforts of technical services, N.G.O.s and municipalities have been committed to the protection of livestock routes and regulating the use of the remaining pastoral resources within and between communities (Bonnet et al. 2005; Ibrahim et al. 2014).

6.5.3 Adapting the Pastoral Economy

There is a contradiction between a supporting and expanding market for young animals, especially for small ruminants, and locally for dairy products, and the degrading access of pastoralists to rangeland resources, water and social services. Among the possible adaptations the most common consists in diversifying economic activities by adding cropping or trade to pastoralism. In both cases the diversification implies that part of the family settle, which reduces the freedom of movement of the herds. There is a large range of options for diversification, from seasonal activities targeting additional income or the satisfaction of part of the family needs in cereals, with a livestock production system that remains largely pastoral, to a transition toward a sedentary, mixed crop-livestock system. The risk is to progressively worsen access to good quality grazing feed by reducing herding and herd mobility, thereby decreasing herd reproductive performances to marginal viability (Turner et al. 2014). An alternative adaptation strategy consists in investing in equipment and inputs to free pastoral livestock from the resource constraints. This implies investing in the digging of private wells or in tankers to transport water to remote herd camps, and investing in large fodder inputs and the means to bring them to the herds to supplement their diet in the late dry season. Contrary to a persisting paradigm of 'prestige husbandry', the pastoral systems in the Sahel have always been integrated with the regional livestock trade (Wane et al. 2009). The recent spread of cell phone communication networks has improved herder information on market conditions and boosted their reactivity as traders. Furthermore, improvement of the road network, cross-border customs, administrative and financial regulations on the livestock trade, and better adapted insurance and banking systems should enhance the efficiency of the livestock trade in Western Africa.

6.5.4 Adapting the Crop Farming Economy

Agriculture intensification which is targeting a jump in productivity per unit of cultivated land is achieved by investing in inputs, technology, infrastructure and knowledge. Cash crops have locally been catalyzers of such intensification (Dufumier 2005). The development of mixed crop-livestock production systems is a complementary pathway of agricultural intensification (Blanchard 2010). In livestock production, labour is the limiting factor: to conduct the herd to water and grazing, to care for animal health, to supply supplementary feed and and to negotiate grazing rights. Mixed systems are becoming more widely distributed—although unequally between families—but are largely undervalued by the government and the international development agencies. Livestock play many roles in mixed farming systems, for example through the diversification of income earning activities, as recycling agents: organic matter, nutrients, crop by-products (Hiernaux and Diawara 2014), as part of social processes (labor, knowledge, networks) and in financial systems (investment, cash, capital) (Fig. 6.11). Mixed crop-livestock farming is not just a merging of the pastoral with crop farming systems. If most livestock continue to be fed by grazing on rangeland, fallows and stubble, they nevertheless become increasingly fed with harvested crop residues, and agro-industrial feed (groundnut and sesame cakes, cotton seed, and cereals).

Fig. 6.11 Livestock in mixed farming system at Dantiandou. Examples of livestock contributions to agricultural intensification in integrated crop-livestock farms at Dantiandou: *top* cattle corralling in the dry season, *bottom left* morning milking for family consumption, *bottom right* sheep grazing and recycling millet stubble.

6.5.5 Adaptive Regional Integration of the Agricultural Sector

The development of mixed crop-livestock farming offers an opportunity to stratify livestock production at a regional scale. The pastoralists provide young animals, a majority of males, to the settled crop- livestock farmers who either breed them as draught animals or fatten them to trade on the meat market. Less often they breed females for dairy production. This regional stratification establishes a dependence of the crop-livestock farming system on the pastoral system. At the same time, however, it increases the competition for grazing resources. The regional stratification of livestock production is not new (Jahnke 1982) but the socio-economical context is changing. The capacity of the local actors, and also the technical services, national and regional policies, is key to supporting and mediating the integrated development of both systems at local and regional scales. This is a condition for agricultural intensification in the Sahel, to enhance productivity, improve the people's welfare and maintain the resilience of the ecosystem.

Acknowledgements The authors are indebted to many colleagues and collaborators for the data collected and analysed over several decades both in Mali and Niger under a series of research projects and funding agencies listed in the endnotes. Among them they would particularly like to thank the main actors of the long term monitoring and production system studies: late Mohamed Idrissa Cissé, late Lassine Diarra, Youssouf Maiga and Nogmana Soumaguel in Mali, Adamou Kalilou, Oumar Moumouni and Seybou Garba in Niger. Among the colleagues whose works were particularly useful in this chapter, the authors would like acknowledge the works of Matthew Turner from University of Wisconsin, Augustine A. Ayantunde from ILRI, Manuela Grippa from the GET laboratory in Toulouse, Valérie Trichon from the University Paul Sabatier in Toulouse. The remote-sensing analyses are borrowed from Cécile Dardel PhD thesis and Mamadou Diawara contributed to the results on the livestock impact. The authors are grateful to Layne Coppock for his thorough review and to Mike Mortimore who has kindly edited the text in proper English. This publication was made possible through the support provided by the research project 'Environmental and social changes in Africa: past, present and future (ESCAPE, ANR-10-CEPL-005) funded by ANR (French National Research Agency).

References

Achard, F., Hiernaux, P., & Banoin, M. (2001). Les jachères fourragères naturelles et améliorées en Afrique de l'Ouest. In C. Floret & R. Pontanier (Eds.), *La jachère en Afrique tropicale. De la jachère naturelle à la jachère améliorée. Le point des connaissances* (pp. 201–240). Paris: John Libbey Eurotext.

Affholder, F. (1995). Effect of organic matter input on the water balance and yield of millet under tropical dryland condition. *Field Crops Research, 41*, 109–121.

Ag Mahmoud, M. (1992). In E. Le Floc'h (Ed.), *Le haut Gourma Central* (2nd Edn.). Montpellier, France: CEFE/CNRS.

Akponikpe, P. B. I. (2008). *Millet response to water and soil fertility management in the Sahelian Niger: Experiments and modelling* (168p). Ph.D. Dissertation, Université catholique de Louvain, Louvain-la-Neuve, Belgium.

Akponikpe, P. B. I., Minet, J., Gérard, B., Defourny, P., & Bielders, C. L. (2010). Spatial fields' dispersion as afarmer strategy to reduce agro-climatic risk at the household level in pearl millet based systems in the Sahel: A modeling perspective. *Agricultural and Forest Meteorology*. doi:10.1016/j.agrformet.2010.10.007.

Ali, A., Lebel, T., & Amani, A. (2003). Invariance in the spatial structure of Sahelian rain fields at climatological scales. *Journal of Hydrometeorology, 4*, 996.

Anthelme, F., Waziri Mato, M., de Boissieu, D., & Giazzi, F. (2006). Dégradation des ressources végétales au contact des activités humaines et perspectives de conservation dans le massif de l'Aïr (Sahara, Niger). *VertigO, 7*, 2.

Anyamba, A., & Tucker, C. J. (2005). Analysis of Sahelian vegetation dynamics using NOAA-AVHRR NDVI data from 1981-2003. *Journal of Arid Environment, 63*(3), 596–614.

Aubréville, A. (1949). Climats, Forêts et Désertification de l'Afrique Tropicale. Paris, Société d'Editions Géographiques, Maritimes et Coloniales.

Ayantunde, A. A., Hiernaux, P., Fernandez-Rivera, S., van Keulen, H., & Udo, H. M. J. (1999). Selective grazing by cattle on spatially and seasonally heterogeneous rangelands in the Sahel. *Journal of Arid Environment, 42*, 261–279.

Bationo, A., Waswa, B., Kihara, J., Adolwa, I., Van Lauwe, B., & Koala, S. (2012). *Lessons learned from long-term soil fertility management experiments in Africa*. Dordrecht: Springer.

Begue, A., Vintrou, E., Ruelland, D., Claden, M., & Dessay, N. (2011). Can a 25-year trend in Soudano-726 Sahelian vegetation dynamics be interpreted in terms of land use change? A remote sensing approach. *Global Environmental Change-Human Policy Dimensions, 21*, 413–420.

Bielders, C. L., Rajot, J.-L., & Amadou, M. (2002). Transport of soil and nutrients by wind in bush fallow land and traditionally managed cultivated fields in the Sahel. *Geoderma, 109*, 19–39.
Bielders, C. L., Rajot, J.-L., & Michels, K. (2004). L'érosion éolienne dans le Sahel nigérien: influence des pratiques culturales actuelles et méthodes de lutte. *Sècheresse, 15*(1), 19–32.
Blanchard, M. (2010). *Gestion de la fertilité des sols et rôle du troupeau dans les systèmes coton-céréales-élevage au Mali sud* (306p), Université Paris-Est Créteil, PhD, Paris, France.
Bonnet, B., Marty, A., & Demante, M.-J. (2005). *Hydraulique et sécurisation des systèmes pastoraux au Sahel. Appui à la gestion locale* (28p). Paris: IRAM, démarches et méthodes.
Boudet, G. (1972). Désertification de l'Afrique t r o p i c a l e sèche. *Adansonia, sér . 2, 12*(4), 505–524.
Boudet, G. (1979). Quelques observations sur les fluctuations du couvert végétal sahélien au Gourma Malien et leurs conséquences pour une stratégie de gestion sylvo-pastorale. *Bois et Forêts des Tropiques, 184*, 31–44.
Boudet, G. (1990). Désertification ou remontée biologique au Sahel. *Cahiers ORSTOM, série biologie, 12*, 293–300.
Boutrais, J, 1990. Derrière les clôtures… Essai d'histoire compare de ranchs africains. *ORSTOM Cahiers des sciences humaines*, 26(1–2), 73–95.
Brandt, M., Verger, A., Diouf, A. A., Baret, F., & Samimi, C. (2014). Local vegetation trends in the Sahel of Mali and Senegal using long time series FAPAR satellite products and field measurements (1982-2010). *Remote Sensing, 6*, 2408–2434.
Breman, H., & de Ridder, N. (1991). *Manuel sur les pâturages des pays sahéliens*. Paris: Karthala.
Brooks, N. (2004). *Drought in the African Sahel: Long term perspectives and future prospects*. Tyndall Centre Working Papers. Anglia, UK: U. O. E.
Brouwer, J., Fussell, L. K., & Herrmann, L. (1993). Soil and crop growth micro-variability in the West African semi-arid tropics: A possible risk reducing factor for subsistence farmers. *Agriculture, Ecosystem & Environment, 45*, 229–238.
Cappelaere, B., Descroix, L., Lebel, T., Boulain, N., Ramier, D., Laurent, J.-P., et al. (2009). The AMMA-CATCH experiment in the cultivated Sahelian area of south-west Niger—Investigating water cycle response to a fluctuating climate and changing environment. *Journal of Hydrology, 375*(1–2), 34–51.
Carrière, M. (1989). *Les communautés végétales sahéliennes en Mauritanie (Région de Kaédi); analyse de la reconstitution annuelle du couvert herbacé*. Maisons-Alfort: IEMVT-CIRAD.
Chauveau, J. P., Colin, J.-P., Jacob, J.-P., Lavigne Delville, P., & Le Meur, P.-Y. (2006). *Modes d'accèsz à la terre, marchés fonciers, gouvernance et politiques foncières en Afrique de l'Ouest* (97p). Résultats du projet de recherche CLAIMS. London: IIED.
Chudeau, R. (1921). Le problème du desséchement en Afrique Occidentale. Bulletin du Comite d'Etudes Historiques et Scientifiques de l'AOF, *4*, 353–369.
Cissé, A. M. (1986). *Dynamique de la strate herbacée des pâturages de la zone sud-sahélienne*. PhD Thesis, Agricultural University. Wageningen, The Netherlands.
d'Herbès, J. M., Valentin, C., Tongway, D., & Leprun, J. C. (2001). Banded vegetation patterns and related structures. In *Banded vegetation patterning in arid and semiarid environments: Ecological processes and consequences for management* (pp. 1–19), Ecological studies 149, New York, USA: Springer.
Dardel, C., Kergoat, L., Hiernaux, P., Grippa, M., Mougin, E., Ciais, P., & Nguyen, C. C. (2014a). Rain-Use Efficiency: What it tells about conflicting Sahel greening and Sahelian paradox. *Remote Sensing, 6*, 3446–3474.
Dardel, C., Kergoat, L., Mougin, E., Hiernaux, P., Grippa, M., & Tucker, C. J. (2014b). Re-greening Sahel: 30 years of remote sensing data and field observations (Mali, Niger). *Remote Sensing of Environment, 140*, 350–364.
Dawalak, A. (2009). *Effet des crises climatiques sur le cheptel bovin et mise au point d'une méthode de calcul sur le taux de croît dans la commune de Hombori (Mali)* (69p). Katibougou, Mali: Mémoire IPR/IFRA.
De Gironcourt, G. (1912). Le sommet de la boucle du Niger. *La Géographie XXV, 3*, 153–172.

De Wispelaere, G. (1980). Les photographies aériennes témoin de la dégradation du couvert ligneux dans un géosystème sahélien sénégalais. Influence de la proximité d'un forage. *Cah. OSRTOM, série Sc. Hum., 17*(3–4), 155–166.

Dick, R. P., Sene, M., Diack, M., Khouma, M., Badiane, A., Samba, S., et al. (2010). The Native Shrubs Philiostigma reticulatum and Guiera senegalensis: The unrecognized potential to remediate degraded soils and optimize productivity of sahelian agroecosystems. In *Le projet majeur africain de la Grande Muraille Verte: Concepts et mise en œuvre*. Montpellier, France: IRD Éditions.

Dodd, J. L. (1994). Desertification and degradation in Sub-Saharan Africa—The role of livestock. *Bioscience, 44*(1), 28–34.

Dufumier, M. (2005). *Étude des systèmes agraires et typologie des systèmes de production agricole dans la région cotonnière du Mali*. Paris: INAPG Paris-Grignon.

Dumont, R. (1962). *L'Afrique noire est mal partie*. Paris: Editions du Seuil.

Eklundh, L., & Olsson, L. (2003). Vegetation index trends for the African Sahel 1982-1999. *Geophysical Research Letters, 30*, 1430–1433.

Ellis, J. E., & Swift, D. M. (1988). Stability of pastoral ecosystems: Alternate paradigms and implications for development. *J. Range Management, 41*, 450–459.

Fensholt, R., & Rasmussen, K. (2010). Analysis of trends in the Sahelian 'rain-use efficiency' using GIMMS NDVI, RFE and GPCP rainfall data. *Remote Sensing of Environment, 115*, 438–451.

Fensholt, R., Sandholt, I., &Stisen, S. (2006). Evaluating MODIS, MERIS, and VEGETATION—Vegetation indices using in situ measurements in a semiarid environment. *IEEE Transactions on Geoscience & Remote Sensing, 44*, 1774–1786.

Fernandez-Rivera, S., Hiernaux, P., Williams, T. O., Turner, M. D., Schlecht, E., Salla, A., et al. (2005). Nutritional constraints to grazing ruminants in the millet-cowpea-livestock farming system of the Sahel. In A. A. Ayantunde, S. Fernandez-Rivera, & G. McCrabb (Eds.), *Coping with feed scarcity in samllholder livestock systems in developing countries* (pp. 157–182). Nairobi, Kenya: ILRI.

Frappart, F., Hiernaux, P., Guichard, F., Mougin, E., Kergoat, L., Arjounin, M., et al. (2009). Rainfall regime over the Sahelian climate gradient in the Gourma, Mali. *Journal of Hydrology, 375*(1–2), 128–142.

Gallais, J. (1967). Le Delta intérieur du Niger, étude géographique régionale, Dakar: *Mémoire de l'IFAN, 79*, 621.

Gallais, J. (1984). *Hommes du Sahel. Espaces-temps et pouvoirs* (289p). Paris: Flammarion.

Gandah, M., Bouma, J., Brouwer, J., Hiernaux, P., & Van Duivenbooden, N. (2003). Strategies to optimize allocation of limited nutrients to sandy soils of the Sahel: A case study from Niger, west Africa. *Agricultural Ecosystem & Environment, 94*(3), 311–319.

Gardelle, J., Hiernaux, P., Kergoat, L., & Grippa, M. (2010). Less rain, more water in ponds: A remote sensing study of the dynamics of surface waters from 1950 to present in pastoral Sahel (Gourma region, Mali). *Hydrology and Earth System Sciences, 14*, 309–324.

Giannini, A., Saravanan, R., & Chang, P. (2003). Oceanic forcing of Sahel rainfall on interannual to interdecadal time scale. *Science, 302*(5647), 1027–1030.

Goutorbe, J. P., Lebel, T., Tinga, A., Bessemoulin, P., Brouwer, J., Dolman, A. J., et al. (1994). HAPEX-Sahel—A large-scale study of land-atmosphere interactions in the semiarid tropics, *Annalae Geophysicae—Atmospheres Hydrospheres and Space Sciences, 12*(1), 53–64.

Grimaud, J.-L., Chardon, D., Beauvais, A., & Christophoul, F. (2011). Cenozoic drainage evolution of West Africa: Spatial and temporal constraints from the lateritic record. *Geophysiacal Research Abstract, 13*, 4308.

Grove, R. H. (1995). *Green imperialism: Colonial expansion, tropical island edens and the origins of environmentalism, 1600–1800*. Cambridge, UK: Cambridge University Press.

Guengant, J. –P., Banoin, M., & Quesnel, A. (2002). *Dynamique des populations, disponibilité en terres et adaptation des régimes fonciers: le cas du Niger* (155p). Roma: FAO.

Hampshire, K. (2002). Fulani on the move: Seasonal economic migration in the Sahel as a social process. *The Journal of Development Studies, 38*(5), 15–36.

Hanan, N. P., Prevost, Y., Diouf, A., & Diallo, O. (1991). Assessment of desertification around deep wells in the Sahel using satellite imagery. *Journal of Applied Ecology, 28*, 173–186.
Harroy, J. P. (1944). Afrique, terre qui meurt. La dégradation des sols africains sous l'influence de la colonisation. Marcel Hayez Office International de la Librairie, Bruxelles.
Hein, L., & de Ridder, N. (2006). Desertification in the Sahel: A reinterpretation. *Global Change Biology, 12*, 751–758.
Hein, L., de Ridder, N., Hiernaux, P., Leemans, R., de Wit, A., & Schaepman, M. (2011). Desertification in the Sahel: Towards better accounting for ecosystem dynamics in the interpretation of remote sensing images. *Journal of Arid Environment, 75*, 1164–1172.
Hellden, U., & Tottrup, C. (2008). Regional desertification: A global synthesis. *Global & Planet. Change, 64*, 169–176.
Hérault, B., & Hiernaux, P. (2004). Soil seed bank and vegetation dynamics in Sahelian fallows: The impact of past cropping and current grazing treatments. *Journal of Tropical Ecology, 20*, 1–9.
Herrmann, S. M., Anyamba, A., & Tucker, J. C. (2005). Recent trends in vegetation dynamics in the African Sahel and their relationship to climate. *Global Environmental Change-Human & Policy Dimensions*, *15*(4), 394–404.
Herrmann, S., & Tappan, G. G. (2013). Vegetation impoverishment despite greening: A case study from central Senegal. *Journal of Arid Environments, 90*, 55–66.
Heumann, B. W., Seaquist, J. W., Eklundh, L., & Jonsson, P. (2007). AVHRR derived phenological change in the Sahel and Soudan, Africa, 1982-2005. *Remote Sensing of Environment, 108*(4), 385–392.
Hickler, T., Eklundh, L., Seaquist, J. W., Smith, B., Ardo, J., Olsson, L., et al. (2005). Precipitation controls Sahel greening trend. *Geophysical Research Letters, 32*, L21415.
Hiernaux, P. (1988). Vegetation monitoring by remote sensing: Progress in calibrating a radiometric index and its application in the Gourma, Mali. *ILCA Bulletin*, *32*, 14–21. Available from http://www.ilri.org/InfoServ/Webpub/fulldocs/Bulletin32/vegeta.htm.
Hiernaux, P. (1998). Effects of grazing on plant species composition and spatial distribution in rangelands of the Sahel. *Plant Ecology, 138*, 191–202.
Hiernaux, P., & Ayantunde, A. A. (2004). The Fakara: A semi-arid agro-ecosystem under stress. Report of research activities, First phase (July 2002-June 2004) of the DMP-GEF Program (GEF/2711-02-4516), ILRI, Nairobi, Kenya.
Hiernaux, P., Ayantunde, A. A., Adamou, K., Mougin, E., Gérard, B., Baup, F., et al. (2009a). Resilience and productivity trends of crops, fallows and rangelands in Southwest Niger: Impact of land use, management and climate changes. *Journal of Hydrology, 375*(1–2), 65–77.
Hiernaux, P.,Cissé, M. I., Diarra, L., & de Leeuw, P. N. (1994). Fluctuations saisonnières de la feuillaison des arbres et des buissons sahéliens. Conséquences pour la quantification des ressources fourragères. *La Revue d'Elevage et de Medecine Vétérinaire des Pays Tropicaux*, *47* (1), 117–125.
Hiernaux, P., Diarra, L., Trichon, V., Mougin, E., Soumaguel, N., & Baup, F. (2009b). Woody plant population dynamics in response to climate changes from 1984 to 2006 in Sahel (Gourma, Mali). *Journal of Hydrology, 375*, 103–113.
Hiernaux, P., & Diawara, M. O. (2014). Livestock: Recyclers that promote the sustainability of smallholder farms. *Rural, 21*(04), 9–11.
Hiernaux, P., & Gérard, B. (1999). The influence of vegetation pattern on the productivity, diversity and stability of vegetation: The case of 'brousse tigree' in the Sahel. *Acta ecologica, 20*, 147–158.
Hiernaux, P., & Le Houérou, H. N. (2006). Les parcours du Sahel. *Sècheresse, 17*(1–2), 51–71.
Hiernaux, P., Mougin, E., Diarra, L., Soumaguel, N., Lavenu, F., Tracol, Y., & Diawara, M. O. (2009c). Rangeland response to rainfall and grazing pressure over two decades: Herbaceous growth pattern, production and species composition in the Gourma, Mali. *Journal of Hydrology, 375*(1–2), 114–127.
Hiernaux, P., Mougin, E., Diawara, M., Soumaguel, N., & Diarra, L. (2012). *How much does grazing contribute to herbaceous decay during the dry season in Sahel rangelands?* (23p)

Toulouse, France: ECliS Deliverable 3.4, GET. http://eclis.get.obsmip.fr/index.php/eng/scientificproductions/delivrable.
Hiernaux, P., Mougin, E., Soumaguel, N. ,Diawara, M. O., Auda, Y., & Le Dantec, V. (2013). *Do seasonal maximum growth rates explain the large variations in annual yields of the herbaceous in Sahel rangelands?* Toulouse, France: ECliS Deliverable 2.7. http://eclis.get.obsmip.fr/index.php/eng/scientificproductions/delivrable.
Hiernaux, P., & Turner, M. D. (1996). The effect of clipping on growth and nutrient uptake of Sahelian annual rangelands. *Journal of Applied Ecology, 33*, 387–399.
Hiernaux, P., Turner, M. D/. (2002). The influence of farmer and pastoralist management practices on desertification processes in the Sahel. In J. F. Reynolds & M. D. Stafford Smith (Eds.), *Global desertification: Do humans cause deserts?* (pp. 135–148). Berlin: Dahlem University Press.
Hubert, H. (1920). Le dessèchement progressif en Afrique occidentale. *Bulletin du Comité d'études historiques et scientifiques de l'Afrique occidentale française, 3*, 401–467.
Hutchinson, J., & Dalziel, J. M. (1954–1972). In R. W. J. Keay & F. N. Hepper (Eds), *Flora of west Tropical Africa* (2nd Edn.). London: Crown Agents for Overseas Government and Administration.
Hutchinson, C. F., Herrmann, S. M., Maukonen, T., & Weber, J. (2005). Introduction: The "Greening" of the Sahel. *Journal of Arid Environment, 63*, 535–537.
Ibrahim, H., Mormont, M., & Yamba, B. (2014). Local innovations and initiatives in the management of conflicts and health risks in agro-pastoral zones: The SAREL and PASEL projects in the Dakoro department in Niger. *Fields Actions Science Reports, 11*, 8p.
IPCC. (2013). *Climate Change 2013: The Physical science basis. Summary for policymakers* (36p). IPCC 5th assessment report.
Jahnke, H. E. (1982*). Livestock production systems and livestock development in tropical Africa* (253p). Kiel, Germany: Kieler Wissenschaftsverlag.
Jones, B. (1938). Desiccation and the West African Colonies. *The Geographical Journal, 91*, 401–423.
Le Barbé, L., & Lebel, T. (1997). Rainfall climatology of the HAPEX-Sahel region during the years 1950-1990. *Journal of Hydrology, 189*(1–4), 43–73.
Le Drézen, Y., Lespez, L., Rasse, M., Garnier, A., Coutard, S., Huysecom, E., & Ballouche, A. (2010). Hydrosedimentary records and Holocene environmental dynamics in the Yamé Valley (Mali, Sudano-Sahelian West Africa). *Comptes Rendus Geoscience, 342*, 244–252.
Le Houérou, H. N. (1989). *The grazing land ecosystems of the African Sahel* (282p). Ecological studies 75. Berlin: Springer.
Le Houérou, H. N. (2009). *Bioclimatology and Biogeography of Africa.* Berlin Heidelberg: Springer-Verlag.
Lebel, T., & Ali, A. (2009). Recent trends in the Central and Western Sahel rainfall regime (1990-2007). *Journal of Hydrology, 375*(1–2), 52–64.
Lebel, T., Cappelaere, B., Galle, S., Hanan, N., Kergoat, L., Levis, S., et al. (2009). AMMA-CATCH studies in the Sahelian region of West-Africa: An overview, *Journal of Hydrology*, *375*(1–2), 3–13.
Leprun, J. C. (1992). Etude de quelques brousses tigrées sahéliennes: structure, dynamique, écologie. In E. Le Floc'h, M. Grouzis, A. Cornet & J. C. Bille (Eds.), *L'aridité une contrainte au développement* (pp. 221–244). Paris: Editions de l'ORSTOM.
Lesnoff, M., Corniaux, C., & Hiernaux, P. (2012). Sensitivity analysis of the recovery dynamics of a cattle population following drought in the Sahel region. *Ecological Modelling, 232*, 28–39.
Ludwig, J. A., & Tonway, D. J. (1995). Spatial organization of landscapes and its function in semi-arid woodlands, Australia. *Landscape Ecology, 10*, 51–63.
Michels, K., Lamers, J. P. A., & Buerkert, A. (1998). Effects of windbreak species and mulching on wind erosion and millet yield in the Sahel. *Experimental Agriculture, 34*, 449–464.
Morel, R, (1991). *Atlas agroclimatique des pays de la zone du CILSS. Notice et commentaire* (161p). Niamey, Niger: Centre Agrhymet.

Mougin, E., Hiernaux, P., Kergoat, L., De Rosnay, P., Grippa, M., Timouk, F., et al. (2009). The AMMA Gourma observatory site in Mali: Relating climatic variations to changes in vegetation, surface hydrology, fluxes and natural resources. *Journal of Hydrology, 375*(1–2), 14–33.

Mougin, E., Lo Seen, D., Rambal, S., Gaston, A., & Hiernaux, P. (1995). A regional Sahelian grassland model to be coupled with multispectral satellite data. I: Model description and validation. *Remote Sensing of Environment, 52*, 181–193.

Myneni, R. B., Hall, F. G., Sellersand. P. J., & Marshak, A. L. (1995). The interpretation of spectral vegetation indexes. *IEEE Transactions on Geoscience & Remote Sensing, 3*, 481–486.

Nicholson, S. E. (2013). The West African Sahel: A review of recent studies on the rainfall regime and its interannual variability. *ISRN Meteorology, 2013*, Article ID 453521. doi10.1155/2013/453521.

Ntaré, B. R., & Williams, J. H. (1992). Response of cowpea cultivars to planting pattern in intercrops with pearl millet in Niger. *Expl. Agric., 28*, 41–48.

Olsson, L., Eklundh, L., & Ardo, J. (2005). A recent greening of the Sahel—Trends, patterns and potential causes. *Journal of Arid Environment, 63*(3), 556–566.

Osbahr, H. (2001). *Livelihood strategies and soil fertility at Fandou Beri, southwestern Niger* (414p). PhD thesis, University of London.

Peacock, C. P. (1983). *Phase exploratoire d'une étude des systèmes de production animale dans le Gourma malien: données de base sur le cheptel caprin et ovin*. Document de programme N° 92c. Bamako, Mali: CIPEA.

Penning de Vries, F. W. T., & Djiteye, M. A. (Eds.). (1982). *La productivité des pâturages sahéliens, une étude des sols, des végétations et de l'exploitation de cette ressource naturelle*. Wageningen, The Netherlands: Pudoc.

Pieri, C. (1989). *Fertilité des terres de savanes. Bilan de trente ans de recherche et de développement agricol au Sud du Sahara* (444p). Montpellier: Min.de la Coopération et CIRAD-IRAT.

Powell, J. M., Fernandez-Rivera, S., Hiernaux, P., & Turner, M. D. (1996). Nutrient cycling in integrated rangeland/cropland systems of the Sahel. *Agricultural System, 52*, 143–170.

Prince, S. D., de Colstoun, E. B., & Kravitz, L. L. (1998). Evidence from rain-use efficiencies does not indicate extensive Sahelian desertification. *Glob Change Biology, 4*, 359–374.

Prince, S. D., Wessels, K. J., Tucker, C. J., & Nicholson, S. E. (2007). Desertification in the Sahel: A reinterpretation of a reinterpretation. *Glob Change Biology, 13*, 1308–1313.

Raynaut, C. (2001). Societies and nature in the Sahel: Ecological diversity and social dynamics. *Global Environmental Change, 11*, 9–18.

Redelsperger, J. L., Thorncroft, C. D., Diedhiou, A., Lebel, T., Parker, D. J., & Polcher, J. (2006). African monsoon multidisciplinary analysis—An international research project and field campaign. *Bulletin of the American Metereological Society, 87*(12), 1739–1746.

Rockström, J., Barron, J., Brouwer, J., Galle, S., & de Rouw, A. (1999). On-farm spatial and temporal variability of soil and water in pearl millet cultivation. *Soil Science Society of America Journal, 63*, 1308–1319.

Rockström, J., & de Rouw, A. (1997). Water, nutrients and slope position in on-farm pearl millet cultivation in the Sahel. *Plant Soil, 195*, 311–327.

Saqalli, M., Bielders, C. L., Defourny, P., & Gérard, B. (2013). Reconstituting family transitions of Sahelian western Niger 1950-2000: An agent-based modelling approach in a low data context. *Cybergeo: European Journal of Geography, 634*, 1–19.

Saqalli, M., Gérard, B., Bielders, C. L., & Defourny, P. (2010). Simulating rural environmentally and socio-economically constrained multi-activity and multi-decision societies in a low-data context: A challenge through empirical agent-based modeling. *Journal of Artificial Society & Social Simulation, 13*, 2.

Schlecht, E., Hiernaux, P., Achard, F., & Turner, M. D. (2004). Livestock related nutrient budgets within village territories in western Niger. *Nutrient Cycling in Agroecosystems, 68*, 199–211.

Seghieri, J. (1996). Rythmes saisonniers d'une savane soudano-sahélienne en relation avec l'hydrodynamique du sol. *Ecologie, 27*(2), 67–78.

Stebbing, E. P. (1935). The encroaching Sahara: The threat to the west African Colonies. *Geographical Journal, 85*, 506–524.

Tabutin, D., & Schoumaker, B. (2004). La demographie de l'Afrique au sud du Sahara des années 1950 aux années 2000. Synthèse des changements et bilan statistique. *Population, 59*(3–4), 521–622.

Thébaud, B., & Batterbury, S. (2001). Sahel pastoralists: Opportunism, struggle, conflict and negotiation. A case study from eastern Niger. *Global Environmental Change, 11*, 69–78.

Toulmin, C. (1987). Drought and the Farming Sector: Loss of animals and post-drought rehabilitation. *Development Policy Review, 5*(2), 125–148.

Touré, I., Ickowicz, A., Wane, A., Garba, I., & Gerber, P. (2012). *Atlas of trends in pastoral systems in Sahel*. Montpellier, France: FAO & Cirad. 36p.

Tracol, Y., Mougin, E., Jarlan, L., & Hiernaux, P. (2006). Testing a sahelian grassland functioning model against herbage mass measurements. *Ecological Modelling, 193*, 437–446.

Trichon, V., Walcker, R., Hiernaux, P., & Mougin, E. (2009). *50 years of woody plant population dynamics in Sahel (Gourma, Mali): A survey based on aerial photographs*. In AMMA International Conference, Ouagadougou, July 20–24.

Trochain, J. (1940). Contribution à l'étude de la végétation du Sénégal (433p.), Larose, Paris.

Tucker, C. J. (1979). Red and photographic infrared linear combinations for monitoring vegetation. *Remote Sensing of Environment, 8*, 127–150.

Tucker, C. J., Dregne, H. E., & Newcomb, W. W. (1991). Expansion and Contraction of the Sahara Desert from 1980 to 1990. *Science, 253*, 299–301.

Tucker, C. J., Pinzon, J. E., Brown, M. E., Slayback, D. A., Pak, E. W., Mahoney, R., et al. (2005). An extended AVHRR 8-km NDVI dataset compatible with MODIS and SPOT vegetation NDVI data. *International Journal of Remote Sensing, 26*, 4485–4498.

Tucker, C. J, Vanpraet, C. L., Sharman, M. J., & van Ittersum, G. (1985). Satellite Remote-Sensing of 741 total herbaceous biomass production in the Senegalese Sahel—1980-1984. *Remote Sensing of Environment, 17*, 233–249.

Turner, M. D. (1998). Long-term effects of daily grazing orbits on nutrient availability in Sahelian West Africa: 1. Gradients in the chemical composition of rangeland soils and vegetation. *Journal of Biogeography, 25*, 669–682.

Turner, M. D. (1999). Merging local and regional analysis of land-use change; the case of livestock in the Sahel. *Annals of the Association of American Geographers, 89*(2), 191–219.

Turner, M. D., & Hiernaux, P. (2002). The use of herders' accounts to map livestock activities across agropastoral landscapes in Semi-Arid Africa. *Landscape Ecology, 17*, 367–385.

Turner, M. D., Hiernaux, P., & Schlecht, E. (2005). The distribution of grazing pressure in relation to vegetation resources in semi-arid west Africa: The role of herding. *Ecosystems, 8*, 668–681.

Turner, M. D., McPeak, J. G., & Ayantunde, A. A. (2014). The role of livestock mobility in the livelihood strategies of rural peoples in semi-arid west Africa. *Human Ecology, 1*, 1–17.

UNEP. (1992). *World Atlas of desertification, Middleton NJ and DSG Thomas*. London: Edward Arnold.

Valenza, J. (1984). *Surveillance continue des pâturages naturels sahéliens sénégalais. Résultats de 10 années d'observation*. Agrostologie 44. Dakar: ISRA-LNERV.

Vetter, S. (2005). Rangeland at equilibrium and non-equilibrium recent developments in the debate. *Journal of Arid Environment, 62*, 321–341.

Vintrou, E., Desbrosse, A., Begue, A., Traore, S., Baron, C., & Lo Seen, D. (2012). Crop area mapping in West Africa using landscape stratification of MODIS time series and comparison with existing global land products. *International Journal of Applied Earth Observation and Geoinformation, 14*(1), 83–93.

Voortman, R. L., Brouwer, J., & Albersen, P. J. (2004). Characterization of spatial soil variability and its effect on Millet yield on Sudano-Sahelian coversands in SW Niger. *Geoderma, 121*(1–2), 65–82.

Walker, B. H., Carpenter, S. R., Rockstrom, J., Crépin, A. S., & Peterson, G. D. (2012). Drivers, "slow" variables, "fast" variables, shocks, and resilience. *Ecology & Society, 17*(3), 30.

Wane, A., Touré, I., & Ancey, V. (2009). Pastoralisme et Recours aux marchés - Cas du Sahel sénégalais (Ferlo). *Cah. de l'Agric., 19*(1), 14–20.

Warren, A. (1995). Changing understandings of African pastoralism and environmental paradigms. *Transactions of the Institute of British Geographers NS, 20*(2), 193–203.

Warren, A., Batterbury, S. P. J., & Osbahr, H. (2001). Soil erosion in the West African Sahel: A review and an application of a 'local political ecology' approach in South West Niger. *Global Environmental Change—Human and Policy Dimensions, 11*, 79–95.

Chapter 7
Land and Natural Resource Governance: Development Issues and Anti-Desertification Initiatives in Niger

Yamba Boubacar

Abstract Desertification in Niger is set in a context of poverty and rapid population growth, but significant advances have been achieved in natural resource management, in particular the wide and growing adoption of assisted natural regeneration of woody vegetation on farmland (régénération naturel assisté, or RNA). Some 250,000 ha of degraded land have been rehabilitated, from 3 to 5 million hectares may be within the reach of improved management, and earth satellite images show 'regreening'. The evolution of Niger's anti-desertification strategy since 1960 is summarized. The 1980s were pivotal years in which power was decentralized to local communities and institutions. The governance of natural resources in central southern Niger is reviewed and the results of surveys in Zinder Region are presented. The linkages between RNA and development are explored and the findings of the study contextualized in the dynamics of regional growth. It is concluded that while the new techniques have potential for reducing poverty and for sustainable natural resource management (NRM), they cannot alone solve the challenge of a rapidly growing population.

Keywords Desertification · Governance · Niger · Regeneration

7.1 Introduction

The African continent can only move forwards if there is synergy between its development plans and efforts to take account of environmental issues. 'Combating desertification' is a common thread linking a range of serious political, socio-economic and cultural problems. The governance of natural resources is a key element.

Y. Boubacar (✉)
Université Abdou Moumouni, Niamey, Niger
e-mail: byamba@refer.ne

R.H. Behnke and M. Mortimore (eds.), *The End of Desertification?*
Springer Earth System Sciences, DOI 10.1007/978-3-642-16014-1_7

According to one estimate, there were 22 million km^2 of usable land in Africa in 1992: 15 million km^2 arable and 7 million km^2 non-arable (Pontié and Gaud 1992). According to the World Atlas of Desertification, the soils of 320 million hectares in the arid, semiarid and dry subhumid zones of Africa are lightly, moderately or highly degraded (Thomas and Middleton 1992).

Niger's land law (Code Rural Niger 2008) defines desertification as "the degradation of land in arid, semi-arid and dry sub-humid areas as a result of various factors, including climatic variations and human activity". Human pressure on land is intensifying as demographic growth (which stood at 3.9 % in 2012) increases the demand for cultivable land and fuelwood, which is the only source of energy for the vast majority of the population. The loss of forest lands, which official statistics estimate shrank by around 8 million hectares between 1970 and 1999, is undermining biological diversity, genetic resources and soil fertility, and reducing the ability to sequester carbon. Areas vulnerable to drought, desertification and the most intense human use of natural resources coincide with areas of primary concern regarding food security and domestic energy needs. Official figures claim that several hundreds of thousands of hectares of land lose their productive potential each year due to various forms of erosion. Natural forest shrank by 60 % between 1981 and 1994; 40 % of officially reserved forests disappeared between 1970 and 2003.

This chapter focuses first on environmental policy, which centred on forest and tree management, seen as the first priority in combating desertification especially after the impact of the 1972–73 drought cycle. This policy was fundamentally revised when a second major drought cycle hit the region in 1983–84, in favour of devolved governance and the promotion of local practices. After examining the underlying reasons for success in reversing the desertification process in central southern Niger, this chapter identifies the main constraints and prospects for the future.

7.2 Evolution of an Anti-Desertification Strategy in Niger Since 1960

Niger's harsh climatic conditions include very low levels of rainfall (700 mm in the south of the country but less than 100 mm in the north) and average annual temperatures of 27–29 °C. The country had 17 million inhabitants in 2012. If the population continues to grow at its current rate, it will reach nearly 78 million by 2050 (Guengant and Banoin 2003). Nearly 80 % live in rural areas, with just under 75 % occupying 12 % of the national territory. This puts intense pressure on an environment where degradation is a well-known and longstanding phenomenon.

Nearly 90 % of the population rely on agriculture and livestock rearing for their livelihoods, although Niger produces uranium and recently moved into oil. Only 12 % of the 1,267,000 km^2 of land in Niger is usable for agriculture, and many regions with high population densities have reached saturation point as agricultural areas have expanded (where ecological conditions allow).

Early strategies. Environmental protection strategies have evolved over time. The strategy during the colonial period and early years of independence in the 1960s was strongly influenced by the legacy of the Centre Technique Forestier Tropical, a colonial structure whose programmes, guidelines and actions were characterized by a coercive style of forest administration. The newly established public sector, which was supposed to drive the modernization of the economy, paid little attention to environmental considerations and pushed groundnut production to such extremes that the entire State budget was covered by revenues from groundnut exports at the end of the 1960s (Hamadou 2000).

Combating desertification first became a national priority in the early 1970s as severe climatic conditions and successive droughts left the population highly vulnerable to food insecurity. The State had hoped that revenues from its new uranium export sector could be used to moderate food insecurity, but was tipped into a cycle of debt by falling prices on the international market.

Consequently, as part of a new policy on food self-sufficiency, the State allocated thousands of hectares to large traders and senior civil servants in the state apparatus, working on the assumption that placing agriculture in the hands of those with the resources to invest in it would produce enough surplus grain to affect the price of basic foodstuffs. In reality, however, this policy led to land-grabbing and poor agricultural practices (such as using tractors on fragile soils) that destroyed the environment and increased erosion.

Power to the people. A landmark debate in Maradi in May, 1984 broke with the technical approach of the 1970s by supporting an environmental policy whose main strategy was to make local people responsible for anti-desertification initiatives. A consensual document known as the 'Maradi Commitment' was adopted, setting out a programme of actions structured into 11 sub-programmes (République du Niger 1984). This dynamic would subsequently result in a national policy, which would not only emphasize improving production systems to achieve food self-sufficiency, but would also entail multisectoral and multidisciplinary interventions by all rural development actors, with interdependent components to address the forestry, agricultural and pastoral dimensions of the process.

Until 1984, state policies for managing rural areas in Niger looked like an extension of the colonial forestry strategies that had focused on plantations, maintaining and restoring soils, and protecting forest resources by classifying gum (*Acacia senegal*) and kapok (*Ceiba pentandra* or *Bombax* spp.) forests and hunting reserves. This line of action was strengthened as a succession of droughts between 1972 and 1984 weakened rural areas and previously unaffected regions of Niger. Many forest plantations were established as interventions by NGOs, the state forest administration and numerous forestry programmes which were extended into public woodlands.

The droughts disrupted ecological balances across the country, forcing the State to rethink its environmental policy. In order to protect natural resources it modified the Forest Code in 1974, and legislative and regulatory texts were formulated to create new classified forests (especially gum trees), planted trees in urban centres (roadsides, public buildings), and in rural and urban concessions, schools, markets and other public spaces. These so-called ‘first generation’ projects included the Forest Project, the Gum Project, the first projects to create green belts around urban areas, promotion of trees on farms, including gums and Borassus palms, creation of village woodlots, dune stabilization, and soil and water conservation.

The results of reforestation initiatives were disappointing. This led to criticism of ‘overly expensive and unsustainable’ sectoral policies. However the relative success of grassroots agroforestry operations suggested that the way forward was to mobilize the general population and encourage them to participate in reforestation initiatives.

The severe drought of 1984 showed the limitations of these state forestry policies and raised questions about accepted forestry dogma. As new guidelines were gradually introduced, the focus switched from protecting forest reserves to the overall management and use of plant resources. The drive to raise awareness and mobilize communities around anti-desertification efforts was helped by the fact that most rural people’s living standards had been seriously affected by this drought, and actions to develop woodland resources were therefore designed to take account of local needs. This important turning point was marked by a new style of environmental project, which managed woodland resources through forestry cooperatives and local organizations, and started to address the problem of fuelwood consumption in urban centres.

Fuelwood markets. These interventions were structured around an approach that encouraged local people to manage local fuelwood resources themselves, and which promoted alternative sources of energy. In 1991 this Household Energy Strategy led to the introduction of rural wood markets as part of the effort to meet the energy needs of large urban centres, and to reorganize the supply chain so that it benefited rural areas, fed into the tax base and was controlled by local structures that organized the collection and marketing of fuelwood (Noppen et al. 2004; World Bank 1988).

The creation of rural fuelwood markets was partly driven by the desire to involve local people in woodland management strategies, and partly to help reverse the environmental damage caused by the uncontrolled exploitation of timber resources. This would be done through selective cutting, introducing quotas for given parcels, and using woodlands in rural concessions registered in the name of local management structures in accordance with their capacity for natural regeneration. Another reason why the State wanted to encourage popular participation in programmes to combat desertification was that it no longer had the resources to tackle this problem itself, due to the draconian spending cuts imposed by the structural adjustment programme and the worsening economic recession triggered by the devaluation of the franc CFA (1994).

Projects and interventions. Conservation and development are understood to be interdependent. Recent studies show that Niger has made considerable efforts in the last few decades.[1] The statistics in Niger's national anti-desertification and natural resource management action plan show that international, non-governmental and private organizations as well as public sector projects have conducted a range of initiatives including 40 forest management projects, 35 soil and water conservation (SWC) and soil defence and restoration (SDR) projects, 26 live hedge planting projects, 26 windbreak planting projects, 18 village woodland projects, 36 natural regeneration projects, 17 river bank (*kori*) stabilization projects, 44 forest plantation projects, 13 bushfire prevention initiatives, 13 dune stabilization actions, 21 set-aside projects and 14 improved stove projects.

But it is essential to look beyond the statistics of technical interventions to consider how natural resources are governed and managed. In Niger it is common for natural resources in a particular space to be used for different, overlapping purposes, with various rights determining the diverse ways in which each resource may be exploited. These systems have to take account of many factors, including social changes whose depth and breadth vary from region to region. While some cultural elements remain stable, even apparently immutable social systems and institutions evolve over time and adapt to socio-economic change, as local communities constantly develop rules and adapt their production strategies to the problems they need to resolve.

Taking account of context. Niger's strategy links combating desertification with drought mitigation, in order to (i) prevent and reduce land degradation; (ii) rehabilitate partially degraded land; (iii) restore desertified land; (iv) strengthen drought early warning systems; (v) strengthen the mechanisms for preventing and managing drought situations; and (vi) put in place and strengthen food security systems.

In addition, Article 88 of Law No. 98-056 on environmental management states that protecting forests, livestock corridors and pastures helps combat the desertification caused by overexploitation, overgrazing, unauthorized clearances, fires, slash-and-burn and the introduction of inappropriate species. The survival of many rural households now hinges on the implementation of this directive as Niger faces the challenge of combating desertification in order to safeguard its future. Effective solutions are needed to tackle the serious constraints that desertification presents to development, to implement the national strategy for poverty reduction, and to improve people's living conditions.

This strategy now needs to address a number of major social, spatial, environmental, economic and financial issues that have arisen since it was first introduced. Efforts to combat desertification are no longer seen in isolation, but treated as part of the overall quest for sustainable development. Therefore, the drive to create favourable conditions to improve food security, resolve the domestic energy crisis and improve local health and economic development ultimately involved strengthening existing initiatives to reduce poverty.

---

[1]Undertaken by the Universities of Niamey and Amsterdam in collaboration with the CILSS (Comité Inter-Etats de Lutte contre la Sécheresse dans le Sahel) and other partners.

This realization marked a break with the old habit of regarding it as a purely technical issue requiring a technical solution. Combating desertification requires concerted efforts by different rural actors, and consensus around shared rules of management. This is particularly challenging in rural communities that have undergone profound social as well as physical changes in recent decades, and have seen their traditional social frameworks for managing and using land and its resources start to lose their relevance as productive logics and relationships, becoming increasingly individualized (Boubacar 2000).

Over the last four decades environmental strategies have been strongly influenced by the international and national context: at the international level by United Nations conferences on the Human Environment (at Stockholm in 1972) and on Environment and Development (at Rio de Janeiro in 1992); and at the national level by the formulation and subsequent adoption (in January, 2000) of the Strategic Framework for Poverty Reduction and the National Environmental Plan for Sustainable Development, as well as the promulgation of various texts relating to the environment, most notably the framework law on environmental management. In 1996, Niger set up the National Environmental Council for Sustainable Development to implement, monitor and evaluate the national environmental policy, which would be put into effect through a Poverty Reduction Strategy Programme (PRSP), Niger's National Environmental Plan for Sustainable Development. This was to include (i) strengthening grassroots participation; (ii) developing partnerships; (iii) improving the articulation and synergy between intervening agencies; and (iv) putting in place effective and sustainable funding mechanisms.

7.3 The Challenge of Land Resource Governance in Central Southern Niger

Land in central southern Niger used to be freely available, and agriculture and livestock rearing were well integrated until the 1970s, when ecological systems were disrupted by a combination of climatic, socio-economic and demographic factors. These imbalances (which sometimes resulted from agricultural practices intended to tackle the growing pressure on land from humans and livestock) were exacerbated by successive droughts, often with dramatic social consequences.

Agricultural transformation. The performance of most production systems in Niger largely depends on farmers' capacity to maintain the fertility of their naturally very fragile soils, and their ability to make the most of what little rain falls on their fields. Precipitation is low and highly variable in both spatial and temporal terms.

In the past producers could restore soil fertility by leaving land fallow for periods that varied according to how much land the household had at its disposal. Fallow served the dual purpose of restoring soil fertility and providing fodder for the livestock. Livestock produce manure, and soils were replenished through customary contracts that allowed transhumant herds to graze the crop residues on cultivated land.

Nowadays land saturation in densely populated areas has reached unprecedented levels, and demand for land is so high that crop rotation is impossible and fallow virtually non-existent. Plots have been extended wherever possible, often encroaching onto pastoral lands. This not only creates conflicts over land, but also reduces the quantity and quality of fodder resources, affecting farmers' ability to maintain and restore soil fertility. Yields have declined so much that many households now only produce enough cereal to cover their needs for 4–6 months of the year. Continuous cultivation has reduced soil fertility, and there is insufficient manure to compensate for the nutrients removed by cropping. This has disrupted the efficiency of the agricultural production system and upset its previously balanced relationship with the physical environment.

The severe droughts of the 1970s and 1980s are burned into the memories of producers all over the Sahel, and especially those in Niger. The death of so many trees and livestock left them extremely poor and vulnerable, eliminating most of their productive capital and contributing to the impoverishment of soils on agricultural land.

Woodfuel resources. The fact that rural communities in Niger are highly dependent on natural resources for their livelihoods has serious implications for the environment and people's capacity to produce and cook food. Data gathered by the environmental services show that the need for fuelwood is rising each year due to growing demand in urban and semi-urban areas. The statistics derived from sectoral consultations on the environment and anti-desertification initiatives show that every person in Niger requires 250–300 kg of wood to meet their annual fuel requirements. Given the low natural productivity of forested areas, this means that the country has already been eating into its resource capital for some time. Some people have been reduced to using crop residues and animal manure for fuel rather than as soil improvers, causing soils to degrade more quickly and reducing the productivity of their land.

Soil productivity. There is no doubt that the expansion of cultivated lands and high demand for fuelwood is impoverishing soils and undermining efforts to protect natural resources, forcing rural communities to find other means of ensuring that agricultural and livestock production systems can co-exist (such as aerial fodder (tree browse) and crop residues).

Producers identified farm size and the extent to which agriculture and livestock rearing are integrated as the two main differentiating factors between farms. Maintaining soil fertility is a key concern for most farmers, who rely on organic manure to maintain a minimum level of productivity. The fact that soils with a low organic matter content are less able to retain moisture has become increasingly problematic as rainfall has declined over the last few decades, making production systems less resilient and increasingly vulnerable to climatic hazards. Agricultural production can no longer fulfil the dual role of covering the food needs and facilitating the social and material reproduction of most rural households.

7.4 Transformation in Central Southern Niger

The critical situation just described has prompted numerous initiatives to tackle declining soil fertility, with joint efforts by the authorities, technicians and producers to find alternative solutions that can reverse the process of desertification and safeguard their survival. One of the more successful methods promoted by extension agents is assisted natural regeneration (RNA), which involves improved clearing techniques and maintaining and protecting the natural regeneration of certain species on farmland. Many producers in central southern Niger have adopted these practices in order to rehabilitate the environment and create more favourable ecological conditions in their locality.

Although it is generally agreed that deforestation leads to the decline and possibly irreversible loss of soil fertility, there is a need for a more nuanced understanding of the process that takes account of recovery periods. The evidence shows that reforestation can reverse the loss of tree cover and help restore soil fertility, and that the medium- and long-term resilience of natural ecosystems depends on certain types of human intervention that can enhance the recovery process in reforested or set-aside areas.

Despite the pessimistic assessment of natural resource management (NRM) in Niger, a series of studies undertaken in 2006[2] made the astonishing finding that despite a huge drive to plant trees in Niger over the past three decades, producers' efforts to protect natural regeneration in their fields far outperformed artificial plantations in terms of their coverage and impact (Bonkano 2006; Nassirou 2006; Boubacar and Nassirou 2010). Numerous reports on environmental degradation have overlooked some remarkably positive developments, despite drought years such as 2004/2005. For example:

– at least 250,000 ha of severely degraded land in the study area have been rehabilitated and regeneration is taking place over three million hectares. Satellite images show that a re-greening process is now well under way in areas affected by long-term wind and/or water erosion after being cleared for crops. This is a unique achievement, and is particularly significant in the densely populated areas of central and southern Niger (Reij et al. 2009);
– a significant increase in irrigated fields and the amount of land under flood-recession crops has resulted in a spectacular upsurge in market garden produce, which provides both food and incomes;
– after the ecological crisis in the 1970s and 1980s, producers started encouraging natural regeneration in their fields and developing more complex production systems that are less vulnerable to drought;
– investments in NRM by producers and the government and its technical and financial partners have had an impact on several Millennium Development Goals, particularly Objective 1 (Reduce extreme poverty and hunger) and Objective 7

[2]Undertaken by the Universities of Niamey and Amsterdam in collaboration with the CILSS (Comité Inter-Etats de Lutte contre la Sécheresse dans le Sahel) and other partners.

(Ensure environmental sustainability). People from villages that have invested in NRM report that household food security has improved since these initiatives began.

- to the uninitiated observer, producers have an astonishing capacity to adapt to ecological and economic crises. Their hard work and creativity in protecting and managing natural regeneration has not only increased their output, but also improved their environment. Various projects provided vital technical and/or material support in rehabilitating degraded land and extending the area under irrigated crops;
- from an economic viewpoint, it is demonstrably rational to invest in natural resource management in order to reduce rural poverty, improve the environment and accelerate economic growth.

7.5 Assisted Natural Regeneration (RNA)

Tree felling has been a major contributing factor to desertification in Niger as it exposes the already fragile soils to all kinds of erosion. The good news is that it is possible to halt this process by planting trees in degraded areas, and that most indigenous woody species in the Sahel have strong regenerative capacities, producing vigorous regrowth after cutting. Woodlands help enrich local ecosystems and improve the performance of local agrarian systems by enhancing soil fertility and providing firewood and timber.

In the early 1990s, local people started restoring woody vegetation that had virtually disappeared as a result of the droughts of 1973 and 1984 and the over-exploitation of resources (in the sense of exceeding the ecosystems' natural capacity to regenerate). With support from various projects and the state technical services, simple improved clearing techniques were used to encourage natural regeneration.

RNA is a simple, inexpensive technique that consists of protecting any regrowth that appears on cultivated land. It entails *not* uprooting, burning or cutting bushes or shrubs when preparing or hoeing fields where crops are grown, and managing regrowth at the base of woody vegetation that has been protected—in other words, allowing them to reproduce naturally from seed or the regrowth from stumps.

RNA did not become widespread, supported by government, until the late 1980s, when farmers were struggling to adapt their production systems to cope with the ecological crisis triggered by the latest drought. Having previously regarded trees as an exploitable common resource, farmers started retaining them in their fields to act as windbreaks, help stabilize the soil, provide fuelwood and timber and enable them to recapitalize their operations. The appropriation of trees started as an individual response to the local environmental conditions, the disappearance of common lands and increasing demographic pressure, and developed into a local dynamic that was subsequently taken up by extension agents as part of the drive to re-green Niger.

This began with a large agro-forestry initiative that was implemented as part of the national RNA strategy. Agro-forestry parklands covered 157,950 ha in Zinder Department (now Region) in 1990 (Energie 2 Project). By 2006 they covered an estimated one million hectares, dominated by *Faidherbia albida* (Larwanou and Adam 2006). The scale of these operations is clearly visible in satellite images, which can be used to monitor the increase in vegetative cover and changes in land use (but not to assess the quality of the vegetation).

In the past, farmers would remove all trees from their fields in order to stop their shade from hindering crop growth, deprive seed-eating birds of perches and to increase their social capital, as clearing large expanses of land demonstrated their capacity to work hard and mobilize a large labour force. They usually only retained trees whose fruit had some nutritional value, or which marked the edges of their fields. Nowadays, farmers use naturally occurring woody species to fulfil a range of objectives, selectively planting, cutting and pruning them in order to create mixed stands of young and mature trees in single fields or across several plots.

This new practice of using a certain density of trees to reconstitute and maintain the vegetative cover on cultivated land has created significant ligneous resources in several parts of Niger, especially in southern Maradi and Zinder, where record densities of nearly 150 different-sized trees per hectare have been observed. Such a high density is confined to *Faidherbia albida*, which is well known for its ability to improve soil fertility, enabling producers to double or even triple their output. In addition to this, it has a reverse seasonal cycle (leafing in the dry season) that does not impede crop growth (in the rains), produces pods that can be used as animal fodder in the hungry season, and provides timber. Surveys of 197 farmers/households conducted in 2010 in four villages in southern Zinder Region show that the practice is widespread (Fig. 7.1; Table 7.1). Species selection is largely determined by the availability of land. Farmers with little land (the majority) try to diversify the species on each plot in order to increase biodiversity, while those with large fields grow more *Faidherbia albida* in order to maintain and improve soil fertility—which is always a problem as organic and mineral fertilizers are in short supply.

Parcels with lower densities of *Faidherbia albida* tend to contain a wider range of species, especially in smaller and more vulnerable holdings where producers have to exploit every opportunity by allowing several types of tree to grow on limited land. Some farmers have reintroduced species that had previously disappeared, especially those that can be used for nutritional purposes. They can also use this more intensive form of production to increase their income by planting species with a high economic value, such as baobab (*Adansonia digitata*). This helps reduce household food insecurity even when rainfall is very low, as they can sell timber products and sub-products that are less vulnerable to climatic hazards.

Figure 7.2 shows satellite images of terroirs (village lands) north and south of the Niger-Nigeria border. The contrasting densities of tree canopies indicate the effect of farmers' differing priorities under their respective agricultural extension and regulatory regimes. The southern terroir represents a control case for estimating the impact of RNA in Niger.

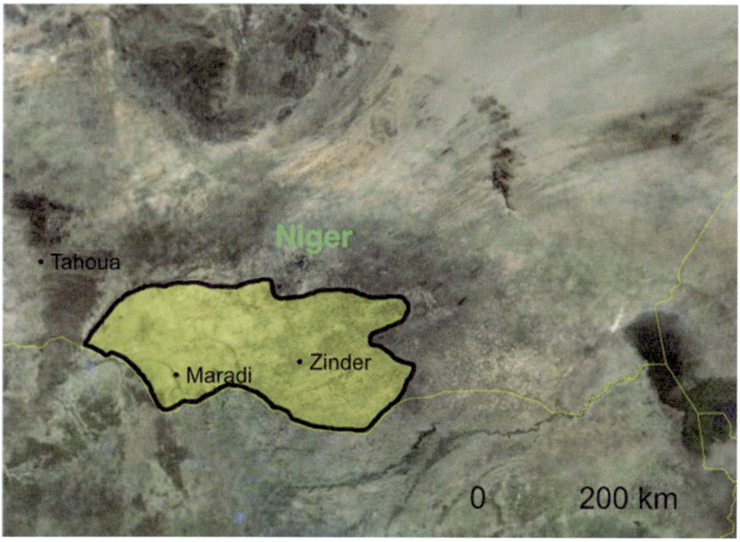

Fig. 7.1 Location of 12 terroirs used for comparing tree density across the Niger-Nigeria border.

Table 7.1 Density of *Faidherbia albida* (Gao) on village lands in the region of Zinder

Number of *Gao* per hectare	Number of respondents	Percentage
0	5	2.5
1–10	79	40.1
11–20	42	21.3
21–30	21	10.7
31–40	16	8.1
41–50	11	5.6
51 and over	23	11.7
Total	197	100.0

Source Boubacar and Nassirou (2010)

Producers have also developed various techniques to ensure that young plants in badly degraded areas have sufficient water, using anti-erosion measures to collect and minimize runoff and encourage rainwater absorption. Central southern Niger can be seen as a symbol of successful synergy between research and development operations that can help reverse desertification.

Figure 7.3 shows (albeit on different soils and at different times of the year) the contrasting landscapes at Illela in 1984 and in 2004, after two decades of tree protection on farmlands.

RNA fulfils multiple functions. In ecological terms, it protects the soil from wind erosion and prevents young plants from being dried out by the wind, traps particular organizms, reduces the risk of attacks by crop pests and helps reconstitute the soil

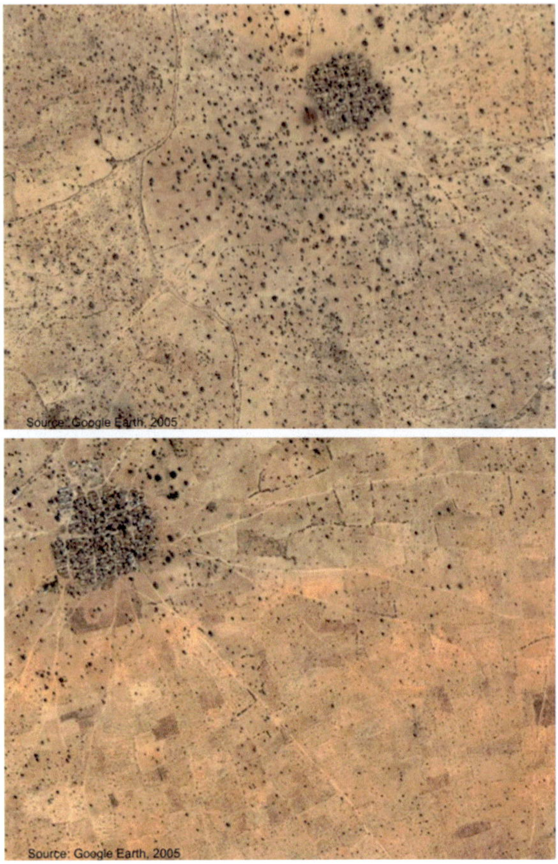

Fig. 7.2 Terroirs 8.5 km N (*upper*) and 1.5 km S (*lower*) of the Niger-Nigeria border.

organic matter content. It also changes the chemical composition of the soil as trees increase or maintain the soil fertility of cultivated plots, improve their carrying capacity, act as windbreaks, create a favourable micro-climate for crops, and enrich the plant composition and thus the biodiversity of village lands (Table 7.2).

In addition to improving food security, nutrition and human and animal health, RNA also has an economic impact as it helps households broaden their sources of income. Researchers in the Zinder region found that nearly 40 % of households use the fruit and leaves of certain trees as an additional source of food; 11 % rely on them in the hungry season; and nearly 39 % see them as a key element of their food security strategy, using the income generated to help cover their food needs throughout the year. All in all, nearly 90 % of the households surveyed depend directly or indirectly on RNA, either through direct consumption or by using the income it generates to buy foodstuffs.

While the primary motivation for RNA is to improve soil fertility (82 % of users), it also plays a very important role in improving household food security by increasing crop yields, providing products and sub-products for domestic

Fig. 7.3 Illela (near Birnin Konni on the Niger-Nigeria border) in 1984 (early rains—clay soils pre-planting) and 2004 (after the rains—sandy soils post-harvest).

consumption, and generating other sources of income (Table 7.2). Different parts of the tree are used as human and animal food complements or supplements, according to local habits and the vulnerability of the households concerned.

Women sometimes keep the pods, leaves or fruit of certain species (such as baobab) and sell them in the hungry season when prices are usually higher. This can provide an extra 10,000–50,000 francs CFA (approximately \$20–\$100 USD in 2010). Beekeeping is another potentially lucrative source of revenue. Farmers who were interviewed for one study (Boubacar and Nassirou 2010) reported that RNA generates enough money to enable them to cover certain costs and avoid having to sell their grain, which would make them more food insecure. They also save on their medical costs by treating themselves with products of the natural

Table 7.2 The intended objectives of RNA users

Intended objectives of RNA (most protected species)	Number of users	Percentage of users
Human consumption	17	8.6
Animal feed	7	3.6
Soil fertility	160	81.2
Fuelwood	19	9.6
Traditional medicine	6	3.0
Timber and lumber	13	6.6
Combating desertification	6	3.0
Sale (income generation)	9	4.6
Shade	5	2.5
Other	6	3.0

Source Boubacar and Nassirou (2010)

pharmacopoeia. All agree that RNA has become an essential part of their lives and they would be worse off without it. It plays a key role in sustaining and improving their production systems, enables them to engage in various economic activities (Table 7.3), and improves resilience to food insecurity by directly and indirectly contributing to the household economy.

Table 7.3 shows that most farmers (93.9 %) make well under 40,000 francs CFA (approximately $80 USD in 2010) from RNA, which represents about a third of an average farmer's annual income. Many households get through bad years by selling timber and non-timber forest products (NTFP), and RNA has become an important factor in economic differentiation. It enables producers to cope with the increasing scarcity of cultivable land (Tables 7.4 and 7.5).

Table 7.6 shows the importance of RNA within the wider framework of responses to food insecurity, cited by respondents in the study sites.

Table 7.3 Total revenues generated from woody species by RNA users

Francs CFA	Number of users	Percentage of users
0–19,999	172	87.3
20,000–39,999	13	6.6
40,000–59,999	4	2.0
60,000–79,999	4	2.0
80,000–99,999	3	1.5
100,000 and above	1	0.5
Total	197	100.0

Note $1.00 US dollar equalled approximately 500 francs CFA in 2010
Source Boubacar and Nassirou (2010)

Table 7.4 The role of RNA in a year of normal rainfall

Role of RNA	Number of respondents	Percentage
No response	4	2.0
Additional source of food	78	39.6
Hungry season food	21	10.7
Food security	76	38.6
Meet other needs	18	9.1
Total	197	100.0

Source Boubacar and Nassirou (2010)

Table 7.5 Impact of RNA on household living standards following poor rains

Estimated improvement in living standards (percent)	Number of respondents	Percentage
No response	27	13.7
10–20	17	8.6
21–30	37	18.8
31–40	73	37.1
41–50	41	20.8
51 and over	2	1.0
Total	197	100.0

Source Boubacar and Nassirou (2010)

Table 7.6 Solutions for coping with cereal shortages

Solutions	Number of respondents	Percentage
No reply	22	11.2
Use RNA products	66	33.5
Leave the area	34	17.3
Income-generating activities	57	28.9
Reduce the number of livestock (sale)	18	9.1
Total	197	100.0

Source Boubacar and Nassirou (2010)

Livestock rearing practices have shifted away from extensive grazing as rangeland has disappeared. Many farmers keep their livestock in pens for much of the year, supplementing cut and carried herbage with the seeds, pods, branches, leaves and shoots of naturally regenerating trees. Some 98.5 % of respondents cited this practice. This 'aerial fodder' is important in maintaining the nutritional quality of their diet before the grass develops and during critical hungry periods. It is supplemented with crop residues after the harvest. A growing number of small ruminants are tethered and reared on ANR products all year round. The good condition of the animals observed in the study sites is a testimony to this intensive feeding system.

People rely on livestock to help reduce their vulnerability, and to maintain them in good condition they protect certain species in order to compensate for shortages of crop residues and grass. They also note that the herbaceous plants they use for fodder during the rainy season grow better near trees.

These plants become increasingly scarce as the dry season progresses, making livestock vulnerable to malnutrition. This is when the fruits and leaves of species such as *Guiera senegalensis, Piliostigma reticulatum* and *Faidherbia albida*, which is one of the very few trees in the Sahel whose leaves stay green during the dry season, come into their own. In addition to enabling herders to get through the hungry season when plant biomass disappears, these trees also play a strategic role in transhumance, creating favourable conditions for stopping-off points in the herds' seasonal movements. Overall, RNA is reported to have reduced the distances that cattle cover on transhumance.

7.6 Assisted Natural Regeneration (RNA) in Sustainable Land Management (SLM) and Development

Niger's environment and sustainable development policy aims to create the favourable conditions needed to improve food security, resolve the domestic fuel crisis and improve local sanitation and economic development. Much of the current thinking on combating desertification focuses on sustainable land management (SLM). This willingness to combine theory with practice represents a significant policy development, since treating RNA as a key component of SLM can enable households to turn their situation round from being in a structural deficit to producing more cereal—although this is still extremely difficult for the very poor in areas where land is increasingly fragmented.

The Government of Niger and its partners agreed on a roadmap for SLM, which was developed on the basis of several in-country missions undertaken by the TerrAfrica platform and its partners (www.terrafrica.org) at the government's request. The two main activities identified are (i) setting up a national SLM committee, and (ii) defining a national vision and strategy for implementing the SLM agenda, with a strategic framework for investment that identifies priorities, strengthens the strategy, and defines mechanisms for coordination, consultation and evaluation.

Niger has gained a good deal of experience in rehabilitating agro-sylvo-pastoral lands in recent decades. Some communities have significantly increased their available land. Yet the SLM activities have only benefited 8 % of the villages in Niger, mainly in the regions of Dosso, Tillaberi and Tahoua (Pender and Ndjeunga 2007).

The most commonly used land management practices are *tassa* or *zaï* (planting pits), semi-circular bunds, stone dikes, tree planting and mulching. These techniques are most widespread in project and programme intervention areas, especially among households that have been given technical training and support. Comparative results from identical plots show that using *zaï* can increase agricultural production by an average of 24 %. Composting and plain or enclosed ridging are also simple, inexpensive methods of managing sandy soils, but their use is highly localized. More needs to be done to disseminate these effective practices and extend them beyond a limited number of villages or producer groups.

The many agricultural development programmes implemented over the last decade have had mixed results. This is often because they did not target the poorest and most vulnerable sectors of the population, improve the technical conditions for production, or create a strong enough dynamic to encourage beneficiary communities to take charge of their own development. In order to improve uptake of research-driven technologies, a collaborative agro-forestry research programme, run in parallel with development operations, is testing participatory techniques that build on grassroot agro-forestry initiatives. This has proved successful, and has involved far more rural people in RNA practices that help restore and protect the productive potential of the land.

The spontaneous uptake of RNA referred to earlier (Larwanou and Adam 2006), even though no funded project has ever involved plantations on this scale, can be incorporated into agro-forestry systems, partly due to changing attitudes and partly to a change in the status of naturally regenerated woody vegetation, which is now appropriated for exclusive private use (CT/PIIP 2003). The increasingly private nature of RNA initiatives is shown by holding inter-village meetings, setting up surveillance brigades to protect trees and introducing rules to sanction offenders. In order to make better use of their natural spaces and build on their investments, local people have rehabilitated livestock corridors, extended the mandate of existing structures to include the management of sylvo-pastoral areas, and invited their customary and administrative authorities to local meetings of grassroot institutions in order to give them a certain legitimacy.

One of the most innovative aspects of this process has been the creation of a local institutional base as a catalyst for organized action, rather than as a mechanism for distributing the benefits of State- or project- managed improvements. This marks a shift from the logic of 'benefit-sharing', long used by central governments to obtain local people's cooperation, to 'power sharing' that aims to transfer responsibility to local communities.

This approach appeals to local people who no longer want to be treated as the passive recipients of project assistance. It uses a flexible procedure with no fixed guidelines, major financial costs or fixed quantitative objectives, which local actors can adapt and shape to their particular circumstances. The effectiveness of the strategy is largely due to the emphasis on making the best use of existing resources and local creativity, rather than identifying problems and trying to use pre-existing solutions to resolve them. This marks a fundamental change in both the technical and socio-organizational aspects of the approach, which encourages villagers to ask why and how actions are being taken and how best to organize themselves.

This dynamic of experimentation and reflection is conducive to development in the sense that it is not simply a matter of improving difficult situations, but also of making changes at the individual level, in people's capacities, attitudes and collective and individual behaviour.

Rather than being carried out in isolation, efforts to combat desertification are seen as part of the overall drive to reduce poverty. Niger has one of the lowest Human Development Index ratings in the world, and its poor economic performance is compounded by a chronic food deficit. The official statistics indicate that nearly 20 % of the rural population are in a critical food situation in normal (i.e. drought-free) years. RNA techniques enable people to use the land in ways that help combat desertification and manage their natural resources sustainably. As part of its decentralization policy and in order to promote sustainability, the government is giving communes the right to manage natural resources at the local level. Various local planning tools have been developed to help them do this, such as Communal Development Plans. These include assessments by elected local officials, which consistently identify environmental degradation as a major problem—yet environmental issues are rarely prioritized because mayors tend to focus on developing infrastructures in order to keep voters onside. There have been cases where State land has been allocated to city dwellers who are more concerned about production levels than about tackling desertification.

7.7 New Issues and Challenges in Combating Desertification

How does this environmental dynamic fit into current efforts to find the best way of managing natural resources? Although RNA has flourished in central southern Niger, the major socio-economic changes that have taken place in this part of the country in recent years have created a new set of challenges. There has certainly been an overall improvement in natural resource management practice, but the prospects for the future will be affected by a range of positive and negative factors, not least the capacity for conceptual planning and practical and financial forecasting.

Increasing individualization in production systems has triggered a crisis in family models, and created households whose members may have very different economic objectives and trajectories, and greater freedom to pursue their own activities and initiatives. The desire to "retain social networks and their key functions of social protection and facilitating access to resources, while trying to reshape them to accommodate the constant trade-off between personal projects and social obligations" (Réseau ENDA 2005) inevitably creates tension. Increased opportunities for trade and growing interdependence with other actors have altered people's frames of reference, as they engage in joint projects and develop functional relations that often transcend family structures, which gradually become less relevant and effective.

New forms of local, regional and international interdependence have emerged with the advent of new structures, such as agricultural supply chains, the NEPAD environmental programme (New Partnership for Africa's Development), UMOA (West African Monetary Union) and the Integrated Ecosystem Management of Shared Catchments between Nigeria and Niger (NNJC 2006). Rural communities are also having to adapt to increasing appropriation of natural resources, and to think in broader terms of common land that requires different management structures and rules, and new forms of consensus around locally shared land and resources.

Working on the basis that the fight for development can only succeed if beneficiaries are actively involved in the process, Niger started a process of administrative reform to improve its social and economic development programmes. The chosen way forward was decentralization, through consultation and participation, taking account of local needs and giving communes the power to consult with stakeholders and make decisions about the environment and natural resource management.

This reconfiguration of public spaces creates a new set of problems and challenges, as the current legislative and regulatory arrangements do not take full account of the state powers that will be transferred to local governments, or of their land-related competences in the decentralization process. Article 12 of Law No. 2002-013 of 11 June 2002, regarding the transfer of powers to the regions, departments and communes, specifies the environmental and NRM responsibilities that are to be transferred to the communes, namely: (i) environmental protection and conservation; (ii) preparing local environmental and NRM action plans (in accordance with departmental plans); and (iii) assessing and authorizing all developments in the commune that are classified as dangerous, unhealthy or inconvenient, and continuing previous initiatives to set up monitoring committees to tackle bush fires.

Most new communes are unclear about the exact extent of the land and land management powers that have been transferred to them. Reflections at the national level, led by the Permanent Secretariat for the Rural Code in 2007, identified: (1) the need for land law specifically designed to deal with decentralization; (2) a need to establish an agreed system of land management; and, above all, (3) a need to define and put to use the modalities of land development. National and local land administrations are now in place, and there is growing consensus over the need for land use plans as a tool for sustainable land management, as well as tools to regulate the productive use of resources and security of tenure.

7.8 Conclusion: The Need to Build on Experience in Niger

The diverse experiences and ongoing constraints to environmental management in Niger make it more important than ever to develop multi-faceted strategies that will help safeguard what has been achieved, while redefining environmental policies in general and efforts to combat desertification in particular. These studies show the need to invest in environmental conservation and management as a sound basis for development.

This will involve rethinking the underlying procedure for actions in certain regions of Niger, to ensure that environmental management policies value and build on local practices and contribute to rural development. Successful policies require clear guidelines, appropriate technical choices and, crucially, popular support. It is worth noting that the most innovative feature of the highly effective RNA strategy was the creation of a local institutional base as a catalyst for organized actions and a mechanism for transferring responsibilities to local communities.

The results so far suggest that anti-desertification policies should focus on encouraging farmers to promote natural regeneration in their fields as this is cheaper and more effective than planting trees. However, it is important to recognize that while RNA techniques have proved highly effective, they will be of limited value in addressing the worrying demographic trends in Niger. The latest data from the National Institute for Statistics show that the population is increasing by 3.9 % each year, and will double in less than 20 years if it continues at this rate. In this context, it is legitimate to ask how much room for manoeuvre local people will have, even with RNA. It is true that this practice has helped increase the productivity of agricultural and pastoral systems, but it cannot be relied upon as the sole solution to their problems. There is an urgent need to significantly increase investment in agriculture, livestock rearing and NRM in order to create the conditions to tackle desertification effectively. Ultimately, this can only be done by accelerating economic growth, reducing rural poverty, sustainably increasing the productivity of natural resources and reducing living costs, finally winning the fight against desertification.

References

Bonkano, BH. (2006). *Etude de la régénération naturelle dans les champs cultivés: cas du village de Dan Gamdji*. Mémoire Ingénieur des Techniques Agricoles. Niamey: Université Abdou Moumouni.

Boubacar, B. (2000). *Evolution des régimes de propriété et d'utilisation des ressources naturelles dans la région de Maradi*. Drylands Research Working Paper 29. Crewkerne, UK: Drylands Research. (www.drylandsresearch.org.uk)

Boubacar, B., & Nassirou, S. M. (2010). *L'étude de la régénération naturelle dans le sud de la région de Zinder*. Niamey: Université de Niamey/Université d'Amsterdam.

Code Rural Niger. (2008). *Rapport Général sur l'Atelier National des Commissions Foncières (ANACO V), Niamey du 1er au 3 juillet 2008*. Niamey: Secrétariat Permanant du Code Rural Niger.

CT/PIIP. (2003). *Expérience du departement d'Aguié (Niger) en matière de protection et de gestion des ressources naturelles*: Contribution de la CT/PIIP à l'atelier sur l'impact de la reglémentation forestière sur les systèmes forestiers, Ségou (Mali) du 8 au 10 décembre 2003. Aguié: CT/PIIP

Guengant, J. P., & Banoin, M. (2003). *Dynamique des populations, disponibilités en terres et adaptation des régimes fonciers: Le Niger*. Rome: UN Food and Agriculture Organisation.

Hamadou, S. (2000). *Evolution à long terme des productions agricoles, du système de commercialisation et des prix des produits dans la zone de Maradi*. Drylands Research Working Paper 32. Crewkerne, UK: Drylands Research (www.drylandsresearch.org.uk)

Larwanou, M., & Adam, T. (2006). *Impacts de la régénération naturelle assistée au Niger- Etudes de quelques cas dans les régions de Maradi et Zinder*. Niamey/London: Centre Régional d'Etude spécialisée en Agriculture Niamey/International Institute for Environment and Development

Nassirou, M. S. (2006). *Impacts de la régénération naturelle assistée des ligneux sur la réduction de la vulnérabilité des ménages: cas du terroir de Kirou Haussa dans la commune urbaine de Matameye (région de Zinder)*. Mémoire DEA: Université de Niamey.

NNJC. (2006). *Integrated Ecosystem Management of Shared Catchments between Nigeria and Niger*. EGEF Council Document, MFA Regional Annex. Niamey: Niger-Nigeria Joint Commission for Development

Noppen, D., Kerkhof, P., & Hesse, C. (2004). *Rural fuelwood markets in Niger: An assessment of Danish support to the Niger household energy strategy 1989–2003*. Copenhagen: Danida.

Pender, J., & Ndjeunga, J. (2007). *Evaluation des impacts des programmes de gestion durable des terres sur la Gestion des Terres et la Pauvreté au Niger*. Washington DC: TerrAfrica (World Bank).

Pontié, G., & Gaud, M. (1992). *L'environnement en Afrique*. Afrique Contemporaine, Trimestriel n°161 janvier-mars 1992

Reij, C., Tappan, G., & Smale, M. (2009). *Agroenvironmental transformation in the Sahel; another kind of "Green Revolution"*. IFPRI Discussion Paper 00914. Washington, DC: International Food Policy Research Institute

République du Niger. (1984). *Le débat National sur la Lutte contre la Désertification (Maradi, 1984)*. Niamey: Ministère de l'environnement.

Réseau ENDA Graf Sahel. (2005). *Changement politique et social: Eléments pour la pensée et l'action*. Dakar: ENDA intermonde.

Thomas, D., & Middleton, N. (1992). *World atlas of desertification*. London/Nairobi: Edward Arnold for United Nations Environment Programme.

World Bank. (1988). *Programme Conjoint PNUD/Banque Mondiale d'Assistance à la Gestion du Secteur de l'Energie Rapport d'Activité No 082/88*. Washington DC: The World Bank.

Annex: Additional Sources

Adam, T., & Reij, C. (2006). *Impacts des investissements dans la GRN: Etude Sahélienne*. Niamey/Amsterdam: Université de Niamey/Free University of Amsterdam.

Adamou, D. (1998). *Contribution à l'étude des potentialités et contraintes d'exploitation des ressources forestières de la vallée de Goulbi N'Kaba dans le canton de Kanan Bakaché (Mayahi). Mémoire Ingénieur des Techniques Agricoles*. Niamey: Eaux et Forêts/Université Abdou Moumouni.

Amoukou, A. I., Marichatou, H., Yamba, B., & Yayé, A. D. (2007). *Vulnérabilité et innovations paysannes: L'expérience d'Aguié au Niger*. Louvain/Niamey: Presse Universitaire de Louvain/Université Abdou Moumouni.

Banoin, M., Aristid, J., & Jouve, P. (1998). Le défrichement amélioré au Sahel. Une pratique agro forestière adoptée par les paysans. *Bois et forêt des tropiques N° 225*(1), 31–43

Boubacar, O. A. (1995). *Caractérisation et gestion des ressources naturelles dans la zone d'intervention du projet de développement rural—Mayahi*. Niamey: Université Abdou Moumouni.

Ciblé (PIC). (2007). Consolider les relations entre recherche universitaire et opérations de développement en renforçant les synergies entre savoirs scientifiques et savoirs paysans. *Projet Interuniversitaire Ciblé, 2B*, 102–220.

Dadi, C. (2000). *La protection de la régénération naturelle dans la zone du développement communautaire de Baban Rafi: Pratiques paysannes et impacts socio économiques. Mémoire Centre Régional d'Etude spécialisée en Agriculture*. Niamey: Université Abdou Moumouni.

De Miranda, E. (1979). *Etude des déséquilibres écologiques et agricoles d'une région tropicale semi-aride au Niger: le problème de l'utilisation des ressources naturelles dans trois villages haoussa*. Bordeaux: DGRST/ACC, Université de Bordeaux II

Direction Départementale de l'Agriculture Niger. (2009). *Rapport d'activité agricole 2008*. Niamey: Direction départementale de l'Agriculture.

Djiwa, O. (2003). *Pratique agricoles et gestion de l'arbre dans les Parcs agro forestières en zone ouest et la région des savanes au Togo*. Mémoire du Centre Régional d'Etude spécialisée en Agriculture. Niamey: Université Abdou Moumouni

Hamadou, G. I. (2005). *Etude de la filière des ressources du doum (Hyphaene thebaica) dans la perspective de mise en place d'un marché dans la grappe de El-Geéza (département d'Aguié). Mémoire Ingénieur des Techniques Agricoles/Protection de Végétaux*. Niamey: Université Abdou Moumouni.

Hamissou, A. (2005). *Etude de faisabilité technique, économique et organisationnelle d'un marché de bois issu de la RN dans la grappe de Dan Saga à Aguié. Mémoire Centre Régional d'Etude spécialisée en Agriculture*. Niamey: Université Abdou Moumouni.

Laouali, A. (1998). *Etude de la croissance des ressources Forestières dans le cadre de l'aménagement intégral du territoire (UTE de Karkamatt). Centre Régional d'Etude spécialisée en Agriculture*. Niamey: Université Abdou Moumouni.

Larwanou, M., Abdoulaye, M., & Reij, C. (2009). *Etude de la régénération naturelle assistée dans la région de Zinder (Niger): Revue*. Washington, DC: United States Agency for International Development.

Malam Kime, M. (2000). *Impact de la régénération naturelle sur la protection de l'environnement et la production agricole dans la zone d'intervention du projet PRIVAT. Mémoire Centre Régional d'Etude spécialisée en Agriculture*. Niamey: Université Abdou Moumouni.

N'Dao, S. (2001). *Pratique Agricoles et gestion de l'arbre dans les Parcs Agro forestier en milieu soudano sahélien (Sine- Saloum, Sénégal). Centre Régional d'Etude spécialisée en Agriculture*. Niamey: Université Abdou Moumouni.

N'Garbaroum, T. (1994). *Evaluation de l'effet de Faidherbia albida sur les sols: Les incidences pour les cultures des zones tropicales semi- aride (cas du Mil, Sorgho, et du Mais)*. Centre Régional d'Etude spécialisée en Agriculture. Niamey: Université Abdou Moumouni

PDRAA. (2000). *Evaluation interne finale du PDRAA: la régénération des ressources naturelles, base de production agropastorale*. Aguié, Niger: Projet de Développement Rural de l'Arrondissement d'Aguié.

PNUE. (2003). *Consultation sectorielle sur l'environnement et la lutte contre la désertification*. Rapport final. Niamey: Ministère de l'Environnement Niger/Programme des Nations Unies pour l'Environnement

Raynaut, C. (1980). *Recherches multidisciplinaires sur la région de Maradi*. Rapport de synthèse: Lutte contre l'aridité en milieu tropical. Bordeaux: DGRST/ACC, Université de Bordeaux II

Raynaut, C. (Ed.). (1997). *Sahels. Diversité et dynamiques des sociétés-nature*. Paris: Kartala

Salissou, A. (2004). *Valorisation des produits et sous produits ligneux dans la partie Nord du département d'Aguié: Cas du terroir villageois de Dan Saga*. Niamey: Université Abdou Moumouni. Mémoire Ingénieur des Techniques Agricoles/Faculté d'agronomie.

Soko, D. (2003). *Caractérisation de la biodiversité agro forestière dans les terroir de Dakala et Boussourou, Segou, Mali. Centre Régional d'Etude spécialisée en Agriculture*. Niamey: Université Abdou Moumouni.

Boubacar, B. (1993). *Ressources ligneuses et problèmes d'aménagement forestier dans la zone agricole du Niger*. Thèse de doctorat, Tome I, Université Michel Montaigne, Bordeaux.

Part II
Global Issues

Chapter 8
Deserts and Drylands Before the Age of Desertification

Diana K. Davis

Abstract This chapter analyses the deep and complex history of western thinking about deserts and desertification in the centuries before the word desertification was coined in 1927 by a French colonial forester. It shows that a relatively benign view of the drylands dominated in western thinking until the early colonial period when notions of deforestation causing desiccation began to take shape. During the period of 19th century colonialism, particularly British and French colonialism in Africa, deserts, drylands, and their degradation became a particular focus of colonial scientific research as well as practical policy formulation. It was during this period that indigenous peoples, primarily but not only nomads, were blamed most often for what later came to be called desertification. French colonial experiences, first in North Africa, and later in West Africa were especially influential in the formation of much of our contemporary mainstream understandings of desertification, and thus our management of the drylands, today. A great deal of our thinking about drylands, as well as many policies for developing them, derive from the colonial period and were carried into the contemporary mainstream in large part by several UN agencies. The chapter concludes by suggesting that desertification is a (neo)colonial concept that would benefit from careful reconsideration.

Keywords Desiccation · Colonialism · Development · Nomads · Environmental history

The word desertification (désertification) was first used by the French colonial forester Louis Lavauden in 1927, and has since been associated strongly with notions of spreading deserts and of human culpability for deforestation, overgrazing and arid land degradation. Lavauden, following an old Anglo-European tradition of thinking about arid lands and nomads, wrote that "desertification is uniquely the act of humans…[T]he nomad has created what we call the pseudo-desert zone"

D.K. Davis (✉)
UC Davis, Davis, CA, USA
e-mail: geovet@ucdavis.edu

R.H. Behnke and M. Mortimore (eds.), *The End of Desertification?*
Springer Earth System Sciences, DOI 10.1007/978-3-642-16014-1_8

(Lavauden 1927, p. 267). From its very first use, then, the word desertification has been politically and ideologically laden with many preconceptions and misconceptions that date back long before the "Age of Desertification" during the 20th century. This chapter traces this long history of thinking about desertification in the centuries before the word desertification was coined and demonstrates the key importance of the colonial period, especially of French colonial experiences in Africa, for the formation of many of the ideas that buttress the mainstream concept of desertification today. In doing so, it suggests that this long history of thinking about deserts and indigenous peoples contributes significantly to the tenacity of the perception of a "crisis of desertification" despite the convincing evidence to the contrary in much of the world.

8.1 Desert Wastes Take Shape: Exploration, Desiccation and Improvement

Most deserts including the Sahara, the Namib, the Kalahari and the Gobi have been where they are today for about 65 million years, shrinking and expanding with global climate conditions (Goudie 1986). They were largely formed in the current locations and extent by the time Greek civilization began to produce geographical and historical works.

From the classical period to the middle ages, deserts were not conceived as ruined or deforested spaces, nor regions where crops should be grown. That is, deserts were not considered the degraded or undeveloped wastes that they have been for the last couple of centuries. From antiquity to the age of European exploration, deserts and their edges were primarily considered by Europeans to be naturally dry, sometimes sandy features that presented difficulties for traveling and were thought to produce strange and exotic people like the Ethiopians who were believed to be "scorched" by the torrid climate and its burning sun. Although parts of the Bible identified the desert as the home of the devil, many early Christians held the desert in Egypt sacred as the site of ascetic perfection where the Desert Saints went to prove their piety. For some, then, the desert represented a heaven on earth. For travelers, like Marco Polo, deserts such as those in Central Asia, were strange and wonderful, but also sometimes home to threatening peoples and frightening spirits.[1]

It is after the so-called "Columbian Exchange" that Western perceptions and understandings of deserts and drylands begin to change in significant ways. The Age of Exploration, when Europeans "discovered" the New World, brought new knowledge of very different environments, especially the tropics and tropical islands. The 16th century witnessed the development of early capitalism and

[1]See, for example, Hugh Murray, *The Travels of Marco Polo* (New York: Harper & Brothers, Publishers, 1858). For more details and further analysis of the history of western thinking about and action in the drylands, see (Davis forthcoming).

colonialism, and also the beginnings of an articulation of desiccation theory which forged the long-lasting link between deserts and forests. Desiccation theory included the idea that deforestation causes the climate to dry out and diminishes rainfall and also, the corollary, that reforestation restores lost rainfall and creates a generally more humid and healthful atmosphere (Grove 1995).

The earliest scientific work on desiccation theory, like that of Edmund Halley, John Woodward, and Stephen Hales, primarily focused on the beneficial effects of deforestation and desiccation.[2] This late 17th and early 18th century work tended to highlight the benefits to health and agriculture of clearing forests in cool, humid, continental climates like eastern North America. It did not mention any dangers of deforestation causing desert-like conditions. By the mid-18th century, however, this began to change and by the 1760s and 1770s connections began to be drawn by colonial officials like Pierre Poivre and Joseph Banks between deforestation, apparent desiccation, and the creation of desert-like conditions, particularly in tropical island settings such as Mauritius and St. Helena. Poivre, a physiocrat, was named Commissaire-Intendant of Mauritius in 1766 and his work and writings on this island have been very influential (Grove 1995). Shortly after his arrival, he outlined his plans for reforestation and for creating forest reserves in order to address the problems he perceived of deforestation and desiccation. He warned that if the deforestation were not stopped, in just a few years Mauritius "will not be habitable any longer; it would have to be abandoned" (Poivre 1797, p. 211).

European experiences in the 17th and 18th centuries mostly of colonial plantation agriculture in tropical island settings including the Caribbean, the South Pacific, and the Indian Ocean, had revealed serious consequences in the wake of deforestation such as soil erosion (Grove 1995, 1997). As Grove has argued, the image of the tropical island with its lush vegetation and association with the Garden of Eden, came to represent an idealized landscape during the 18th century with very positive moral connotations. The fear that anthropogenic deforestation and overgrazing was causing desiccation in such settings therefore took on an urgency it might not have otherwise. Those concerned with the effects of deforestation were worried not only about desiccation and erosion but also, and perhaps more significantly, about these Edenic islands becoming worthless desert wastes.[3] In the words of Joseph Banks, who visited some of these tropical islands in 1771, "the paradise" of St. Helena had been allowed, through deforestation and overgrazing, to "become a desert" (Grove 1997, p. 52). After the late eighteenth century, desiccation theory became dominant in Europe, especially in Britain and France.

During the 18th century and into the 19th, another transformation took place that has direct bearing on Western notions of drylands and their value. With the further development of capitalism and the rise of economic liberalism and the related

---

[2]See John Woodward, "Some Thoughts and Experiments Concerning Vegetation," *Philosophical Transactions* 21, no. 1 (1699): 193–227, for example.

[3]About this time, deserts came to be more strongly associated with "evil" just as forests were being strongly associated with "good" (Grove 1997).

influence of the agricultural improvement movement, the meaning and use of commons and other "waste land" changed in profound ways. Before the late 18th century, these common lands, governed by a variety of customary land use rules, were primarily perceived as "a productive remainder," as land that was not being used for direct cultivation but that was still valuable in a number of different ways to many different people (Goldstein 2013). By the end of the 18th century, conceptions of agricultural improvement had progressed to such a degree, influenced strongly by the rise of liberal economic thought during the century, that by the beginning of the 19th century, waste or common land was no longer perceived as the "productive remainder," or uncultivated but still valuable land. Rather common and waste land came to be widely understood as *wasted* land that had the potential to be profitable but that was either not being used efficiently or was actually being degraded by improper use or overgrazing, etc. (Goldstein 2013).

The notion that all land, even common or waste land, held potential agricultural value bolstered the enclosure movement already well underway in England and parts of Europe. Enclosure, the process of removing common rights to land, essentially privatizing it, was strengthened by this transformation because it helped to make the argument that "the entire edifice of common right was itself a wastage of the improvers' economic right—presented as a natural [divine] right—to realize the maximum productive potential of all things, at all times, and in all ways" (Goldstein 2013, p. 12).

In France, despite the Physiocrat movement which advocated agricultural improvement, enclosure of waste land took much longer by nearly a century than it had in England for complicated reasons. One of these was that due to the system of agriculture and land tenure in France at this time, the rights of common pasture (vaine pâture) and right of way (parcours) strongly shaped agricultural production and prevented enclosure on any significant scale in most of the country (Rozental 1956). For this reason, the French improvers tried repeatedly to enact legislation to curtail and abolish common pasturage and right of way. Opposition was intense from many quarters and no success was achieved until 1889 when right of way, but not common pasturage, was finally abolished (Rozental 1956, p. 70). The long battle of the improvers led to the development of very negative perceptions of grazing livestock in many sectors of French society (Davis 2007; Whited 2000). Accusations of environmental degradation, overgrazing, deforestation, torrents and desiccation grew increasingly common in France throughout the late 18th and through the 19th century.

These changing conceptions of waste land, commons, and herding/grazing influenced the ways in which British and French colonial agents interpreted and understood arid regions and pastoral peoples in their overseas territories and thus shaped colonial agricultural and environmental policies for development (improvement) in significant ways during the 19th century. For the agricultural improvers, deserts were the ultimate wastes and the drylands and their inhabitants needed the most reform and improvement. France was one of the earliest colonial powers to put these ideas into action at a significant scale.

8.2 The Desert Within: France and Arid Lands at Home and Abroad

Of the three most important Western colonial powers active in the drylands of Africa and Asia, England, France and Germany, only France contained semi-arid regions. Thus, only France had any experience with trying to manage semi-arid, dryland environments, primarily in the southern, Mediterranean region of the country, but also along parts of the Atlantic coast. This likely has much to do with the fact that the French also have one of the longest histories of thinking and writing about arid lands and their presumed ruination. Influenced by the still powerful ideas of environmental determinism, for example, Charles Louis de Secondat, the Baron de Montesquieu, wrote about "Oriental despotism" in the arid lands of what we now call the Middle East in his influential 1748 *Spirit of the Laws*. In this volume, though, he also states that the nomadic Tartars, a "nation of herdsmen and shepherds, … have destroyed Asia from India even to the Mediterranean; and all the country which forms the east of Persia they have rendered a desert" (Montesquieu 1949 [1748], pp. 279–280).[4] This is one of the earliest iterations of what would become common wisdom in France regarding arid lands and the impact of nomads over the following two centuries.

Not long after Montesquieu condemned the Tartars for rendering much of SW Asia a desert, Voltaire pronounced the French Landes region comparable to the Sinai desert in his 1769 *Dictionnaire philosophique* (Traimond 1986, p. 221). This is the sandy coastal region in southwestern France known as the Landes de Gascogne (the wastelands of Gascony). It was commonly believed during the 18th century and later that the area had been deforested due to fire and overgrazing.[5] With roughly 400 square miles of littoral producing mobile sand dunes that had engulfed fields and villages, it was widely feared from the late 18th century that France itself would be invaded and turned to desert by the menacing dunes of the Landes. It was also frequently compared to the desert of Arabia and to the Sahara desert, and called "the Sahara of France" during the early 19th century (Traimond 1986). Important for long-term French thinking about how to deal with sand dunes and spreading deserts, the Landes dunes were stabilized with massive plantings of maritime pines, a project begun under the direction of Nicolas Bremontier in 1787 (Blanchard 1926). Within a century, the dunes were stabilized and the Landes had been successfully transformed into a 2,000,000 acre pinery and become a major producer of turpentine.

Notions of deforestation, threatening sand dunes and deserts had therefore been combined in the French mind since the late 18th century with the home-grown example of the Landes. Such thinking began to be combined with notions of

[4] By Tartars he is referring to the Turkic and Mongol peoples who inhabited central and northern Asia.

[5] More recent research indicates, however, that there may have been a severe freezing period in the early 18th century that lasted for 5 months and killed up to 80,000 ha of forest. See Traimond (1986, p. 222).

deforestation and desiccation during the late 18th century. This is reflected in the influential writings of George Louis Leclerc, le Comte de Buffon. Buffon believed that the climate could be modified by humans, especially that it could be warmed and dried by clearing forests. Although Buffon noted that it was easier to warm the climate than to cool it by human action, he believed that deserts could be transformed, could be cooled, humidified, and made more habitable. He wrote in 1778, for example, that in Arabia "a single forest [planted] in the middle of these burning deserts would be adequate to temper them, by leading the rain to them, [and] give to the earth all the elements of its fertility" (Buffon 1865, p. 100). Buffon thus brought together key elements of desiccation theory and the notion of improvement and applied them directly to what he considered emblematic desert wastes.

Within a decade, the romantic writer and utopianist, Jacques Henri Bernardin de Saint Pierre had published his influential *Studies of Nature* (1784–88). Bernardin de Saint Pierre had worked closely with Pierre Poivre on Mauritius for several years. Much of Poivre's desiccationist thinking is strongly reflected in Bernardin de St. Pierre's works, especially ideas of deforestation and desiccation leading to desert-like conditions. In the closing years of the 18th century, a constellation of events focused the attention of many in France on questions of climate change, deforestation, desiccation and the decline of civilizations. The French revolution of 1789 had loosened state control over the forests leading to an increase in forest clearance about the time the Landes were beginning to be planted with pines. Napoleon's campaign in Egypt from 1798 to 1801 maintained French attention on deserts and dunes. Indeed, the French experience in Egypt as chronicled in the monumental *Description de l'Egypte* (1809–1829) reveals a fascination for the many centuries of "devastation" and the burial in sand of so many ancient monuments.

In the middle of Napoleon's invasion of Egypt, during the summer of 1800, a severe heat wave and drought hit France which further stoked the fears that were growing over deforestation and desiccation in the country. The prominent French agronomist Cadet de Vaux, for instance, wrote during that hot, dry summer, in the *Moniteur Universel*, that due to deforestation "we are devoured by drought.... By altering Earth's surface we have changed the atmosphere and seasons ... [our existence] is linked to the fate of the forests."[6] The late 18th century publications of Poivre, Buffon and Bernardin de Saint-Pierre, in combination with works like those of Horace-Bénedict de Saussure on deforestation and the lowering of Lake Geneva (Saussure 1779) and Jean Antoine Fabre on deforestation and torrents in the French Alps (Fabre 1797) set the stage for desiccation theory to develop more rapidly and widely in France than anywhere else in Europe at this time.

In 1802, the influential French geographer and engineer, F. A. Rauch, brought these ideas together to proclaim that "in the provinces of Asia, anciently covered with people and woods, which assured the balance of the elements, and the destruction of which has since transformed [the provinces] into vast deserts, ...

[6]Cadet de Vaux quoted in (Fressoz 2007, p. 338).

emanates to the present all the elements of a rapid and premature death" (Rauch 1802, p. 47). He was discussing here the "antique plains of Ninevah, Babylon, Chaldee, and nearly all of Asia Minor and old Egypt … the Orient" which had, with the "disappearance of their protective forests," lost their rains, their undulating rivers and the rich fertility of their soils to become nothing but scorched deserts (Rauch 1802, p. 44). Rauch had read extensively in the literature on desiccation and related subjects and cited Poivre, Bernardin de St. Pierre, Hales, Halley, and others in his provocative discussions.

Over the course of the seventeenth, and into the 18th century, Europeans' notions of the "decline" of the Ottoman empire, and its "despotism," weakness and deterioration, had grown to be quite common. Many French thinkers in particular began to worry that France might decline in very similar ways and it stimulated much writing on the subject (Lockman 2009, pp. 45–48). Rauch, however, appears to be one of the first to put together these common fears of political/economic decline with fears of environmental decline. Thus he may be one of the earliest thinkers to develop the declensionist narrative of civilizational decline, deforestation and desertification.

A few years later, in 1819, Alexander von Humboldt published the fourth volume of his *Personal Narrative of Travels to the Equinoctial Regions of the New Continent During the Years 1799–1804*, in which he concluded that the level of Lake Valencia (Venezuela) had dropped precipitously due in large part to deforestation of the hills and mountains that surrounded the lake's basin (Humboldt et al. 1819, pp. 364–379). Having lived in Paris from 1808, Humboldt had read widely in contemporary French literature on deforestation and desiccation and was clearly influenced by the writings of Bernardin de Saint Pierre and others whom he cites in his many publications. Humboldt's work on Lake Valencia is widely accepted as "proving" the link between deforestation and desiccation, but he was clearly strongly influenced by a long tradition of French thinking about the subject.[7] These ideas were further extended and popularized by the work of Jean-Baptiste Boussingault, the agricultural chemist who elaborated on Humboldt's work on deforestation and desiccation (Boussingault 1837).[8]

By mid-century, notions of deforestation/desiccation and desertification were well established in France and elsewhere in Europe, including in Germany and England. Major figures from Jean-Baptiste Lamarck to Edward Balfour to Matthias Schleiden subscribed to these ideas and wrote about them (Balfour 1849; Lamarck 1820; Schleiden 1848). The eminent French scientist Antoine Becquerel summed up the prevailing logic in 1853 when he answered the question: "to what cause do we attribute the formation of these vast deserts of the interior of Africa? Deforestation" (Becquerel 1853, p. 254).

[7]The increase of irrigation using water from the streams which fed this lake, however, likely had much more to do with the lowering of Lake Valencia that did deforestation in the region. See Davis (2007, p. 217 note 173) for details.

[8]Boussingault's paper was translated and published in English and had a significant impact in India where it was read by influential actors including Edward Balfour (Grove 1995, pp. 328, 379, 443).

French agronomists and foresters were particularly struck with the idea that deforestation had caused desiccation and the creation of deserts in the Middle East and North Africa. The influential agronomist Jean-Baptiste Rougier de la Bergerie lamented, in 1817 for instance, the deforestation of Mount Lebanon, the "pride of the Orient" and the "fertility of the Euphrates," that had become "nothing more than the king of ruins and of deserts," its famous cedars having disappeared (Rougier de la Bergerie 1817, p. 10). Similarly, the powerful forester Jacques-Joseph Baudrillart cautioned in 1823 that due to deforestation France would become a "vast desert comparable to that which is today Asia Minor, Judea, Egypt, Greece and many other countries formerly flourishing and only recognizable today by their ruins" (Baudrillart 1823, p. 27). It was Baudrillart's plan that led to the creation of the French Forestry School in 1824 and this way of thinking was written into the forestry curriculum from its earliest days (Guyot 1898).

When France invaded Algeria in 1830, then, French thinking about deforestation and desertification was already well formed and it took only a short period of time for these notions to be applied to the new territorial acquisition. By the time Algeria was declared an official province of France in 1848, no longer a mere colony, the French had learned that their new territory was largely desert. Algeria thus became "the desert within" France with very interesting, far-ranging, and complex consequences.

8.3 Colonial Crisis Narratives and the Threatening Sahara

The French arrived in Algeria armed with the belief, derived primarily from reading ancient classical sources, that North Africa had been the granary of Rome, a lush environment with thick forests and rich soils, expecting a fertile and productive land. The arid and semi-arid landscapes they encountered in 1830s Algeria, especially beyond the more humid coastal and mountain zones, though, soon made them realize that they had acquired mostly desert territory and their perceptions began to change rapidly. Within just a few years of the occupation, pastoralists, especially nomads, were blamed for deforesting Algeria over the centuries and for creating desert-like conditions, a trend that continued for over a century (Davis 2007).

Nomads began to be blamed for destroying the environment as early as 1834, the year the French government fully committed to colonial occupation, following a French tradition of thinking dating back to at least Montesquieu. M. Poinçot, the leader of a cavalry troop in Algeria explained that "during all the time that you have occupied the soil, it [the soil] has been destroyed because you are nomads. If we want to, we can take your lands" (Nouschi 1961, p. 161). Such accusations were not limited to government personnel or "experts" but were also common in the popular press. The new popular French magazine, *L'Illustration*, reported in 1843, for instance, that Algeria was deforested and largely desert, and that "the nomads who range over the country are the cause of this desolation. The Arabs … constantly destroy [it] by the grazing of their livestock and the burning of pastures" (Anon 1843, p. 124). Such claims of degradation and desiccation were used to justify expropriating land and forests, and for sedentarizing nomads from this point onward during the colonial period. Although contested by several figures throughout the colonial period, the desiccation/degradation narrative became increasingly influential as the colonial period wore on, finally becoming dominant by the 1870s.

In the first decade of the colonial period, the view of the land as ruined by the "natives" began to be written into laws and policies governing agricultural, forest and environmental management in Algeria (and later Tunisia and Morocco). A prime example is the 1838 law outlawing "the burning, for any cause, of wood, scrub, copse, brush, hedges … [and] standing grasses and plants" (Marc 1916, p. 208) passed the same year as the creation of the Algerian Forest Service. Burning had been an important livelihood tool for the Algerians in managing pastures and fields. As the colonial period progressed, the condemnations of Algerians, and especially nomads, became more vitriolic and the accusations of deforestation/desertification became more pointed and widespread. In 1865, for example, a powerful French doctor and politician in Algeria, August Warnier, voiced the common opinion that due to the Arabs' destruction, Algeria, "long ago a sort of terrestrial paradise covered in groves…is [now] a sterile desert, bare and without water, that all call the *land of thirst*" (Warnier 1865, p. 28). Warnier was responsible for writing and passing into law extensive legislation which contained this narrative in the 1870s that destroyed indigenous communal (tribal) land tenure and instituted private property rights. These land tenure laws severely handicapped nomadic pastoralism as they dramatically reduced grazing territories, which was part of their purpose.

From the 1870s to the end of the colonial period, increasingly draconian forestry laws and calls for reforestation were justified with the deforestation/desiccation narrative. Indeed, this desertification narrative was actually written into many of these laws and related policies. The most important of these was the 1903 Algerian Forestry Law (Davis 2007). The influential physician, Dr. Paulin Trolard, was the primary force leading to the development and passage of this law. Trolard blamed the desert on deforestation and overgrazing by the indigenous populations, fiercely advocated the criminalization of their traditional land uses and the implementation of aggressive reforestation (Trolard 1891, 1893). If no action were taken, he warned, "the Sahara, this hearth of evil, stretches its arms toward us every day; it will soon enclose us, suffocate us, annihilate us" (Reboisement 1883, p. 2). Fear of the northward spread of the Sahara was widespread in French Algeria for many decades. In passing the comprehensive 1903 forestry law, the French institutionalized the desiccation/desertification narrative with far-reaching effects, primarily to the detriment of the indigenous populations.[9] Combined with the land laws of the 1840s

[9]The Algerian Forest Code formed the basis of the Forest code in Tunisia and in Morocco where it had many similar negative effects on the indigenous populations, especially nomads and other pastoralists. See Davis (2007) for details.

which defined land not being "productively used" as vacant/available and the property laws of the 1870s, these restrictions on the use of areas defined as forest (including "deforested" areas deemed reforestable), pastoralists were forced to sedentarize in large numbers.

Contributing to the plight of the pastoral nomads were the range management and veterinary policies developed outside of the official forest and agricultural areas. Much of Algeria south of the coastal region was arid and semi-arid rangeland. From the first decades of French occupation in Algeria, a programme of encouraging sedentarization and the 'rationalization' of livestock production was implemented (Davis 2008). This included the genetic improvement of livestock, the restoration of pastures and the provision of more water points (Bernis 1855, 1856). In his influential 1906 book, *L'Évolution du Nomadisme in Algérie*, the eminent geographer Augustin Bernard made clear why such measures were necessary when he explained that "nomadism, with its herds, tends to enlarge its domain unceasingly, to sterilize bigger and bigger regions... the nomad must disappear before the agriculturalist" (Bernard and Lacroix 1906, p. 5, 7).

French veterinarians were responsible for these programmes and they developed sedentary, ranching-like livestock projects in many places. Most of these measures discouraged mobility as they required inputs of time, money and/or work in a single location. They encouraged, and sometimes enforced, sedentarization because these men, too, perceived the nomads as causing overgrazing, burning, and deforestation. Even more vigorous policies to enforce sedentarization and limit pastoral migration were implemented in Morocco during the French Protectorate (1912–1956) where the government stated in 1939 "that the goal to attain is the progressive extinction of transhumance" (Ruet 1952, p. 6). In fact, controlling nomads was a serious concern across the French colonial world and the former chief of the central livestock and range service for the French colonies has explained that many such programmes in zones of pastoral nomadism were designed specifically to obtain "a reduction in the amplitude of seasonal migrations, the progressive sedentarization of errant populations" (Larrat 1965, p. 483).[10]

A substantial part of the reason that so many people working in colonial settings in the drylands, including veterinarians, botanists, and foresters, believed that the environment was deforested, degraded and desertified is found in 19th and early 20th century plant ecology. Beginning with Humboldt, the dominant mode of

---

[10]As a result, in Algeria the nomadic population, which had comprised over 60 % of the total population in 1830, was reduced to only 5 % of the population by the early 20th century (Boukhobza 1992). In Tunisia basically all the nomads were sedentarized and in Morocco they were reduced to about 16 % of the population by independence. However, the standard prescriptions for "restoring" dryland environments that emerged from the French experience in Algeria (limiting or eliminating grazing and burning, destocking and sedentarization of nomads, pasture improvement and reforestation) had negative impacts on the majority of indigenous Algerians, not only on nomads. Most of these policies were later applied to Tunisia and Morocco as they were occupied with very similar results. By the mid-twentieth century, the vast majority of North Africans had been disenfranchised of their lands, forests and livestock and many were reduced to wage labour and deep poverty (Davis 2007).

determining plant associations was by identifying the one or two "characteristic" species of the region.[11] Called phytosociology in France, this was a subjective exercise and depended on the expertise of the botanist conducting the study. This approach defined natural (or potential) vegetation by a dominant species (nearly always trees) that was presumed to have existed in an earlier "natural," undisturbed, state. It became the dominant approach throughout Europe, was closely related to both Tansley's work on ecosystems and Clement's work on succession and "climax" vegetation. It was codified by Josias Braun-Blanquet and his "relevé method" of deducing the natural or "potential" vegetation of a region by expert opinion.

In the French Maghreb, Louis Emberger used this method to construct his influential potential vegetation maps of Morocco and Algeria.[12] Actual landscapes that did not look like these maps were assumed to be degraded: deforested and desertified. Analyses utilizing paleoecological research have demonstrated just how problematic and ideologically informed this type of plant ecology and its maps can be (Davis 2007). Although criticized by many over the years for being too subjective, this phytosociological relevé method remained the most widely used, except in North America, until the close of the 20th century (Barbour et al. 1999). It is in the complex context of this long legacy of scientific and cultural thinking about and action in deserts and drylands that ideas of desertification should be considered.

Most colonial officials, French and otherwise, by the turn of the 20th century, understood desertification as tightly allied to deforestation and as a problem caused primarily by indigenous populations, most of all nomads and other migratory groups. Most did not regard the expansion of agriculture into marginal areas or over-irrigation to be significant contributors to desertification. This resulted in fairly standardized policies for arid lands "development" that restricted and criminalized native grazing, burning and agriculture, promoted reforestation, encouraged dry-farming, and built irrigation works for sedentary farming, that spread across the colonial world from the Maghreb to S. Africa to the Middle East to India. The environmental and social damage of many of these policies is made clear elsewhere in this volume.

Britain and France were the two colonial powers who controlled the most territory in arid and semi-arid lands during the 19th and early 20th centuries. The understandings of arid lands that developed in France and Britain while administering their imperial territories, have been, therefore, especially influential in global thinking about drylands and desertification. During the 20th century, this Anglo-European dominance of knowledge and policy in the drylands grew. It came to bear in particularly important and influential ways on Sub-Saharan Africa, especially in the Sahel region.

---

[11]For details, see Davis (2007, pp. 144–149).

[12]Emberger defined all five of his vegetation zones for Morocco by their "dominant" trees and he also later based his bioclimatic map of the Mediterranean basin on Morocco which he considered to be the best and most complete example of Mediterranean climate and vegetation (Davis 2007). His work thus significantly influenced later work on the entire Mediterranean basin.

8.4 Desertification Comes of Age: The 20th Century

The southern edge of the Sahara desert, the Sahel, has been a focal point for desertification fears for much of the 20th century. Ideologies of desertification and the policies they produced, though, have rarely been analysed in the long-standing Anglo-European context detailed above. The French imperial context of North Africa was, however, especially important for both thinking about and action in the drylands of the Sahel. It was within this context that the French colonial forester Louis Lavauden, working in southern Tunisia, coined the term "desertification" in 1927 and pronounced that desertification was the result of the improvidence of nomads and the overgrazing of their livestock (Lavauden 1927).[13] This French tradition of thinking about drylands and their inhabitants strongly influenced environmental policies in the roughly 3/4 of the Sahel controlled by the French in the first half of the 20th century. It also likely influenced the eastern zone under British control due to the close relationships amongst French and British colonial scientists and administrators.

For most French colonial officials, as they were conquering and occupying western Africa at the turn of the 20th century, the Sahel was frequently envisioned and understood as part of a single zone that also included the northern edges of the Sahara, namely, south central Algeria and southern Tunisia. The northern zone and the southern zone together were called the "Saharan confines" (confins du Sahara) as explained by the influential colonial botanist Auguste Chevalier (Chevalier 1932, p. 677).[14] Although the vegetation and climate regimes were recognized to be somewhat different, their similarities were highlighted and they were both believed to contain good regions for dry-farming. Both zones received between 250 and 500 mm annual average rainfall by these accounts (Chevalier 1932). The steppes and Hauts Plateaux of Algeria and sections of southern Tunisia had long been considered as forming a transition zone with the northern Sahara, but the Sahel region (zone sahélienne) was defined only in 1900 as the southern edge of the Sahara forming a transition zone with the more tropical areas further south (Chevalier 1932, p. 676).

The French aimed to unite their North African and West African possessions and many of the experts sent to French West Africa (AOF, l'Afrique Occidentale Française) were well versed in the environmental literature on the French Maghreb, they frequently cited the most prominent botanists, foresters and other experts working in Algeria, and they were actively looking for similarities between the two

---

[13]I am grateful to the late Henry Noel Le Houérou for sharing his knowledge of arid lands and desertification with me and for pointing out that it was Lavauden who first used the word desertification. See Le Houérou (2002).

[14]Auguste Chevalier was trained as a botanist. He rose to the very influential position of deputy director (1907) and then director (1912) of the Colonial Agriculture Laboratory (Laboratoire d'agronomie coloniale). In 1929 he was made professor at the Museum of Natural History and was later elected to the French Academy of Sciences.

regions. In the words of one outspoken military advocate of African colonization, "is it not natural to extend to the Sahara [and Sahel] the conclusions taken from the analysis of Algeria and neighboring countries?" (Lahache 1907, p. 182). In fact Algeria, and later Morocco, were held up as models of colonial installation and emulated in many ways (Chevalier 1930). Colonial officials were also very familiar with the desiccationist literature of the last 150 years, all the way back to Pierre Poivre (Chevalier 1909, p. 44). Despite some heated debate during the first two decades of the 20th century over whether desiccation was climatic or anthropogenic (Ballouche and Taibi 2013), by the late 1920s when Lavauden first used the word desertification, the consensus had grown among French colonial officials and many British colonial officials that desiccation was taking place and that humans and their livestock were to blame.[15]

Chevalier summed this up succinctly in 1932 when he wrote that "all the Sahara is nothing more than a vast pasture devastated by man and mowed down by his animals, above all the camel… not only is Saharan vegetation the residue of an age when the region was a steppe, it represents at the same time the last stage of degradation of the plant cover caused by the abuse of chaotic exploitation" (Chevalier 1932, p. 682).[16] Moreover, the long-standing concern that the Sahara was encroaching to the north was expanded to include the fear that it was also growing to the south. One of the earliest to raise the alarm was Chevalier who was convinced before 1907 that the Sahara was spreading south towards French Soudan (Ballouche et al. 2013, p. 5; Lahache 1907, p. 160). This concern about the spread of the Sahara to the south was shared by many French colonial officials and later also by many British colonial officials (Aubréville 1938, 1949; Harroy 1944; Lavauden 1927, 1931). British authors who rang alarm bells about the Sahara spreading southwards, such as Edward Bovill and Edward Stebbing, interestingly, relied heavily on French sources and clearly were influenced by the discussions in the French colonial and metropolitan literature on deforestation and anthropogenic versus climatic desiccation.[17] Moreover, Bovill esteemed the French example of

[15]The late 19th century and early 20th century debates about whether non-anthropogenic climatic desiccation was occurring were global in scope with evidence marshalled from every continent and included notable figures such as prince Kropotkin and Ellsworth Huntington. For an excellent overview from the time, see Gregory (1914a, b). There were also vigorous debates about whether desiccation was anthropogenic or natural or whether the real question was anthropogenic degradation and not desiccation. Amidst all this, there were also those who dissented with the prevailing views, as there had been in French Algeria, and championed much of the local populations' knowledge and environmental management skills. For more detailed discussion of the Sahelian context, see Benjaminsen and Berge (2004).

[16]Chevalier had been trained in phytosociology and thus he deduced "natural" vegetation from what he considered dominant species, even when these were "relicts." This approach is very evident in many of his publications and was just as problematic in the AOF as it had been in the French Maghreb. See Chevalier (1920) for a representative example which contains a copy of the botanical, agricultural and forestry map of AOF which he had made in the late 1890s.

[17]More than half (16/28 or 57 %) of the sources that Bovill cited, for example, were French while Stebbing also discussed many French authors and colonial officials he had interviewed on his tour

dealing with nomads (Bovill 1921, p. 268) and Stebbing also admired much of the work done by the French in AOF (Stebbing 1937).

Just as they had looked to the example of North Africa in their analysis of the vegetation and climate of the Sahel, many French colonial officials also looked in large part to forestry, pastoral management, and agricultural policies used Algeria (and later Morocco) for inspiration.[18] Deforestation and overgrazing along with the use of fire in agricultural and pastoral settings were identified as the biggest problems leading to desertification in the Sahel, just as they had been in North Africa. Chevalier suggested as early as 1909 that the forest administration in Algeria was the model that would achieve success in stopping deforestation in French West Africa (Chevalier 1909, p. 49). This was followed in 1924 with a detailed analysis of the problem and plan of action by Maurice Mangin, a prominent French forest inspector who had been charged by the French government to conduct a study on AOF (AIGREF 2001, p. 203). Mangin extolled both the 1903 Algerian Forest Code as a legal model to be followed in AOF and Lyautey's Moroccan Forest Service as a bureaucratic and research model to be emulated (Mangin 1924, pp. 634, 636, 647, 653).[19]

Meaningful and effective forest legislation, however, was very slow to come to AOF and the most important legal instrument did not appear until the 1935 Forest Law was passed on 4 July 1935 (AIGREF 2001; Montagne and Bertrand 2006; Ribot 2001). This forest law was modeled on the Madagascar forest law of 1930 (GGAOF 1935, p. 599) which itself was modeled on the Algerian Forest Code of 1903 (Madagascar 1913, p. 1203).[20] The resemblances between the AOF 1935 forest law, the Algerian Forest Code and the Moroccan Forest Code are striking (Davis 2007; Montagne and Bertrand 2006). In this legal way, too then, long-standing French ideas regarding deforestation, desiccation and desertification as well as the solutions to these perceived problems were carried from French North Africa to French West Africa.

Livestock production, l'élevage, was considered one of the riches of AOF, although not as rich as timber production in the southern, more tropical regions. In the Sahel, where timber production was impossible, improved livestock production

(Footnote 17 continued)

(Bovill 1921; Stebbing 1935, 1937). The influence of French thinking about environmental change during the 19th and 20th century on British thinking and action in the colonial environmental realm has not yet been well explored and deserves further academic scrutiny especially given that 84 British foresters sent to India between 1867 and 1885 (4.5 per year) had been trained at the French forestry school at Nancy, France (AIGREF 2001, p. 195).

[18]In the more tropical parts of West and Central Africa, the French looked to other examples like those in India, Indochina and Madagascar for models of forest management.

[19]The Moroccan Forest Law of 1917 was based closely on the Algerian 1903 Forest Code (Davis 2007).

[20]The Madagascar forest law of 1930 was a revision of the 1913 forest law which had been deliberately based on the 1903 Algerian forest Code. In both cases, there were other influences: for the AOF forest law of 1935, inspiration was also derived from the Indochina forest law and for the 1913 Madagascar forest law, metropolitan French forest law was also influential.

was one the main goals of French administration. Early in the colonial period, improving livestock production was perceived primarily as a civilizing mission and an effort to raise the level of development of the indigenous populations. Raising the number of animals and their condition was seen as an answer to both the problem of slavery (animals could provide a capital investment with good returns as slavery had done) and also the problem of cannibalism which was alleged to occur (animals could provide meat for human consumption) (Pierre and Monteil 1905, pp. 54–55). Working towards the improvement of livestock production would also, in this colonial view, eventually lead to production for market export and improve the environment if done "rationally." It was widely believed that the Sahelian environment needed improvement and hard work or it would revert "again to denudation and the sterility of the desert" (Pierre 1906, pp. 16–17).

The "traditional" techniques of raising livestock by the indigenous populations, especially the nomads, were usually seen as primitive, wasteful, and destructive by French pastoral technicians. They explained that the long migrations exhausted the livestock, made them more susceptible to disease, prevented them from gaining weight properly, and led to overgrazing. Moreover, the livestock were hardy but scrawny and did not gain weight well or quickly, two attributes needed to market meat. The transformations sought by the French were oriented to market production of meat and later of wool, whereas the indigenous peoples were oriented more towards subsistence production and hardy stock that could better survive environmental changes wrought by the frequent droughts and unreliable rainfall. French Algeria was often held up as the model to follow, especially for sheep production (Pierre 1906, pp. 168–169).

The policies for pastoral management in AOF mirrored those developed in Algeria during the 19th century quite closely. This "rational" management included, on the animal side, selection of the best stock according to French standards, mandatory castration of inferior stock, cross-breeding with superior animals imported from France or Algeria (already "improved"), and provision of veterinary care (l'Armée 1931, p. 63). On the environment side, rational pastoral improvement included the creation of forage reserves, increasing the plane of nutrition with certain forage crops and agricultural crop residues, providing more frequent watering points and the creation of model pastoral farms where better techniques could be demonstrated to the locals (l'Armée 1931, p. 64). All of these techniques made it more likely if not necessary for livestock movements to be shortened and lessened, which was one of the primary goals of the French administration.

The twin sets of policies adopted in AOF, in the sectors of forestry and pastoral development were emblematic of colonial policies in most of the drylands of the world in the first half of the 20th century. Sedentarizing nomads, controlling or eliminating grazing, livestock and pasture improvement, fire suppression, forest reserves and reforestation, and increasing dry-farming and irrigated agriculture in various combination were found in the great majority of colonial territories with drylands, and in the USA. Alarmist publications in the 1930s by Stebbing and others about the spread of the Sahara and the international debates raised with the Anglo-French Forestry Commission, which concluded that the Sahara was not spreading to the south, drew international attention to the problems of drylands.[21] The American Dust Bowl of the 1930s further heightened knowledge of and concern with what was becoming known as desertification, and with it, focused attention on soil erosion in dryland environments.

The events of the 1930s thus led to the internationalization of the idea of a "crisis of desertification" that raised the profile of environmental change in the drylands to a global level. The American Dust Bowl raised concern with erosion which was taken up in many Western colonial territories in Africa and Asia with American-inspired policies to curb erosion in ways that were often ecologically inappropriate. The new erosion policies, along with more well-established anti-desertification policies, though, frequently, led to socially repressive programmes and the execution of political goals in the name of environmental protection (Adams 2009; Anderson 1984; Showers 2005). In Mandate Palestine, for instance, such thinking led the British-run forest department to report, incorrectly, that half of the habitable area of Palestine had become an artificial desert due to overgrazing, and to implement reforestation programmes and grazing restrictions with the help of Zionist interest groups (Davis 2011).

Momentum for believing in a crisis of anthropogenic desertification continued to grow and become popularized in the late 1930s and throughout the 1940s with publications like *The Rape of the Earth* and *Climate, Forests and Desertification in Tropical Africa* (Aubréville 1949; Jacks and Whyte 1939). André Aubréville's book on forests and desertification in tropical Africa was particularly influential and he has been frequently cited as the first to use the word desertification, although he had actually read and cited Lavauden's (1927) article where the word was first used and almost certainly got it from him (Aubréville 1938, 1949). Aubréville was trained as a botanist and as a forester and he attained the high and influential rank of Inspecteur Général des Eaux et Forêts des Colonies by the time he published his 1949 tome on forests and desertification. He was most concerned with the savannah zones, though, rather than the more arid northerly zones of West Africa. The American Walter C. Lowdermilk helped to spread fears of desertification to the heart of the Middle East with his 1944 book *Palestine: Land of Promise* which was replete with the deforestation, desiccation, desertification narrative and pointedly blamed Arab nomads for most of the presumed environmental destruction down through the centuries (Lowdermilk 1944).

Lowdermilk was later chair of UNESCO's natural sciences department's committee of experts which established the Arid Zone Program in 1949 (Selcer 2011). Following WWII, as many former colonial territories were gaining, or about to gain, independence, the newly created United Nations framed their Arid Zone Program as "Man's battle against the desert." The programme grew in size, funding, and influence and by 1956 it had been renamed the Arid Lands Major Project, with permanent seats held by Great Britain, France and the United States (Selcer 2011).

---

[21]The Commission did, however, agree that anthropogenic environmental degradation was a serious concern in the regions studied (Niger and Nigeria) (Mortimore 1998, p. 20).

This programme incorporated the vast majority of earlier Anglo-European thinking and policies on arid lands and threatening deserts and further spread these ideas and policies globally with conferences, training and education programmes, and with publications.

The programme targeted the belt of arid and semi-arid lands from Morocco to India for research and training, using a neo-Malthusian rationale that aimed to make arid lands productive for more people in a crowded world. It recruited and incorporated many European "experts" from key positions in former colonies, like the phytosociologist Louis Emberger and Saharan expert Théodore Monod. It also brought together scholars and scientists from post-colonial countries with Anglo-European experts, including Soviet and American, to create an international network of scholars working on arid lands around the world. Although the more applied part of the programme itself (especially training seminars) targeted North Africa, the Middle East to Pakistan, and India, the basic science research and recommendations developed from this work became widely accepted, disseminated, and later applied to many different arid lands around the globe.

Although it had, at it's end in 1962, "neither shrunk the deserts nor stopped erosion," (Batisse 2005, p. 77) the Arid Lands Program had succeeded in spreading, promoting and popularizing the old (colonial) Anglo-European notion of desertification of the drylands "which man himself has reduced to a desert-like state by irrational exploitation of the soil and vegetation, but whose present economy could be improved."[22] It also trained many officials and scholars from the target countries in methods like phytosociology which spread the very scientific methods which had helped to form and reinforce the declensionist narrative.

Perhaps more problematically, the Arid Lands Program also helped to legitimize the sedentarization of nomads in the Middle East in the name of environmental protection while elites in these countries were actually interested in social control and nation building (breaking down tribal affiliations and controlling nomads).[23] Many policy recommendations issued by UNESCO included the sedentarization of nomads and/or "rationalization" of pastoral production. In fact, the great majority of policy prescriptions in UNESCO literature strongly reflected colonial policy prescriptions for arid lands and included destocking and sedentarization, fire suppression, provision of water and irrigation of fodder crops, pasture amelioration including pasture reserves and reseeding, and reforestation (White 1956, pp. 179–188). Although it was subsumed under the International Hydrological Program in the 1960s, and focused primarily on the Middle East and North Africa, the Arid Lands Program set the stage in several ways for the UN Conference on Desertification (UNCOD) a decade later.

[22]Final quote from "UNESCO and the Arid Zone," 24 January 1958, UNESCO/NS/AZ/334, Paris, p. 1. See also the chapter on "Evolution of Land Use in South-Western Asia" by R. O. Whyte in Stamp (1961, pp. 57–118) for a 50 page treatise on how humans and livestock have desertified the Middle East (Stamp 1961, pp. 57–118).

[23]See, for example the chapter by the Egyptian Mohammed Awad, chairman of UNESCO's executive board in (UNESCO 1962, pp. 325–340). See also (Selcer 2011, pp. 232–233).

8.5 Conclusion

When the prolonged drought in the early 1970s hit the Sahel with spectacularly high human and animal mortality, global concern with desertification exploded and resulted in a bureaucracy, research agenda and policy prescriptions for desertification still influential (and problematic) today, as detailed in Camilla Toulmin and Karen Brock's chapter in this volume. Many involved in UNCOD were familiar with and drew information and inspiration from UNESCO's Arid Lands Program. UNESCO had published 30 volumes in its *Arid Zone Research Series* with contributions from 600 authors totaling 7600 pages (Batisse 2005, p. 77). One of the primary organizers of the UN's World Conference on Desertification held in Nairobi in 1977 was Mohamed Kassas who "repeatedly confirmed that all or nearly all had already been studied and stated on the subject of desertification in the series published by UNESCO" (Batisse 2005, p. 78).

Thus many of the old, colonial ideas about deserts and desertification were carried over from UNESCO's Arid Lands Program to UNCOD and later to UNCCD. The dominant perceptions of deserts and desertification embraced by UNESCO, UNCCD, and many other institutions, governments and scholars, though, is deeply flawed. As is now widely recognized by a large number of scholars, "it seems to have triggered many problems, of understanding and of action, that manifested themselves in the ways in which desertification [has been] conceptualized, represented and approached as an environmental, social and political issue" (Thomas and Middleton 1994, p. 49). A substantial number of these misunderstandings have been caused by the long and influential history of Western thinking about deserts and desertification detailed above, particularly during the colonial period. This includes not only our flawed notions of deserts as ruined former forests that are inexorably spreading, but also our beliefs that arid lands should be made (more) productive. Why should the desert bloom? The answers to this question are at once scientific, economic, ideological and political and the question deserves much more scrutiny. Desertification is in many ways a (neo)-colonial concept in need of revision.

References

Adams, W. M. (2009). *Green development: Environment and sustainability in a developing world* (third ed.). London: Routledge.

AIGREF. (2001). *Des Officiers royaux aux ingénieurs d'état dans la France rurale (1219-1965).* Paris: Éditions Technique et Documentation.

Anderson, D. M. (1984). Depression, dust bowl, demography and drought: The colonial state and soil conservation in East Africa during the 1930s. *African Affairs, 83*, 321–344.

Anon., (1843). Algérie: Description géographique. *L'Illustration, 1*(8), 124–125.

Aubréville, A. (1938). *La Forêt coloniale: Les forêts de l'Afrique Occidentale Française. Annales d'Académie des Sciences Coloniales* (Vol. IX). Paris: Société d'Éditions Géographiques, Maritimes et Coloniales.

Aubréville, A. (1949). *Climats, Forêts et Désertification de l'Afrique Tropicale*. Paris: Société d'Éditions Géographiques, Maritimes et Coloniales.
Balfour, E. (1849). Notes on the influence exercised by trees in inducing rain and preserving moisture. *The Madras Journal of Literature and Science, 15*(36), 400–448.
Ballouche, A., & Taibi, A. N. (2013). Le "Dessèchement"de l'Afrique sahélienn. Un Leitmotiv du discours d'expert revisité. *Autrepart*, 65(1), 47–66.
Barbour, M. G., Burk, J. H., Pitts, W. D., Gilliam, F. S., & Schwartz, M. W. (1999). *Terrestrial plant ecology* (3rd ed.). Menlo Park, CA: Addison, Wesley, Longman.
Batisse, M. (2005). *The UNESCO water adventure from desert to water, 1948–1974*. Paris: UNESCO.
Baudrillart, J.-J. (1823). *Traité général des eaux et forêts, chasses et pêches* (10 vols. Vol. 1). Paris: Chez Madame Huzard.
Becquerel, A. C. (1853). *Des Climats et de l'influence qu'exercent les sols boisées et non boisées*. Paris: Firmin Didot Frères.
Benjaminsen, T., & Berge, G. (2004). Myths of Timbuktu: From African El Dorado to desertification. *International Journal of Political Economy, 34*(1), 31–59.
Bernard, A., & Lacroix, N. (1906). *L'Évolution du nomadisme en Algérie*. Alger: Adolphe Jourdan.
Bernis, J. (1855). Sur l'espèce ovine de l'Algérie. *Bulletin de la société impériale zoologique d'acclimatation, 2*, 597–613.
Bernis, J. (1856). Sur l'espèce ovine de l'Algérie, deuxième et troisième lettre. *Bulletin de la société impériale zoologique d'acclimatation, 3*(63–76), 108–123.
Blanchard, W. O. (1926). The landes: Reclaimed waste lands of France. *Economic Geography, 2* (2), 249–255.
Boukhobza, M. H. (1992). *Monde rural: Contraintes et mutations*. Alger: Office des Publications Universitaires.
Boussingault, J.-B. (1837). Mémoire sur l'influence des défrichements dans la diminuition des cours d'eau. *Annales de Chimie et de Physique, 64*(1), 113–141.
Bovill, E. W. (1921). The encroachment of the Sahara on the Sudan. *Journal of the Royal African Society, 20*(79), 174–185, 259–269.
Buffon, C. (1865). *Oeuvres choisies de Buffon* (Vol. premier). Paris: Librairie de Firmin Didot Frères, Fils et Cie.
Chevalier, A. (1909). *Les Végétaux utiles de l'Afrique tropical française: Première étude sur les bois de la Côte d'Ivoire. Études Scientifiques et Agronomiques* (Vol. fascicule V). Paris: A. Challamel.
Chevalier, A. (1920). *Exploration botanique de l'Afrique Occidentale Française*. Paris: Paul Lechevalier.
Chevalier, A. (1930). *L'Agronomie coloniale et le Muséum National d'Histoire Naturelle: Premieres conférences du cours sur les productions coloniales végétales & l'agronomie tropicale*. Paris: Laboratoire d'Agronomie Coloniale.
Chevalier, A. (1932). Les Productions végétales du Sahara et de ses confins Nord et Sud: Passé - présent - avenir. *Revue de Botanique Apliquée & d'Agriculture Tropicale, 12*(133–134), 669–710.
Davis, D. K. (2007). *Resurrecting the granary of Rome: Environmental history and French colonial expansion in North Africa*. Athens: Ohio University Press.
Davis, D. K. (2008). Brutes, beasts and empire: Veterinary medicine and environmental policy in French North Africa and British India. *Journal of Historical Geography, 34*(2), 242–267.
Davis, D. K. (2011). Scorched Earth: The problematic environmental history that defines the middle East. In A. Amanat, M. Bonine, & M. Gasper (Eds.), *Is there a middle East: The evolution of a geopolitical concept*. Stanford: Stanford University Press.
Davis, D. K. (forthcoming). *The Arid Lands: History, Power, Knowledge*. Cambridge, MA: MIT Press.
Fabre, J. A. (1797). *Essai sur la théorie des torrens et des rivières*. Paris: Chez Bidault.

Fressoz, J.-B. (2007). Beck back in the 19th century: Towards a genealogy of risk society. *History and Technology, 23*(4), 333.
GGAOF, G. G. d. l. A. O. F. (1935). Arrêté promulguant en Afrique occidentale française le décret du 4 juillet 1935, sur le régime forestier de l'Afrique occidentale française. *Journal Officiel de Sénégal,* 24 July 1935:599–606.
Goldstein, J. (2013). *Terra Economica:* Waste and the production of enclosed nature. *Antipode, 45* (2), 357–375.
Goudie, A. S. (1986). *The search for Timbuktu: A view of Deserts*. Oxford: Clarendon Press.
Gregory, J. W. (1914a). Is the Earth drying up? Part I. *The Geographical Journal, 43*(2), 148–172.
Gregory, J. W. (1914b). Is the Earth drying up? Part II. *The Geographical Journal, 43*(3), 293–313.
Grove, R. H. (1995). *Green imperialism: Colonial expansion, tropical Island Edens and the origins of environmentalism, 1600–1860*. Cambridge: Cambridge University Press.
Grove, R. H. (1997). *Ecology, climate and empire: Colonialism and global environmental history 1400–1940*. Cambridge: The White Horse Press.
Guyot, C. (1898). *L'Enseignement forestier en France: L'école de Nancy*. Nancy: Crépin-Leblond.
Harroy, J.-P. (1944). *Afrique terre qui meurt: La dégradation des sols africains sous l'influence de la colonisation*. Bruxelles: Marcel Hayez.
Humboldt, A. D., & Bonpland, A. (1819). *Personal narrative of travels to the equinoctial regions of the new Continent, 1799–1804* (Vol. IV). Translated by Williams, Helen Maria. London: Printed for Longman, Hurst, Rees, Orme and Brown.
Jacks, G. V., & Whyte, R. O. (1939). *The rape of the Earth: A World survey of soil erosion*. London: Faber and Faber.
Lahache, J. (1907). Le Dessechement de l'Afrique française est-il Démontré? *Bulletin de la Société de Géographie et d'Études Coloniales de Marseille, 31*(1), 149–185.
Lamarck, Ml C. (1820). *Système analytique des connaissances positives de l'homme*. Paris: A. Belin.
l'Armée, L. I.d. S. V.d. (1931). *Le Service vétérinaire et le service de la remonte aux colonies*. Paris: Imprimerie Nationale.
Larrat, R. (1965). L'Oeuvre des vétérinaires français dans le monde: Introduction. In J.-G. Tridon (Ed.), *Vétérinaires de France*. Paris: Service de Propagande et Édition, Information.
Lavauden, L. (1927). Les Forêts du Sahara. *Revue des Eaux et Forêts, 65*(6), 265–277, 329–341.
Lavauden, L. (1931). *Le Problème forestier colonial*. Paris: Éditions Berger-Levrault.
Le Houérou, H. N. (2002). Man-made deserts: Desertization processes and threats. *Arid Land Research and Management, 16*(1), 1–36.
Lockman, Z. (2009). *Contending visions of the middle East: The history and politics of orientalism* (2nd ed.). Cambridge: Cambridge University Press.
Lowdermilk, W. C. (1944). *Palestine: Land of promise*. New York and London: Harper & Brothers Publishers.
Madagascar, G. G. D. (1913). Arrêté promulguant dans la colonie de Madagascar et Dépendences le décret du 28 août 1913, établissant le régime forestier applicable à Madagascar. *Journal Officiel de Madagascar et Dépendences,* 18 Octobre 1913:1202–1209.
Mangin, M. (1924). Une Mission forestière en Afrique Occidentale Française. *La Géographie*, 42 (4), 449–483, 629–654.
Marc, H. (1916). *Notes sur les forêts de l'Algérie*. Alger: Typographie Adolphe Jourdan.
Montagne, P., & Bertrand, A. (2006). Histoire des politiques forestières au Niger, au Mali et à Madagascar. In A. Bertrand (Ed.), *L'État et la gestion locale durable du forêts en Afrique francophone*. Paris: CIRAD.
Montesquieu, B. D. (1949 [1748]). *The spirit of the laws* (T. Nugent, Trans.). New York: Hafner Publishing Co.
Mortimore, M. (1998). *Roots in the African dust: Sustaining the Sub-Saharan drylands*. Cambridge: Cambridge University Press.
Nouschi, A. (1961). *Enquête sur le niveau de vie des populations rurales constantinois de la conquête jusqu'en 1919*. Paris: Presses Universitaires de France.

Pierre, C. (1906). *L'Élevage dans l'Afrique Occidentale Française*. Paris: A. Challamel.
Pierre, C., & Monteil, C. (1905). *L'Élevage au Soudan*. Paris: A. Challamel.
Poivre, P. (1797). *Oeuvres complètes de P. Poivre*. Paris: Chez Fuchs.
Rauch, F. A. (1802). *Harmonie hydro-végétale et météorologique, ou recherehces sur les moyens de recréer avec nos fôrets la force des températures et la régularité des saisons, par des plantations raisonnées*. Paris: Chez les frères Levrault.
Reboisement, Ld. (1883). *La Forêt: Conseils aux indigènes*. Alger: Imprimerie de l'Association Ouvrière P. Fontana.
Ribot, J. (2001). *Science, use rights and exclusion: A history of forestry in francophone West Africa*. London: IIED.
Rougier de la Bergerie, J.-B. (1817). *Les Forêts de la France*. Paris: Chez Arthus Bertrand.
Rozental, A. A. (1956). The enclosure movement in France. *The American Journal of Economics and Sociology, 16*(1), 55–71.
Ruet, C. (1952). *La Transhumance dans le moyen Atlas et la haute Moulouya*. CHEAM: Unpublished report.
Saint-Pierre, J. H. B. D. (1786). *Études de la nature* (2 vols. Vol. 2). Paris: Chez P. F. Didot.
Saussure, H. -B. D. (1779). *Voyage dans les Alpes*. Neuchatel: Chez Samuel Fauche.
Schleiden, M. J. (1848). *The plant; A biography*. London: Hippolyte Bailliere.
Selcer, P. (2011). *Patterns of science: Developing knowledge for a World community at UNESCO*. Dissertation, History of Science Department, University of Pennyslvania.
Showers, K. B. (2005). *Imperial gullies: Soil erosion and conservation in lesotho*. Athens: Ohio University Press.
Stamp, D. L. (Ed.). (1961). *A history of land use in Arid regions*. Paris: UNESCO.
Stebbing, E. P. (1935). The encroaching Sahara: The threat to the West African colonies. *Geographical Journal, 85*(6), 506–524.
Stebbing, E. P. (1937). *The forests of West Africa and the Sahara: A study of modern conditions*. London: W. & R. Chambers.
Thomas, D. S. G., & Middleton, N. (1994). *Desertification: Exploding the myth*. West Sussex: John Wiley & Sons Ltd.
Traimond, B. (1986). Le Voyage dans les Landes de Gascone ou la traversée du Sahara français. *Études Rurales, 103–104*, 221–234.
Trolard, P. (1891). *La Colonisation et la question forestière*. Alger: Imprimerie Casabianca.
Trolard, P. (1893). *Bréviaire du ligeur*. Alger: Imprimerie Casabianca.
UNESCO. (Ed.). (1962). *The problems of the Arid Zone: Proceedings of the Paris symposium*. Paris: UNESCO.
Warnier, A. (1865). *L'Algérie devant l'empereur, pour faire suite à L'Algérie devant le sénat, et à L'Algérie devant l'opinion publique*. Paris: Challamel Ainé.
White, G. F. (Ed.). (1956). *The future of Arid lands: Papers and recommendations from the international Arid lands meetings*. Washington, DC: American Association for the Advancement of Science.
Whited, T. L. (2000). *Forests and peasant politics in modern France*. New Haven: Yale University Press.

Chapter 9
Where Does Desertification Occur? Mapping Dryland Degradation at Regional to Global Scales

Stephen D. Prince

Abstract To make sense of the controversies about Sahelian desertification, we must be able to assess objectively the processes, their location and level of threat to drylands and those who inhabit them. Surprisingly, after the United Nations Conference on Desertification (UNCOD) was set up in 1977, there still are no reliable maps or means of monitoring desertification at the sub-national to global scales, even in the iconic Sahel region. These are the scales needed to formulate institutional policies for prevention and remediation and for Earth system science. While there are a few maps that do extend to larger areas, they mostly have serious shortcomings. This void is partly because suitable metrics have not been available and those that have been proposed can be difficult or impossible to apply over large areas. One of the most basic problems is the lack of comparison with non-desertified sites and an absence of validation, without which statements about more or less desertification are misleading. Furthermore, existing maps are mostly based on subjective assessments by experts, and therefore cannot be applied elsewhere or by different observers, nor can they be used in future for monitoring. Consequently, unsupported statements about the extent and severity of desertification abound. This chapter reviews the existing global maps and sets out principles for more rigorous mapping and proposes methods that adhere to these. Decades of work have improved our understanding of individual components of the desertification processes, but have not answered the three fundamental questions at the regional to global scales: *What is degraded*? *Where does it occur*? *How severe is the degradation*?

Keywords Methods for mapping desertification · Drylands · Global · GLASOD · GLADA · USDA NRCS

S.D. Prince (✉)
University of Maryland, College Park, USA
e-mail: sprince@umd.edu

R.H. Behnke and M. Mortimore (eds.), *The End of Desertification?*
Springer Earth System Sciences, DOI 10.1007/978-3-642-16014-1_9

9.1 Introduction

Desertification processes have causes and impacts at country, provincial and county scales (administrative divisions 1, 2 and 3) in addition to the household to community scales which have been studied in depth (e.g. Mortimore and Adams 2001; Reynolds et al. 2011). Data covering large areas are needed for many purposes (Table 9.1). For example Vogt et al. (2011) list 11, large user communities with needs for data on degradation beyond the local scale on which to base investments for prevention and remediation. Other applications include the effect of degradation on the declining agricultural land base needed for an increasing population (Gibbs and Salmon 2015); to determine the size of investment required—for example Spain spent €12 $\times$ 10^6 on actions against soil erosion from 2004 to 2007 (Martín-Fernández and Martínez-Núñez 2011); and policy support systems for degradation control (Mulligan 2015). Several new United Nations agencies with global spheres of activity have been established (Safriel 2007) culminating in the United Nations Convention to Combat Desertification (UNCCD 1994) with 194 signatory states. Furthermore, the need for information at the larger scale is now recognized by the Conventions on Biodiversity and Climatic Change, Agenda 21 (UNCED 1992), the Millennium Goals, and the Sustainable Development Goals (United Nations General Assembly 2013).

Yet, after 40 years of research and despite the existence of UNCCD, comprising a Committee on Science and Technology (CST), the Group of Experts (GoE), National Action Programmes (NAP), Action Programmes for Sub-regional (SRAP), the Regional groups (RAP), a large and accelerating output of scientific journal and popular articles, and particularly unpublished reports ("grey literature" Lobell 2010), there is a dearth of reliable observations on the spatial extent, location and severity, and current monitoring of degraded and susceptible lands is manifestly inadequate for sub-national to global needs (Herrick et al. 2010; Verón et al. 2006).

Assessments are critically dependent on objective data, but development of appropriate measurements has proved difficult. Scaling-up local data to larger areas is often impossible because some phenomena are restricted to particular scales (Prince 2002). For example, field observations of gulley erosion around cattle watering points due to local factors such as livestock number cannot be applied directly to the national scale because of lack of data and anyway different issues, such as policy, become the most influential factors, however, at the local scale, policy is an invariant state-variable (UNCED 1992, parag. 12.28). There are also differences in timescales: most existing studies of degradation are of processes that manifest themselves over one or a few years, but there are also changes that transform land to the extent that long-term loss of productivity occur (Prince 2002). Long-term losses require very large investments to reverse (Abella et al. 2015), and are therefore effectively permanent, at least within one generation. Clearly data collected over a few years cannot detect this condition.

Table 9.1 Examples of applications of current maps to different aspects of global degradation

Topic	Application	Findings	Source of data on degradation	Author(s)
Sustainability	Land degradation in the developing world: implications for food, agriculture, and the environment to 2020	Land degradation is a serious threat to food production and rural livelihoods with potentially severe effects by the year 2020	GLASOD (in part)	Scherr and Yadav (1996)
Soil conservation	Assessment of degradation-induced loss of global soil productivity	Average degradation-induced loss of global soil productivity was approximately 0.1–0.2 % $year^{-1}$ during the period 1945–1990. 1.6 billion hectare of land are affected by erosion globally, 1.1 billion by water and 0.5 billion hectare by wind erosion	GLASOD	Dregne and Chou (1992), Oldeman and van Lynden (1996), Batjes et al. (2012)
	Global soil water erosion	A model of degradation validated with GLASOD	GLASOD	Batjes (1996)
Soil carbon	Soil carbon sequestration in overgrazed grassland	Widespread rehabilitation of overgrazed grasslands could sequester approximately 45 Tg C $year^{-1}$, most of which can be achieved by cessation or moderating grazing	GLASOD	Conant and Paustian (2002)
Salinization	Assessment of significance of human-induced salinization	Anthropogenic salinization accounts for only 10 % of Africa's non-irrigated saline soils: nutrient depletion is a far greater constraint on food production	GLASOD	Thomas and Middleton (1993)
Carbon sequestration	Impact of human-induced degradation on net primary production in Zimbabwe	Only 16 % of Zimbabwe was at its potential production and the total loss in productivity due to degradation was about 17.6 Tg C $year^{-1}$, that is 13 % of the entire national potential	LNS	Prince et al. (2009)
Agriculture	Effects of soil erosion on crop yields	Production loss of 0.3 % $year^{-1}$ by six crops at the global level. Economic loss \$523.1 million $year^{-1}$	NRCS	den Biggelaar et al. (2004)
	Number of people affected by global desertification	11.9 million square kilometre with 1.4 billion inhabitants at very high risk of desertification	NRCS	Eswaran et al. (2001)

(continued)

Table 9.1 (continued)

Topic	Application	Findings	Source of data on degradation	Author(s)
	Global agricultural productivity and soil degradation	Productivity of some lands has declined 50 % due to soil erosion and desertification. In Africa, past soil erosion has reduced yield 2–40 % (mean 8.2 %). In South Asia, annual loss in productivity is 36×10^6 tons of cereal valued at US$5.4 $\times 10^9$ by water and US$1.8 $\times 10^9$ by wind erosion. Total annual cost of yield loss in USA is $44 $\times 10^9$ per year and $247 per ha of cropland and pasture. At the global-scale annual loss is 75×10^9 tons of soil, costing US$400 $\times 10^9$ per year, or approximately US$70 per person per year	NRCS	Beinroth et al. (2001)
	Losses in global productivity	Declining rain-use efficiency-adjusted NDVI was approximately 24 % of the global land area, mainly in Africa south of the equator, South-East Asia and south China, north-central Australia, the Pampas and swaths of the Siberian and north American taiga	GLADA	Bai et al. (2008)
	Global net primary production losses caused by human-induced dryland degradation	Global dryland degradation is 4–10 % of the potential NPP. Losses amount to 20–40 % of the potential NPP on degraded agricultural areas and >55 % in some world regions. Total human appropriation of NPP in drylands is close to the overall annual harvest	GLASOD	Zika and Erb (2009)
Carbon sequestration	Changes in global productivity	NE Europe, N China, N India, the mid-western part of USA and southern Australia tend to have been positively influenced by human interventions despite the fact that weather conditions reduced the productivity	GLADA	Bai et al. (2012)

(continued)

Table 9.1 (continued)

Topic	Application	Findings	Source of data on degradation	Author(s)
Climate	Link between desertification and global warming	Soil degradation has had a significant impact on local water resources over a 30 year period (1950–1980)	GLASOD	Feddema (1999)
Policy	Management of land	Understanding of land use at the landscape level is essential	GLASOD	Bossio et al. (2010)

GLADA Global Assessment of Land Degradation and Improvement; *GLASOD* Global Assessment of Soil Degradation; *LNS* Local Net Production Scaling; *NRCS* US Natural Resource Conservation Service; *NPP* Net Primary Production; *NDVI* Normalized Difference Vegetation Index, which is remotely-sensed and often used as a surrogate for *NPP*

At present, there are just four maps of degradation with global coverage. There are others (Safriel 2007) but their spatial resolutions are too low for use at a regional scale. The four maps are actually suites, since they include maps of different themes. They are: the Global Assessment of Soil Degradation (GLASOD, Oldeman et al. 1991); the United States Natural Resource Conservation Service, Major Land Resource Stresses and Conditions (NRCS, Beinroth et al. 2001); Global Assessment of Land Degradation and Improvement (GLADA, Bai et al. 2008); and the Global Land Degradation Information Systems (GLADIS, Nachtergaele et al. 2011). The inputs, methodologies for assessment of the extent, trends, and severity of desertification, and the results, shown by these maps differ dramatically (Table 9.2). Furthermore there has been strong criticism of their data and methodologies, so it is not surprising that they have led to skepticism and, ultimately, to a delay of effective actions (Verón et al. 2006; Wessels 2009).

Making maps and global assessments is difficult, partly due to the lack of recognition that "desertification" is not a single phenomenon (Gibbs and Salmon 2015), and is therefore incapable of simple measurement (e.g. Adeel et al. 2005; Vogt et al. 2011). Another source of confusion is whether maps show existing degradation, degradation that is in progress, or risk of degradation. While there is a plethora of broad, but non-specific definitions, propositions, and conceptual diagrams (Reynolds and Stafford Smith 2002), they do not lead to measurable variables. A recent special issue of a well-regarded, scientific journal on dryland degradation has 12 papers filling 167 pages in which there are 17 conceptual diagrams but no mention of actual data! The seemingly endless semantics and diagrams published on desertification have had little demonstrable, practical, influence on research, prevention or remediation.

Table 9.2 Areas of degraded land in selected countries, Africa and world-wide from three global degradation maps; GLASOD, NRCS, and GLADA. Each map has different metrics of degradation

	Land area ($km^2 \times 10^6$)	GLASOD		NRCS		GLADA	
		Modified from Nachtergaele et al. (2011)		Reich et al. (2001)		Bai et al. (2008)	
		Total severe and very severe ($km^2 \times 10^3$)	% severe and very severe	High and very high vulnerability ($km^2 \times 10^3$)	% high and very high vulnerable	Affected area ($km^2 \times 10^3$)	% area affected
Country							
Burkina Faso	0.273	33.45	12	136.0	50	9.255	3
Chad	1.251	12.00	1	392.0	31	52.735	4
Mali	1.220	18.93	2	267.0	22	35.637	3
Mauritania	1.030	9.58	1	67.0	7	6.301	1
Niger	1.266	27.67	2	217.0	17	22.563	2
Nigeria	0.910	26.38	3	289.0	32	91.443	10
Senegal	0.192	20.17	11	126.0	66	34.655	18
Sudan	2.376	17.71	1	480.0	20	166.031	7
Africa							
Not degraded	–	–	83	–	54	–	
Light-moderate	–	–	6	–	14	–	24
Moderate	–	–	–	–	16	–	
High	–	–	–	–	11	–	
Very high	–	–	11	–	5	–	
Global							
Not degraded	–	–	87	–	33	–	
Light-moderate	–	–	5	–	11	–	15
Moderate	–	–	–	–	10	–	
High	–	–	–	–	5	–	
Very high	–	–	8	–	6	–	

9.2 Chapter Aims

The aim of this chapter is to assess the current knowledge about degradation at Sahelian and global scales and from years to decades. Answers to three fundamental questions are explored: *What* is degraded? *Where* does it occur? *How severe* is it? While the discussion is limited to global maps, there are numerous regional and local scale maps, but few of these are for the Sahel which is the main focus of this book. The sources, strengths and limitations of the maps are considered with respect to the uses for which they were intended and for which they are often used. The basic principles of mapping degradation are discussed and some suggestions are made that should lead to much improved answers to the three fundamental questions.

The terms "desertification" and "dryland degradation" are frequently used synonymously, as is implicit in the UNCCD definition (UNCCD 1994). The two terms, however, are used differently here since one theme of this chapter is that "desertification", is more a concept than a single, measurable condition. It consists of different aspects of land degradation, operating at different scales, and with different effects, for example loss of soil organic carbon, loss of productivity, reduced cover of vegetation, loss of palatable species, wind and water erosion. Thus "desertification" is used here for the overall concept, while "degradation" refers to individual processes.

9.3 Current Maps

9.3.1 Global

Before embarking on an examination of the strengths and weaknesses of current mapping and monitoring desertification, a selection of existing maps are described briefly in order to illustrate some of their characteristics and to provide a context for further discussion (Gibbs and Salmon 2015). The maps selected include all four global maps (but one, GLADIS, is still in preparation) that provide information at sub-national scales. Safriel (2007) analysed the degree and severity of degradation shown in four global maps, including some not discussed here. The maps are considered with an emphasis on the "*where*" question. While this chapter is confined to the global and Sahel scales, there are maps of smaller areas but, unfortunately, many of these are not available other than in the original figure in the publication or, worse, only in grey literature (Lobell 2010). Beyond the maps discussed here, there are few archives or data-bases, still fewer in digital formats. Moreover, the mapping methods are not always specified, and most are based on subjective assessment by experts, and are therefore effectively non-repeatable and not useful for the purposes listed in Table 9.1 or similar applications.

1. *GLASOD and the World Atlas of Desertification*

The most widely used global map of types of desertification, locations and severity, is the World Map of Desertification (Oldeman et al. 1991) also known as GLASOD, later published in modified form (UNEP 1997) (Figs. 9.1 and 9.2). Given the acknowledged need for a map of desertification, both now and especially at the time of its publication, the title itself, its global coverage, and lack of alternatives, have led to its widespread but often uncritical use.

Four types of degradation were mapped in GLASOD; water erosion, wind erosion, chemical deterioration and physical deterioration, each shown by a different color in Fig. 9.3. Thus GLASOD consists of much more than the presence or absence of "desertification", incorporating, as it does, information on the type, the area affected, its severity and causes, albeit subject to severe criticisms that are discussed later in this Chapter. Areas that were degraded by natural processes or were considered stable were omitted.

2. *U.S. Natural Resource Conservation Service (NRCS)*

The U.S. NRCS published a series of global maps of themes related to degradation (Beinroth et al. 2001) (Figs. 9.1, 9.2, and 9.4). These consist of maps of "Vulnerability" to degradation, salinization, physical damage, water and wind erosion. Vulnerability is the estimated cost of remediation of each of the stresses. Each of these conditions is coupled with a map of "Risk", calculated by weighting the Vulnerability by human population density (Deichmann 1994).

3. *Global assessment of land degradation and improvement (GLADA)*

The International Soil Reference and Information Center (ISRIC) and FAO have published global suites of maps of degradation as a part of the Global Land Degradation Assessment (GLADA) programme (Figs. 9.1 and 9.2) (Bai et al. 2008). The maps consist of the means of 25-years of satellite measurements of annual normalized vegetation index (NDVI) which were used as a surrogate for net primary production (NPP). In addition to the temporal trends of NDVI, a ratio of annual NDVI and rainfall (RUE), as first introduced for dryland degradation by Prince et al. (1998), and residual trends (RESTREND) (Wessels et al. 2007b) in which the gradient of annual deviations (residuals) of each pixel's value from the regression of annual NDVI on rainfall were plotted in a time-series. In all three cases the coefficients of a linear regression of the time-series were mapped in GLADA: negative values indicating a downward trend, interpreted as degradation, and a positive value, recovery. Three of the several maps derived from these data are considered here (Figs. 9.1 and 9.2).

4. *Global Land Degradation Information System (GLADIS)*

GLADIS is being developed by FAO using the concept of "ecosystem goods and services" to assess degradation (Nachtergaele et al. 2011). Because services are more concepts than specific metrics, indices of biomass, soil, water, biodiversity, economics and social factors are calculated from existing data. The basic data used

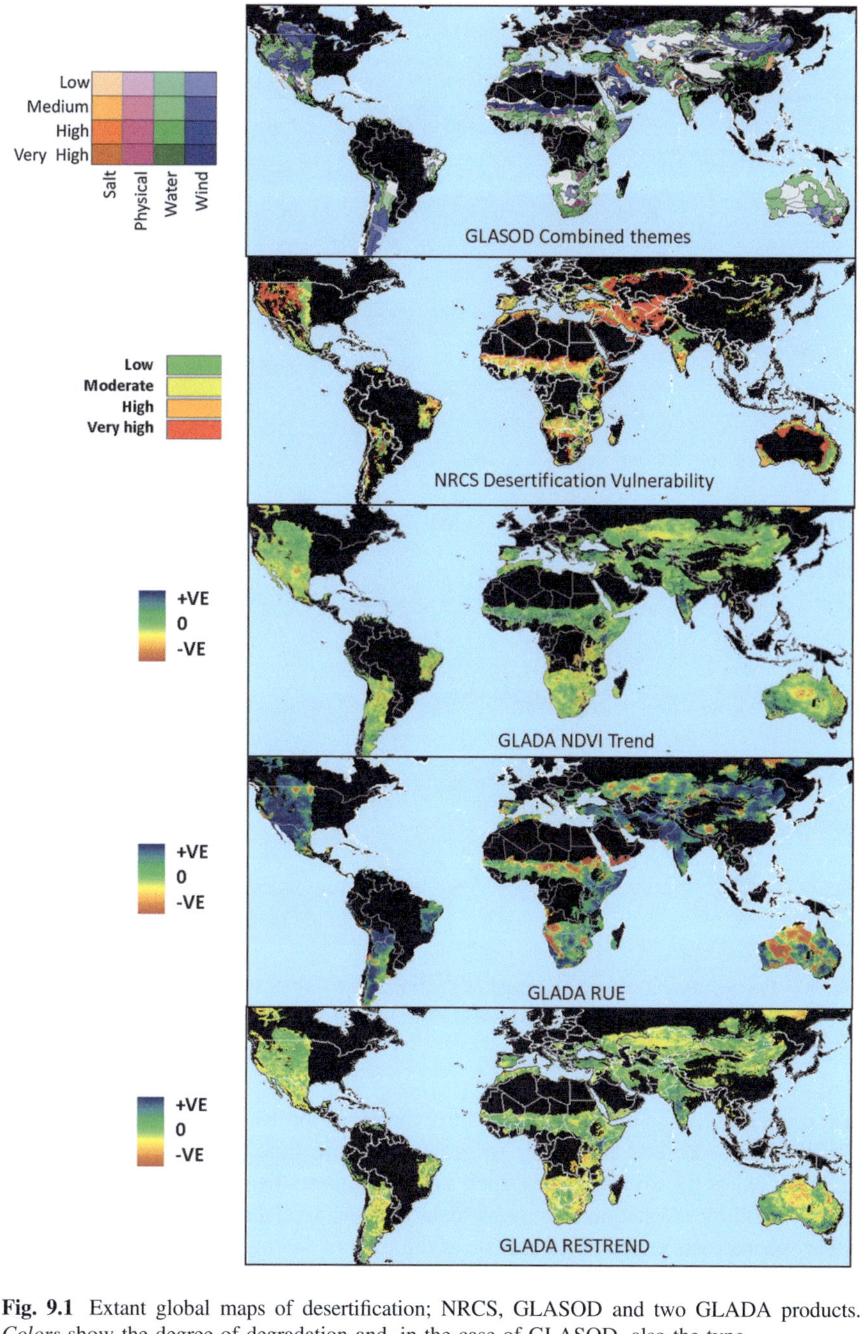

Fig. 9.1 Extant global maps of desertification; NRCS, GLASOD and two GLADA products. *Colors* show the degree of degradation and, in the case of GLASOD, also the type.

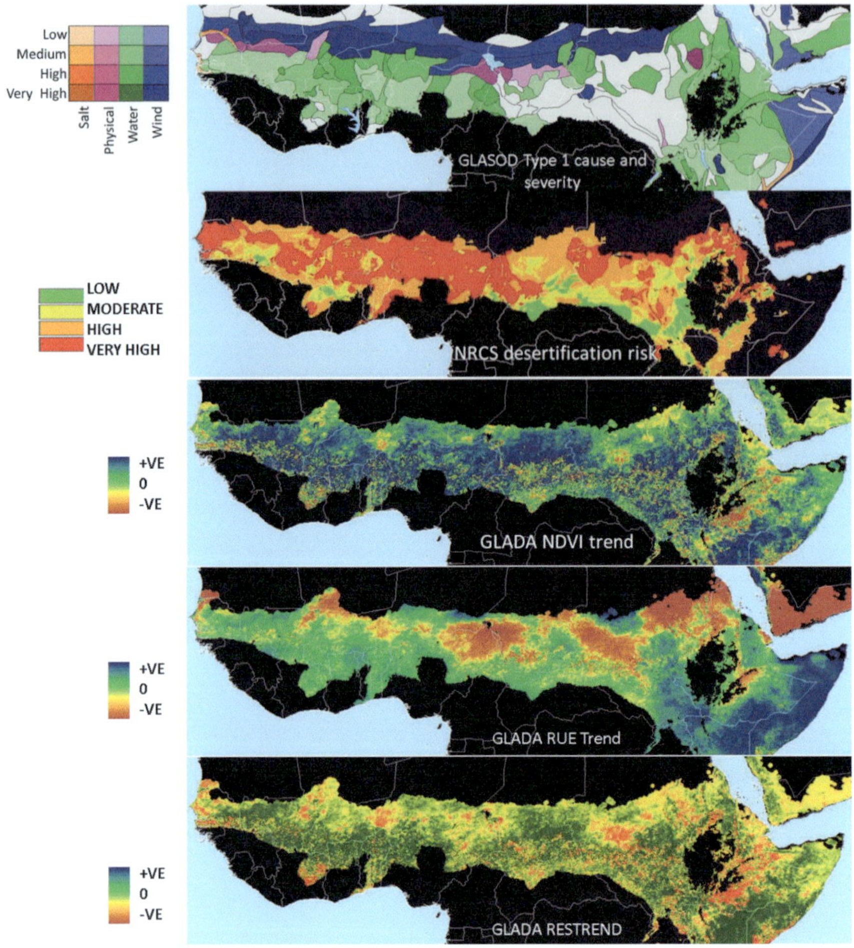

Fig. 9.2 Extracts from three global maps of desertification (Fig. 9.1) for the Sahelian, Sudanian and Dry-Subhumid eco-climatic zones. *Colors* show the degree of degradation and, in the case of GLASOD, also the type.

in the indices can be difficult to measure, therefore indirect modeling was used to translate stressors and processes to indices. At the time of writing, no suitable indicator exists for some services, such as economic, so the system is designed to be ready to add these as soon as suitable data become available. GLADIS outputs are under reconstruction and not available at the time of writing but they will include a database of global maps of the status and trends of the six ecosystem goods and services. Thus the GLADIS database will contain a compendium of data for the biophysical and socio-economic factors that are judged to be relevant and are available.

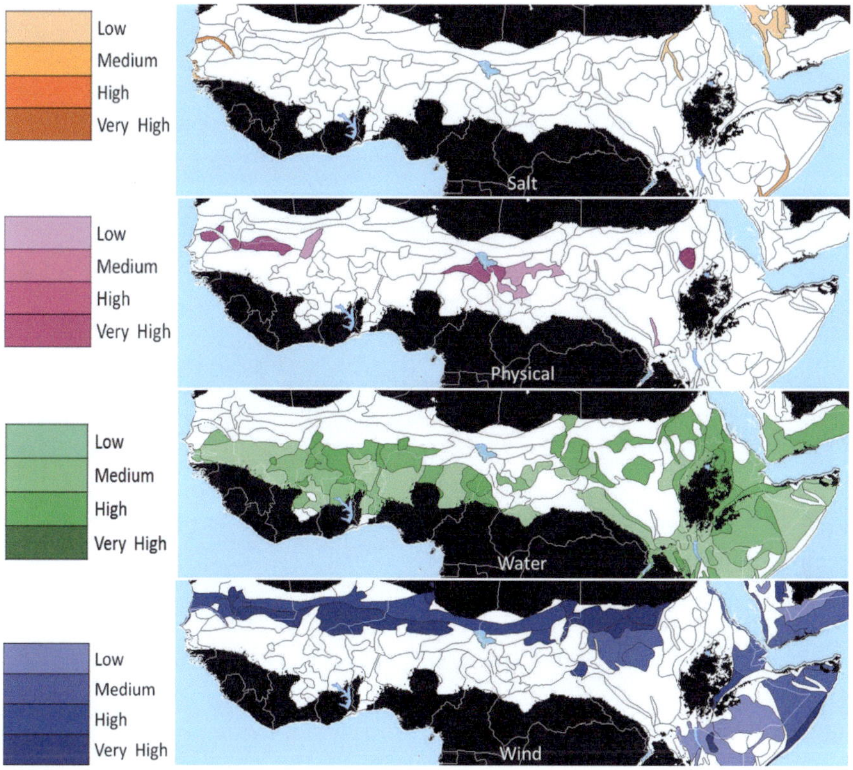

Fig. 9.3 GLASOD, its four components and severity (Oldeman et al. 1991).

9.4 Assessment of Techniques Used in Mapping

The global maps outlined above and many of their regional counterparts raise critical issues that have not previously been considered in respect of mapping degradation. Moreover, none of the existing maps meet all or, in fact, many of these criteria.

1. *Fundamental issues*

There are some basic principles and issues that apply to most degradation mapping that are all too often overlooked. Some of these are listed with brief explanations in Table 9.3.

2. *Subjective and objective data and interpretation*

Subjective assessments, especially by multiple observers (e.g. GLASOD), even if they are experts, are subject to errors in consistency and reproducibility. Monitoring using repeated expert assessments is subject to even larger errors which can easily be greater than any real changes that might have occurred. Sonneveld and

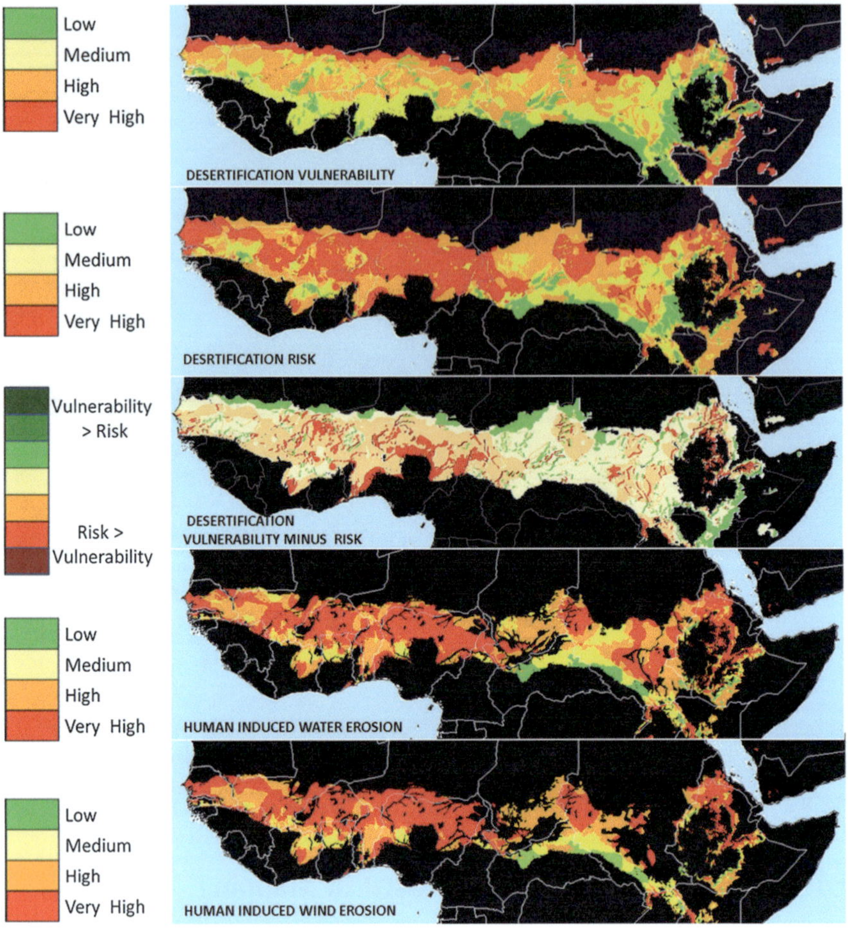

Fig. 9.4 Five NRCS themes, each showing four levels of Vulnerability (Beinroth et al. 2001).

Dent (2009) found that the GLASOD maps were only moderately consistent and hardly reproducible, and concluded that they are "not very reliable". The NRCS maps avoid some of the worst consistency problems of GLASOD by using a single, small group of experts, but assignment of values of severity to the soils and climate data was still subjective and therefore could not be repeated by others with adequate accuracy for monitoring.

3. *Meteorological and environmental normalization*

Removing the effects of variations caused by short-term fluctuations in the environment, such as interannual variation in rainfall (Holm et al. 2003), is essential to detect underlying degradation (Prince et al. 1998). Where NPP is only limited by annual rainfall, a simple index such as the rain use efficiency (RUE), that is the ratio

Table 9.3 Four fundamental principles of degradation mapping

1. Five different degradation states that are often confused	Comments	Citations
(i) Appearance of degradation	Land with low resource availability in its natural state often appears to be similar to degraded land	Safriel (2009), Castro et al. (1980)
(ii) Susceptible to degradation	Susceptible land owing to its environment, but not actually degraded	UNEP (1997), Beinroth et al. (2001)
(iii) Land recovers when stressors removed	Land apparently degraded, but within its range of resilience. When stressors removed (e.g. drought, overstocking), the land returns to its initial, non-degraded condition	Olsson et al. (2005), Tucker et al. (1991)
(iv) Temporal trend of increase in degradation	The degradation persists when stressors (e.g. drought, overstocking) are removed—and there is a temporal trend of increasing degradation	Wessels et al. (2007a)
(v) Stable, degraded state	Degraded land in static condition that does not change when stressors (e.g. drought, overstocking) are removed	Milton et al. (1994)
2. Degradation is not absolute, rather it is determined by the expectation of the land manager	Ecologically potential vegetation is not the baseline for assessment of degradation, rather the expectations of the land manager; i.e. the "utilitarian approach" of the Millennium Ecosystem Assessment. Comparisons with parks and other areas that are protected are not appropriate	Millennium Ecosystem Assessment (2005), Warren (2002)
3. Reference sites, in a non-degraded condition, are essential before land can be said to be degraded	A logical necessity for the identification of degraded land is an explicit, standard, base-line, or reference condition with which the site in question can be compared. Otherwise naturally low potential land could be mapped as degraded. Qualitative judgment of degradation by experts usually has an implicit notion of a reference, non-degraded condition, albeit subjective, based on the opinion of the assessor	Boer and Puigdefábregas (2003)
4. Point data from field plots cannot be used to make maps without explicit spatial sampling design	If spatial continuity of observations cannot be achieved, appropriate sampling schemes are needed to interpolate between field samples	Herrick et al. (2010)

of NPP (i.e. NDVI) to rainfall, can identify degraded sites since their NPP per unit rainfall is lower than non-degraded sites. However, NPP is usually affected by more than rainfall alone so additional factors (e.g. temperature, humidity) must be accounted for (Rishmawi 2013). Furthermore, if applied over several years, RUE is subject to wild fluctuations since productivity, and any other environmental factors that may be included, usually change much more slowly than rainfall, but annual rainfall in drylands characteristically varies markedly between years; thus the ratio of productivity and rainfall varies even without any change in degradation. In the short term, RUE says more about rainfall fluctuation than about land degradation (Wessels 2009). RUE calculated for averaged productivity and rainfall over a number of years (e.g., 5 years, Prince et al. 1998) reduces, but does not eliminate, this problem but at the cost of reduced temporal resolution.

4. *Comparison with non-degraded reference sites*

Fundamentally, the term "degradation" implies a comparison, and no measure of "degradation" is useful unless the condition in the absence of degradation is known with which to compare putative degraded sites (Table 9.3). Without this comparison, the causes of differences in condition could be differences in factors not related to degradation (e.g. soil fertility, climate). However, rarely is it known which sites are not degraded. The US Department of Agriculture's (USDA) Natural Resources Conservation Service (NRCS) (Herrick et al. 2010) used the data from field study plots which were rated by Natural Resource Inventory (NRI) local experts and land managers by comparison with close-by reference plots selected by them which had similar soil, climate, plant communities and NPP, however this is only possible at a field site scale and could never be applied globally. The communal lands of southern Africa (Zimbabwe, South Africa) provide an unusual circumstance where degraded and non-degraded areas are adjacent and easily identified on the ground (Prince et al. 2009; Wessels 2009; Wessels et al. 2008).

A method designed to be used for large areas has been developed called Local Net production Scaling (LNS) (Prince et al. 2009) which detects reference sites objectively. So far, results have been published for the whole of Zimbabwe (Prince et al. 2009), northern South Africa (Wessels et al. 2008) and the entire northern and southern Great Plains of the U.S.A. (Reeves and Baggett 2014). However, there is no limitation on the size of area of application, only practical considerations. Bastin et al. (1993) and Pickup and Chewings (1994) devised a technique similar in concept to LNS called the grazing gradient method. The measure of degradation was the steepness of the gradient of vegetation cover from more distant, less grazed parts of a paddock towards the typically bare area at a watering point. However, grazing gradients only exist in single paddocks. In theory, process modelling could also be used to estimate the non-degraded condition (Bai et al. 2012), but current limitations of the models, including the absence of the necessary inputs, restrict their use.

5. *Synthesis of multiple measurements of degradation*

Anthropogenic land degradation generally consists of multiple conditions and so assembly of a number of different measurements is often used (e.g. Symeonakis and Drake 2004; Zucca and Biancalani 2011). To derive indices of degradation, UNCCD uses 11 key conditions (Berry et al. 2009), WOCAT 57 (Liniger et al. 2008) and GLADA 132 (Nachtergaele and Licona-Manzur 2008). Such indices, however, have several flaws. A single index hides which of the component processes are affected—it may be several or just one—and so the condition in need of remediation cannot be known. Furthermore, combination of measurements, either implicitly or explicitly, weights the measurements of the properties by different amounts, thus emphasizing some aspects more than others (Kosmas et al. 2012). Cai et al. (2010) optimized the combinations of conditions, but the result is still an index, based on the subjective selection of inputs. Weightings based on some knowledge of the processes would be better, but this is rarely done and indices are often more recipes than objective summarizations based on the process (e.g. GLADIS).

6. *Process modeling and monitoring*

Mechanistic models simulate the processes of interest using mathematical representations. Many such models exist, relevant to different aspects of degradation (e.g. Izaurralde et al. 2007; Kirkby et al. 2008; Tamene and Le 2015). Process models are attractive since they are designed to behave according to the same rules that determine the underlying production and degradation processes. Model results can be very accurate when the fundamental vegetation characteristics, the local environment, and the biophysical processes are known. However, the more realistic models are, the greater their complexity and their need for data. The demand for data and parameters can be prohibitive and often default values have to be used with consequent reduction of accuracy. On the other hand, simple process models sacrifice realism (e.g. Bai et al. 2012). As with all modeling, calibration with field data is critical. Hybrid observation/model systems exist, such as a national programme on soil erosion in Spain. Remote sensing can also be used to add the spatial dimension to point process models (e.g. Izaurralde et al. 2007).

7. *Explicit data on socioeconomics*

In many cases mapping degradation is a step towards understanding the drivers of anthropogenic processes. At the global scale, socioeconomic and institutional indicators are available, although often highly aggregated. A number of maps relevant to human livelihoods (e.g. agriculture, population, land use) exist that have the necessary characteristics of resolution, coverage, accuracy and repeatability, but many based on field surveys are not frequently repeated. Nevertheless, progress is beginning to be made by linking socioeconomic models with remote sensing (e.g. GLADIS, Nachtergaele et al. 2011) thereby enabling spatial representation of the human causes and effects of degradation.

8. *Inventories*

Many inventories of factors presumed to be caused by degradation have been created by extracting country- or regional-scale data from global maps (e.g. Reich et al. 2001; Bai et al. 2008). However, those derived from desertification maps can be no better than the individual maps used to create them. Others are based on country statistics (e.g. den Biggelaar et al. 2004).

9. *Syndromes*

Syndromes of degradation that describe "archetypical, dynamic, coevolutionary patterns of human-environment interactions" have been proposed and applied at limited scales (Petschel-Held et al. 1999; Geist 2005). They are derived from qualitative studies of the physical and human aspects of selected desertification case studies. Syndromes have been used in relation to degradation and its socioeconomic effects (Ibáñez et al. 2008) and in a predictive model (Sietz et al. 2006). While not applied to global mapping, Geist (2005) has provided an inventory of types of desertification. While attractive as summaries of the nature of specific degradation processes, the selection of types of syndromes was not based on any objective scheme.

10. *Remote sensing*

Land cover mapping is frequently undertaken using a wide range of types of satellite data and analytical methods. Suitable remote sensing data are now available over most of the Earth, at regular intervals, consistently and with defined accuracy. Some measurements at scales of <1 m are available thus spanning the range between local anthropogenic, sub-national to global scales. Furthermore, instruments carried on satellites can measure more properties than those used for land cover (Hill 2001; Ustin et al. 2009; Bai et al. 2012) and there are many emerging techniques (Dewitte et al. 2012) that have application to drylands (Table 9.4) (Zucca et al. 2012; Ustin et al. 2009) such as wildfires, actual evapotranspiration, woody biomass, species types, grassland and crop production, standing dead vegetation, dust generation, human population density, crop yields and others. However, remote sensing cannot measure directly many of the traditional indicators of degradation, such as changes in vegetation species composition (Mbow et al. 2013) but, for some indicators, they can provide proxies—that is spectral correlates of the desired metric (e.g. the use of NDVI as a proxy for NPP, Seaquist et al. 2009). Most remote sensing techniques yield quantitative measurements that allow severity to be assessed objectively. The high temporal frequency enables the use of time-series of vegetation properties that can measure, for example, productivity of natural vegetation and crop yields and the timing of fires, distinguishing late season fires which are more destructive and those earlier in the dry season. Remote sensing data are useful in their own right but more often as an input to analyses that include other types of data (e.g. Tamene and Le 2015). Recent developments are expanding the range and several are being developed that incorporate socioeconomic conditions (Brown et al. 2014).

Table 9.4 New and emerging remote sensing technologies for mapping land degradation

(i) Optical	Addition of wave bands, improved calibration, corrections for atmosphere and bidirectional reflectance effects, enhanced accuracy
(ii) High spatial resolution optical data	20 to <1 m spatial resolution is appropriate for observation of anthropogenic surface features. Many systems have a limited spectral range, but this is changing. Google Earth is publically available but only suitable for visual analysis. Data are currently expensive but some companies will provide data free for research and development
(iii) Hyperspectral (imaging) spectrometers	Increasing number of surface conditions can be measured using more wavelengths or finer spectral band widths. Mostly on aircraft, few and limited coverage with satellite-borne sensors at this time
(iv) Thermal	Land surface temperature and energy balance are closely related to the water balance and soil moisture including actual evapotranspiration. There are many satellite-borne thermal sensors and new instruments and several analytical techniques, with more under development
(v) RADAR	Topography, surface roughness, surface soil moisture, vegetation cover
(vi) LiDAR	Topography, slope and aspect, vegetation height (accuracy is very high—often much less than 1 m), biomass and canopy structure. Elevation
(vii) Integrated field data collection and computer-aided analysis	A powerful combination of portable instruments are now available which increase the quality and quantity of field data and its numerical analysis. These include ruggedized computers that can run GIS and image processing software on remotely sensed data in the field, linked to Global Positioning Systems, digital cameras and digital entry of field observations

Remote sensing, however, is no panacea; the data themselves may suffer from uncontrolled factors, such as cloud cover and haze that may prevent acquisitions on critical dates, also technical issues that degrade the signal, such as differences in the illumination and view angles in successive orbits. Notwithstanding its limitations, remote sensing remains the only available way to address the sub-national to global scales (Prince 2002), despite the limited range of types of measurements. Surprisingly, remote sensing methods for mapping and monitoring degradation are still resisted by some (Reynolds pers.com.). This opinion would, of course, be understandable where the variables that can be measured by remote sensing are not relevant to a specific application, such as some socio-economic processes, although these too are beginning to use remotely sensed inputs (Brown et al. 2014).

9.5 Assessments of Maps of Global Dryland Degradation

9.5.1 Overview

As discussed above, there are just four sets of maps of degradation or major aspects of degradation that have global coverage and are available in digital form (GLADIS was under review at the time of writing). Unfortunately independent data and especially field observations with which to assess the accuracy of the maps are almost completely lacking, thus most maps of degradation are not validated and their value is therefore unknown. GLADA has been assessed qualitatively in three countries (China, Kenya, South Africa) with mixed results (e.g. Wessels 2009), but detailed assessments are not available for NRCS or GLASOD.

9.5.2 Methods Used to Make Maps

An assessment of the validity and hence utility of the desertification maps must start with an examination of the methods and data that were used in their creation.

1. *GLASOD*

GLASOD maps (Figs. 9.1, 9.2, and 9.3) were created using "structured informed opinion": "structured", in that a unified, global scheme of description was used and "informed", in that the information was provided by regional experts. The world was divided into 21 regions, each of which had its own team of experts led by a "Correlator". In the first step in each region the main physiographic units were established by the regional team using their knowledge of geology, topography, soils, climate and vegetation. These 21 units were compiled into a single world map. The physiographic units differed greatly in size, which affects the appropriate scale of use of the maps. According to the authors, they should only be used for continental—global applications (Oldeman and van Lynden 1996), nevertheless, there are many cases of country-scale uses.

The percentage of each mapping unit affected was simplified into five intervals and the degree of degradation in four levels, giving 20 classes for each of the four degradation types (water erosion, wind erosion, chemical deterioration and physical deterioration). These 20 classes were summarized into five which combine area and severity, each shown by a different density of the same color as in Figs. 9.1, 9.2 and 9.3. For each polygon only the color of the class with the highest score is shown and, if a second is also present, its score is given in text annotations. In addition, each unit has notes that summarize the main characteristics of the degradation. All polygons except areas considered stable were placed in one of the four types, even if as little as 1–5 % of the polygon's area was actually affected, so the extent of degradation shown on the map is visually exaggerated.

2. *NRCS*

The U.S. NRCS maps (Figs. 9.1, 9.2, and 9.4), as noted above, show "Vulnerability" to degradation, salinization, physical damage, and water and wind erosion (Beinroth et al. 2001). Each of these conditions is coupled with a map of "Risk", calculated by weighting the Vulnerability by human population density (Deichmann 1994). The sources of the classification units were the Suborders of the FAO/UNESCO Soil Map of the World (FAO/UNESCO 1991), a pedoclimate map compiled by the authors using a water-balance model and data from 25,000 weather stations. The quality of the climate data is unknown since the data base has not been published or assessed against other climatologies. The 1991 U.N. Food and Agriculture Organization (FAO) soils map shows only taxonomy, not physical properties, and so to assign Vulnerability, the authors used the available field measurements of erosivity, extrapolated to entire soil units.

The criterion of land quality was its ability to sustain grain production—as a measure of food security—and the estimated cost of remediation of each of the stresses. The main factors considered were the coefficient of variability in rainfall, depth, extreme levels of chemical and physical condition, resilience and general information on each soil type. The locations of serious human conflicts as defined by the International Peace Research Institute (IPRI) were also included. 24 stress conditions were defined, based mainly on properties associated with the soil types. These conditions were summarized into nine land quality classes and each of these into four levels of Vulnerability to wind, water, physical damage and salinity. The maps were gridded at 0.033° (16 km^2 at the equator). The authors compared their map with GLASOD and plot data from India at specific sites but not the class boundaries.

3. *GLADA NDVI, RUE and trends*

The GLADA programme (Bai et al. 2008) used NOAA AVHRR NDVI satellite data to estimate the amount of vegetation in each pixel. 25 years (1981–2006) of NDVI data were obtained from the NASA GIMMS group (Tucker et al. 2005). The data (pixels) have 0.05° × 0.05° resolution (approximately 36 km^2 at the equator) but several technical characteristics of the sensor system (not noted by the authors) reduce this to about 324 km^2. Monthly rainfall data at 0.5° resolution (3136 km^2 at the equator) were obtained from VASClimO and CRU TS (Harris et al. 2014), and temperature from CRU alone. Pixels were eliminated from the analyses if their rainfall was not correlated with NDVI.

The principal analytical technique was to examine temporal trends of annual NDVI, RUE, and RESTREND (Wessels et al. 2007b; Bai et al. 2008) to detect negative trends (degradation) and positive trends (recovery). The difference between RUE and RESTREND is that the RESTREND results are expressed in the time dimension so trends can be detected, whereas RUE displays changes in the rainfall dimension, giving only a snap-shot.

The main drawback of NDVI alone is that it is not normalized for interannual meteorological variations, which therefore dominate the results. In the case of RUE and RESTREND another problem is that it is not known a priori which sites are degraded and which are not, so there is no standard with which to compare. Usually the average (or best–fit of a regression of NPP on rainfall) of all RUE values is used as an estimator of the non-degraded RUE but, since the average or regression lines include both degraded and non-degraded pixels, the relationship is usually an underestimate. Other difficulties have also been noted (Rishmawi 2013; Wessels et al. 2012). However, RUE and RESTREND have an important advantage, which is the ratios are for individual pixels so the effects of between-pixel variation in the environment (e.g. soil, meteorological variables, topography) are eliminated.

Elaborations of both RUE and RESTREND have been developed, including the use of improved methods of estimating the potential NPP, such as using additional meteorological variables (e.g. temperature, humidity) which, together, give a more accurate measure of short-term meteorological variation (Rishmawi 2013) and simple models of production (Bai et al. 2012), however, estimation of the potential NPP still remains a source of inaccuracy.

Pickup and Chewings (1994) suggested that degradation should be assessed in wet years when the contrast between reference and other locations are most evident. However, Wessels et al. (2008) found that the proportional difference between degraded and a non-degraded reference remained the same irrespective of the rainfall in their study region.

9.5.3 *Geographical Review of Maps and Their Relationships with Environment*

Strong correlations of areas of degradation and environmental factors related to degradation would be expected in any map of degradation. Since GLASOD and NRCS were published, many data sets of environmental factors and aspects of human utilization have become available. Here the relationships are analysed qualitatively and quantitatively.

Digital versions of the global and Sahel GLASOD, NRCS, GLADA NDVI trend and GLADA RUE maps were coregistered with five maps showing environmental and human factors (Table 9.5). The GLADA RUE trend map was omitted since it is essentially the same as the NDVI trend. Masks were used to remove areas where either of each pair of maps had no data. A numerical comparison of the degradation and environmental maps was carried out using a "fuzzy numerical" version (Hagen-Zanker et al. 2006) of the original kappa (κ) test (Cohen 1960) (Table 9.6). Kappa varies from 0, if there is no agreement, to 1 indicating complete agreement. The similarity of the 25 year mean environmental variable for single pixels (κ) were mapped allowing visual interpretations of the spatial patterns of agreement and disagreement. The overall similarity of each map pair ($\bar{\kappa}$) was also calculated.

Table 9.5 Sources of environmental data shown in Figs. 9.5 and 9.6 and discussed in the text

Environmental data used in Figs. 9.5 and 9.6	
(i) Aridity, the ratio (dimensionless) of precipitation to potential evapotranspiration	Trabucco and Zomer (2009)
(ii) Net primary production	USGS (2015)
(iii) Ecoregion	Olson et al. (2001)
(iv) Population	CIESIN (2005)
(v) Land use	Ellis and Ramankutty (2008)
Data discussed in text	
(vi) Salinity	Batjes et al. (2012)
(vii) Human influence index (HII)	Wildlife Conservation Society (2005)
(viii) Ratio of cattle density to NPP	Robinson et al. (2007)
(ix) Soil nutrient retention	Batjes et al. (2012)
(x) Soil nutrient availability	Batjes et al. (2012)

Table 9.6 Comparison of three degradation maps with five environmental variables using the fuzzy-numerical kappa test ($\bar{\kappa}$) (Hagen-Zanker et al. 2006)

Global	Aridity	Land use	Primary production (NPP)	Population	Ecoregion
NRCS	0.73	0.68	0.57	0.43	0.36
GLASOD	0.70	0.54	0.56	0.53	0.31
GLADA NDVI trend	0.18	0.13	0.17	0.17	0.10
GLADA rain use efficiency (RUE)	0.17	0.12	0.18	0.25	0.10
Sahel					
NRCS	0.73	0.76	0.47	0.35	0.28
GLASOD	0.59	0.58	0.54	0.58	0.19
GLADA NDVI trend	0.06	0.05	0.07	0.07	0.04
GLADA rain use efficiency (RUE)	0.38	0.28	0.49	0.61	0.10

0 indicates no correspondence and 1 complete agreement. Sources of the environmental data are shown in Table 9.5

1. *GLASOD*

There are some studies where GLASOD has been compared with one or more environmental variables, occasionally finding degradation where a specific range of values of an environmental variable occurs. However, correlations were mostly one-directional, that is, although an environmental factor may frequently be found where degradation is shown by GLASOD, the reverse is not true: areas with the same environmental values were not always degraded and in no cases could they predict degradation. This point is very clear in the case of salinity which is more easily recognized in the field than most others degraded conditions. While salinity shown in GLASOD generally agreed with the Harmonized World Soil Database (HWSD, Batjes et al. 2012) map, many saline areas in the HWSD were not shown in the GLASOD salinity map.

In the GLASOD maps severity was higher in more arid areas (Table 9.6) but generally stopped short of the Hyperarid zone. By far the largest contributions to overall degradation were wind and water erosion; salinity and physical deterioration dominated in only a very small parts of the maps. Many of the GLASOD maps had areas in which the types of degradation and their severity coincided with putative causal factors (compare Figs. 9.1 and 9.5) and had moderate to high $\bar{\kappa}$ (Table 9.6). Thus it seems that GLASOD coincides with a set of environmental conditions, not distinguishing these from actual degradation. For example, in drylands with low (but not zero), productivity, low nutrient retention, low nutrient availability, used for livestock production, and where human impacts are large, had greater severity, suggesting that the maps are more a statement of susceptibility than of actual degradation. There were, however, some notable exceptions, such as the large area shown as degraded in Argentina, in South Africa and in E. China that differed from the environmental patterns with which they were compared.

GLASOD mostly shows degradation in the zone between Hyperarid and Dry Sub-humid (except for the most northerly steppes) but many large areas within this aridity range are not mapped as degraded (e.g. much of South America). Some correlations with soil characteristics might have been expected, given that GLASOD was largely based on soils, but no strong ones were found—at least with soil nutrient status. Only moderate correlations with population were found (but the dates of GLASOD and population differ). The productivity in areas mapped with high severity was almost always in the 0–0.3 gm^{-2} year^{-1} range, except in temperate regions where it was sometimes above this range. Surprisingly the similarity with ecoregion was lower.

When limited to the Sahel region (Figs. 9.2, 9.3 and 9.6), the patterns were similar to and more clear than at the global scale. $\bar{\kappa}$ values were very similar for all factors ($0.59 - 0.54$, Table 9.6) except ecoregion.

2. *NRCS*

NRCS Vulnerability was highly correlated with aridity and land use ($\bar{\kappa} = 0.73 - 0.68$, Table 9.6), moderately with NPP ($\bar{\kappa} = 0.57$) and least with ecosysyem ($\bar{\kappa} = 0.36$). In the Arid and Dry-Subhumid zones, the map (Figs. 9.1 and 9.5) was essentially the same as aridity zones and low NPP (Table 9.6). In-between, in the Semi-Arid zone, the NRCS map showed mainly High with smaller invaginations of the Moderate class. The Risk maps had a higher correlation with population, which is not surprising because population is one of its ingredients. The Mediterranean, Central Asia, India and the Sahel had largely High to Very High Risk, of which much was related to human activities, while drylands in the rest of the world were mostly Moderate to Low. The strong correlation between population and Risk is probably also responsible for the positive correlations with the Human Influence Index (HII) and cattle density per unit NPP (Table 9.5).

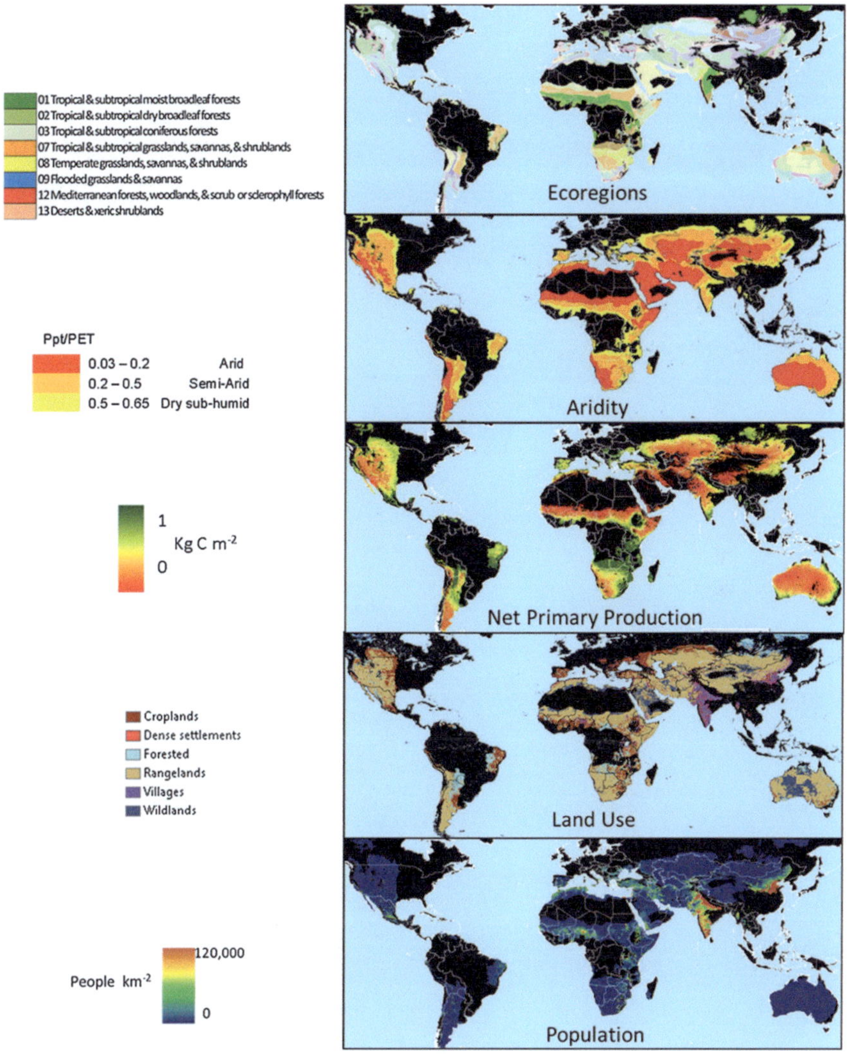

Fig. 9.5 Global patterns of five variables related to degradation. See Table 9.5 for sources.

High degradation in tropical Arid and Semi-Arid zones was almost exclusively in grasslands, dry woodland, savannas and scrub. At a regional scale, however, there were some unexpected features. For example, in USA, it was shown as High in the wetter, eastern Great Plains, but declined westwards to Low even as aridity increased. Beyond the Front Range of the Rocky Mountains, degradation was again positively related to aridity. A similar pattern occurred in E. Australia and China. If the USA feature has any meaning, it may reflect the more intense land utilization in the least arid (eastern) part of these drylands associated with higher degradation,

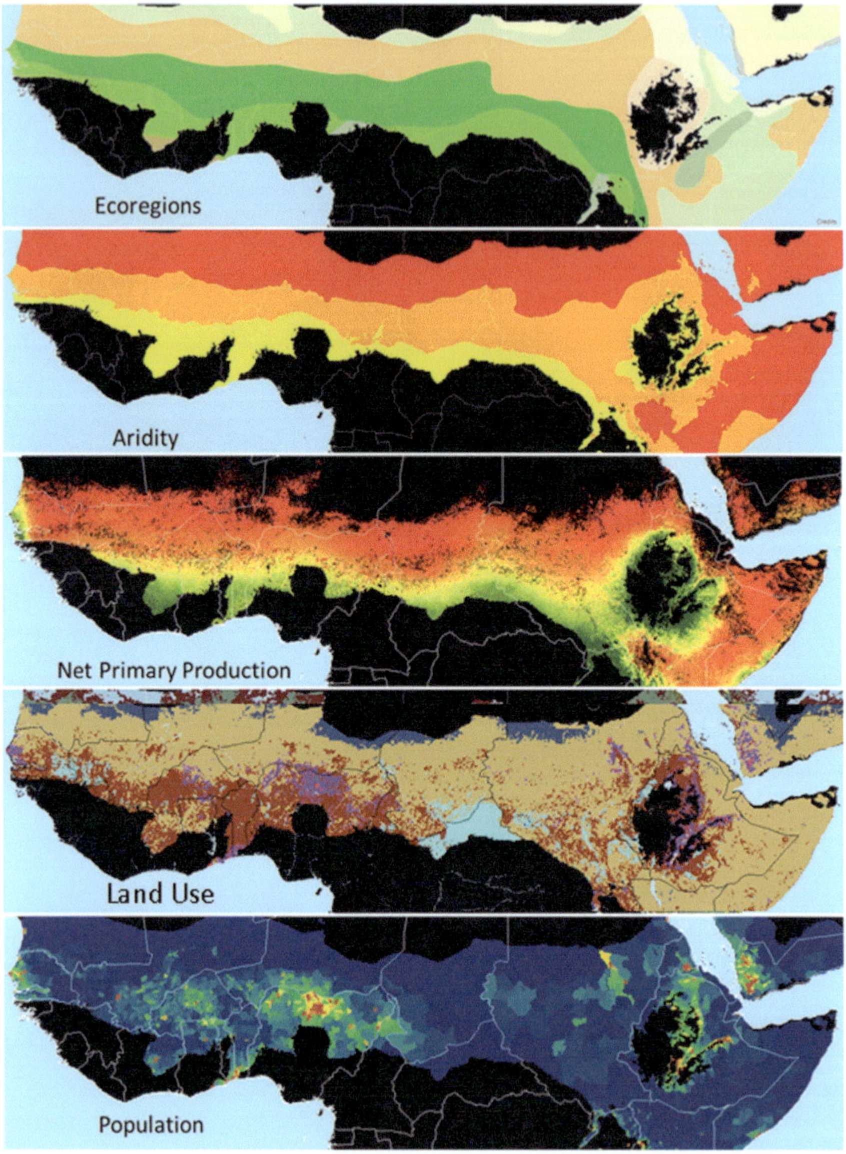

Fig. 9.6 Sahel, Sudanian and Dry-Subhumid zone patterns of five variables related to degradation. See Fig. 9.5 for *color* scales and Table 9.5 for sources.

followed by reduced degradation in the rangelands of the High Plains to the west and high again in the mountainous West.

The NRCS maps of overall desertification were closely related to NPP. In India, Moderate was associated with human activities almost exclusively, but it is surprising that the densely populated Ganges valley was shown High, rather than Very

High; maybe this reflects the high NPP which offsets heavy utilization. In N. Africa, while Vulnerability increased southwards, Risk was the reverse, presumably owing to the higher population along the Mediterranean coast. The highest values of saline-induced degradation in the NRCS map correlated quite closely with the HWSD salinity map.

Wind erosion was largely zonal, like overall degradation, increasing with aridity up to the boundary of the Semi-Arid and Arid zones, and then decreasing into the Arid zone. However, large areas within the drylands are not shown in the NRCS map. In China human causes were mainly at the wetter margins, while in India, vast areas were attributed to human action. Central Asia is shown as Very High and most of this, too, can be associated with human action. Only small areas of anthropogenic wind erosion were shown in Australia and South America. The Dust Bowl region of the USA was mapped with a Low Risk of wind erosion in spite of signs of a return of the erosion of the 1930s (Romm 2011). In the Sahel, where all the erosion was attributed to human action, most of the high values of wind erosion were in the Semi-Arid zone and decline to the north and south.

Most parts of the world's drylands are mapped as High and Very High for water erosion. High values occurred particularly in the Arid zone, but not in the most Arid areas, except for the high plains of the USA where, once again, along the east to west gradient of increasing aridity, water erosion was first high, then fell in the High Plains and increased into the western mountains. Southern Africa and Australia were generally Low to Moderate for water erosion.

In the Sahel, the north-south zonal patterns dominated the NRCS maps and the environment. Once again, the differences between Vulnerability and Risk reflected the population distribution, shown clearly by the similarity of their arithmetic difference and population. The measures of similarities (Table 9.6) were high for aridity and land use ($\bar{\kappa} = 0.73 - 0.68$, Table 9.6), but only moderate for NPP and population ($\bar{\kappa} = 0.47 - 0.35$, Table 9.6). Ecoregion was low, as with GLASOD, possibly because of the coarse resolution of the ecoregion map compared with the degradation maps.

3. *GLADA*

GLADA (Bai et al. 2008) undertook detailed analyses of NDVI trends and found negative values in most Arid zones and also widespread, but less-coherent, patches in Dry-Subhumid zones. Large positive trends were mainly in the Semi-Arid zone in-between, but all three zones had small, scattered areas with positive trends. However, when the patterns were compared quantitatively there were effectively no similarity with any of the environmental factors ($\bar{\kappa} = 0.10 - 0.18$, Table 9.6) and even less so when restricted to the Sahel ($\bar{\kappa} = 0.07 - 0.04$, Table 9.6). This suggests that the NDVI trend was insensitive to any of the environmental factors associated with degradation—a surprising finding. In the time-series of single pixels, both negative and positive trends occurred, and so it must be questioned whether, in view of such rapid changes and reversals, they constitute degradation in the sense where the time scale of recovery from true desertification is often longer than 25 years.

The global pattern of RESTREND was similar to the NDVI trend but more intense and widespread. Exceptions occurred, for example the Gedaref (Al Qadarif) province of eastern Sudan which was positive with NDVI trend and negative negative with RESTREND. The Sahel Semi-Arid belt had large positive NDVI trends, but smaller areas of RESTREND were positive and all had low values. About half of the Sahel had negative RESTREND values. Thus RESTREND, with its rainfall normalization, suggests large areas of degradation even though NDVI was increasing. The RUE pattern was similar to RESTREND, but more coherent. Taken together GLADA suggest that the well-known "greening" of the Sahel (Olsson et al. 2005) is masking underlying degradation of productivity, in contrast to Rishmawi (2013) who used similar methods and found only isolated patches of degradation in the region.

Unlike NRCS and GLASOD, both of which used qualitative, expert judgment, GLADA products are derived from numerical measurements. However, a serious drawback is the uncertainty in interpretation—do negative trends indicate degradation? The finding here of effectively zero correlation with causative factors is not encouraging. Assessments of GLADA products carried out in partner countries (Dijkshoorn et al. 2008) gave mixed results (Wessels 2009).

9.5.4 Inter-comparisons of Degradation Maps

Since there were no strong correlations of any of the maps with the selected environmental variables, a comparison of each map with the others was carried out to see if there is any consistency of mapping of degraded areas. While agreements between maps are of interest, of course they do not indicate validity: on the contrary, given that GLASOD has been found to have serious shortcomings (Sonneveld and Dent 2009; Nachtergaele et al. 2011; UNCED 1992), any agreements with other maps could be because they agree in their invalidity.

Both the overall global and Sahel comparison results (Figs. 9.7, 9.8 and 9.9) revealed two groups of maps that were similar amongst themselves, but very different between. The first set consisted of the GLADA maps (NDVI Trend, RESTREND, and RUE) all of which had $\bar{\kappa}$ values greater than 0.6. The similarity was greatest for the NDVI Trend *v* RESTREND comparison, probably because both are linear fits to temporal trends of the satellite data. However, the level of agreement suggests that the rainfall normalization had rather little effect on RESTREND, at least at a global scale. The lesser agreement with RUE is probably because RESTREND expresses the changes in NDVI through time, while RUE shows the degree of degradation with respect to rainfall. The maps of κ for each pixel (Figs. 9.8 and 9.9) were not random but had geographically coherent areas of similar values.

The second group of maps with internal similarity was GLASOD and NRCS. Unlike the NDVI-derived maps which are global, less of the world is covered by these, and an even smaller area jointly, but they do include most of the Arid,

Fig. 9.7 Comparisons of all 10 pairs of five global degradation maps. *Kappa* ($\bar{\kappa}$) indicates the degree of agreement of map pairs. Global results are shown in *black* and Sahelian, Sudanian and Dry-Subhumid zones (labelled Sahel+) in *grey*.

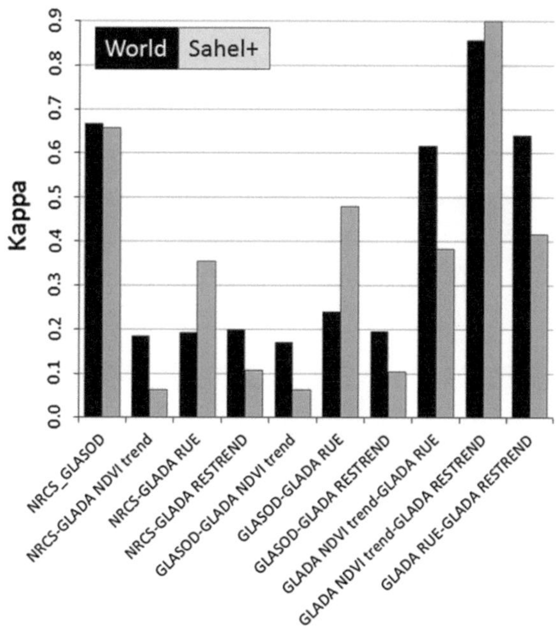

Semi-Arid and Dry-Subhumid regions. Their relatively high similarity ($\bar{\kappa} = 0.667$) may be because properties related to soils were the most important inputs used in both maps. Again, the areas of agreement and disagreement tended to be in distinct patches with sharp boundaries.

Although the three maps derived from NDVI have overall low similarities with GLASOD (also reported by Gibbs and Salmon 2015) and NRCS, the kappa for both NRCS and GLASOD with RUE in the Sahel were higher than the other comparisons between the two groups. This is most likely a result of the similarity of GLASOD and NRCS, so both have similar correlations with RUE. Gibbs and Salmon's (2015) results for two of the maps included here were very different, probably because they included all parts of the Earth, not just drylands, and areas where one or more maps had no data.

9.6 Discussion

The general lack of agreement between the degradation maps and the lack of adherence to basic principles of detection leads to the conclusion that the questions, "*What*?", "*Where*?" and "*How severe*?", at the global and Sahel scales, are largely unanswered, and the maps that exist are unsuitable for most of the applications for which they are used. There are many reasons for this, including the following issues.

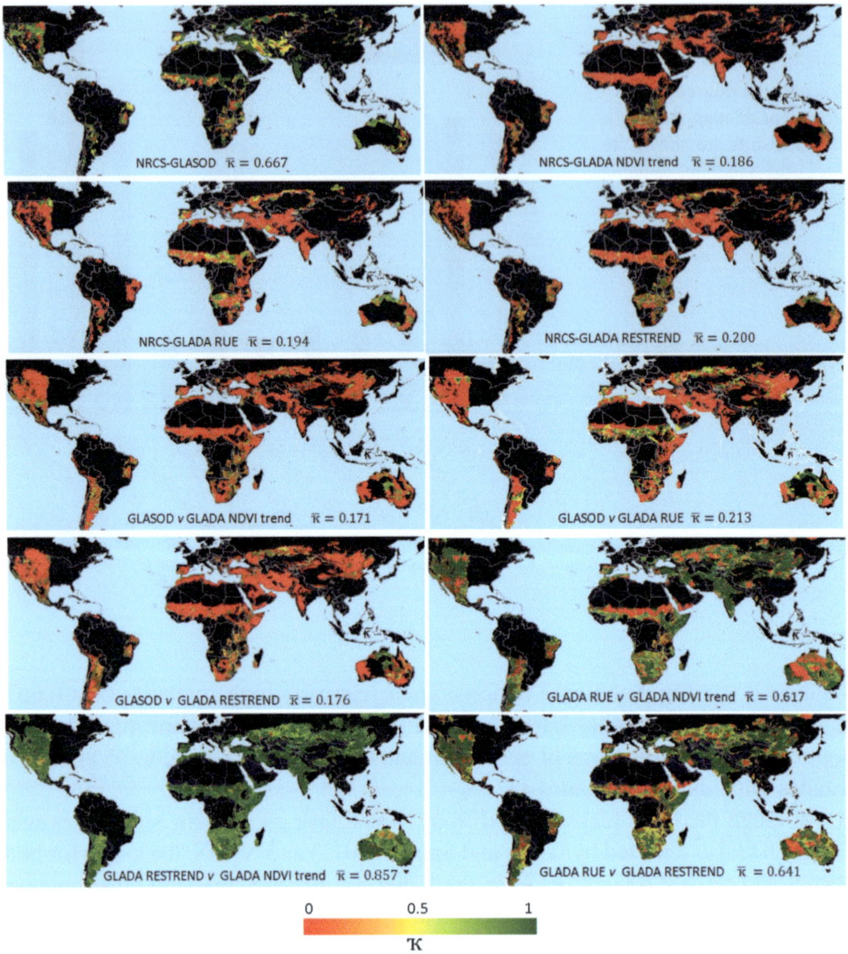

Fig. 9.8 Comparisons of all 10 pairs of five global degradation maps. Individual pixel Kappa (κ) shown in *color*: *1* total agreement, *0* total disagreement. Global summary comparisons ($\bar{\kappa}$) are given on each panel. *Black areas* indicate areas that are not Arid, Semi-Arid or Dry-Subhumid or not shown in one or both maps.

9.6.1 Pursuit of the Undefinable "Desertification"

It was only in the immediate aftermath of the Sahelian droughts that maps were published of a unitary property of "desertification" (e.g. Dregne 1983), and these were, in fact, oriented to a particular set of conditions (similar to the "Sahelian" syndrome of Geist 2005) rather than the overall concept enshrined in the "World Atlas of Desertification". It is surprising, therefore, that the idea of a singular property of "desertification" is still held. In the special case of GLASOD, because

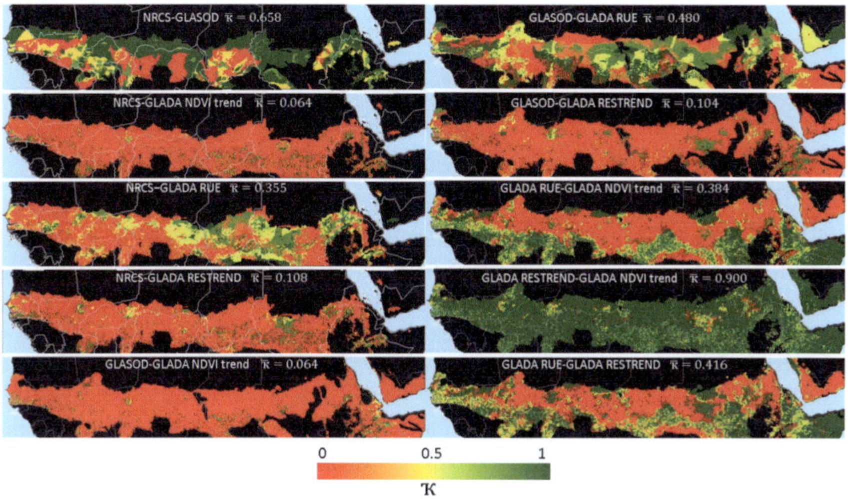

Fig. 9.9 Comparisons of all 10 pairs of five Sahel degradation maps. Individual pixel Kappa (κ) shown in *color*: *1* total agreement, *0* total disagreement. Summary comparisons for the whole maps ($\bar{\kappa}$) are given on each panel. *Black areas* indicate areas that are not Arid, Semi-Arid or Dry-Subhumid or are not shown in one or both maps.

there is no overlap of causes and severities between mapped units (Fig. 9.3), the maps of each type of degradation can be shown together in one image (Figs. 9.1 and 9.2), but this does not imply that it is a map of desertification in the holistic sense. Similarly, the several NRCS maps each show the Risk and Vulnerability of a separate condition, not all together. The GLADA maps are all based on a surrogate for NPP, again, not an overall "desertification" metric. It has been proposed by Prince (2002) that, because several factors associated with degradation result in loss of productivity, NPP can be regarded as a surrogate for these in addition to productivity itself; however, this is an integration of processes that control productivity, not a candidate for an all-encompassing metric of desertification. Whether the use of production as an integrator is adequate or not, the result is a study of impacts on productivity, not on "desertification".

The notion of a process of "desertification" has recently returned in the rejection of individual degradation processes in favor of an index that attempts to link all factors that lead to human loss (Bridges and Oldeman 1999; Reynolds and Stafford Smith 2002; Nachtergaele et al. 2011). However, there are good reasons not to combine metrics: all variables cannot be measured with equal accuracy; an individual process may be of interest, such the soil carbon balance, rather than "desertification"; and a synthetic index requires the use of arbitrary weightings of the individual processes. Nevertheless, it has been suggested that the degrees of interactions vary, some are stronger than others, and so some simple variables may be adequate in some cases (Reynolds et al. 2007).

Progress will be difficult unless the unitary concept of desertification is abandoned in favor of mapping the drivers, processes and outcomes.

9.6.2 Limited Measurement Techniques

While a large number of objective metrics of degradation can be used in the field (DESIRE 2008), upscaling of most of these is generally not possible beyond the local site scale because spatial variations between field sites cannot be taken into account. More fundamentally, local measurements often cannot be used for larger areas because of the incongruity of data at different scales. Similarly, qualitative expert opinions, while not to be dismissed, particularly in view of the critical role of humans in the designation of degradation, are severely limited when spatially and temporally extensive coverage is needed. Furthermore, maps based on opinion, whether "expert" or not, are qualitative and therefore cannot be used for monitoring.

Satellite data and their derivatives, while expanding rapidly in type, precision and resolution, are not substitutes for field measurements. They can provide only a few of the properties often measured in the field, moreover, even measurements of the same variable (e.g. vegetation cover) differ owing to the differences in measurement techniques. Nonetheless, there is an expanding inventory of remotely-sensed variables that can be used directly or as proxies, and the fact remains that there is no substitute for the intrinsic spatial- and temporal-continuity of remotely-sensed measurements. New satellite measurements with local to global coverage, high spatial resolution and long-term archives are emerging and these, coupled with rapidly improving global meteorological data and continuing improvements of maps of critical properties such as soil fertility, warrant some optimism.

9.6.3 Lack of Comparison with Non-degraded Sites

An incontrovertible reality for any measurement system is that it requires a reference standard. Without this, statements about more or less degradation or "desertification" are meaningless. Despite obvious difficulties, comparisons within areas that are not degraded are essential.

9.6.4 Slow Degradation and Fast Changes in Weather

In most cases, degradation is a gradual process and its presence is often obscured by higher temporal frequency variations in the environment. The most obvious case is rainfall; no two growing seasons are the same and neither are the responses of the vegetation. Without being able to normalize these factors, any supposed measurement of degradation is only suggestive at best. Where there are recognizable cycles of these other factors, degradation may be revealed by averaging over the period of the cycle (Prince et al. 1998). However, such regularities are rare (Wessels et al. 2007b).

9.7 Conclusions

The *What*? question, if in reference to "desertification", is meaningless. On the other hand, if applied to specific conditions and processes, such as losses of productivity, biomass, soil fertility, or erosion, and can be measured quantitatively, "degradation" can be defined rigorously. Thus, if specific degradative processes are selected, rather than a nebulous, all-encompassing concept of "desertification" and objective measurement techniques exist, progress can be made.

The *Where*? question can only be answered for a specific "*What*" answer. Even if the condition is well-defined, a limitation can be the lack of an appropriate method for mapping and monitoring at scales useful to the application.

How severe? Given the current narrow selection of measurable variables (the *What*?) and the constraints of adequate spatial coverage (the *Where*?), assessments of severity are limited and generally not adequate beyond the local, and certainly not the global scale. An exception is the precision of satellite-borne instruments but, with these, the challenge is the interpretation of the data to obtain information on the specific conditions and processes of degradation.

The current situation is well summarized by Nachtergaele et al. (2011)

> Global assessments of soil and land degradation have started more than 35 years ago but had until now not achieved a clear answer on where land degradation takes place, what impact it has on the population and what the cost to governments and land users would be if the decline in soil, water and vegetation resources continued unabated. Although institutional, socio-economic and biophysical causes of land degradation have been identified locally in many case studies, these have not been inventoried systematically at district, national or regional level. Much of the investment in land reclamation and rehabilitation during recent years has been driven by donor interest to fund action rather than research and understanding of the problem. Even the impact of what works and what does not in combating land degradation is hardly known nor investigated (modified from Nachtergaele et al. 2011).

But what of the future? Is the objective of mapping *what*, *where* and *severity* beyond the local scale, totally quixotic? On the contrary—there are some reasons for optimism, including the following.

(i) While all methods for mapping degradation have intrinsic limitations, remote sensing not only has important capabilities with its present data and analytical methods but also is in a phase of rapid expansion. Some highlights of near-term developments of satellite remote-sensing include (Table 9.4): radar measurement of soil water and rainfall; LiDAR data of biomass; high spectral-resolution radiometers (hyperspectral) for detection of plant species and functions (Ustin et al. 2009); and increasing accuracy and spatial resolution of global weather data.

(ii) While digital versions of degradation maps and their input data are rarely made available, access via Internet to data on Earth properties is advancing rapidly. For example: the Oak Ridge National Laboratory Distributed Active Archive Center (ORNL DAAC 2015) has over 1134 data sets for biogeochemical dynamics; the Global Change Master Directory (GCMD 2015) has six relevant categories, containing over 18,000 records, including comprehensive data relevant to dryland degradation; the Socioeconomic Data and Applications Center (SEDAC 2015) which focuses on human interactions with the environment, has data in 163 subject areas; and the FAO Global Information and Early Warning System (GIEWS 2015) which issues reports on the world food situation and provides early warnings of impending food crises.

(iii) Notwithstanding these new tools and improved data access, a fundamental need is for independent organizations at national, regional and global scales to undertake routine monitoring. The United Nations Environment Program (UNEP 2015) and FAO GLADIS have limited, but very relevant activities of this sort. However, at present, even existing data are not easily available—a glaring example is the paucity of free, easily available, basic meteorological station data such as rainfall. However these issues are being overcome by some agencies, for example, the Centre de Suivi Ecologique in Senegal (CSE 2015) and the USAID Famine Early Warning System (FEWS Net 2015; Brown 2008).

(iv) Finally, technically credible leadership, free from external pressures, is needed to address the issues of degradation at international-global scales and to harmonize policy-relevant research, monitoring and interpretation. Unfortunately governmental agencies are often not well-informed of appropriate techniques, they are subject to political pressures and have no incentive to participate at the international scale. While GIEWS has an on-going programme of harmonization of data collection, the interpretation step is not included. In the case of climate change, the Intergovernmental Platform on Climate Change (IPCC 2015) provides a blueprint and it is to be hoped that the new Intergovernmental Panel on Biodiversity and Ecosystem Services (IPBES 2015), which includes the UNCCD, will develop into this role.

Credible answers to the *what*, *where* and *how severe* questions are possible if better fundamental principles are developed and applied. There is hope that such progress will not be long in coming.

Acknowledgments Dr. K. Rishmawi, Jeremy Stanfield and Hasan Jackson assisted with the map comparisons.

References

Abella, S. R., Chiquoine, L. P., Newton, A. C., & Vanier, C. H. (2015). Restoring a desert ecosystem using soil salvage, revegetation, and irrigation. *Journal of Arid Environments, 115*, 44–52.

Adeel, Z., Safriel, U., Niemeijer, D., & White, R. (Eds.). (2005). *Millennium ecosystem assessment. Ecosystems and human well-being: Desertification synthesis*. Washington, D.C.: World Resources Institute.

Al-Bakri, J. T., Taylor, J. C., & Brewer, T. R. (2001). Monitoring land use change in the Badia transition zone in Jordan using aerial photography and satellite imagery. *The Geographical Journal, 167*, 248–262.

Asner, G. P., & Heidebrecht, K. B. (2002). Spectral unmixing of vegetation, soil and dry carbon cover in arid regions: Comparing multispectral and hyperspectral observations. *International Journal of Remote Sensing, 23*, 3939–3958.

Bai, Z. G., Conijn, J. G., Bindraban, P. S., & Rutgers, B. (2012). *Global changes of remotely sensed greenness and simulated biomass production since 1981. Towards mapping global soil degradation Report 2012/02*. Wageningen: ISRIC.

Bai, Z. G., Dent, D. L., Olsson, L., & Schaepman, M. E. (2008). Proxy global assessment of land degradation. *Soil Use and Management, 24*, 223–234.

Bastin, G., Pickup, G., Chewings, V., & Pearce, G. (1993). Land degradation assessment in central Australia using a grazing gradient method. *Rangeland Journal, 15*, 190–216.

Batjes, N. H. (1996). Global assessment of land vulnerability to water erosion on a 1/2 by 1/2 grid. *Land Degradation and Development, 7*, 353–365.

Batjes, N., Dijkshoorn, K., van Engelen, V., Fischer, G., Jones, A., Montanarella, L., et al. (2012). *Harmonized world soil database: Version 1.2*. FAO, Rome, Italy and IIASA, Laxenburg, Austria: FAO/IIASA/ISRIC/ISSCAS/JRC.

Beinroth, F. H., Eswaran, H., & Reich, P. F. (2001). Global assessment of land quality. In D. E. Stott, R. H. Mohtar & G. C. Steinhardt (Eds.), *Sustaining the global farm: Selected papers from the 10th international soil conservation organization meeting* (pp. 569–574). Purdue University and the USDA-ARS National Soil Erosion Research Laboratory.

Berry, L., Abraham, E., & Essahli, W. (2009). *UNCCD recommended minimum set of impact indicators*. UNCCD.

Boer, M. M., & Puigdefábregas, J. (2003). Predicting potential vegetation index values as a reference for the assessment and monitoring of dryland condition. *International Journal of Remote Sensing, 24*, 1135–1141.

Bossio, D., Geheb, K., & Critchley, W. (2010). Managing water by managing land: Addressing land degradation to improve water productivity and rural livelihoods. *Agricultural Water Management, 97*, 536–542.

Bowker, M. A., Belnap, J., & Miller, M. E. (2006). Spatial modeling of biological soil crusts to support rangeland assessment and monitoring. *Rangeland Ecology and Management, 59*, 519–529.

Bridges, E. M., & Oldeman, L. R. (1999). Global assessment of human-induced soil degradation. *Arid Soil Research and Rehabilitation, 13*, 319–325.

Brown, M. E. (2008). *Famine early warning systems and remote sensing data*. Berlin: Springer.

Brown, M. E., Grace, K., Shively, G., Johnson, K. B., & Carroll, M. (2014). Using satellite remote sensing and household survey data to assess human health and nutrition response to environmental change. *Population and Environment, 36*, 48–72.

Cai, X., Zhang, X., & Wang, D. (2010). Land availability for biofuel production. *Environmental Science and Technology, 45*, 334–339.

Castro, J. M., Salomone, J. M., & Reichart, R. N. (1980). Estudio de los focos de erosión en el SO de la Provincia de Chubut. In *Informe Técnico* (p. 68). Trelew, Argentina: Instituto Nacional de Tecnología Agropecuaria.

Cerdan, O., Govers, G., Le Bissonnais, Y., Van Oost, K., Poesen, J., Saby, N., et al. (2010). Rates and spatial variations of soil erosion in Europe: A study based on erosion plot data. *Geomorphology, 122*, 167–177.

CIESIN. (2005). Gridded Population of the World (GPW), v3. Socioeconomic Data and Applications Center (SEDAC), http://sedac.ciesin.columbia.edu/. Accessed February 7, 2015.

Cohen, J. (1960). A coefficient of agreement for nominal scales. *Educational and Psychological Measurement, 20*, 37–46.

Conant, R. T., & Paustian, K. (2002). Potential soil carbon sequestration in overgrazed grassland ecosystems. *Global Biogeochemical Cycles, 16*, 1143.

Cook, B., & Pau, S. (2013). A global assessment of long-term greening and browning trends in pasture lands using the GIMMS LAI3 g dataset. *Remote Sensing, 5*, 2492–2512.

Coppin, P., Jonckheere, I., Nackaerts, K., Muys, B., & Lambin, E. (2004). Digital change detection methods in ecosystem monitoring: A review. *International Journal of Remote Sensing, 25*, 1565–1596.

CSE. (2015). Centre de Suivi Ecologique (Ecological Monitoring Center), http://www.cse.sn. Accessed February 6, 2015.

Dawelbait, M., & Morari, F. (2012). Monitoring desertification in a Savannah region in Sudan using landsat images and spectral mixture analysis. *Journal of Arid Environments, 80*, 45–55.

Deichmann, U. (1994). *A medium resolution population database for Africa: Technical paper and digital database*. Santa Barbara: National Center for Geographic Information and Analysis.

del Valle, H. F., Blanco, P. D., Metternicht, G. I., & Zinck, J. A. (2010). Radar remote sensing of wind-driven land degradation processes in Northeastern Patagonia. *Journal of Environmental Quality, 39*, 62–75.

den Biggelaar, C., Lal, R., Wiebe, K., Eswaran, H., Breneman, V., & Reich, P. (2004). The global impact of soil erosion on productivity II: Effects on crop yields and production over time. *Advances in Agronomy, 81*, 1–48.

DESIRE. (2008). *Manual for describing land degradation indicators*. Athens: Agricultural University of Athens.

Dewitte, O., Jones, A., Elbelrhiti, H., Horion, S., & Montanarella, L. (2012). Satellite remote sensing for soil mapping in Africa: An overview. *Progress in Physical Geography, 36*, 514–538.

Dijkshoorn, J. A., van Engelen, W. W. P., & Huting, J. R. M. (2008). *Global Assessment of Land Degradation Soil and landform properties for LADA partner countries (Argentina, China, Cuba, Senegal and The Gambia, South Africa and Tunisia). ISRIC report 2008/06 and GLADA report 2008/0*. Wageningen: ISRIC—World Soil Information Center and FAO.

Dregne, H. E. (1983). Desertification of arid lands. In F. El-Baz & M. H. A. Hassan (Eds.), *Physics of desertification* (p. 242). Chur, Switzerland; New York: Harwood Academic Publishers.

Dregne, H. E., & Chou, N. T. (1992). Global desertification dimensions and costs. In H. E. Dregne (Ed.), *Degradation and restoration of arid lands*. Lubbock, Texas: International Center for Arid and Semiarid Land Studies, Texas Tech. University.

Ellis, E. C., & Ramankutty, N. (2008). Putting people in the map: Anthropogenic biomes of the world. *Frontiers in Ecology and the Environment, 6*, 439–447.

Eswaran, H., Almaraz, R., van den Berg, E., & Reich, P. (1997). An assessment of soil resources of Africa in relation to productivity. *Geoderma, 77*, 1–18.

Eswaran, H., Lal, R., & Reich, P. F. (2001). Land degradation: an Overview. In E. M. Bridges, I. D. Hannam, L. R. Oldeman, F. W. T. Pening de Vries, S. J. Scherr, & S. Sompatpanit (Eds.), *Responses to land degradation*. Khon Kaen, Thailand: Oxford Press.

FAO/UNESCO. (1991). The digitized soil map of the world. In *World Soil Resources Report*, Rome.

FAOSTAT. (2015). *Food and agriculture organization of the United Nations, statistics division*. http://faostat.fao.org. Accessed February 7, 2015.

Feddema, J. J. (1999). Future African water resources: Interactions between soil degradation and global warming. *Climatic Change, 42*, 561–596.

FEWS Net. (2015). *Famine early warning network*. http://www.fews.net. Accessed February 7, 2015.

Furby, S., Caccetta, P., & Wallace, J. (2010). Salinity monitoring in Western Australia using remotely sensed and other spatial data. *Journal of Environmental Quality, 39*, 16–25.

Gao, J., & Liu, Y. (2010). Determination of land degradation causes in Tongyu County, Northeast China via land cover change detection. *International Journal of Applied Earth Observation and Geoinformation, 12*, 9–16.

GCMD. (2015). *Global change master directory*. http://gcmd.nasa.gov. Accessed February 6, 2015.

Geist, H. (2005). *The causes and progression of desertification*. Abingdon Oxon, UK: Ashgate Publishing.

Gibbs, H. K., & Salmon, J. M. (2015). Mapping the world's degraded lands. *Applied Geography, 57*, 12–21.

GIEWS. (2015). *Global information and early warning system*. http://www.fao.org/giews/english/index.htm. Accessed February 15, 2015.

GLADIS. (2015). *Global land data information system*. http://www.fao.org/nr/lada. Accessed February 8, 2015.

Hagen-Zanker, A., Engelen, G., Hurkens, J., Vanhout, R., & Uljee, I. (2006). *Map Comparison Kit (MCK). 3.0*. Maastricht, The Netherlands: Research Institute for Knowledge Systems (RIKS bv).

Harris, I., Jones, P. D., Osborn, T. J., & Lister, D. H. (2014). Updated high-resolution grids of monthly climatic observations—the CRU TS3.10 dataset. *International Journal of Climatology, 34*, 623–642. doi:10.1002/joc.3711.

Herrick, J. E., Lessard, V. C., Spaeth, K. E., Shaver, P. L., Dayton, R. S., Pyke, D. A., et al. (2010). National ecosystem assessments supported by scientific and local knowledge. *Frontiers in Ecology and the Environment, 8*, 403–408.

Hill, J. (2001). Remote sensing of surface properties. In S.-W. Breckle, M. Veste & W. Wucherer (Eds.), *The key to land degradation and desertification assessments. Sustainable land use in deserts* (pp. 243–254). Springer, Berlin.

Holm, A. M., Cridland, S. W., & Roderick, M. L. (2003). The use of time-integrated NOAA NDVI data and rainfall to assess landscape degradation in the arid shrubland of Western Australia. *Remote Sensing of Environment, 85*, 145–158.

Ibáñez, J., Martínez Valderrama, J., & Puigdefabregas, J. (2008). Assessing desertification risk using system stability condition analysis. *Ecological Modelling, 213*, 180–190.

IPBES. (2015). *Intergovernmental panel on biodiversity and ecosystem services*. http://www.ipbes.net/. Accessed June 9, 2015.

IPCC. (2015). *Intergovernmental panel on climate change*. http://ipcc.ch/. Accessed February 7, 2015.

Izaurralde, R. C., Williams, J. R., Post, W. M., Thomson, A. M., McGill, W. B., Owens, L. B., & Lal, R. (2007). Long-term modeling of soil C erosion and sequestration at the small watershed scale. *Climatic Change, 80*, 73–90.

Kirkby, M. J., Irvine, B. J., Jones, R. J. A., & Govers, G. (2008). The PESERA coarse scale erosion model for Europe. I.—Model rationale and implementation. *European Journal of Soil Science, 59*, 1293–1306.

Kosmas, C., Karavitis, C., Kairis, O., Kounalaki, A., Fasouli, V., & Tsesmelis, D. (2012). *Using indicators for identifying best land management practices for combating desertification. DESIRE Scientific reports. Deliverable 2.2.3*. Agricultural University of Athens.

Lambin, E. F., & Ehrlich, D. (1997). Land-cover changes in sub-saharan Africa (1982–1991): Application of a change index based on remotely sensed surface temperature and vegetation indices at a continental scale. *Remote Sensing of Environment, 61*, 181–200.

Liniger, H., Schwilch, G., Gurtner, M., Studer, R.M., Hauert, C., van Lynden, G., et al. (2008). *WOCAT degradation categorization system*. WOCAT—World Overview of Conservation Approaches and Technologies.

Lobell, D. B. (2010). Remote sensing of soil degradation: Introduction. *Journal of Environmental Quality, 39*, 1–4.

Martín-Fernández, L., & Martínez-Núñez, M. (2011). An empirical approach to estimate soil erosion risk in Spain. *Science of the Total Environment, 409*, 3114–3123.

Mbow, C., Fensholt, R., Rasmussen, K., & Diop, D. (2013). Can vegetation productivity be derived from greenness in a semi-arid environment? Evidence from ground-based measurements. *Journal of Arid Environments, 97*, 56–65.

Millennium Ecosystem Assessment. (2005). *Millennium ecosystem assessment synthesis report. Ecosystems and human well-being: A framework for assessment. Concepts of ecosystem value and validation approaches.* Washington, DC: Island Press and World Resources Institute.

Milton, S. J., Dean, W. R. J., du Plessis, M., & Siegfried, W. R. (1994). A conceptual model of arid rangeland degradation. *Bioscience, 44*, 70–76.

Mortimore, M. J., & Adams, W. M. (2001). Farmer adaptation, change and 'crisis' in the Sahel. *Global Environmental Change-Human and Policy Dimensions, 11*, 49–57.

Mulligan, M. (2015). Computational policy support systems for understanding land degradation effects on water and food security for and from Africa. Land restoration: Reclaiming landscapes for a sustainable future, Elsevier (p. 400). ILLP Publication.

Mulligan, M., Burke, S., & Ogilvie, A. (2015). Much more than simply "desertification": understanding agricultural sustainability and change in the Mediterranean. In R. Benkhe (Ed.), (pp. x–xx). doi:10.1007/978-3-642-16014-1_8. Berlin: Springer.

Nachtergaele, F. O. F., & Licona-Manzur, C. (2008). The land degradation assessment in drylands (LADA) project: Reflections on indicators for land degradation assessment. In C. C. Lee & T. Schaaf (Eds.), *The future of drylands* (pp. 327–348). Rome, Italy: UNESCO, Food and Agriculture Organization of the United Nations (FAO).

Nachtergaele, F. O., Petri, M., Biancalani, R., van Lynden, G., van Velthuizen, H., & Bloise, M. (2011). An Information database for land degradation assessment at global level. In *Global land degradation information system* (*GLADIS*).

Oldeman, L. R., Hakkeling, R. T. A., & Sombroek, W. G. (1991). *World map of the status of human-induced soil degradation: An explanatory note. Global Assessment of Soil Degradation* (*GLASOD*) (2nd ed.). Wageningen Winand Staring Center, International Society for Soil Science, FAO, International Institute for Aerospace Survey and Earth Science.

Oldeman, L. R., & van Lynden, G. W. J. (1996). *Revisiting the GLASOD methodology. ISRIC report 1996/03*. Wageningen: ISRIC World Soil Information.

Olson, D. M., Dinerstein, E., Wikramanayake, E. D., Burgess, N. D., Powell, G. V. N., Underwood, E. C., et al. (2001). Terrestrial ecoregions of the world: A new map of life on earth. *Bioscience, 51*, 933.

Olsson, L., Eklundh, L., & Ardö, J. (2005). A recent greening of the Sahel—trends, patterns and potential causes. *Journal of Arid Environments, 63*, 556–566.

ORNL DAAC. (2015). Oak Ridge National Laboratory Distributed Active Archive Center, http://daac.ornl.gov/index.shtml. Accessed February 15, 2015.

Petschel-Held, G., Block, A., Cassel-Gintz, M., Kropp, J., Lüdeke, M. K. B., Moldenhauer, O., et al. (1999). Syndromes of global change: A qualitative modelling approach to assist global environmental management. *Environmental Modeling and Assessment, 4*, 295–314.

Pickup, G., & Chewings, V. (1994). A grazing gradient approach to land degradation assessment in arid areas from remotely-sensed data. *International Journal of Remote Sensing, 15*, 597–617.

Prince, S. D. (1991). Satellite remote sensing of primary production: Comparison of results for Sahelian grasslands 1981–1988. *International Journal of Remote Sensing, 12*, 1301–1311.

Prince, S. D. (2002). Spatial and temporal scales of measurement of desertification. In M. Stafford-Smith & J. F. Reynolds (Eds.), *Global desertification: Do humans create deserts?* (pp. 23–40). Berlin: Dahlem University Press.

Prince, S. D., Becker-Reshef, I., & Rishmawi, K. (2009). Detection and mapping of long-term land degradation using local net production scaling: Application to Zimbabwe. *Remote Sensing of Environment, 113*, 1046–1057.

Prince, S. D., De Colstoun, E. B., & Kravitz, L. L. (1998). Evidence from rain-use efficiencies does not indicate extensive Sahelian desertification. *Global Change Biology, 4*, 359–374.

Reeves, M. C., & Baggett, L. S. (2014). A remote sensing protocol for identifying rangelands with degraded productive capacity. *Ecological Indicators, 43*, 172–182.

Reich, P. F., Numbem, S. T., Almaraz, R. A., & Eswaran, H. (2001). *Land resource stresses and desertification in Africa. Responses to Land Degradation. Proceedings of the 2nd International Conference on Land Degradation and Desertification, Khon Kaen, Thailand.* New Delhi, India: Oxford Press.

Reynolds, J. F., Grainger, A., Stafford Smith, D. M., Bastin, G., Garcia-Barrios, L., Fernández, R. J., et al. (2011). Scientific concepts for an integrated analysis of desertification. *Land Degradation and Development, 22*, 166–183.

Reynolds, J. F., & Stafford Smith, M. (Eds.). (2002). *Global desertification: Do humans create deserts?* Berlin: Dahlem University Press.

Reynolds, J. F., Stafford-Smith, D. M., Lambin, E. F., Turner, B. L, I. I., Mortimore, M., Batterbury, S. P. J., et al. (2007). Global desertification: Building a science for dryland development. *Science, 316*, 847–851.

Rishmawi, K. (2013). Spatial patterns and potential mechanisms of land degradation in the Sahel. Unpublished PhD, University of Maryland.

Robinson, T. P., Francescini, G., & Wint, W. (2007). The Food and Agriculture Organization's gridded livestock of the world. *Veterinaria Italiana, 43*, 745–751.

Romm, J. (2011). The next dustbowl. *Nature, 478*, 450–451.

Safriel, U. (2007). The assessment of global trends in land degradation. In M. V. K. Sivakumar & N. Ndiang'ui (Eds.), *Climate and land degradation* (pp. 1–38). New York: Springer.

Safriel, U. (2009). Deserts and desertification: Challenges but also opportunities. *Land Degradation and Development, 20*, 353–366.

Scherr, S. J., & Yadav, S. (1996). Land degradation in the developing world: Implications for food, agriculture, and the environment to 2020. In *Food, agriculture, and the environment*. International Food Policy Research Institute.

Seaquist, J. W., Hickler, T., Eklundh, L., Ardo, J., & Heumann, B. W. (2009). Disentangling the effects of climate and people on Sahel vegetation dynamics. *Biogeosciences, 6*(3), 469–477.

SEDAC. (2015). Socioeconomic Data and Applications Center, http://sedac.ciesin.columbia.edu/about. Accessed February 15, 2015.

Sietz, D., Untied, B., Walkenhorst, O., Ludeke, M. K. B., Mertins, G., Petschel-Held, G., & Schellnhuber, H. J. (2006). Smallholder agriculture in Northeast Brazil: Assessing heterogeneous human-environmental dynamics. *Regional Environmental Change, 6*, 132–146.

Singh, G., Bundela, D. S., Sethi, M., Lal, K., & Kamra, S. K. (2010). Remote sensing and geographic information system for appraisal of salt-affected soils in India. *Journal of Environmental Quality, 39*, 5–15.

Sonneveld, B. G., & Dent, D. L. (2009). How good is GLASOD? *Journal of Environmental Management, 90*, 274–283.

Symeonakis, E., & Drake, N. (2004). Monitoring desertification and land degradation over sub-Saharan Africa. *International Journal of Remote Sensing, 25*, 573–592.

Tamene, L., & Le, Q. B. (2015). Estimating soil erosion in sub-Saharan Africa based on landscape similarity mapping and using the revised universal soil loss equation (RUSLE). *Nutrient Cycling in Agroecosystems, 101*, 1–15. doi:10.1007/s10705-10015-19674-10709.

Thomas, D. S. G., & Middleton, N. J. (1993). Salinization: new perspectives on a major desertification issue. *Journal of Arid Environments, 24*, 95–105.

Trabucco, A., & Zomer, R. J. (2009). *Global Aridity Index (Global-Aridity) and Global Potential Evapo-Transpiration (Global-PET) Geospatial Database*. Published online, available from the CGIAR-CSI GeoPortal at: http://www.csi.cgiar.org

Tucker, C. J., Dregne, H. E., & Newcomb, W. W. (1991). Expansion and contraction of the Sahara desert from 1980 to 1990. *Science, 253*, 299–301.

Tucker, C. J., Justice, C. O., & Prince, S. D. (1986). Monitoring the grasslands of the Sahel 1984–1985. *International Journal of Remote Sensing, 7*, 1571–1581.

Tucker, C. J., Pinzon, J. E., Brown, M. E., Slayback, D. A., Pak, E. W., Mahoney, R., et al. (2005). An extended AVHRR 8-km NDVI dataset compatible with MODIS and SPOT vegetation NDVI data. *International Journal of Remote Sensing, 26*, 4485–4498.

UNCCD. (1994). Elaboration of an international convention to combat desertification in countries experiencing serious drought and/or desertification, particularly in Africa (p. 58). United Nations General Assembly, 93rd plenary meeting, New York.

UNCED. (1992). Managing fragile ecosystems: combating desertification and drought. In *Agenda 21: Earth summit*. Rio de Janerio: United Nations Conference on Environment and Development.

UNEP. (1997). *World Atlas of Desertification*. (2nd ed.). London, New York: Arnold & Wiley, on behalf of UNEP.

UNEP. (2015). United Nations Environment Programme, http://www.unep.org/. Accessed February 7, 2015.

United Nations General Assembly. (2013). Future we want—Outcome document. In *Resolution adopted by the General Assembly*. United Nations Department of Economic and Social Affairs (DESA).

USGS. (2015). The Land Processes Distributed Active Archive Center (LP DAAC), https://lpdaac.usgs.gov/. Accessed February 7, 2015.

Ustin, S. L., Palacios-Orueta, A., Whiting, M. L., Jacquemoud, S., & Li, L. (2009). Remote sensing based assessment of biophysical indicators for land degradation and deserlification. In R. Hill (Ed.), *Recent advances in remote sensing and geoinformation processing for land degradation assessment* (pp. 15–44). London: Taylor & Francis Group.

Verón, S. R., Paruelo, J. M., & Oesterheld, M. (2006). Assessing desertification. *Journal of Arid Environments, 66*, 751–763.

Vogt, J. V., Safriel, U., Von Maltitz, G., Sokona, Y., Zougmore, R., Bastin, G., & Hill, J. (2011). Monitoring and assessment of land degradation and desertification: Towards new conceptual and integrated approaches. *Land Degradation and Development, 22*, 150–165.

Warren, A. (2002). Land degradation is contextual. *Land Degradation and Development, 13*, 449–459.

Wessels, K. J. (2009). Comments on 'Proxy global assessment of land degradation' by Bai et al. (2008). *Soil Use and Management, 25*, 91–92.

Wessels, K. J., Prince, S. D., Carroll, M., & Malherbe, J. (2007a). Relevance of rangeland degradation in semiarid northeastern South Africa to the nonequilibrium theory. *Ecological Applications, 17*, 815–827.

Wessels, K. J., Prince, S. D., Malherbe, J., Small, J., Frost, P. E., & VanZyl, D. (2007b). Can human-induced land degradation be distinguished from the effects of rainfall variability? A case study in South Africa. *Journal of Arid Environments, 68*, 271–297.

Wessels, K. J., Prince, S. D., & Reshef, I. (2008). Mapping land degradation by comparison of vegetation production to spatially derived estimates of potential production. *Journal of Arid Environments, 72*, 1940–1949.

Wessels, K. J., van den Bergh, F., & Scholes, R. J. (2012). Limits to detectability of land degradation by trend analysis of vegetation index data. *Remote Sensing of Environment, 125*, 10–22.

Whitlow, R. (1988). *Land degradation in Zimbabwe*. Harare: Department of Natural Resources, Government of Zimbabwe/Department of Geography, University of Zimbabwe.

Wildlife Conservation Society (WCS) and Center for International Earth Science Information Network (CIESIN). (2005). *Last of the Wild Project, (LWP-2): Global Human Influence Index (HII) Dataset*. Palisades, NY: NASA Socioeconomic Data and Applications Center (SEDAC), Columbia University.

Zika, M., & Erb, K. H. (2009). The global loss of net primary production resulting from human-induced soil degradation in drylands. *Ecological Economics, 69*(2), 310–318.

Zucca, C., & Biancalani, R. (2011). Guidelines for the use of the LADA QM DB in the frame of the "National Piloting of provisional UNCCD impact indicators" (p. 24), LADA FAO.

Zucca, C., Della Peruta, R., Salvia, R., Sommer, S., & Cherlet, M. (2012). Towards a world desertification atlas. Relating and selecting indicators and data sets to represent complex issues. *Ecological Indicators, 2012*, 157–170.

Chapter 10
40 Years of Climate Modeling: The Causes of Late-20th Century Drought in the Sahel

Alessandra Giannini

Abstract This chapter reviews the evolution of our understanding of the causes of the late 20th century Sahel drought in the context of global climate model development. In describing, in coherent and holistic fashion, physical arguments for local and global influences on regional rainfall variability and change, it aims to make sense of current trends and future projections in the context of the recent past.

Keywords Sahel · Drought · Climate modeling · Seasonal prediction · Climate change · Attribution

10.1 Introduction

The interpretation of drought in the Sahel as a consequence of local environmental degradation fundamentally rooted in rapid population growth arose in the 1970s—the time of the "population bomb" (Ehrlich 1968) and "limits to growth" (Meadows et al. 1972). It may have been intuitive in that context, and with little direct experience of local conditions, to embrace the notion that mismanagement of natural resources could have an influence beyond local land degradation, and lead to self-perpetuating drought and occasional widespread famine. Perhaps the conception persisted because an explanation centered on local causes also meant that local solutions could be implemented. However, in the decades since the early 1970s, understanding of the global interconnectedness of physical climate has advanced, as is apparent in the development of dynamical models to predict the El Niño-Southern Oscillation (Cane et al. 1986), and in the demonstration that humans are indeed capable of influencing the global climate system through emissions from fossil fuel burning (Intergovernmental Panel on Climate Change 2007, 2014). Therefore, the popular notion of the influence of local land degradation on rainfall

A. Giannini (✉)
International Research Institute for Climate and Society, The Earth Institute, Columbia University, 61 Route 9W, Palisades, NY 10964-8000, USA
e-mail: alesall@iri.columbia.edu

R.H. Behnke and M. Mortimore (eds.), *The End of Desertification?*
Springer Earth System Sciences, DOI 10.1007/978-3-642-16014-1_10

and the more recent realization of global anthropogenic influence on climate demand a review to make sense of the scientific debate that has surrounded the cause of drought in the Sahel over the last 40 years.

In broad climatic terms, the Sahel represents the fluctuating semi-arid grassland-savanna boundary between desert and tropical rainforest, running across Africa from the Atlantic coast of Mauritania and Senegal to the Red Sea coast of Sudan and Eritrea. In reality there is a gradation in the ecosystems that span annual mean rainfall between ~200 and ~1000 mm from north to south, and from desert to tropical rainforest, usually referred to as Sahelian, Sahelo-Sudanian and Sudanian (Nicholson 1979), as well as great variation in population density, in patterns of land use, in human settlement and in livelihood strategies, not only in the north-south, but also in the east-west direction (Raynaut 1997; Cour and Snrech 1998; Leach and Mearns 1996). However, the coarse resolution of global climate models, where one spatial unit or "grid box" has a dimension of ~100 km, justifies referring to the Sahel as the strip of land occupying the latitudinal belt between 10° N and 20°N, i.e. covering ~1000 km in the north-south direction, and ~6000 km in the east-west direction.

In this latitudinal belt climate is defined by the northernmost annual reach of the African monsoon, which is attained in northern hemisphere summer. This climatic regime is characterized by the alternation of a relatively brief rainy season with the longer dry season. Climate is susceptible to significant variation in precipitation on all time scales, from years to centuries to millennia and longer (Nicholson 2000a; Shanahan et al. 2009; de Menocal et al. 2000; Petit-Maire 2002; Gasse et al. 1990; Gasse 2000; Braconnot et al. 2007a, b). Conversely, temperature does not vary as dramatically as in temperate latitudes, for well understood dynamical reasons (Polvani and Sobel 2002). However, on the time scale of a person's lifetime nothing quite compares, in abruptness, magnitude and severity of human impact, with the shift to persistently dry conditions that manifested itself in the Sahel at the end of the 1960s (Lamb 1982; Nicholson 1983; Katz and Glantz 1986; Hulme 2001; Dai et al. 2004; Trenberth et al. 2007; Greene et al. 2009). This shift was an extreme phenomenon physically—worldwide, no region of similar extent has witnessed such magnitude of change during the 20th century—as well as societally—with recurrent food security crises and talk of irreversible land degradation, or desertification. For this reason, the cause of drought in the Sahel has been hotly debated since the inception of its last episode, in the late 1960s, with repercussions in development practice, as further explored in this book.

In this chapter I seek to reconcile through an examination of physical factors the alternative views on the cause of the persistence of late 20th century drought[1] in the Sahel, and to define the context in which to situate 21st century climate change in

[1]For definitions of drought—meteorological, hydrological, agricultural—the reader is referred e.g. to Glantz (1994) and Wilhite (2001). Here drought is synonymous with deficient, or below-average rainfall, where the average is computed on a sufficiently long period, with 30 years being the standard implemented by the UN World Meteorological Organization.

the region. In simplified terms, the debate in the climate science community pitted two alternative interpretations of the genesis of drought against each other, which, by nature of one being internal, the other external to the region (Brooks 2004) appeared to be mutually exclusive. In fact, I will show that elements of the physical arguments underlying the two interpretations—one to do with the impact of local human activity on the physical properties of the land surface and its purported regional or global consequences on climate, the other with the remote influence of the oceans on the climate of this margin of the monsoon—are recurrent in explanations of past change, e.g. stretching to the paleo-climate reconstructions and modeling interpretation of the "African humid period" which ended around 6000 years ago (Kutzbach and Liu 1997; Patricola and Cook 2007), and of future change, e.g. the apparent disagreement among global climate models in projections of regional rainfall. For this reason, it is important to make sense of them holistically.

Therefore, the goal of this chapter, which is organized in two sections and conclusions, is to review the competing explanations of drought to highlight the possibility of consistency between them, and to find convergence in explanations of climatic change in the Sahel in a way that is meaningful to the development of resilient livelihood strategies. The first section is about the past. It reviews the internal/local/land and external/remote/ocean interpretations of physical climate change in the Sahel, highlighting their strengths and weaknesses, and proposes a theoretical framework, largely based on the work of Held, Neelin, Sobel and collaborators (e.g. Held and Soden 2006; Neelin 2007; Zeng and Neelin 1999; Neelin et al. 2003; Chou and Neelin 2004; Chiang and Sobel 2002; Sobel 2007) in which to make sense of the parts. The second section is about the present and future—how do we interpret current trends and projections of future change in a way that is consistent with the recent past, i.e. the last century of direct, instrumental records, to begin to communicate meaningfully policy and practice implications?

10.2 The Climate Science Behind Understanding Drought in the Sahel

10.2.1 An Interpretive Lens: Bottom–Up and Top–Down Processes Affecting Rainfall of Relevance to Local and Remote Influences

> Why does it rain? If a parcel of air rises, it expands in the lower pressure, cools, and therefore condenses moisture in the parcel, producing cloud, and ultimately, rainfall – or perhaps snowfall. So a key ingredient is certainly the many and varied mechanisms for causing air to rise. (Trenberth et al. 2003)

A *parcel of air* will rise against gravity if it has the energy, if it is buoyant, or vertically unstable, with respect to the *environment air* surrounding it (Fig. 10.1). Therefore, one way to cause instability and rain is to make a parcel lighter, hence more buoyant or unstable than the air surrounding it, by making it warmer and/or

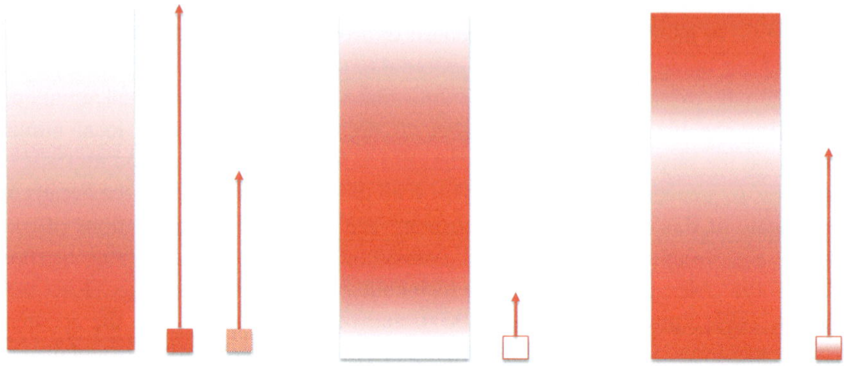

Fig. 10.1 The stability or instability of a parcel of air with respect to the environment air surrounding it. *White* to *red* hues indicate increasing energy content, in the vertical column representing the environment air, as well as in the small square representing a parcel of air near the surface.

moister[2] near the surface. As the parcel of air rises, it cools and condenses the moisture within until it reaches a point where its temperature and moisture content is the same as the surrounding air, hence it is no longer buoyant. Given the local environment's vertical profile of stability, how far a parcel can ascend, whether high, i.e., up to an altitude between 15 and 20 km in the troposphere,[3] likely causing deep convection and heavy precipitation, or only a few kilometers, as in dry/shallow convection, is largely determined by its initial condition, its temperature and humidity in the *boundary layer* near the surface, before it initiates ascent (Emanuel et al. 1994; Polcher 1995).

Charney's hypothesis for a bio-geophysical feedback between albedo, rainfall and vegetation cover can be cast as a theory of Sahel drought that relates vertical instability and rainfall to near-surface conditions over land. Studies that relate rainfall variability in the Sahel to sea surface temperature anomalies in the tropical Atlantic (e.g. Lamb 1978a, b; Eltahir and Gong 1996) also make use of this framework—changes in vertical stability and rainfall are determined from the bottom up, by perturbing near-surface atmospheric conditions.

Another way to perturb rain-producing processes is to make the air surrounding the parcel more or less stable, so that it becomes less or more difficult for that parcel of air to reach a threshold beyond which it becomes unstable and wants to ascend, or continue to ascend deep into the troposphere. The analogue for this type of influence is the El Niño-Southern Oscillation (ENSO) "teleconnection", i.e. ENSO's global influence on apparently remote regions of the world, especially in

[2]Moist air is lighter than dry air, because a molecule of water vapor is lighter than a molecule of nitrogen or oxygen, the most prevalent constituents of the Earth's atmosphere.

[3]The troposphere is the layer of atmosphere closest to the surface, where all human activity takes place. It is ~15 km deep in the tropics and ~10 km deep at high latitudes, and contains 90 % of the atmosphere's mass.

the tropics. During a warm ENSO or El Niño event, ocean and atmosphere in the central equatorial Pacific interact in a positive feedback loop that extracts energy from the system: a reduction in wind speed reduces evaporation, warming the surface of the ocean, and the atmosphere responds with increased instability, increased convection and rainfall, which further weakens the winds and warms the oceans (Cane 1991; Bjerknes 1969). In the source region of the central equatorial Pacific increased energy extraction from the surface of the ocean and more abundant rainfall go hand in hand. This local increase in energy is first communicated upward by rising parcels of air in deep convection, and then around the global tropics by atmospheric waves that spread the increase in temperature at upper levels of the troposphere, i.e. at $\sim$15–20 km of altitude (Polvani and Sobel 2002). Around the tropics, outside of the tropical Pacific, warmer environmental air at upper levels translates into a higher instability/energy threshold for a parcel that wants to ascend, and a more stable vertical profile. Consequently the rain-producing process is inhibited, and tropical drought spreads wide in the months that follow the onset of a warm ENSO episode in regions remote from the ENSO source, e.g. in the South Asian and West African monsoons (Chiang and Sobel 2002; Yulaeva and Wallace 1994; Lyon 2004).

Similar arguments have recently been invoked in studies that discuss the role of warming of the Indian Ocean in the persistence of drought in the Sahel and Horn of Africa (Held et al. 2005; Funk et al. 2008), or of generalized warming of the oceans in future projections of drought (Neelin et al. 2003, 2006), with an increase in greenhouse gases leading to a transient warming of upper levels in the troposphere (Chou and Neelin 2004; Yang et al. 2003; Sugi and Yoshimura 2004).

10.2.2 Complementary Theories on the Cause of Late-20th Century Sahel Drought

10.2.2.1 Charney's Local/Land Bio-Geophysical Feedback for the Persistence of Drought

Charney's original formulation of a bio-geophysical feedback (Charney et al. 1975) states that:

> A decrease in plant cover may be accompanied by an increase in the surface albedo.[4] This would lead to a decrease in the net incoming radiation and an increase in the radiative cooling of the air. As a consequence, the air would sink to maintain thermal equilibrium by adiabatic compression, and cumulus convection and its associated rainfall would be suppressed.

---

[4]The albedo is the reflectivity of the Earth's surface to solar radiation. It is a unitless number between 0 and 1. Surfaces that reflect a high proportion of the incoming insolation have high albedo.

The fundamental elements of Charney's mechanism to explain the persistence of drought in a semi-arid desert border region such as the Sahel are: (1) the relationship between increase in surface albedo, decrease in precipitation, and consequent further decrease in vegetation cover and increase in albedo—a reinforcing, or positive feedback, (2) the role of surface albedo in defining the local net surface and atmospheric energy budgets, and (3) though not explicitly stated in this specific paragraph, the role of human activity in disturbing surface albedo through systematic modification of the land cover.

Charney and collaborators (1975, 1977) were the first to test this mechanism in general circulation models of the atmosphere (Schneider and Dickinson 1974), the same class of computational tools in use to make weather forecasts and projections of future climate change, albeit in their primordial stages.[5] Using the model developed at NASA's Goddard Institute for Space Studies (Somerville et al. 1974), they compared "control" and "perturbation" simulations. In the control, the value of surface albedo in the Sahel and in similar semi-arid margins of monsoons in North America and South Asia is prescribed to be the same as all other land points, i.e. 0.14. For comparison, albedo is set to a low value, equal to 0.07, for oceanic grid points, which reflect very little incoming solar radiation, and to a high value, equal to 0.70, for ice points, which reflect a lot (Charney et al. 1977). In the perturbation, the land value for albedo is doubled to 0.35 only in semi-arid margins of monsoons such as the Sahel. In the comparison, the perturbed simulations, those with increased albedo values at the desert border, always display drier conditions in these regions, and significantly so, despite the small sample size.

Charney et al. (1977) comment on two points of interest to ensuing developments. First of all, they acknowledge that "As better general circulation models are developed, the experiments described in this article [should] be repeated". This comment stems out of a specific concern about their model's simulation of evaporation, which, whether too high or too low, was unrealistic. Secondly, Charney et al. note that "the perturbation will remain until external counteracting climatological forces become large enough to overcome the local feedback effect and return the system toward the old equilibrium". In other words, while the impact of land

---

[5]In this general class of models, also known as GCMs, for "general circulation models" or "global climate models", the surface of the Earth is divided into squares, or grid boxes, of dimensions currently of the order of ~100 km. In each grid box the evolution of the atmosphere only, initially, and of the coupled ocean-atmosphere system, more recently, is derived from the equations of motion—the motion of fluids, the atmosphere and ocean, on a rotating sphere heated by an external source of energy. In the atmosphere only case, the surface temperature of the oceans is prescribed, either to monthly climatological values, or to monthly varying values derived from observations [or from seasonal predictions], while the exchange between land surface and atmosphere is described by empirical relations referred to as parameterizations. In the simplest coupled ocean-atmosphere case, the sole external source of energy is the sun, insolation at the top of the atmosphere is kept constant in time, and the system is let to adjust to the spatial variation in heating that derives from the planet's own geometry and rotation around its axis, and that of its revolution around the sun. The circulations that ensue in the atmosphere and oceans are manifestations of the inherent "internal" variability in the system as it redistributes heat from warmer to cooler regions.

surface change is not negligible, Charney et al. fully expect that in longer model runs the dominant role of the large-scale circulation should be restored.

Follow-up studies to Charney's seminal work focused on his recommendation to reproduce their perturbation simulations as the complexity and internal consistency of the physical processes described by models improved. Sud and Fennessy (1982) repeated the same numerical experiments and came to the same conclusions with the GLAS (Goddard Laboratory for Atmospheric Sciences) model. Laval and Picon (1986) did the same with the model developed at the Laboratoire de Météorologie Dynamique (LMD) in Paris. Xue and Shukla (1993) introduced a vegetation parameterization in the COLA (Center for Ocean-Land-Atmosphere Studies) model, which permitted them to characterize the sensitivity of the model to vegetation change, through its effect on net surface energy budget and hydrology. By changing vegetation cover in a more realistic way compared to the idealized increase in albedo used in prior studies, Xue and Shukla recovered the decrease in precipitation, as well as the changes in the regional atmospheric circulation consistent with a weakened monsoon. These modeled changes described a decrease in the south-westerly flow of moist, monsoonal air, a weakening in the tropical easterly jet aloft, and in the mid-level African easterly jet—all features described by Newell and Kidson (1984) to be typical of dry years. They also highlighted a ~2-week delay in the onset of the rainy season in their 'desertification' integrations. This facet of Xue and Shukla's simulations is particularly interesting vis-à-vis the current state-of-the-art in interpretation of climate change projections. Studies have consistently shown a delay in the onset of the West African (Biasutti and Sobel 2009; Biasutti 2013) and global monsoons (Seth et al. 2011, 2013) in response to global warming, despite the uncertainty in the direction of the seasonal or annual average change in precipitation.

Many more studies have been carried out since Charney et al.'s simulations to check the consistency of improved model versions, e.g. moving from the simplistic consideration of a uniform surface albedo change to the more involved consideration of soil moisture feedbacks and dynamic vegetation. The work of Taylor et al. (2002) provides an end point, because it summarizes the modeling sophistication achieved, and at the same time epitomizes the limits reached by sensitivity studies. These are studies in which the hypothetical cause of a change in the real world is identified. In this case, the change in land surface conditions is hypothesized to be the cause of the change in observed precipitation between wet and dry decades. Simulations are then run with and without the cause, and the two sets, a "perturbed" and a "control" case, are compared. In Taylor et al. (2002) land use change is computed based on an independent equilibrium model developed by Stephenne and Lambin (2001). This model takes into account the competing human activities modifying the rural landscape—agricultural and livestock production, and fuelwood extraction—going so far as to postulate that extensification of agricultural production should occur before intensification. Land use change is then translated into a change in the parameters that describe vegetation cover in a general circulation model. The end result is the expected decrease in rainfall between conditions as simulated appropriate for the 1960s and 1990s. However, the simulated decrease,

4.6 %, is only a fraction of the estimated ~10–20 % reduction in regional rainfall observed between the wet and dry decades of the 20th century.

In sum, we can synthesize as follows. Over the past ~40 years model development has led to increased sophistication in the treatment of land-atmosphere interaction processes, which started with imposing a simple albedo change, and evolved through consideration of interactive soil moisture and dynamic vegetation, and more realistic patterns of land use change. It has become increasingly clear that the processes by which anthropogenic land use change can reduce precipitation are physically plausible, and consistent with a natural positive feedback response of the land surface to precipitation. On the other hand, prescribing a change in local land surface conditions without allowing for the system to react, to verify whether the dynamics of the global climate system will allow the amplification of the initial perturbation, is only half of the story.

10.2.2.2 The Remote Influence of the Oceans on the Climate of the Sahel

The other half of the story is the influence of climatic factors external to the region, most notably the oceans, the source of most moisture in this semi-arid region. Oceanic influence on Sahelian drought was first quantified empirically, in studies that pointed to the coincidence of warm sea surface temperature (SST) anomalies in the equatorial Atlantic Ocean during dry years in the Sahel, and related them to anomalies in the atmospheric circulation (Hastenrath and Lamb 1977; Lamb 1978a, b; Lamb and Peppler 1992).

The first modeling studies were conducted at the UK Meteorological Office (Folland et al. 1986; Palmer 1986), in a control-perturbation set-up similar to that utilized to demonstrate sensitivity to land cover change. In this case, a control simulation, run with climatological sea surface temperature conditions, was compared to a perturbation simulation, where SST anomalies typical of dry years are superimposed on the same climatology (Fig. 10.2). The response was a decrease in rainfall in the Sahel in the perturbation compared to the control simulation. These studies were the first to dynamically demonstrate the sensitivity of Sahel rainfall to worldwide sea surface temperatures. Why, then, were the climate science community and the broader community of scholars and practitioners of drylands development, not convinced that the cause of Sahel drought lay in the oceans, not over land?

Almost 10 years later, Rowell et al. (1995) confirmed Folland et al.'s (1986) results. They also reported that Owen and Folland (1988) had not been able to reproduce them as cleanly and convincingly with the subsequent version of the UK Met Office model, pointing out that the difference between the two generations of models lay in the representation of clouds. The model employed by Folland et al. (1986) prescribed cloud cover to follow climatology at each location, avoiding the complexity of dynamically modeling the evolution in cloud cover. In contrast, the subsequent version employed a prognostic cloud scheme, i.e. attempted to model the evolution of cloud cover within the model. As had already been noted by

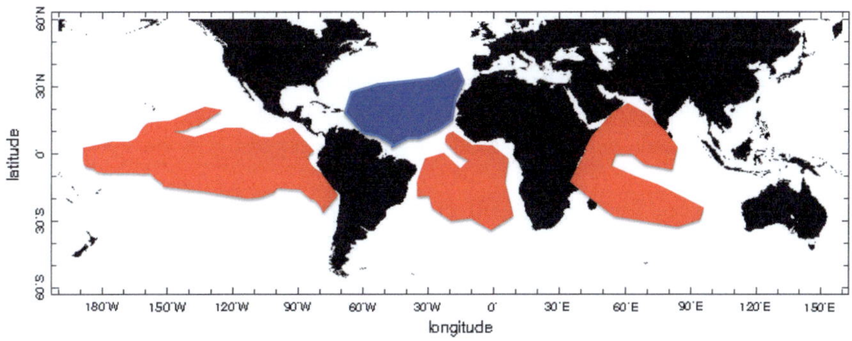

Fig. 10.2 The pattern of sea surface temperature associated with late 20th century Sahel drought. Notable features are: the warming of the tropical and southern oceans, including the South Atlantic and Indian Oceans, and the relative cooling of the northern hemisphere mid-latitude oceans, especially the North Atlantic Ocean.

Charney et al. (1977), cloud cover is a potential negative feedback on the hypothesized bio-geophysical feedback, i.e. it can potentially weaken or disable it. An increase in surface albedo causes a reduction in solar radiation absorbed at the surface and in energy potentially available to the atmospheric column. If this is followed by a decrease in precipitation, such a decrease also implies a decrease in cloud cover and in cloud albedo, hence an increase in solar radiation that can reach the surface. Therefore, in terms of atmospheric energy budget, the cloud albedo response can potentially cancel the initial surface albedo anomaly, and re-equilibrate the system. In Folland et al.'s model, cloud cover would not have changed as a consequence of the albedo-precipitation change, therefore eliminating the potential to cancel out the effect of the initial surface albedo change. In the subsequent set-up, where clouds varied in concert with precipitation, it could have cancelled the effect of surface albedo change. It should be noted that the dominance of this potential negative feedback is physically unrealistic in semi-arid regions such as the Sahel. In such regions, a reduction in precipitation translates into a reduction in soil moisture. Regardless of how much radiation reaches the surface, evaporation is limited by water availability, not by the energy required to extract it through evaporation (Koster et al. 2003, Koster et al. 2006). In the real world, if the balance between surface and cloud albedo were to result in an increase in radiation into the surface, in the absence of soil moisture that increase in radiation would not translate into an increase in evaporation that could fuel an increase in humidity and vertical instability.

We may ask why the Folland et al. results did not have the broader impact warranted by their significance to the dominant discourse of human-induced desertification (Bullock et al. 1995). While it may have been reasonable for the development community to wait until climate scientists had reached a consensus, why did the latter take so long to agree amongst themselves? Ten years after Folland et al. (1986), Sud and Lau (1996) expressed skepticism at the Rowell et al.

(1995) results, which had replicated Folland et al. (1986), but could not in turn be replicated with the simultaneous effort of the Atmospheric Model Intercomparison Project (AMIP)—the first of many inter-comparisons of atmospheric models that used observed SST run over a short period (1979–88), due to computing limitations at the time. Sud and Lau (1996) went so far as to suggest that the Rowell et al. simulation of the relationship between sea surface temperature and rainfall anomalies between 1949 and 1990 could be "success by chance". A similar inter-comparison of models attempted more recently, in the context of the CLIVAR Climate of the 20th Century project (Scaife et al. 2009), similarly documented the failure of many models to reproduce the amplitude of Sahelian climate fluctuations.

These results, however, do not call into question the conclusion that oceans were the primary cause of Sahel drought. Since Giannini et al. (2003), various state-of-the-art models have successfully reproduced the sensitivity of Sahel rainfall to global-scale sea surface temperature changes, including anomalous years, and its multi-decadal evolution. These multi-decadal shifts include the anomalously wet 1950s and 1960s, persistent drought in the 1970s and 1980s, and partial recovery since the 1990s. They can be reproduced in simulations based on inclusion of the observed evolution of global sea surface temperatures only as the prescribed boundary condition. Whereas Sud and Lau (1996) could argue that a model could by chance correctly simulate a few random years, this argument cannot reasonably be supported in the face of ensembles of decades-long simulations that reproduce the alternation of wet and dry epochs as described by Giannini et al. (2003), Lu and Delworth (2005), and Caminade and Terray (2010) using different models.

10.2.2.3 Limitations of Both Explanations

Counter-arguments to either explanation have to do with inconsistencies between the real and modeled worlds, either in the magnitude of the perturbations to the system or in the magnitude of the response to a perturbation.

Scientists seeking to refute the role of Charney's bio-geophysical feedback as the sole explanation for the persistence of drought appealed to the much smaller observed surface albedo change compared to that used in models to provoke a significant response in precipitation (e.g. Nicholson 2000b). With regards to the observed difference in surface albedo between dry and wet years, Govaerts and Lattanzio (2008) have the last word: in Meteosat observations that compare the very dry year of 1984 to the wetter 2003, they found a maximum surface albedo change of 0.15 along 16ºN latitude, and an average regional change of 0.06, i.e. significantly smaller than the change from 0.14 to 0.35 imposed by Charney et al. (1977) and similar subsequent studies.

Simulations based on the dominant role of the oceans also inadequately represent the amplitude of rainfall variability (Scaife et al. 2009), especially when ocean and atmosphere are coupled as in the model systems in use for the IPCC Assessment Reports (Biasutti 2013), albeit with exceptions (e.g. Lu and Delworth 2005). In other words, if one were to hold simulations of the response of Sahel rainfall to oceanic change to the same standard applied by Taylor et al. (2002) in the case of the land use/land cover change, one would fail in many instances to recover the amplitude of the observed change, a 15–20 % decrease in rainfall with respect to the long-term average. However, as discussed later in this chapter, the strength of the explanation based on a remote/oceanic forcing of Sahel drought lies in the faithfulness of the simulation of the century-scale evolution of regional rainfall anomalies (Giannini et al. 2003) more than in its magnitude.

10.2.3 Synthesis: Oceans Lead and Land Amplifies

In the 1990s and early 2000s most attempts made to compare the relative roles of sea surface temperature anomalies or land use change did not yield firm conclusions (e.g. Diedhiou and Mahfouf 1996; Douville et al. 2001; Douville 2002). Maynard and Polcher (2003) correctly pointed to the complexity of balancing remote oceanic influence with continental response mediated by land surface-atmosphere feedbacks that could either enhance, through evaporation, or cancel out, through clouds, the initial ocean-forced perturbation. However, when Zeng et al. (1999) and Giannini et al. (2003) revisited the issue, the time was ripe for stronger conclusions. In an idealized modeling context, and in a global general circulation model context, respectively, they showed that the slow, but steady refinement of climate models had enabled them to conclusively identify the oceans as the primary cause of drought. State-of-the-art models similar to those used for climate change projections can reproduce the 20th century evolution of Sahelian climate given only the historical evolution of sea surface temperatures (Fig. 10.3, from Giannini et al. 2003; Lu and Delworth 2005). Inclusion of vegetation dynamics in an idealized model (Zeng et al. 1999), and of land surface-atmosphere coupling in a global general circulation model (Giannini et al. 2003) lead to amplification of the ocean-driven rainfall anomalies, testament to the fact that a positive bio-geophysical feedback is indeed in place.

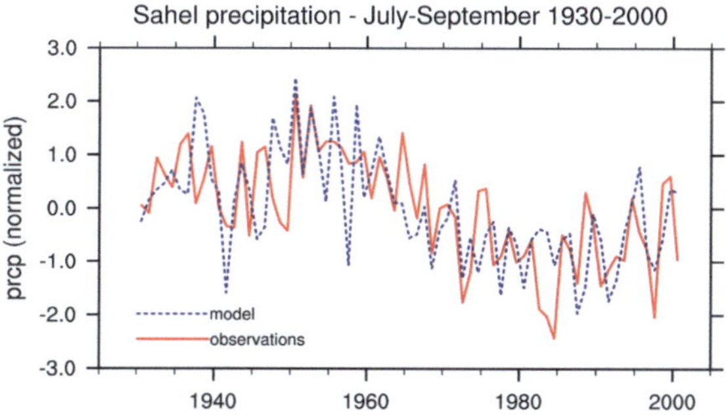

Fig. 10.3 The evolution of Sahel rainfall, as averaged from station records, in *red*, and from the ensemble of simulations analysed in Giannini et al. (2003) (reproduced from Giannini et al. 2003).

Giannini et al. (2003) took their inspiration from seasonal climate forecasts that predict shifts in the probabilities of a season being warmer or wetter than normal a few months ahead of time. Seasonal climate prediction relies on the influence that slowly evolving surfaces of the earth, such as the oceans or soil moisture, have in nudging the atmosphere into preferred states, affecting the probability of their recurrence. Using the model developed by Suarez and collaborators at NASA's Goddard Space Flight Center (Bacmeister et al. 2000), they described Sahel rainfall variability as a result of well-defined global SST anomalies, and hence as potentially predictable given a prediction of the SST anomalies.

The novelty in the Giannini et al. (2003) study with respect to the Folland et al. (1986) and Rowell et al. (1995) was what it tried to simulate. What was simulated was not the magnitude of the Sahel rainfall response to sea surface temperature anomalies associated with drought. Rather, it was the 20th century evolution of Sahel rainfall patterns that was reproduced accurately in its interannual and interdecadal modulations. As in Folland et al. (1986), the analysis highlighted the sensitivity of Sahel rainfall to global tropical SST. The El Niño-Southern Oscillation (ENSO) dominates on interannual time scales, but it is warming of the oceans around Africa, especially the equatorial Indian Ocean and south Atlantic, that can be directly related to the multi-decadal persistence of drought.

10.2.3.1 How the Influences of the Atlantic and Indian Oceans Work

The Atlantic and Indian Oceans influence Sahel rainfall on longer time scales than interannual, because their variations occur on these longer time scales, as in the Atlantic Multidecadal Oscillation (Kushnir 1994; Enfield and Mestas-Nuñez 1999; Mann and Emanuel 2006) and the pronounced warming trend in the Indian Ocean since the 1970s (Du and Xie 2008). In addition to the literature discussed in greater detail below, the patterns of oceanic influence affecting the Sahel have been thoroughly dissected in the modeling component of the recent Euro-African research endeavor known as AMMA—African Monsoon Multidisciplinary Analyses (Hourdin et al. 2010; Ruti et al. 2011; Rodríguez-Fonseca et al. 2011; and references therein).

The influence of the Atlantic Ocean on Sahel rainfall has historically been conceived of as a continental extrapolation of the association between maximum sea surface temperature (SST) and deep convection over the adjacent Atlantic Ocean (Zeng 2003). Over the oceans, the latitudinal location of the Intertropical Convergence Zone (ITCZ), the east-west belt of deep convection and precipitation, coincides with the warmest surface temperatures (Lindzen and Nigam 1987). The north-south, or meridional gradient in SST, i.e. the temperature difference between the north and south tropical Atlantic, affects the displacement of the ITCZ (Hurrell et al. 2006; Knight et al. 2006). In the Atlantic basin, a warmer south than north would attract the ITCZ southward, compared to its climatological mean location (Moura and Shukla 1981).

Eltahir and collaborators extended these considerations to view the ocean-atmosphere-land as a coupled system via the "meridional gradient in boundary-layer entropy" (Eltahir and Gong 1996). Entropy is a variable directly linked to the meridional overturning circulation in the atmosphere, also known as the Hadley cell. A positive north-south gradient in near-surface entropy is associated with more vigorous deep convection and precipitation when the ascending branch of the Hadley cell is in the northern hemisphere (Kang et al. 2009). In the atmosphere entropy is synonymous with equivalent potential temperature—a measure of the temperature and humidity contributions to the energetic content of a parcel of air near the surface. Considering the geography of ocean and land in West Africa, the meridional gradient in boundary layer entropy is enhanced if land heats up, as in traditional theories of the monsoon, leading to a vigorous on-shore circulation, and attendant increase in moisture convergence. It is weakened if the equatorial Atlantic Ocean heats up, as in the observational work of Lamb (1978a, b) and Nicholson (1980, 1981), and modeling work of Fontaine et al. (1998), Vizy and Cook (2001, 2002), and Losada et al. (2010) which link drought in the Sahel to warming along the Gulf of Guinea.

The influence of the Indian Ocean had already been noted by Shinoda and Kawamura (1994), but it is with Giannini et al. (2003) and Bader and Latif (2003) that warming of the equatorial Indian Ocean took on prominence in explanations of Sahel drought (e.g. Kerr 2003). The influence can be conceptualized as a vertical stability argument akin to that behind the influence of ENSO on the global tropics (Chou et al. 2001; Neelin et al. 2003). A warmer than normal equatorial Indian Ocean favors oceanic deep convection. It is this oceanic deep convection that sets the energy threshold for vertical instability by warming the upper troposphere and raising the threshold for convection. If the threshold is too high for a continental parcel of air to attain it, e.g. because over land the moisture required to make the parcel lighter is not readily available, then land gets drier. A more involved dynamic explanation invokes the westward propagation of Rossby waves to carry over Africa the atmospheric adjustment to enhanced deep convection in the equatorial Indian Ocean (Hagos and Cook 2008; Lu 2009). This atmospheric adjustment bears all the circulation features typical of dry years in the Sahel, starting from north-easterly near-surface wind anomalies, which oppose the climatological south-westerly flow, and express a weakening of the moist, on-shore monsoonal flow (Giannini et al. 2005).

10.2.4 Conclusion on Understanding the Recent Past of Sahelian Climate

Regarding the cause of 20th century drought in the Sahel, the explanation linking drying to changes in the oceans is more inclusive than the alternative explanation which links drought to human-induced baring of the soils. If one accepts that the 20th century drought had its origin in the oceans, this conclusion does not exclude the possibility that an ocean-forced reduction of rainfall may be amplified by natural land surface-atmosphere feedbacks, and exacerbated (or countered) by human-induced land use change. Conversely, at present there is no evidence that corroborates the purported global influence of the climatic consequences of land cover/land use change. To the contrary, the latest model comparison of the scales of influence of sea surface temperature anomalies and land cover change supports the conclusion that the latter is local (Findell et al. 2009).

10.3 Current and Future Change

10.3.1 Simulating Climate Change with Coupled Models

Climate change projections are made with global, coupled ocean-atmosphere-land models—the direct descendants of the models that began to be developed in the late 1960s (Le Treut et al. 2007) starting with the atmosphere only. To simulate the climate of the 20th century, the whole suite of "external influences" that can affect the internal variability of the ocean-atmosphere system is described according to historical reconstruction. These influences include natural factors such as volcanic eruptions and variations in insolation at the top of the atmosphere related to cycles in solar activity, and anthropogenic factors, such as concentrations of pollutants—gases like carbon dioxide and methane, or aerosols like sulfur dioxide. In some instances land use change is also prescribed in addition to natural variations in vegetation due to climate. The goal here is to faithfully reproduce the response of the climate system to all the external influences it has experienced, natural or anthropogenic. This exercise has been most successful in explaining global mean temperature. The observed evolution of global mean temperature falls within the uncertainty envelope of the multi-model average of simulations that include anthropogenic influences. In late-20th century, it falls outside of the multi-model range of simulations that only include natural influences (Stott et al. 2006; Hegerl et al. 2007 and references therein). Attribution of the warming to anthropogenic causes follows.

In these global coupled models of the climate system the equations of motion are derived directly from first principles. They constitute the "dynamical core" of models. They are complemented by parameterizations, which describe processes that take place at scales—from molecular to tens of kilometers—that cannot be represented explicitly given the coarse spatio-temporal resolution of the models. Many of these processes—surface-atmosphere exchange of energy and water, cloud development, and especially convection—exert a stronger influence on the climate of the tropics than on that of middle and high latitudes, where the large-scale dynamics of the circulation dominate. It is therefore likely that the significant variation, indeed the discrepancy, in projections of regional precipitation change over tropical land is caused by the inter-model variation in the representation of these processes.

The Sahel is a case in point. In the face of remarkable agreement in the models' simulation of late 20th century drought, 21st century projections of precipitation change diverge, i.e. some models project a drier future, others a wetter future, and perhaps the majority anticipate no significant change (Biasutti and Giannini 2006; Cook and Vizy 2006; Douville et al. 2006; Joly et al. 2007; Biasutti 2013).

10.3.1.1 Climate Change and the Late 20th Century Sahel Drought

The magnitude, and the spatial and temporal extent of late 20th century Sahel drought naturally raise questions about its potential association with global climate change. Section 10.2.3 detailed the history of the successful simulation of Sahel drought in atmosphere-only models, when the observed evolution of global sea surface temperatures is prescribed as a boundary condition. It also provided an indirect attribution argument, whereby Sahel drought could be tied to anthropogenic emissions from fossil fuel burning via their influence on sea surface temperatures. Late 20th century drying is also reproduced among the approximately 20 coupled ocean-atmosphere models contributed to the 4th Assessment Report of the Intergovernmental Panel on Climate Change (IPCC AR4), where anthropogenic influences are directly included. Here, no model simulates wetter Sahelian conditions at the end of the 20th century, that is, with the external influences typical of the end of the 20th century, compared with a "pre-Industrial" control, that is, simulations where external influences are constant and fixed to "pre-Industrial" values. Under these constraints, a considerable number of models actually simulate statistically significantly drier conditions (Biasutti and Giannini 2006). This agreement is remarkable because in the IPCC models the oceanic influence effecting late 20th century drought is one step removed compared to the simulations that initially detected this relationship. In the IPCC simulations, the 20th-century evolution of sea surface temperature is not directly imposed; rather it is simulated in response to external influences applied to the coupled ocean-atmosphere-land system. It is these external influences that consistently cause the sea surface temperature change associated with drought in the Sahel. This is further proof that the protracted 20th century climate anomaly in the Sahel was indeed a manifestation of influences at play much larger than the Sahel itself.

Biasutti and Giannini (2006) fall short of attributing the Sahel drought solely to global anthropogenic influences from industrialization, because most global change simulations combine natural and anthropogenic influences. When anthropogenic influences are isolated from natural factors, as in Zhang et al. (2007), there is a late 20th century reduction in precipitation in the latitudinal belt that comprises the Sahel, but the results obtained from such "single-forcing" simulations are not fully convincing. For example, the model currently in use by the UK Met Office/Hadley Centre simulates drying of the Sahel in response to aerosols, but wetting in response to greenhouse gases (Ackerley et al. 2011).

Global anthropogenic causes of Sahelian drought can nonetheless be demonstrated in a more general sense, by the link between greenhouse gases and sulfate aerosols with changes in the surface temperatures of the Indian and Atlantic and Oceans respectively. The pattern of sea surface temperature change associated with Sahel drought in IPCC simulations broadly reproduces observed changes (Fig. 10.2). The warming of the Indian Ocean is recognized to be part of the general warming trend in the oceans, and has been attributed to the direct influence of greenhouse gases on the heat content of the ocean (Barnett et al. 2005). Specifically, Indian Ocean warming has been shown to evolve as a small imbalance between the direct warming influence of greenhouse gases augmented by the water vapor feedback, and cooling through an increase in evaporation (Du and Xie 2008). In contrast, the North Atlantic does not warm as much as the tropical and southern oceans, including the equatorial Indian Ocean. The relative lack of warming of the North Atlantic Ocean compared to the South Atlantic is likely a combination of natural and anthropogenic effects. A multi-decadal "oscillation" in the surface temperature of the North Atlantic has been shown to be a manifestation of internal variability of the coupled ocean-atmosphere in IPCC models (Ting et al. 2009), distinguishable from the externally forced warming trend. Separate from this multi-decadal variability, it is possible to distinguish a secular trend that is associated with northern hemisphere sulfate aerosol emissions—increasing until the mid-1980s and abating since the introduction of legislation aimed at controlling pollution and acid rain in North America and Europe (Chang et al. 2011; Booth et al. 2012). Sulfate aerosols cool the surface by increasing the albedo of the atmosphere (Liepert 2002; Stanhill and Cohen 2001; Rotstayn and Lohmann 2002). The sulfate aerosol particles themselves reflect solar radiation. They also act as cloud condensation nuclei, and by facilitating cloud formation, indirectly increase the portion of solar radiation that is reflected by the atmosphere. The influence of aerosols on surface temperature is more localized in the northern hemisphere because their lifetime is much shorter than that of greenhouse gases.

In sum, warming of the Indian Ocean and reduced warming of the North Atlantic compared to the South Atlantic both had a part in late 20th century Sahel drying, and are consistent with a partial attribution of Sahel drought to greenhouse gases and aerosols, respectively.

10.3.1.2 Projections of Future Precipitation Change in the Sahel

The dominant role of the oceans in effecting late 20th century drought suggests that the same sea surface temperature-precipitation relationships should emerge in the models used for IPCC assessments (Meehl et al. 2007; Taylor et al. 2012; Biasutti et al. 2008; Rowell 2013), and that differences in projected Sahel precipitation change may be the consequence of differences in projected sea surface temperature change. Biasutti et al. (2008) examined this possibility using two oceanic indices, one for the tropical Indo-Pacific and the other for the Atlantic. This approach successfully reproduced the 20th century behavior of Sahel rainfall, but failed in predicting end of 21st century patterns.

Giannini et al. (2013) modified this approach by representing oceanic influence in one index—the relative warming of the sub-tropical North Atlantic with respect to the global tropical oceans—instead of the two used in Biasutti et al. (2008). This re-interpretation also distinguished the oceanic Inter-tropical Convergence Zone (ITCZ) in the Atlantic from the continental rainbelt over Africa (Nicholson 2013). This re-formulation was sufficient to both predict the future and reconcile past rainfall patterns with the theory of precipitation change due to global warming (Fig. 10.4). It was also consistent with current interpretations of trends in hurricane activity in the Atlantic sector (Vecchi et al. 2008). These hypothesize that warming of the tropical oceans in response to greenhouse gases "ups the ante" for deep convection, in an expression popularized by Neelin et al. (2003; also see Chou and Neelin (2004) and Sobel (2010)), and that the higher threshold to trigger vertical instability, deep convection and precipitation can be met if sufficient moisture is supplied from the Atlantic, in monsoonal flow. Assuming no change in the strength of the atmospheric circulation, i.e. in near-surface winds and vertical velocity, the role of a warmer North Atlantic Ocean is seen not so much as attracting convergence northward, rather as effecting an increase in atmospheric moisture content, and in the moisture that can be converged over land (Pu and Cook 2012). As warming of the tropical oceans began to emerge in the 1970s, the reduced warming of the North Atlantic, in part due to the cooling effect of sulfate aerosols (Rotstayn and Lohmann 2002; Chang et al. 2011; Chiang et al. 2013), in part due to natural variability in the oceanic circulation (Knight et al. 2006; Ting et al. 2009), starved the African continent of the moisture needed to trigger deep convection, causing persistent drought. Now that the North Atlantic has reversed its trend and is warming, a wetter Sahel has become possible, as manifest in a "partial recovery" of the rains since the mid-1990s (Nicholson 2005; Ali and Lebel 2009; Salack et al. 2011). It is interesting to note that this recovery, to the extent that it is driven by an increase in the mean or median intensity of daily rainfall rather than by an increase in the number of rainy days, is qualitatively different from the driest years, which were characterized by fewer rainy days (Lebel et al. 2003). Further, it is in agreement with farmers' perceptions of a more variable climate, which includes highly disruptive extreme events (West et al. 2008; Lodoun et al. 2013).

10.4 Conclusions

In the 40 years since the devastating Sahelian drought and famine of 1972–73, climate models have developed to a point that simulation of the abrupt shift from the wet 1950s and 1960s to the persistently dry 1970s and 1980s now constitutes a litmus test of our confidence in these models. In response to our best estimates of global sea surface temperature, state-of-the-art atmospheric models require no information on human-induced land cover/land use change to reproduce Sahelian drought. These simulations lead to the conclusion that drought was caused by large-scale, if subtle, shifts in oceanic temperatures, not by local anthropogenic

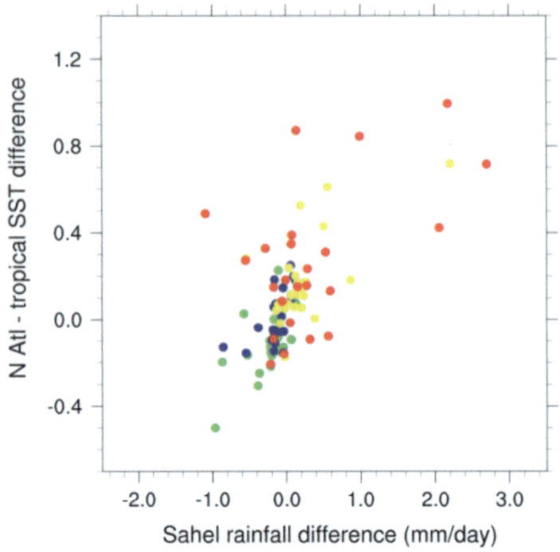

Fig. 10.4 The relationship between change in Sahel rainfall, on the horizontal axis, and change in the difference between sub-tropical North Atlantic and global tropical SST, on the vertical axis. *Green* and *blue dots* refer to the late 20th century change compared to pre-Industrial or beginning of the 20th century, while *yellow* and *red dots* refer to mid- and late-21st century compared to the end of the 20th century (reproduced from Giannini et al. 2013).

pressure on the environment. Even in the face of significant uncertainty in projections of future tropical precipitation change, coupled ocean-atmosphere models of the kind used in Assessment Reports of the Intergovernmental Panel on Climate Change (IPCC), when forced with our best estimate of late 20th century influences on the climate system, natural and anthropogenic, are capable of simulating drying of the Sahel at the end of the 20th century.

Ultimately, it should not come as a surprise that the oceans play such an important role in the climate of this semi-arid, desert margin region, since the only source of moisture for a system with alternation between a short rainy season and a prolonged dry season must be the moisture imported from the surrounding oceans. However, the discourse explaining Sahel drought as a local positive feedback between human-induced baring of the soil and reduction in precipitation has exerted, and still exerts such dominance in environment and development discussions at regional and global levels that it has warranted inclusion of a climate science chapter in a book on desertification, and warrants briefly summarizing the relationship between endogenous and exogenous hypotheses of drought causation once more. In sum, there needs to be no contradiction between the two. The endogenous hypothesis linking baring of the soils to drought fits within the context of the exogenous hypothesis, linking oceanic warming to drought. Land surface-atmosphere interaction has a secondary role, nevertheless an amplifying role, as posited by Charney in his bio-geophysical feedback. On the one hand, while we do not know the exact spatial extent of the human-induced land cover change hypothesized to have triggered persistence of drought, it is highly unlikely that land degradation may have occurred uniformly at the scale required to effect a large-scale climatic change spanning the entire Sahelian belt, from the Atlantic Ocean to the Red Sea. On the other, if oceanic conditions translate into less precipitation than normal, then the consequences of drought for land surface properties, whether increased albedo or reduced soil moisture, act as amplifiers. Anthropogenic disturbance of the land surface can further exacerbate such drought, whereas efforts to remedy desertification may alleviate its local impact.

In the decades since the abrupt onset of persistent drought in the Sahel our understanding of the dynamics of tropical climates has advanced significantly. We now understand that land use/land cover change may be a reasonable way to change the vertical stability of the atmosphere from the surface up, and consequently the occurrence of deep convection and precipitation, as are changes in sea surface temperature in the tropical Atlantic. We also understand that there is an alternative way to change vertical stability and the occurrence of precipitation that invokes changes in upper tropospheric temperature effected by changes in the strength and/or location of deep convection over the oceans. As has become increasingly apparent in the age of El Niño and global warming, the climate is interconnected at a planetary scale, such that changes in remote oceans, particularly the tropical Pacific, can actually have significant impact, through "teleconnections".

It is now widely recognized not only that changes in the oceans caused drought, but that these changes in the oceans are qualitatively consistent with the influence of emissions of aerosols and greenhouse gases. Recognition that emissions from industrialization may have played a role in Sahel drought obviously shifts the blame from local to global anthropogenic influence.

Relating past drought to global anthropogenic influence does not imply that anthropogenic influences will continue to cause drought into the future. Whether making seasonal predictions of relevance to early warning systems and disaster risk reduction, or near-term climate change projections of relevance to adaptation policy and planning, it helps to cast dynamical understanding of regional change in terms of oceanic influence. Warming of the remote tropical oceans—in the case of the Sahel, the Pacific and Indian Oceans—may "up the ante" for convection, but warming of the local oceans—in the case of the Sahel, the Atlantic—may be able to provide sufficient moisture to meet the upped ante. Past, persistent drought lay in an imbalance in warming in the oceans around Africa. As long-term warming of the Indian Ocean emerged, the cooling effect of aerosols greatly reduced warming of the North Atlantic. Now that warming has begun to emerge in the North Atlantic, too, the Sahel has been getting wetter, but with large year-to-year variations in outcome.

What is of greater practical relevance is how the Sahel is getting wetter, and whether these emerging trends are consistent with the expectation that with climate change precipitation events may become more intense, but less frequent. While initial research, qualitative and quantitative (West et al. 2008; Salack et al. 2011; Lodoun et al. 2013), seems to point in this direction, more should be invested in elucidating these trends. At the same time, as we move in the direction of a warmer, and likely more variable climate on all time scales, from intra-seasonal to multi-decadal, it is imperative that we learn from coping with past climatic stress.

As is manifest in other chapters contributed to this volume, strategies in adaptation that spurred individuals and communities to pro-actively re-think environmental management seem to have effected positive change. With the caveat that the role of a recovery in precipitation in the "re-greening" of the Sahel still needs to be separated from the impact of agro-forestry, soil and water conservation, "farmer-managed natural regeneration" etc., mechanisms to favor the uptake of these very same interventions should be given full consideration in global negotiations around adaptation to climate change.

Acknowledgements The author was supported by the National Oceanic and Atmospheric Administration through Office of Global Programs grant NA07GP0213, and by the National Science Foundation through Division of Atmospheric and Geospace Sciences grant 0955372. The author wishes to acknowledge the invaluable comments received from the reviewer (Stephen D. Prince) and the editor (Roy Behnke), which have helped clarify, focus and streamline the original manuscript.

References

Ackerley, D., Booth, B. B. B., Knight, S. H. E., Highwood, E. J., Frame, D. J., Allen, M. R., & Rowell, D. P. (2011). Sensitivity of 20th century Sahel rainfall to sulfate aerosol and CO_2 forcing. *Journal of Climate, 24*, 4999–5014.

Ali, A., & Lebel, T. (2009). The Sahelian standardized rainfall index revisited. *International Journal of Climatology, 29*, 1705–1714.

Bacmeister, J. T., Pegion, P. J., Schubert, S. D., & Suarez, M. J. (2000). *Atlas of seasonal means simulated by the NSIPP1 atmospheric GCM*. NASA Tech. Memo-2000-104606. 100 pp

Bader, J., & Latif, M. (2003). The impact of decadal-scale Indian Ocean sea surface temperature anomalies on Sahelian rainfall and the North Atlantic Oscillation. *Geophysical Research Letters, 30*, 2169. doi:10.1029/2003GL018426.

Barnett, T. P., Pierce, D. W., AchutaRao, K. M., Gleicker, P. J., Santer, B. D., Gregory, J. M., & Washington, W. M. (2005). Penetration of human-induced warming into the world's oceans. *Science, 309*, 284–287.

Biasutti, M. (2013). Forced Sahel rainfall trends in the CMIP5 archive. *Journal of Geophysical Research, 118*, 1613–1623.

Biasutti, M., & Giannini, A. (2006). Robust Sahel drying in response to late 20th century forcings. *Geophysical Research Letters, 33*, L11706. doi:10.1029/2006GL026067.

Biasutti, M., & Sobel, A. H. (2009). Delayed Sahel rainfall and global seasonal cycle in a warmer climate. *Geophysical Research Letters, 36*, L23707. doi:10.1029/2009GL041303.

Biasutti, M., Held, I. M., Sobel, A. H., & Giannini, A. (2008). SST forcings and Sahel rainfall variability in simulations of 20th and 21st centuries. *Journal of Climate, 21*, 3471–3486.

Bjerknes, J. (1969). Atmospheric teleconnections from the equatorial Pacific. *Monthly Weather Review, 97*, 163–172.

Booth, B. B. B., Dunstone, N. J., Halloran, P. R., Andrews, T., & Bellouin, N. (2012). Aerosols implicated as a prime driver of twentieth-century North Atlantic climate variability. *Nature, 484*, 228–232.

Braconnot, P., Otto-Bliesner, B., Harrison, S., Joussaume, S., Peterchmitt, J.-Y., Abe-Ouchi, A., et al. (2007a). Results of PMIP2 coupled simulations of the Mid-Holocene and Last Glacial maximum part 1: Experiments and large-scale features. *Climate of the Past, 3*, 261–277.

Braconnot, P., Otto-Bliesner, B., Harrison, S., Joussaume, S., Peterchmitt, J.-Y., Abe-Ouchi, A., et al. (2007b). Results of PMIP2 coupled simulations of the Mid-Holocene and last Glacial maximum part 2: Feedbacks with emphasis on the location of the ITCZ and mid- and high latitudes heat budget. *Climate of the Past, 3*, 279–296.

Brooks, N. (2004). *Drought in the African Sahel: Long term perspectives and future prospects*. Tyndall Centre for Climate Change Research, Working paper 61.

Bullock, P., Le Houérou, H., Hoffman, M. T., Rounsevell, M., Sehgal, J., & Várallyay, G. (1995). Land degradation and desertification. Vol. Climate change 1995. Impacts, adaptation and mitigation of climate change: scientific-technical analyses. Contribution of Working Group II to the Second Assessment Report of the Intergovernmental Panel on Climate Change. Cambridge University Press. (chapter 4).

Caminade, C., & Terray, L. (2010). Twentieth century Sahel rainfall variability as simulated by the ARPEGE GCM, and future changes. *Climate Dynamics, 35*, 75–94.

Cane, M. A., Zebiak, S. E., & Dolan, S. C. (1986). Experimental forecasts of El Niño. *Nature, 321*, 827–832.

Cane, M. A. (1991) Forecasting El Niño with a geophysical model. pp. 345–369. In M. H. Glantz (Ed.), *Teleconnections linking worldwide climate anomalies* (pp. 345–369). Cambridge: Cambridge University Press.

Chang, C.-Y., Chiang, J. C. H., Wehner, M. F., Friedman, A., & Ruedy, R. (2011). Sulfate aerosol control of tropical Atlantic climate over the 20th century. *Journal of Climate, 24*, 2540–2555.

Charney, J. G., Stone, P. H., & Quirk, W. J. (1975). Drought in the Sahara: A biogeophysical feedback mechanism. *Science, 187*, 434–435.

Charney, J., Quirk, W. J., Chow, S.-H., & Kornfeld, J. (1977). A comparative study of the effects of albedo change on drought in semi-arid regions. *Journal of the Atmospheric Sciences, 34*, 1366–1385.

Chiang, J. C. H., & Sobel, A. H. (2002). Tropical tropospheric temperature variations caused by ENSO and their influence on the remote tropical climate. *Journal of Climate, 15*, 2616–2631.

Chiang, J. C. H., Chang, C.-Y., & Wehner, M. F. (2013). Long-term behavior of the Atlantic interhemispheric SST gradient in the CMIP5 historical simulations. *Journal of Climate, 26*, 8628–8640.

Chou, C., & Neelin, J. D. (2004). Mechanisms of global warming impacts on regional tropical precipitation. *Journal of Climate, 17*, 2688–2701.

Chou, C., Neelin, J. D., & Su, H. (2001). Ocean-atmosphere-land feedbacks in an idealized monsoon. *Quarterly Journal of the Royal Meteorological Society, 127*, 1869–1891.

Cook, K. H., & Vizy, E. K. (2006). Coupled model simulations of the West African monsoon system: 20th century simulations and 21st century predictions. *Journal of Climate, 19*, 3681–3703.

Cour, J. -M., & Snrech, S. (1998). Preparing for the future: A vision of West Africa in the year 2020. West Africa long-term perspective study. Sahel and West Africa Club, Organisation for Economic Cooperation and Development, Paris.

Dai, A., Lamb, P. J., Trenberth, K. E., Hulme, M., Jones, P. D., & Xie, P. (2004). The recent Sahel drought is real. *International Journal of Climatology, 24*, 1323–1331.

de Menocal, P., Ortiz, J., Guilderson, T., Adkins, J., Sarnthein, M., Baker, L., & Yarusinskya, M. (2000). Abrupt onset and termination of the African Humid Period: Rapid climate responses to gradual insolation forcing. *Quaternary science reviews, 19*, 347–361.

Diedhiou, A., & Mahfouf, J. -F. (1996). Comparative influence of land and sea surfaces on the Sahelian drought: A numerical study. *Annales Geophysicae, 14*, 115–130.

Douville, H. (2002). Influence of soil moisture on the Asian and African monsoons. Part II: Interannual variability. *Journal of Climate, 15*, 701–720.

Douville, H., Chauvin, F., & Broqua, H. (2001). Influence of soil moisture on the Asian and African monsoons. Part I: Mean monsoon and daily precipitation. *Journal of Climate, 14*, 2381–2403.

Douville, H., Salas-Melia, D., & Tyteca, S. (2006). On the tropical origin of uncertainties in the global land precipitation response to global warming. *Climate Dynamics, 26*, 367–385.

Du, Y., & Xie, S.-P. (2008). Role of atmospheric adjustments in the tropical Indian Ocean warming during the 20th century in climate models. *Geophysical Research Letters, 35*, L08712. doi:10.1029/2008GL033631.

Ehrlich, P. R. (1968). *The population bomb* (pp. 223). New York: Ballantine Books.

Eltahir, E. A. B., & Gong, C. (1996). Dynamics of wet and dry years in West Africa. *Journal of Climate, 9*, 1030–1042.

Emanuel, K. A., Neelin, J. D., & Bretherton, C. S. (1994). On large-scale circulations in convecting atmospheres. *Quarterly Journal of the Royal Meteorological Society, 120*, 1111–1143.

Enfield, D., & Mestas-Nuñez, A. (1999). Multiscale variabilities in global sea surface temperatures and their relationships with tropospheric climate patterns. *Journal of Climate, 12*, 2719–2733.

Findell, K. L., Pitman, A. J., England, M. H., & Pegion, P. J. (2009). Regional and global impacts of land cover change and sea surface temperature anomalies. *Journal of Climate, 22*, 3248–3269.

Folland, C. K., Palmer, T. N., & Parker, D. E. (1986). Sahel rainfall and worldwide sea temperatures, 1901–85. *Nature, 320*, 602–607.

Fontaine, B., Trzaska, S., & Janicot, S. (1998). Evolution of the relationship between near global and Atlantic SST modes and the rainy season in West Africa: Statistical analyses and sensitivity experiments. *Climate Dynamics, 14*, 353–368.

Funk, C., Dettinger, M., Michaelsen, J., Verdin, J., Brown, M., Barlow, M., & Hoell, A. (2008). Warming of the Indian Ocean threatens eastern and southern African food security but could be mitigated by agricultural development. *Proceedings of the National Academy of Sciences, 105*, 11081–11086. doi:10.1073/pnas.0708196105.

Gasse, F. (2000). Hydrological changes in the African tropics since the Last Glacial maximum. *Quaternary Science Reviews, 19*, 189–211.

Gasse, F., Tehet, R., Durand, A., Gibert, E., & Fontes, J.-C. (1990). The arid-humid transition in the Sahara and the Sahel during the last deglaciation. *Nature, 346*, 141–146.

Giannini, A., Salack, S., Lodoun, T., Ali, A., Gaye, A. T., & Ndiaye, O. (2013). A unifying view of climate change in the Sahel linking intra-seasonal, interannual and longer time scales. *Environmantal Research Letters, 8*, 024010.

Giannini, A., Saravanan, R., & Chang, P. (2003). Oceanic forcing of Sahel rainfall on interannual to interdecadal time scales. *Science, 302*, 1027–1030. doi:10.1126/science.1089357.

Giannini, A., Saravanan, R., & Chang, P. (2005). Dynamics of the boreal summer African monsoon in the NSIPP1 atmospheric model. *Climate Dynamics, 25*, 517–535. doi:10.1007/s00382005-0056-x.

Glantz, M. H. (1994). *Drought follows the plow: Cultivating marginal areas*. Cambridge: Cambridge University Press.

Govaerts, Y., & Lattanzio, A. (2008). Estimation of surface albedo increase during the eighties Sahel drought from Meteosat observations. *Global and Planetary Change, 64*, 139–145.

Greene, A. M., Giannini, A., & Zebiak, S. E. (2009). Drought return times in the Sahel: A question of attribution. *Geophysical Research Letters, 36*, L12701. doi:10.1029/2009GL038868.

Hagos, S., & Cook, K. H. (2008). Ocean warming and late 20th century recovery. *Journal of Climate, 21*, 3797–3811. doi:10.1175/2008JCLI2055.1.

Hastenrath, S., & Lamb, P. J. (1977). Some aspects of circulation and climate over the eastern equatorial Atlantic. *Monthly Weather Review, 105*, 1019–1023.

Hegerl, G., Zwiers, F., Braconnot, P., Gillett, N., Luo, Y., Marengo Orsini, J., Nicholls, N., Penner, J. E., & Stott, P. (2007). Understanding and attributing climate change. In S. Solomon, D. Qin, M. Manning, Z. Chen, M. Marquis, K. B. Averyt, M. Tignor & H. L. Miller (Eds.), *Climate change 2007: The physical science basis. Contribution of working group I to the fourth assessment report of the intergovernmental panel on climate change*. Cambridge, United Kingdom and New York, NY, USA: Cambridge University Press.

Held, I., & Soden, B. (2006). Robust responses of the hydrological cycle to global warming. *Journal of Climate, 19*, 5686–5699.

Held, I. M., Delworth, T. L., Lu, J., Findell, K., & Knutson, T. R. (2005). Simulation of Sahel drought in the 20th and 21st centuries. *Proceedings of the National Academy of Sciences of the United States of America, 102*, 17891–17896. doi:10.1073/pnas.0509057102.

Hourdin, F., Musat, I., Guichard, F., Ruti, P. M., Favot, F., Filiberti, M.-A., et al. (2010). AMMA-Model intercomparison project. *Bulletin of the American Meteorological Society, 91*, 95–104.

Hulme, M. (2001). Climatic perspectives on Sahelian dessication: 1973–1998. *Global Environmental Change, 11*, 19–29.

Hurrell, J. W., & Co-authors. (2006). Atlantic climate variability and predictability: A CLIVAR perspective. *Journal of Climate, 19*, 5100–5121.

Intergovernmental Panel on Climate Change. (2007) Climate Change 2007 Synthesis Report. Available at: http://www.ipcc.ch/pdf/assessment-report/ar4/syr/ar4_syr.pdf.

Intergovernmental Panel on Climate Change. (2014). Climate Change 2014 Synthesis Report. Available at: http://www.ipcc.ch/pdf/assessmentreport/ar5/syr/SYR_AR5_SPMcorr1.pdf.

Joly, M., Voldoire, A., Douville, H., Terray, P., & Royer, J.-F. (2007). African monsoon teleconnections with tropical SSTs: Validation and evolution in a set of IPCC4 simulations. *Climate Dynamics, 29*, 1–20.

Kang, S. M., Frierson, D. M. W., & Held, I. M. (2009). The tropical response to extratropical thermal forcing in an idealized GCM: The importance of radiative feedbacks and convective parameterization. *Journal of the Atmospheric Sciences, 66*, 2812–2827.

Katz, R. W., & Glantz, M. H. (1986). Anatomy of a rainfall index. *Monthly Weather Review, 114*, 764–771.

Kerr, R. A. (2003). Warming Indian Ocean wringing moisture from the Sahel. *Science, 302*, 210–211.

Knight, J. R., Folland, C. K., & Scaife, A. A. (2006). Climate impacts of the Atlantic multi-decadal oscillation. *Geophysical Research Letters, 33*, L17706. doi:10.1029/2006GL026242.

Koster, R. D., Suarez, M. J., & Schubert, S. D. (2006). Distinct hydrological signatures in observed historical temperature fields. *Journal of Hydrometeorology, 7*, 1061–1075.

Koster, R. D., Suarez, M. J., Higgins, R. W., & Van den Dool, H. M. (2003). Observational evidence that soil moisture variations affect precipitation. *Geophysical Research Letters*. doi:10.1029/2002GL016571.

Kushnir, Y. (1994). Interdecadal variations in the North Atlantic sea surface temperature and associated atmospheric conditions. *Journal of Climate, 7*, 141–157.

Kutzbach, J. E., & Liu, Z. (1997). Response of the African monsoon to orbital forcing and ocean feedbacks in the middle Holocene. *Science, 278*, 440–443.

Lamb, P. J. (1978a). Case studies of tropical Atlantic surface circulation patterns during recent sub-Saharan weather anomalies: 1967 and 1968. *Monthly Weather Review, 106*, 482–491.

Lamb, P. J. (1978b). Large-scale tropical Atlantic surface circulation patterns associated with sub-Saharan weather anomalies. *Tellus, 30*, 240–251.

Lamb, P. J. (1982). Persistence of Sub-Saharan drought. *Nature*, 299, 46–48.

Lamb, P. J., & Peppler, R. A. (1992). Further case studies of tropical Atlantic surface atmospheric and oceanic patterns associated with sub-Saharan drought. *Journal of Climate, 5*, 476–488.

Laval, K., & Picon, L. (1986). Effect of a change of the surface albedo of the Sahel on climate. *Journal of the Atmospheric Sciences, 43*, 2418–2429.

Leach, M., & Mearns, R. (1996). *The lie of the land: Challenging received wisdom on the African environment*. Oxford: James Currey, Ltd.

Lebel, T., Diedhiou, A., & Laurent, H. (2003). Seasonal cycle and interannual variability of the Sahelian rainfall at hydrological scales. *Journal of Geophysical Research,*. doi:10.1029/2001JD001580.

Le Treut, H., Somerville, R., Cubasch, U., Ding, Y., Mauritzen, C., Mokssit, A., Peterson, T., & Prather, M. (2007). Historical overview of climate change. In S. Solomon, D. Qin, M. Manning, Z. Chen, M. Marquis, K. B. Averyt, M. Tignor & H. L. Miller (Eds.), *Climate change 2007: The physical science basis. Contribution of working group I to the fourth*

assessment report of the intergovernmental panel on climate change. Cambridge, United Kingdom and New York, NY, USA: Cambridge University Press.
Liepert, B. G. (2002). *Observed reductions in surface solar radiation in the United States and worldwide from 1961 to 1990*. Geophys: Research Letters. doi:10.1029/2002GL014910.
Lindzen, R. S., & Nigam, S. (1987). On the role of sea surface temperature gradients in forcing low-level winds and convergence in the tropics. *Journal of the Atmospheric Science, 44*, 2418–2436.
Lodoun, T., Giannini, A., Traoré, P. S., Somé, L., Sanon, M. M. V., & Millogo-Rasolodimby, J. (2013). Changes in the character of precipitation in Burkina Faso associated with late-20th century drought and recovery in the Sahel. *Environmental Development, 5*, 96–108.
Losada, T., Rodríguez-Fonseca, B., Polo, I., Janicot, S., Gervois, S., Chauvin, F., & Ruti, P. (2010). Tropical response to the Atlantic equatorial mode: AGCM multi-model approach. *Climate Dynamics, 35*, 45–52.
Lu, J. (2009). The dynamics of the Indian Ocean sea surface temperature forcing of Sahel drought. *Climate Dynamics, 33*, 445–460. doi:10.1007/s00382-009-0596-6.
Lu, J., & Delworth, T. L. (2005). Oceanic forcing of late 20th century Sahel drought. *Geophysical Research Letters, 32*, L22706. doi:10.1029/2005GL023316.
Lyon, B. (2004). The strength of El Niño and the spatial extent of tropical drought. *Geophysical Research Letters, 31*, L21204. doi:10.1029/2004GL020901.
Mann, M. E., & Emanuel, K. A. (2006). Atlantic hurricane trends linked to climate change. *EOS, 87*, 233–244.
Maynard, K., & Polcher, J. (2003). Impact of land-surface processes on the interannual variability of tropical climate in the LMD GCM. *Climate Dynamics, 20*, 613–633.
Meadows, D. H., Randers, J., & Behrens, W. W, I. I. I. (1972). *The limits to growth: A report to the club of Rome (1972)*. New York: Universe Books.
Meehl, G. A., Covey, C., Delworth, T., Latif, M., McAvaney, B., Mitchell, J. F. B., et al. (2007). The WCRP CMIP3 multi-model dataset: A new era in climate change research. *Bulletin of the American Meteorological Society, 88*, 1383–1394.
Moura, A. D., & Shukla, J. (1981). On the dynamics of droughts in northeast Brazil: Observations, theory and numerical experiments with a general circulation model. *Journal of the Atmospheric Sciences, 38*, 2653–2675.
Neelin, J. D. (2007). Moist dynamics of tropical convection zones in monsoons, teleconnections, and global warming. In T. Schneider & A. H. Sobel (Eds.), *The global circulation of the atmosphere* (pp. 267–301). Princeton: Princeton University Press. (chapter 10).
Neelin, J. D., Chou, C., & Su, H. (2003). Tropical drought regions in global warming and El Niño teleconnections. *Geophysical Research Letters, 30*, 2275. doi:10.1029/2003GL0018625.
Neelin, J. D., Munnich, M., Su, H., Meyerson, J. E., & Holloway, C. E. (2006). Tropical drying trends in global warming models and observations. *Proceedings of the National Academy of Sciences, 103*, 6110–6115.
Newell, R. E., & Kidson, J. W. (1984). African mean wind changes between Sahelian wet and dry periods. *Journal of Climatology, 4*, 27–33.
Nicholson, S. E. (1979). Revised rainfall series for the West African subtropics. *Monthly Weather Review, 107*, 620–623.
Nicholson, S. E. (1980). The nature of rainfall fluctuations in subtropical West Africa. *Monthly Weather Review, 108*, 473–487.
Nicholson, S. E. (1981). Rainfall and atmospheric circulation during drought periods and wetter years in West Africa. *Monthly Weather Review, 109*, 2191–2208.
Nicholson, S. E. (1983). Sub-Saharan rainfall in the years 1976–1980: Evidence of continued drought. *Monthly Weather Review, 111*, 1646–1654.
Nicholson, S. E. (2000a). Land surface processes and Sahel climate. *Review of Geophysics, 38*, 117–139.
Nicholson, S. E. (2000b). The nature of rainfall variability over Africa on time scales of decades to millennia. *Global and Planetary Change, 26*, 137–158.

Nicholson, S. E. (2005). On the question of the "recovery" of the rains in the West African Sahel. *Journal of Arid Environments, 63*, 615–641.

Nicholson, S. E. (2013) The West African Sahel: A review of recent studies on the rainfall regime and its interannual variability. *ISRN Meteorology*, 453521, 32.

Owen, J., & Folland, C. K. (1988). Modelling the influence of sea surface temperatures on tropical rainfall. In S. Gregory (Ed.), *Recent climate change—a regional approach*. London: Belhaven Press.

Palmer, T. N. (1986). Influence of the Atlantic, Pacific and Indian Oceans on Sahel rainfall. *Nature, 322*, 251–253.

Patricola, C. M., & Cook, K. H. (2007). Dynamics of the West African monsoon under mid-Holocene precessional forcing: Regional climate model simulations. *Journal of Climate, 20*, 694–716.

Petit-Maire, N. (2002). *Sous le sable ... des lacs. Un voyage dans le temps*. Paris: CNRS Editions.

Polcher, J. (1995). Sensitivity of tropical convection to land surface processes. *Journal of the Atmospheric Science, 52*, 3143–3161.

Polvani, L. M., & Sobel, A. H. (2002). The Hadley circulation and the weak temperature gradient approximation. *Journal of the Atmospheric Science, 59*, 1744–1752.

Pu, B., & Cook, K. H. (2010). Dynamics of the West African westerly jet. *Journal of Climate, 23* (23), 6263–6276.

Raynaut, C. (1997). *Societies and nature in the Sahel: rethinking environmental degradation* (pp. xxiii, 351). UK: Routledge/SEI Global Environment and Development Series.

Rodríguez-Fonseca, B., Janicot, S., Mohino, E., Losada, T., Bader, J., Caminade, C., & Voldoire, A. (2011). Interannual and decadal SST-forced responses of the West African monsoon. *Atmospheric Science Letters, 12*(1), 67–74.

Rotstayn, L., & Lohmann, U. (2002). Tropical rainfall trends and the indirect aerosol effect. *Journal of Climate, 15*, 2103–2116.

Rowell, D. P. (2013). Simulating SST teleconnections to Africa: What is the state of the art? *Journal of Climate, 26*, 5397–5418.

Rowell, D. P., Folland, C. K., Maskell, K., & Ward, M. N. (1995). Variability of summer rainfall over tropical North Africa (1906–92): Observations and modeling. *Quarterly Journal of the Royal Meteorological Society, 122*, 669–704.

Ruti, P. M., Williams, J. E., Hourdin, F., Guichard, F., Boone, A., Van Velthoven, P., et al. (2011). The West African climate system: A review of the AMMA model inter-comparison initiatives. *Atmospheric Science Letters, 12*, 116–122.

Salack, S., Muller, B., & Gaye, A. T. (2011). Rain-based factors of high agricultural impacts over Senegal. Part I: Integration of local to sub-regional trends and variability. *Theoretical and Applied Climatology, 106*, 1–22.

Scaife, A. A., Kucharski, F., Folland, C. K., Kinter, J., Brônniman, S., Fereday, D., et al. (2009). The CLIVAR C20C project: Selected twentieth century climate events. *Climate Dynamics, 33*, 603–614. doi:10.1007/s00382-008-0451-1.

Schneider, S. H., & Dickinson, R. E. (1974). Climate modeling. *Review of Geophysics, 12*, 447–493.

Seth, A., Rauscher, S. A., Biasutti, M., Giannini, A., Camargo, S. J., & Rojas, M. (2013). CMIP5 projected changes in the annual cycle of precipitation in monsoon regions. *Journal of Climate, 26*, 7328–7351.

Seth, A., Rauscher, S. A., Rojas, M., Giannini, A., & Camargo, S. J. (2011). Enhanced spring convective barrier for monsoons in a warmer world? A letter. *Climate Change, 104*, 403–414.

Shanahan, T. M., Overpeck, J. T., Anchukaitis, K. J., Beck, J. W., Cole, J. E., Dettman, D. L., et al. (2009). Atlantic forcing of persistent drought in West Africa. *Science, 324*, 377–380.

Shinoda, M., & Kawamura, R. (1994). Tropical rainbelt, circulation, and sea surface temperatures associated with the Sahelian rainfall trend. *Journal of the Meteorological Society of Japan, 72*, 341–357.

Sobel, A. H. (2007). Simple models of ensemble-averaged tropical precipitation and surface wind, given the sea surface temperature. In T. Schneider & A. H. Sobel (Eds.), *The global circulation of the atmosphere* (pp. 219–251). Princeton : Princeton University Press (chapter 8).
Sobel, A. H. (2010). Raised bar for rain. *Nature Geoscience, 3*, 821–822.
Somerville, R. C. J., Stone, P. H., Halem, M., Hansen, J. E., Hogan, J. S., Druyan, L. M., et al. (1974). The GISS model of the global atmosphere. *Journal of the Atmospheric Sciences, 31*, 84–117.
Stanhill, G., & Cohen, S. (2001). Global dimming: A review of the evidence for a widespread and significant reduction in global radiation with discussion of its probable causes and possible agricultural consequences. *Agricultural and Forest Meteorology, 107*, 255–278.
Stephenne, N., & Lambin, E. F. (2001). A dynamic simulation model of land-use changes in Sudano-Sahelian countries of Africa (SALU). *Agriculture Ecosystems and Environment, 85*, 145–161.
Stott, P. A., Mitchell, J. F. B., Allen, M. R., Delworth, T. L., Gregory, J. M., Meehl, G. A., & Santer, B. D. (2006). Observational constraints on past attributable warming and predictions of future global warming. *Journal of Climate, 19*, 3055–3069.
Sud, Y. C., & Fennessy, M. (1982). Study of the influence of surface albedo on July circulation in semiarid regions using the GLAS GCM. *Journal of Climatology, 2*, 105–125.
Sud, Y. C., & Lau, K.-M. (1996). Comments on 'Variability of summer rainfall over tropical north Africa (1906-92): observations and modelling' by D. P. Rowell, C. K. Folland, K. Maskell and M. N. Ward (April A, 1995, 121, 669-704). *Quarterly Journal of the Royal Meteorological Society, 122*, 1001–1006.
Sugi, M., & Yoshimura, J. (2004). A mechanism of tropical precipitation change due to CO_2 increase. *Journal of Climate, 17*, 238–243.
Taylor, C. M., Lambin, E. F., Stephenne, N., Harding, R. J., & Essery, R. L. H. (2002). The influence of land use change on climate in the Sahel. *Journal of Climate, 15*, 3615–3629.
Taylor, K. E., Stouffer, R. J., & Meehl, G. A. (2012). An overview of CMIP5 and the experiment design. *Bulletin of the American Meteorological Society, 93*, 485–498.
Ting, M., Kushnir, Y., Seager, R., & Li, C. (2009). Forced and internal 20th century SST trends in the North Atlantic. *Journal of Climate, 22*, 1469–1481.
Trenberth, K. E., Dai, A., Rasmussen, R. M., & Parsons, D. B. (2003). The changing character of precipitation. *Bulletin of the American Meteorological Society, 84*, 1205–1217.
Trenberth, K., Jones, P. D., Ambenje, P., Bojariu, R., Easterling, D., Klein Tank, A., Parker, D., Rahimzadeh, F., Renwick, J. A., Rusticucci, M., Soden, B., & Zhai, P. (2007). Observations: Surface and atmospheric climate change. In S. Solomon, D. Qin, M. Manning, Z. Chen, M. Marquis, K. B. Averyt, M. Tignor & H. L. Miller (Eds.), *Climate change 2007: The physical science basis. Contribution of working group I to the fourth as- sessment report of the intergovernmental panel on climate change* (pp. 235–336). Cambridge, United Kingdom and New York, NY, USA: Cambridge University Press.
Vecchi, G. A., Swanson, K. L., & Soden, B. J. (2008). Whither hurricane activity? *Science, 322*, 687–689.
Vizy, E. K., & Cook, K. H. (2001). Mechanisms by which Gulf of Guinea and eastern North Atlantic sea surface temperature anomalies can influence African rainfall. *Journal of Climate, 14*, 795–821.
Vizy, E. K., & Cook, K. H. (2002). Development and application of a mesoscale climate model for the tropics: Influence of sea surface temperature anomalies on the West African monsoon. *Journal of Geophysical Research Atmospheres,*. doi:10.1029/2001JD000686.
West, C. T., Roncoli, C., & Ouattara, F. (2008). Local perceptions and regional climate trends of the Central Plateau of Burkina Faso. *Land Degradation and Development, 19*, 289–304. doi:10.1002/ldr.842.
Wilhite, D. (2001). *Drought: A global assessment*. Routledge hazards and disasters series: Routledge.
Xue, Y., & Shukla, J. (1993). The influence of land surface properties on the Sahel climate. Part I: Desertification. *Journal of Climate, 6*, 2232–2245.

Yang, F., Kumar, A., Schlesinger, M. E., & Wang, W. (2003). Intensity of hydrological cycles in warmer climates. *Journal of Climate, 16*, 2419–2423.

Yulaeva, E., & Wallace, J. M. (1994). The signature of ENSO in global temperature and precipitation fields derived from the microwave sounding unit. *Journal of Climate, 7*, 1719–1736.

Zeng, N. (2003). Drought in the Sahel. *Science, 302*, 999–1000.

Zeng, N., & Neelin, J. D. (1999). A land-atmosphere interaction theory for the tropical deforestation problem. *Journal of Climate, 12*, 857–872.

Zeng, N., Neelin, J. D., Lau, K.-M., & Tucker, C. J. (1999). Enhancement of interdecadal climate variability in the Sahel by vegetation interaction. *Science, 286*, 1537–1540.

Zhang, X., Zwiers, F. W., Hegerl, G. C., Gillett, N. P., Solomon, S., Stott, P. A., & Nozawa, T. (2007). Detection of human influence on twentieth-century precipitation trends. *Nature, 448*, 461–466. doi:10.1038/nature06025.

Chapter 11
The Tragedy of the Common Narrative: Re-telling Degradation in the American West

Lynn Huntsinger

Abstract Stories—with a beginning, middle and end, and a moral message, have had a major role in how desertification, and range condition, have been understood on western rangelands in the United States. Stories that attempt to make sense out of vegetation change, whether the bad guy is the self-interested human exploiter or the low-statured ruderal species, take hold in the scientific and public imagination and influence interpretation of policy and management outcomes. The development of policy and management for western grazing lands was shaped by a declensionist narrative of human greed and unrestrained self interest that developed in a parallel and an eventually mutually reinforcing way with a similarly declensionist ecological narrative, creating a story that is deeply embedded in existing institutions for rangelands. This narrative underpins retention of half of the American West in government ownership, how grazing resources are allocated to graziers, and the way that rangeland conditions, including indicators of desertification or degradation, are assessed and monitored. Once such stories take hold, new ideas about ecological dynamics that have a non-linear story and more complex characters have a hard time supplanting or even augmenting old paradigms. This in turn supports policy and management decisions. The reader is warned against charismatic stories–stories encourage and conceal deep-rooted, untested assumptions, simplify complex relationships, and universalize truths that may hold true only in a single time and place.

Keywords Succession · Public land policy · Forestry · Holistic resource management · State and transitions · Equilibrium · Rangeland ecology

The arid rangelands of the western United States include cold desert steppe from eastern California to the base of the Rockies, Mediterranean rangelands along the Pacific Coast, and southwestern deserts down to the border with Mexico (Fig. 11.1). The Rocky Mountains cut the continent in half north to south. Rain-fed agriculture is

---

L. Huntsinger (✉)
University of California, Berkeley, USA
e-mail: huntsinger@berkeley.edu

R.H. Behnke and M. Mortimore (eds.), *The End of Desertification?*
Springer Earth System Sciences, DOI 10.1007/978-3-642-16014-1_11

Fig. 11.1 The continental United States. West of the 100th Meridian, rainfed agriculture is generally no longer feasible. Figure 11.2 presents climate graphs for each town listed.

found mostly to the east of the range; to the west, farming takes place on irrigable plots near water or where irrigation projects bring water to fields, capturing snow melt from the many mountain ranges or dipping into groundwater. The geographer John Wesley Powell, considering the possible development of this western country, recognized in 1878 that here "the lands without water have no value" (Powell 1879). This land with no value has been used for grazing and mining since the 16th century.

Historical concerns about what was termed rangeland degradation or desertification in the nineteenth century gave impetus to the development of the current system of federally administrated and managed western rangelands. Changing theoretical paradigms explaining the causes and processes of degradation and desertification have shaped management and policy on the federal lands since 1900. Livestock grazing was introduced into North America a little over 400 years ago, concurrent with sweeping changes in environment, culture, and economy. Given such a history, how can the role of livestock and other factors in shaping conditions on U.S. rangelands be determined? Introductions of new plants and animals, suppression of traditional peoples and their management practices, political and scientific cultures that have shaped the definition and interpretation of degradation, and the institutions governing arid lands have each played a part.

In his essay "A Place for Stories" the environmental historian Bill Cronon (1992) points out that historians interpret and report the past through stories that have a beginning, middle and end, present a moral message, and feature a change in the main protagonist(s)—whether environmental or human. His argument that stories are innately appealing ways to transmit complex information also sheds light on the development of policy and management for western grazing lands, where a declensionist narrative of human greed and unrestrained self interest developed in a parallel and an eventually mutually reinforcing way with a similarly declensionist ecological story, creating the justification for how rangelands have been allocated, assessed, and managed for more than one hundred years.

Deconstructing the founding narrative of western rangelands, and the use of the concept of degradation or desertification, is important because it underpins retention of half of the American West in government ownership, how grazing resources are allocated to graziers, and the way that rangeland conditions, including indicators of desertification, are assessed and monitored. This in turn supports policy and management decisions. The story is a key component in the education of those specializing in the management and ecology of western arid lands. This chapter explores the origins of the concept of rangeland degradation as a result of livestock grazing, how degradation has been identified, the attempted remedies, and the complications of livestock grazing as simultaneously a cause and a cure. An interplay of ecological science and social norms evolves through time in the arid west. Outcomes are illustrated with the story of fire and invasive species as agents of desertification in the Great Basin.

11.1 The Geographical Setting of Western Rangelands

There are more than 3.25 m km^2 of arid rangelands in the United States (Lal et al. 2004), encompassing a diverse geology, vegetation, topography, and climate. Internal variation in physiography is pronounced: elevations ascend from a low in Death Valley at 86 m below sea level to alpine zone summits and volcanic peaks of some 4000 m. Western montane regions are an important element in livestock mobility patterns (Huntsinger et al. 2010).

The morphology and physiography of the North American west ranges from the glacially-eroded continental shield in central Canada and the slowly westward upward-ramping Great Plains of the United States to a massive cordillera that extends from mid-Mexico through the Rocky Mountains of the United States into the Canadian Rockies (Fig. 11.1). The 100th longitudinal meridian bisects the continent: To the west, the landscape is distinguished by fault-shaped landforms, often-dramatic relief, and aridity (Fig. 11.2). A transect from west to east can help put this geomorphology into perspective (Huntsinger and Starrs 2006) (Fig. 11.3).

Land ownership along the transect is linked to ecological regions and topography. At the Pacific Coast, in California, the land is about 50 % publicly owned, montane areas mostly managed by the federal United States Forest Service (USFS), and lowland deserts by the federal Bureau of Land Management (BLM). There is a significant amount of land in other forms of public ownership, and also in large private ranches. Continuing east, traversing the mountains of the Sierra Nevada to the Great Basin's cold desert steppe, most land is under government management, with small private landholdings along rivers and streams. From the Pacific to the Rocky Mountains, no state in the United States has less than 30 % government

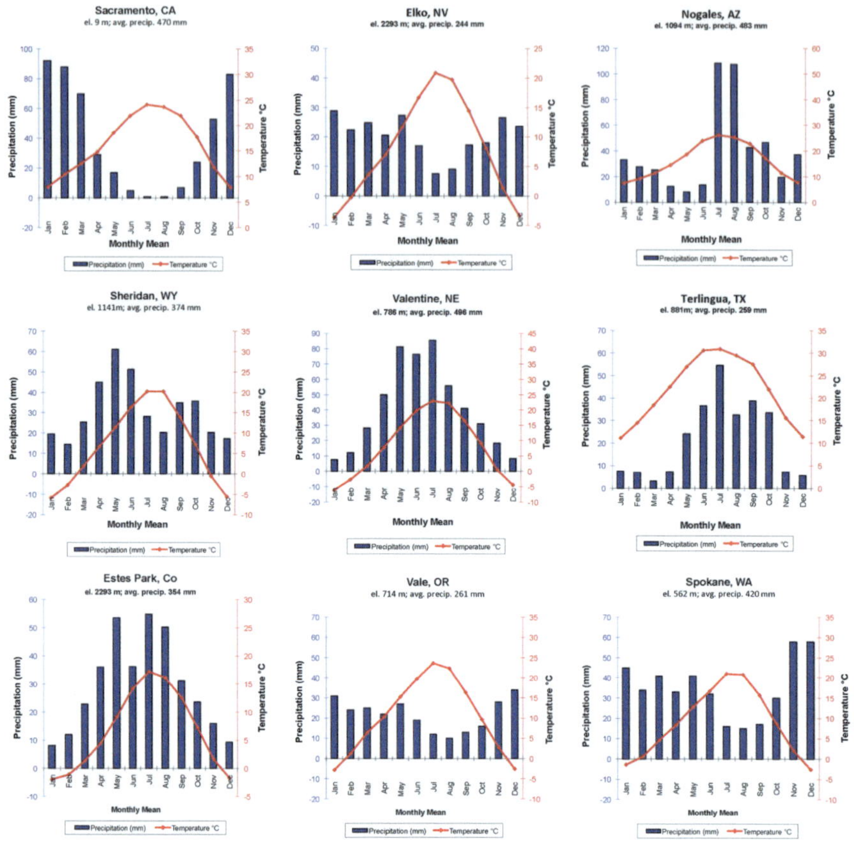

Fig. 11.2 Climate in selected areas identified in Fig. 11.1 (adapted and augmented from Huntsinger and Starrs 2006).

land, and some, like Nevada in the cold desert steppe, are more than 80 % owned and administered by federal agencies. East across the Rocky Mountains are the Great Plains and Midwest, where as the climate becomes wetter and the shortgrass prairie gives way to the mixed and then tall grass prairie, rainfed agriculture prevails and most lands are in private hands. The tallgrass prairie is now almost entirely converted to crop production.

Livestock use of federal lands in the western states today is declining, but still is around 15 million animal unit months annually (AUMs, where one AUM represents forage use by a cow and nursing calf, one horse, or five sheep for one month), controlled through grazing permits that require an annual fee per AUM and specify the timing, amount, and location of grazing. Permitted grazing occurs on approximately 63 m ha managed by the BLM, 38 m ha managed by the (USFS), and on much less of the land managed by the National Park Service (NPS) and the US Fish and Wildlife Service (FWS) (BLM 2015a, BLM 2015b, USFS 2015). Department of Defense lands, Indian Reservations, state lands, and the holdings of municipalities,

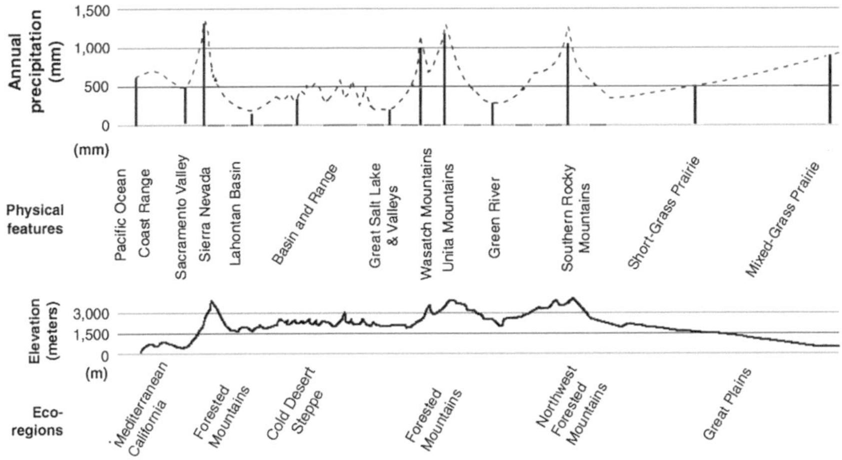

Fig. 11.3 A transect across the continent, west-east on the 40th parallel (adapted from Huntsinger and Starrs 2006). The 100th Meridian runs perpendicular to the transect through the short and mixed-grass prairie.

utility districts, and small public agencies are also sometimes grazed by livestock, on more than an additional 10 million ha.

11.2 Framing the Ecological Setting for Livestock Grazing

In 1998, Perevolotsky and Seligman posited that ecosystem conditions on rangelands adjacent to the Mediterranean Sea were changed sharply with the advent of animal domestication, then reached a relatively stable situation subject to the ups and downs caused by drought, and changes in land use and human populations (Fig. 11.4). Their figure presents the traditional view of degradation in the Mediterranean Basin (1-solid line) compared with the model presented in their paper (2-dashed line). The y-axis represents a descriptive conceptual measure of ecosystem status that "integrates soil characteristics, productivity, and diversity." The authors point out that in the last 50 years, there has been mounting concern among ecologists and range scientists that technological and demographic developments have accelerated the process of environmental degradation on many rangelands worldwide. This view, when applied to the Mediterranean Basin, assumes that local ecosystems were relatively stable until the 1950s and have been intensively degraded since then (model 1).

In contrast, the authors argue that any major changes that may have occurred to the "pristine" ecosystems of the Mediterranean Basin and the Near East are related to the domestication revolution that took place between 5000 and 10,000 years ago (1). Subsequently, the state of local ecosystems declined initially but then began to

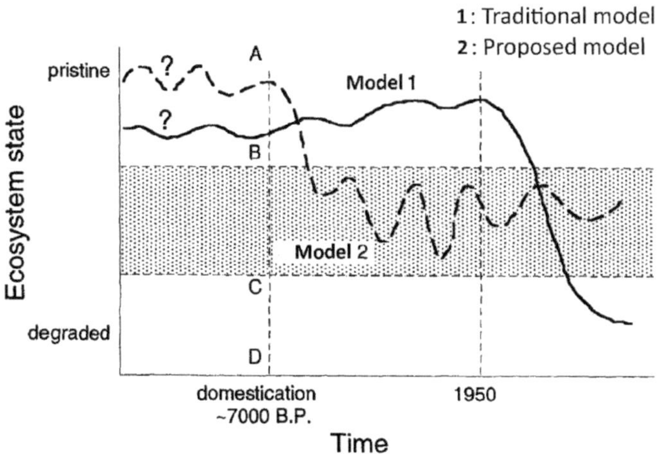

Fig. 11.4 The 2 models contrasted in Perevolotsky and Seligman (1998) for ecosystem change in the Mediterranean Basin.

fluctuate in a cycle between relatively stable limits. The domain A-B indicates the assumed change in ecosystem state following domestication and evolution of pastoralism, while B-C (shaded area) indicates the domain of ecosystem fluctuations within a relatively stable limit cycle.

In North America, livestock were not introduced into the West until 1598 in the Southwest, and 1769 on the Pacific coast. Yet Native populations have been shaping ecosystems since the Pleistocene, primarily by using fire, hunting, and cultivation. Fire was used to open up forests, clear brush, drive game, improve game habitat, reduce pests, and increase the production of foods and other useful plants (Blackburn and Anderson 1993; Anderson 2005). The ecosystem state was heavily influenced by frequent burning, perhaps achieving a B-C domain for anthropogenic fire.

Yet the managed and shaped ecosystem that European colonists confronted is the one long thought of by many ecologists as the "pristine" backdrop against which the settlement of the United States took place. Restoration projects today still look to this ecosystem as the goal, although it was shaped and maintained by human activities beginning thousands of years ago, a product of cultures that no longer have the opportunity to manage most U.S. lands. Rather than thinking in terms of a value-laden directionally downward shift to a "degraded" state on the y-Axis, a "distancing" without inherent positive or negative connotations from the pre-anthropogenic landscape might be a better conceptualization.

With the introduction of livestock, the massive re-allocation and use of water that began in the 19th century, and suppression of fire beginning in the 20th century, a period of huge change in western arid ecosystems was inaugurated. It seems very likely that current ecosystems have as yet not reached any sort of stability in response to these changes, even in the sense of the domain B-C in Fig. 11.4.

Yet it is in this context that government rangeland managers, assigned to manage the arid lands of the United States for grazing and a variety of other uses, have struggled for nearly a hundred years to assess the ecological status of western rangelands, deliberating how grazing should be managed to maintain or improve ecological conditions and forage productivity. Answers to these questions are needed to identify the existence of desertification or degradation, and to understand the human role in it, yet in fact huge gaps in knowledge are often filled by solely theoretical assumptions.

11.3 The Story of Government Management

The federal government manages more than 1.44 m km^2 of public lands in the 11 western states exclusive of Alaska, nearly half the total land area (Gorte et al. 2012). A simple version of the narrative that supported government retention of rangelands has all the characteristics of a story as defined by Cronon. The story begins with a pristine wilderness exploited by rapacious settlers. A failure to privatize most western rangelands, due to inappropriate policies and impracticality, left them open to abuse by avaricious miners, speculative grazing enterprises, and railroads that distributed land-hungry settlers willy-nilly. According to the story, public concern about damage to watersheds and a growing interest in nature made it essential for the government to step in. The story concludes, in the "happily ever after" sense, when the government removes the abuse, restores the land, and maintains productivity in the interest of all the people, unleashing the power of science and technology. A typical statement appears in a 1982 federal report on desertification: "overgrazing in the United States occurred previous to the 1930s, but it has since been largely controlled due to regulatory control and better grazing management systems by livestock producers… in 1977 range condition was improving in 13 of the 17 major range states, remaining static in three, and declining in one" (Sabadell et al. 1982).

That livestock grazing brought about great changes in western ecosystems is likely (Mack and Thompson 1982; Cole et al. 1997; Lal et al. 2004), but the idea that overgrazing and threats to the U.S. natural endowment were driven by irresponsibility and greed, as well as an influx of transient immigrants, fits a bit too conveniently into the founding narrative. In the nineteenth century, much of the West was part of the "public domain," land resulting from the acquisition by the United States of a large part of the North American continent, but as yet not allocated, according to the stated goal of the government, to private citizens for farming and private enterprise. Public domain was used freely by settlers for grazing, mining, and woodcutting. Observing some the of the environmental consequences, in 1895, soon-to-be President Theodore Roosevelt wrote:

> It is almost needless to say that this country needs a thoroughly scientific and permanent system of forest management in the interests of the people of to-day, and, above all, in the interests of their children and grandchildren… Many of the people in these imperiled regions are not permanent inhabitants at all; they are mere nomads, with no intention of remaining for any great length of time in the locality where they happen to be for the moment, and with still less idea of seeing their children grow up there. They, of course, care nothing whatever for the future of the country; they destroy the trees and render the land barren, often from sheer brutal carelessness, often for a pecuniary reward which is absolutely trivial in comparison with the damage done; yet their selfish clamor is allowed to stand in the way of a great measure intended to benefit the whole community. (Bowers et al. 1895)

In 1891, the Forest Reserves began to be set aside for government management in order to provide for the "protection and improvement of forests for the purpose of insuring a permanent supply of timber and the conditions favorable to a continuous waterflow" for the people of the United States (Roth 1901). This came on the heels of a period known for the overstocking of western rangelands, and grazing was initially halted in the Reserves. Stockmen argued that these montane ranges had been grazed for decades and the industry was dependent on them, and after 1897 policy provisions began to be made to allow for livestock grazing where it did not do injury to forest growth or water supply (Roth 1901). Eventually, grazing was constrained to specific districts at specific times, and the number of animals was limited. In getting permits, residents had precedence over "tramp owners" from other states. As is apparent from Roosevelt's remarks, immigration, transience, selfishness, and greed were common arguments at the time for why unrestricted use was degrading the land.

The majority of Forest Reserves were established early in the 20th century. At the time, grazing enterprises were of several kinds, ranging from small family homesteads to massive speculative operations fueled by foreign investment. The development of large herds on western rangelands was abetted by an enormous over-supply of imported financial capital, which until the 1893 trans-Atlantic depression enabled aspiring producers to increase herds rapidly on credit (Sayre and Fernandez-Gimenez 2003). Family ranches on small areas of private land found themselves competing for the surrounding rangelands with widely roaming herds of cattle and sheep. Unable to control grazing on the public domain, many ranchers called for government oversight, with the American National Livestock Association passing resolutions asking that public lands be protected from overgrazing as early as 1884. Congress heard a report from the Public Land Commission in 1906 lamenting overgrazing on the public grazing lands, and a survey of stockmen used for the report found 78 % favored some sort of government control of grazing (Fleischner 2002). The Dust Bowl of the 1930s, though quite different in origins, lent further support to a movement for government oversight and control of arid lands (Cronon 1992; Worster 1979), and in 1934 the Taylor Grazing Act created a system of regulation for the arid rangelands that would eventually come under the aegis of the BLM.

Although generally portrayed in the canon of rangeland science as the savior of rangelands from private greed, closer examination makes it clear that the federal government was far from innocent in creating a system that fostered abuse. As early as the 19th and into the 20th century, government policy contributed to the lack of control of grazing in the arid west by deliberately maintaining an open access system on rangelands (Starrs 1998; Nelson 1995). Two main aspects of this were first, a settlement policy that restricted the allocation of rangelands to unsustainably small parcels for ranching, and second, quashing efforts at the range management and control efforts of ranching communities (Starrs 1998; Nelson 1995). A review of the history of the introduction of grazing and settlement highlights these influences.

11.4 The Introduction of Grazing and the Homestead Act

A general history of livestock grazing in the western United States begins with the implantation of livestock in the Southwest. In 1598, Spanish settlers brought cattle, sheep and goats into what is now New Mexico. For about 200 years, Spanish and Mexican land grants, thousands of hectares in size, were given to individuals and communities for farming, grazing, and woodcutting. Tribes such as the Navajo adopted livestock grazing very early on (Bailey 1980). In California, a short-lived Spanish colonization began in 1769, and as in the Southwest, was superseded by Mexican control in 1822, and finally by the United States in 1848.

In the mid-19th century settlers from eastern regions moved rapidly into the arid western territories, drawn by the California Gold Rush and other mineral finds, and by abundant open land available for settlement. Once Native Americans were displaced, land allocation policies were implemented that, beginning with the 1862 Homestead Act, limited settler land claims to first 64 and then 256 ha. These claims were generally made in the rare areas with decent soils and water for irrigation, leaving arid and mountainous land in the public domain. In the Southwest and California, under American governance, the majority of community and individual grants given out by the Spanish and Mexican governments were abrogated by the courts by the late 1860s, ceded to clever entrepreneurs and lawyers, or returned to the federal or state governments for back taxes, and only rarely remained in the hands of grantees (de Buys 1985). Yet these old claims remain responsible for some of the larger ranches in the arid west.

Powell estimated it would take more than 1000 ha to support a ranching enterprise in the arid regions (1879). He proposed selling irrigable parcels, but allocating rangelands and timberlands to communities based on watersheds to be managed as a common resource. By the time his report made it to Congress, too much was at stake in development and speculation schemes, and in timber interests, to make Powell's suggestions palatable to politicians. In addition, surveying and assessing land potential before allocating it for homesteading would have led to delays that many settlers found unacceptable (Hutchinson 2000).

Instead, the U.S. government pursued a policy of fragmentation based on notions of farmland use from wetter climes. With rangeland allocation limited to small parcels, ranchers throughout the West claimed land near water and then grazed the non-arable surrounding public domain. They created patterns of livestock mobility that suited local geographies, moving stock from arid lowlands in the winter to montane meadows in the summer in some areas, or using crop aftermath or irrigated pastures during forage short seasons in others. In an effort to piece together a reasonably large private ranch, various strategies of subverting the Homestead Act were employed, including hiring people to make false claims, but also, purchasing the homesteads of failed claimants—a not uncommon occurrence (Raban 1996). The majority of ranchers practicing extensive grazing in the arid regions today remain dependent on the use of federal lands that were never successfully transferred to private owners.

11.5 Community Control Attempts

A strategy of "control of the range by control of the water" emerged in much of the West in the nineteenth century. In ranching communities, absent a national programme for managing access to rangelands in the public domain, informal rules and practices evolved that helped control grazing, including legal fencing of private home properties, illegal fencing of public domain range, grazing agreements among members of a community, and extra-legal threats and pressures to fend off outside intruders (Nelson 1995; Starrs 1998). An informal 19th century rule in Arizona held that the owner of a water source had the rights to graze the public domain halfway to the next water source (Sayre 2002). Common gathers where livestock were sorted, with reciprocal labor and herding, and brands to monitor cattle ownership, reflected a nascent pastoral culture as well as Hispano influence (Farquhar 1930; Starrs 1998).

An influx of speculative money in the 19th century fueled the rapid development of a commercial livestock industry based on access to low-cost, uncontrolled land, with few ties to local communities. The patenting of barbed wire in 1867, and its proliferation in the 1870s, changed the face of open range grazing across much of the west. Barbed wire dramatically reduced the cost of restricting free-ranging cattle. Conflicts erupted between sedentary farmers and ranchers, and the graziers of free-ranging herds who came across newly constructed barbed wire fences and cut them. The resulting excess of animals on the range contributed to the impact of severe winters late in the nineteenth century.

In the early 1880s, livestock prices dropped. Livestock producers who had accumulated debt were reluctant or unable to sell off their herds at low prices, because that meant defaulting on their loans (Sayre and Fernandez-Gimenez 2003). This augmented an excess of animals on the range. Following on decades of relatively abundant grass, the mid– to late–1880s had droughts and some particularly severe winters. It has often been reported that in places 85 % of cattle were lost (Fleischner 2002), sometimes found piled up against fences that prevented them from migrating south. While this has become known as "the Big Die-Off" and is part of range lore, economic research has recently challenged the actual severity and impact of these events on the cattle business, arguing that the industry continued to increase into the 1890s when financial upheaval finally constricted the enterprise and rancher net earnings dropped (McFerrin and Wills 2013).

Congress forbade stretching barbed wire across the public domain in 1885. Unfortunately this enforcement of the open access character of government rangelands fostered continued tensions over pasture use between settled communities and 'outsiders' such as widely roaming shepherds, in-migrant homesteaders and speculative cattle enterprises (Nelson 1995). John Muir describes a race to California's Sierra Nevada every year by sheep herders trying to be the first into high elevation range in the late 1800s when the mountains were largely open access public domain (1911). He wrote in 1895:

> Incredible numbers of sheep are driven into the California forest pastures every summer, and their courses are marked by desolation. Not only the moisture absorbing grasses are devoured, but the bushes also are stripped bare. Even the young conifers, which are not eaten by sheep when they can find anything else to stay hunger, are greedily devoured in their famishing condition; and to make destruction doubly sure, fires are set during the dry autumn months to clear the ground of fallen trunks and underbrush in order to facilitate the movements of the flocks and to improve the pastures by letting in the sunshine. (Bowers et al. 1895)

He continues, stating that:

> One soldier in the woods, armed with authority and a gun, would be more effective in forest preservation than millions of forbidding notices. I believe that the good time of the suffering forests can be hastened through the War Department... (Bowers et al. 1895)

A failed settlement policy for open public range that restricted private claims to small areas, and laws that limited community control, were major factors in the heavy livestock grazing of the time. Typical in open access tenure of a common pool resource, when users are prevented from organizing and controlling use, whoever can access the resources first gets them. The chosen solution, well-described by the "leviathan" model defined by Ostrom (1990), was to turn to the government for increased control of grazing lands.

While the open range of the late nineteenth and early 20th centuries was an open access system rather than a community managed resource, Hardin's *Tragedy of the Commons* (1968) is often used to support the need for government retention and management of most of the arid rangelands in the United States. The founding narrative as it has persisted in the history of the rangeland management profession blames a destructive period of overgrazing on the Homestead Act and its subversion; the failure to privatize and divide the range and the subsequent lack of accountability; the influence of outside investment; ignorance or discounting of range carrying capacity; and human greed (Holechek et al. 2010), but neglects the richer story that includes the destruction of emerging local management institutions, the loss of indigenous management, and the impact of financial interests lobbying Congress. This shaping of the founding narrative reflects the formative role of federal management in creating the field of rangeland management itself.

11.6 The Origins and Shaping of Rangeland Management and Science

In 1905 management of the Forest Reserves was transferred from the Department of the Interior to the Department of Agriculture, to be managed by the United States Forest Service. The number of reserved forests grew rapidly during this period. Gifford Pinchot was the first Chief Forester of the agency, under the administration of President Theodore Roosevelt. The Forest Service began charging a fee to graziers in 1906 to support the management of grazing. Grazing "allotments" were created with more specific boundaries than before. They were designated for use by a certain number of animals, for a specified time period, by an individual rancher. It was believed that allotting a well-defined area to each grazier would "would induce the stockman to care for his range, to protect it against fire, and to improve it by seeding or otherwise, and would prevent heedless overgrazing" (Roth 1901, p. 348). A defined a carrying capacity was sought for each allotment, designed to prevent an overabundance of animals that might over-use the range. A set carrying capacity also gave the allotment an indirect market value that could be exchanged in the market along with private lands (Merrill 2002; Sayre and Fernandez-Gimenez 2003). Setting fees, establishing carrying capacities, monitoring range impacts, and mapping allotment boundaries became the tools of a national-scale administration for grazing.

The grazing allocation process on Forest Service lands was highly contentious. Nearby and sedentary ranchers were given precedence in obtaining allotments. Reflective of Roosevelt's 1895 remarks about the abuse of the forests, an implicit goal was the exclusion of foreign-born and "outsider" herders and their "tramp" herds. An explicit social goal was to promote the development of the West by supporting family industry. In addition to determining who would get allotments, deciding on the appropriate carrying capacity was sometimes a contested and difficult process, as it was obviously an issue crucial to the financial welfare of graziers. Ranchers might use different criteria or have a different view of appropriate stocking than foresters, or Forest Service range managers. An authoritative basis for making grazing decisions was needed.

Range science "did not grow out of 'pure' scientific inquiry," but was "dominated from the beginning by government institutions" (Sayre and Fernandez-Gimenez 2003). Study of grasses and forage plants was encouraged by Congress in 1895 through the creation of the Division of Agrostology (Sayre 2002). Early grassland science was adopted for the purposes of the agency within an institutional setting dominated by a cadre of foresters educated in accordance with the vision of Gifford Pinchot, the first chief of the Forest Service, and that of his German-trained mentor Bernhard Fernow. Amid the faith in science and invention characteristic of the Progressive Era, a scientific method was developed for the management of forests. Highly simplified, forest management was driven by a belief in the need for a sustainable, maximized, supply of timber to support the development of the United States (Fairfax et al. 1999). The premise of a coming “timber famine,” which was a driver of the development of practices oriented to “maximum sustained yield,” is well challenged—and indeed the U.S. has yet to experience a timber famine of any sort. However, under scientific management the salient features of policy were an absolute focus on maximizing sustainable production, trees and technology as a solution to social problems, science and technical expertise as the basis for decision making, and large scale comprehensive planning (Fortmann and Fairfax 1989). These rapidly became professional norms in the Forest Service, a lens through which forest related issues were viewed and filtered.

These tenets of scientific management are reflected in the development of rangeland management and science. If timber was considered the goal for forests at the time, it followed that “food and fiber” were the appropriate goals for rangelands —with productivity and efficiency the key to community and national development. Local or traditional goals were pushed aside for those of a central authority, in the interests of protecting the range from ill-informed, self-interested graziers. The role of the Forest Service and BLM in enforcing forest and range policies and regulations is exemplified in the pseudo-military uniforms that were and are standard issue for Forest Service personnel working with the public. Science and technology needed to be developed to replace out-moded and inefficient local practice. However, though the science of silviculture was well developed in France and Germany (Schama 1996) by the nineteenth century, and was promulgated by Fernow and Pinchot in the new Forest Service, there was no similar northern European science to be adopted for arid rangelands. North African and Mediterranean grazing knowledge was ignored by the East Coast-oriented national administration (Davis 2007). A scientific management approach for evaluating grazing and setting a sustainable carrying capacity, close to the heart of Pinchot and his professional management cadre, was called for.

11.7 The Succession Story

In 1916 a professor of botany at the University of Nebraska named Frederick Clements published *Plant succession, an analysis of the development of vegetation*. Based on observations of vegetation patterns on Midwestern tallgrass prairie that had been plowed in different time periods, he created a model for plant community change after a “disturbance” like plowing. This “secondary succession” predicted a linear, predictable development of the plant community through seral stages or states, from post disturbance weedy species to a final, “climax,” assemblage that he suggested was an equilibrial, steady state. The plant community was like a living organism, proceeding from infant to maturation. Some disagreed with the Clementian view (e.g. Gleason 1917), but as is well discussed by the environmental historian Worster (1994) Clements’ account became the plant ecology story of the century, with the vegetation as an organism-like protagonist. First, it allowed the characterization of a disturbance that changes the ecosystem away from the climax assemblage as a setback, and even further, a bad thing. As the plant community develops over time, initial colonizers, or weedy species, are considered subsidiary, temporary interlopers that disappear with "recovery" or restoration to the final state, where nature is in balance. The final state is conferred with positive value by this model. This "story" was not only easily understood, with its beginning, middle, end and moral message, it provided a scientific foundation for rangeland management, and a model that foresters could apply to predict forest growth after harvest or fire.

The norms of scientific forestry were incorporated into Forest Service policy and practice. One of the major "disturbances" foresters and reserve managers sought to control was fire. Because in this view fire reduces climax forest to weedy, early successional species, consumes valuable timber, and can cause erosion, fire was quickly brokered to the role of a major evil, ignoring its valuable place in Native American management, and its use by early ranchers and farmers to clear woody vegetation for cultivation and grazing. Huge fires in the western U.S. in 1910 contributed to an anti-fire sentiment (Pyne 2008). Along with a focus on producing timber, the Clementian model, with its implicit devaluation of "disturbed" vegetation and high valuation of climax stages, fit well with efforts to control and stop all fire in the forest in the 20th century, and to manage post-harvest to maximize forest regrowth. In an 1895 published letter on how to manage Yosemite National Park, where native Californians had burned regularly for centuries, Pinchot wrote that "There is no doubt that forest fires encourage a spirit of lawlessness and a disregard for property rights." Bernhard Fernow explained that "the whole fire question in the United States is one of bad habits and loose morals. There is no other reason for these frequent and recurring conflagrations" (Bowers et al. 1895).

Ranchers using fire to clear the understory on Forest Service ranges were accused not only of damaging the forest, but of being unpatriotic. In 1918 the Shasta Trinity Forest Supervisor sent letters to local stockmen quoting President Wilson:

> Preventable fire is more than a private misfortune. It is a public dereliction. At a time like this of emergency and manifest necessity for the conservation of national resources, it is more than ever a matter of deep and pressing consequences that every means should be taken to prevent this evil. (Morrow 1918; Forero 2002)

The Forest Supervisor goes on to imply that, with World War I ongoing, the crime of burning is especially heinous. He states that it took the equivalent of 400 men working every day for four months to suppress human-caused fires, and these men were needed at the front. It was therefore the patriotic duty of stockmen to prevent fire (Morrow 1918; Forero 2002; Huntsinger et al. 2010).

Yet the role of fire in western ecosystems was not invisible to those looking from outside the perspective of scientific forestry. In 1890, Powell, who spent considerable time exploring the western frontier and spoke several Indian languages, wrote on indigenous use of fire and the impacts of the fire suppression:

> ...[U]nder conditions of civilization, the great forests of the arid lands are being swept from the mountains and plateaus. Before the white man came the natives systematically burned over the forest lands with each recurrent year as one of the great hunting economies. By this process little destruction of timber was accomplished; but, protected by civilized men,

> forests are rapidly disappearing. The needles, cones, and brush, together with the leaves of grass and shrubs below, accumulate when not burned annually. New deposits are made from year to year, until the ground is covered with a thick mantle of inflammable material. Then a spark is dropped, a fire is accidentally or purposely kindled, and the flames have abundant food. (Powell 1890)

Once again, Powell's ideas were badly timed and largely ignored in the face of the Progressive Era conservation mandate.

In fact the widespread suppression of indigenous, rural, and natural burning that began in the 20th century is credited with increasing the density of trees and drought stress to forests. Together with climate change, increased tree density has contributed to pest outbreaks devastating millions of ha of US forests (Taylor 2000; Gallant et al. 2003; Raumann and Cabik 2008), and to increasingly massive and intense wild fires. Compared to the average year in the 1970s, in the past decade there were seven times more fires with an extent greater than 10,000 acres each year (Climate Central 2012). Suppression has also led to the invasion of montane grasslands and meadows by shrubs and trees, drying formerly moist habitats, as trees consume and transpire soil moisture (Raumann and Cabik 2008).

11.8 Development of Range Condition Assessment

On Forest Service and then BLM rangelands, the positive values attributed to the climax state were put to use to evaluate grazing impacts and management. Arthur Sampson, a grassland ecologist with the Forest Service (1909), adapted the model developed by Frederick Clements for understanding grazing effects on rangelands. Like Clements, Sampson accepted the idea that there is a single climax or *equilibrium* state for grasslands of a particular climate (i.e. "climatic climax"), and that this was the ideal state. He explicitly added excessive livestock grazing in the equilibrium-based model as the cause of "range retrogression" away from the climax state, and that the reduction or removal of livestock would allow for competition and other biotic processes internal to the system to drive succession back towards the climax state. The existing state relative to the climax was used to evaluate the "condition" of the range (Spiegal et al. 2015).

In the late 1940s, E.J. Dykesterhuis, a U.S. Soil Conservation Service range scientist who eventually became a professor of range science at Texas A&M University, emphasized the importance of explicitly delineating sites by their inherent edaphic and topographic characteristics (1949). Dyksterhuis posited that a range landscape had a "polyclimax." Each "range site," defined by its climatic and physical features, had its own climax. Grazing remained the primary reason for a range site's departure from climax. This work represented a step forward in defining site potential not with current vegetation, but with soils, climate, and physiographic factors. Plant species were grouped and identified as "decreasers," "increasers," and "invaders" depending on their response to herbivory, with climax species often in the decreaser category, as they are by definition the least adapted to disturbance (Humphrey 1947; Dykesterhuis 1949).

Reinforcing the Clementian model as used in management, the climax species became the presumptively "good" species and those most sensitive to grazing, so the presence of these species could be used to gauge the quality of the livestock management. Too much grazing caused "retrogression" toward the "bad" early successional conditions, while an absence of grazing would lead to recovery of the equlibiral condition of the "climax" vegetation believed to dominate grasslands prior to the introduction of livestock. "Range Condition Classes" were developed based on the proximity of the vegetation on a specific kind of site to the climax community believed to represent its potential. The BLM adopted range condition assessments and allocation practices similar to those of the Forest Service, including the succession based method of assessing range condition.

The ecological story that was forged in the crucible of scientific management proved compelling, stubborn, and adaptable. Despite a long series of ecologists who attempted to modify or better explain vegetation change, the organism-like, beginning-middle-end characteristics of the linear succession model has proved intractable to this day. With this model, managers found that degradation in the biotic sense could be defined as the distance from the steady state, balanced, equilibrium endpoint termed "climax." Like degradation, desertification in this model is considered the distancing of vegetation from the ideal state in arid lands. Maintenance of the climax community was also assumed to protect soils from erosion and maximize biodiversity. Armed with this conceptual tool, government managers believed they now had specific, measurable goals for rangelands that could be used to set numbers of livestock and other parameters "scientifically," evaluate the efficacy of management, and determine where restoration was needed and how to get there. Recovery, or restoration, required the reduction or removal of disturbance, according to the evolving story, and it became gospel.

11.9 Lost in the Story

The convergence of succession-based science, government management and authority needs, and a narrative of heedless destruction by early settlers created a big story with good and bad actors and moral lessons. It starts with a pristine natural world, in balance, with climax plant communities supporting climax wildlife species. Then the settlers arrive on stage, and driven by competition, need, and greed, begin to disturb and upset the virtuous native ecosystems. Bad plants start to increase at the expense of good plants. Soil washes away. Biodiversity declines. Fires destroy native plant communities. Finally, in this story, the government white hats arrive, with the mandate of science and eternal, sustained production of goods for human society.

While it took time to tease out, it has become apparent that there are things wrong with this story. Foremost, perhaps, is that despite continual reductions in stocking, rangeland "recovery" has been slow and unpredictable on U.S. arid lands. The data show that after decades of effort, massive areas of rangeland have not yet recovered to the former, supposedly, pristine state or "potential plant community." The long-held notions of "pristine state" are easily challengeable because the state of rangelands prior to the introduction of livestock was the outcome of indigenous management and native grazers. This context had been lost to the Clementian world.

The preamble to the 1978 Public Lands Improvement Act (3 U.S.C. §§1901–1908) states that "(1) vast segments of the public rangelands are producing less than their potential for livestock, wildlife habitat, recreation, forage, and water and soil conservation benefits, and for that reason, are in unsatisfactory condition;…(3) unsatisfactory conditions on public rangelands present a high risk of soil loss, desertification, and a resultant underproduction…." The legislation defines "range condition" as "the present state of vegetation of a range site in relation to the potential plant community for that site," among other measures, including productivity.

Livestock production on government ranges is in decline, with public lands supporting far fewer livestock than when government management began. After a peak during WWI, the amount of grazing on public range has declined every subsequent decade, due to a combination of thickening trees and other woody plants, removal of livestock for restoration, and changing uses and policies for public lands. Uncounted ranchers and sheep herders have left the business. The deterministic linear succession model has proven a weak predictor of response to grazing, and efforts to remove disturbance and restore the climax remain largely unrealized. The model simply does not fit most rangelands outside of mesic Midwestern grasslands, with ecosystems neither "succeeding" or "retrogressing" reliably in response to grazing management.

It is important to remember that models are not reality, and ecosystems do not conform perfectly to any model. Managers have developed ways of simplifying complex systems using models to identify critical components and processes and to link those processes to possible interventions. Clements' model is one example, but it proved to be of low utility for predicting the consequences of management on arid rangelands.

11.10 New Models for Arid Lands

In the 1930s, A.G Tansley from Oxford took issue with the concept of climax conditions as a management goal, and a measure of environmental quality. He noted that in a given zone there may be many apparently permanent types of vegetation that could be termed climaxes, including those maintained by soils, heavy grazing, or recurrent fires. He contested the underlying assumption that human activities were an external factor that inevitably had the effect of undermining an ideal, pristine, climax and degrading an ecosystem. He promoted the concept of an "anthropogenic" climax instead, one that was not automatically inferior (Tansley 1935; Worster 1994; van der Valk 2013).

In 1988, Ellis and Swift challenged the notion that arid rangelands in Africa can be modeled as equilibrial systems that can be managed by assigning a set "carrying capacity" for grazing animals. Shortly thereafter, Westoby et al. (1989) had major impact on U.S. range management by publishing an article in the *Journal of Rangeland Management* arguing that arid rangelands are not well-described by models based on equilibrium systems, and comparing management using models based in equlibrial versus non-equilibrial ecological dynamics. The non-equilibrial model holds that instead of developing in a linear, progressive pattern, and responding to grazing with directional shifts along a continuum from ruderal to climax conditions as grazing influences plant competitive relations, rangeland vegetation in arid lands at a given site can develop into more than one "stable state" depending on abiotic factors, such as rainfall and temperature, and the history of the site with respect to fire, floods, and so forth. Grazing, as a factor, is under this model often a relatively minor influence compared to the influence of abiotic factors, especially rainfall on arid lands, and removing grazing does not necessarily change the state or change it back to a state that existed before livestock grazed it, a challenge to traditional paradigms for rangeland management worldwide (Behnke et al. 1993). In fact, certain constellations of factors with or without grazing could cause irreversible, or difficult to reverse change, something termed a threshold effect. Based on this, livestock management has to be opportunistic, rather than strategic, to take advantage of irregular rainfall, and vegetation response is to some degree "decoupled" from grazing effects. Rather than focusing on establishing and enforcing a set carrying capacity, in this conceptualization, management should focus on conserving and developing pastoral systems that can cope with and adapt rapidly to unpredictable conditions.

Instead of relying on a linear map to potential vegetation change on a specific type of site following disturbance, a map of this type of model has many stable states. A "states and transition" model identifies such states and lists drivers that cause transitions leading to different states—these drivers are not termed "disturbances," but factors that shape vegetation. In other words, fire is not a disturbance that sets vegetation back, but an occurrence that under some conditions causes transitions or serves to maintain states (Fig. 11.5). For a specific site, livestock grazing can act as a driver, but at moderate levels may not have any affect on the vegetation state, because the state is most strongly influenced by rainfall. Another way to say it is that while deterministic succession is based on competition and other plant-plant interactions as the major driver of vegetation change, on arid lands plants are competing with the environment, rather than each other. There is no single, predictable end state, particularly one that inherently optimizes ecological "goods" like native climax species diversity or protection from erosion. Each state has differing implications for biodiversity, soils, and forage production.

State and transition models take many of the value judgments out of assessing vegetation change. They require the manager to decide what the goal is and why. This is politically less useful than having the decision made for you in the linear

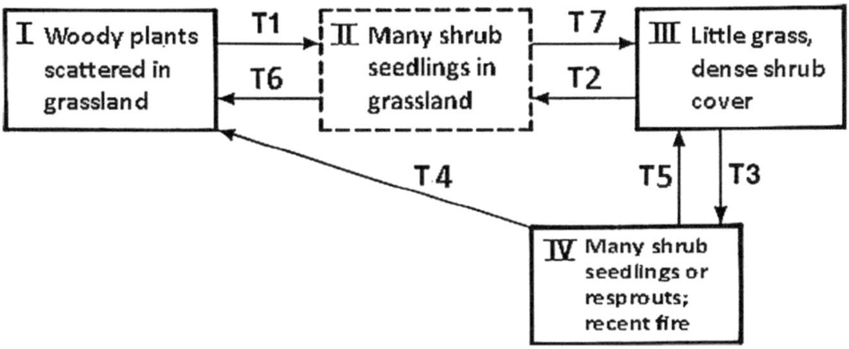

Fig. 11.5 A simplified state and transition model for a semi-arid grassland with potential for increase in shrubs presented by Westoby et al. (1989). Vegetation types stable in a management relevant timeframe and the transitions among them are identified. Researchers, managers, and practitioners work to understand the causes and effects of transitions, the feedbacks that stabilize states, and the nature of the thresholds between them. Thresholds maintain state stability. For example, transition 3 (*T3*) does not occur without fire, while transition 1 (*T1*) has a low threshold, requiring only time to occur. State III might be optimal for some bird species, while states I, II, and IV might be better for livestock and wildlife grazing.

succession paradigm, but it does allow the collection, use, and testing of data in building models for specific sites. Rather than a story that drives how vegetation change is seen, it provides a framework for documenting and eventually explaining how vegetation changes in response to various events. Unfortunately, the "multiple stable states" model of vegetation change does not have the same simple, moralistic and appealing story as linear succession—there really isn't a beginning, end, or moral lesson. There is a site, it rains and things change or it doesn't rain and things change. Changes may be permanent. Rainfall is unpredictable and not influenced by human actors. We need to watch, experiment, and record to learn what is going on. This is in fact very like Cronon's version of a "non-story" (1992).

Non-equilibrium models make it obvious that the terms degradation and desertification are subjective, and based on criteria set by the manager or scientist. Perevolotsky and Seligman (1998) discuss the concept of degradation as follows:

> "degradation" refers to a change of state with a negative value assessment that is related to subjectively chosen criteria. A shift in the nutrient status of soil may increase biodiversity or the persistence of rare species; the result would then be judged as improvement if higher biodiversity or longer persistence of certain species were the criterion. However, if the shifted nutrient status resulted in decreased primary production, the same phenomenon could be judged as degradation. Assessment of vegetation damage also depends on management goals: A forester or environmentalist may regard the domestic ruminant as a pest that degrades the woodland, whereas a herder is more likely to regard the woodland as a source of livestock forage. That is, for the herder, an oak grazed down to a dense dwarf shrub is simply a well-used forage plant.

In the multiple steady states model, the manager must make the decision about which are desirable and undesirable states. This includes being explicit about what constitutes degradation or desertification. It may be change to a state with lower productivity, or with accelerated erosion. To define desertification in this model then, it is necessary to be clear about causes and outcomes, rather than relying on value judgments, or narrative flow. Assumptions can be made visible and tested.

11.11 Degradation/Desertification

As in much of the world, in the U.S. it is difficult to sort out the effects of variations in aridity, climate change, and human influences. The role of livestock grazing, as it interweaves with other factors, is complex. Federal management has led to large reductions in livestock grazing, but there are processes of what could be termed "degradation" that have been exacerbated by government policy and the land tenure institutions that place vast areas of arid lands under "one size fits all" government management, most notably suppression of natural and anthropogenic fires. The relationships among wildfire, invasive plants, and grazing illustrates the interactions that may lead to devastating forms of undesirable vegetation change, how states and transition models are a better fit than linear succession models, and why land allocation and management institutions may ultimately be in large part responsible for causing desertification.

Evidence is accumulating that in the forests of the western United States, as dry material has built up and vegetation thickened due to fire suppression, wildfires are becoming more intense and extensive, converting forests and shrublands to grasslands for an unknown period of time (Hagman et al. 2013; Miller and Safford 2012; Miller et al. 2009; Goforth and Minnich 2008). It is likely that eventually shrubs and trees will regrow on most of these areas, though this is by no means certain. In some cases the new conditions, and unfortunately efforts to reintroduce burning or treat vegetation to reduce the probability of wildfire, benefit invasive plants and contribute to their spread (Keeley 2006). Some of these plants may support more frequent fires, or change moisture dynamics on the site. In the meantime, fire releases methane and carbon in the short term, and over a longer period, carbon storage is reduced, contributing to climate change. There is greater risk of erosion following severe fires, with soil loss potentially reducing the productive capacity of the site—also a form of desertification. Desirable habitats and timber production will be reduced. This could be considered degradation.

Invasive species are new actors on the stage in the fire drama that can interact with fire and grazing in ways that result in what might be termed "desertification." The advent of colonization from Europe began a process of wave after wave of invasive species arriving in North America from around the globe. In addition, with changes in fire and fauna, native species that have undesirable impacts on soils and vegetation may expand their known ranges and become more dominant in regions formerly occupied by grasslands, acting as native invasives. In some rangeland systems, invasives have changed ecosystem dynamics in ways that could be argued to be a form of desertification, with species of less value to wildlife as well as livestock becoming dominant. Occupation by invasive annual species may mean a less complex or shallower root structure that is subject to more erosion, a longer period of dry vegetation conditions that once were buffered by perennial plants drawing moisture from greater depth, and greater risk of wildfire.

The case of the invasive non-native annual grass known as cheatgrass, *Bromus tectorum*, in the cold desert is a useful example of interactions among fire, livestock grazing, and invasive species. The cold desert steppe of the Great Basin region starts in the rain shadow of California's Sierra Nevada, extends east into Utah, north into Idaho, and northeast as far as parts of Wyoming and Montana (Fig. 11.1). Several subspecies of sagebrush (*Artemisa* spp.) intermix with perennial bunchgrasses (*Pseudoroegneria, Achnatherum, Poa, Leymus, Oryzopsis* spp. and others). Much of the region has now been invaded by cheatgrass. Cheatgrass creates an even grass cover, rather than the "bunched" grass cover with spaces between the grass plants that prevailed previously. It is fine stemmed, and dries out every summer as soil moisture declines, creating excellent conditions for the ignition and spread of wildfire. Cheatgrass has substantially altered the regional fire regime, supporting larger and more frequent fires (Balch et al. 2013). Cheatgrass was a major fuel contributor to the six most significant wildfire seasons since 1960, all of which have occurred since 2004 (NIFC 2015). Studies have found a shortening of fire return intervals from 30–110 years to 3–5 years (Brooks et al. 2004; Chambers et al. 2007; Baker 2011).

Accidentally introduced from southwestern Asia in contaminated grain in the 1890s (Mack and Pyke 1983), with seeds that can be carried by livestock, the Bureau of Land Management (BLM 1991) estimated in 1991, that of the lands it managed, 3.6 million hectares contained cheatgrass as the dominant understory, and it had the potential to dominate 16.2 million acres of BLM land in the future (BLM 1991). The National Science and Technology Center mapped 12.4 million hectares of cheatgrass in the Great Basin in 2000 (Menakis et al. 2000). Although grazing and cultivation facilitated the spread of the grass (Young and Clements 2009), rangelands never cultivated and ungrazed since 1944 have been invaded by it (Brandt and Rickard 1994). Daubenmire, in a classic paper in plant community ecology, argued that cheatgrass could invade native perennial grass communities that never had been grazed and were in excellent ecological condition (1940). Once cheatgrass was introduced, for him it was only a matter of time until virtually all sagebrush plant communities in the Great Basin were invaded.

Ironically, in the 1970s and 80s government ecologists were concerned with an increase in sagebrush, and chain dragging to uproot shrubs and herbicides were used to control it (Sabadell et al. 1982). Now cheatgrass invasion, together with increased fire, has led to the loss of woody vegetation and deep-rooted perennial grasses, and reduced forage capacity. Along with declining forage production, frequent fires that cause the loss of perennial grass and shrub species, and the creation of a moisture regime that is annual, with rapid use of water in early spring and senescence and death of the plant life in mid summer and fall, the changes may be seen as a form of desertification. If climate change leads to more aridity, and more frequent wildfire, cheatgrass invasion is likely to expand. Many ecologists believe that a threshold is crossed once cheatgrass is common on a site, and only expensive treatments and reseeding have a chance of restoring the native shrub-grass complex (Young and Clements 2009) (Fig. 11.6).

In one of the few studies that used long term data to examine vegetation changes in the bluebunch wheatgrass-sagebrush steppe of southern Oregon, Allen-Diaz and Bartolome (1998) found that in the absence of fire, grazing practices had no effect on transitions to or away from dominance by cheatgrass. Weather and time seemed to be most closely related to this transition, supporting Daubenmire's observations. Other studies failed to show a link between grazing and vegetation change, except perhaps at very intensive levels of grazing (Laycock 1987; Eckert and Spencer 1987). Bagchi et al. (2013), assessing historical records of cheatgrass invasions, found no relationship with grazing regime and the frequency of transitions among states in the absence of fire. Cheatgrass establishment was related to periods of average to below-average annual rainfall, and its growth and fecundity related to patterns of infrequent rain in fall and early spring. Both Bagchi et al. (2013) and Allen-Diaz and Bartolome (1998) found that transitions away from cheatgrass dominance may occur over time in the absence of fire without any relation to grazing, though in the Bagchi et al. study, reversibility was more common at the site with grazing. Young and Clements (2007) argue that under rest-rotation grazing cheatgrass benefits from deferral of grazing until after seed ripe, or complete rest from grazing. These types of findings lead some to argue that grazing, once a facilitator to invasion, may be a way to control the impacts. The problem has not proved amenable to the default approach of reduced stocking.

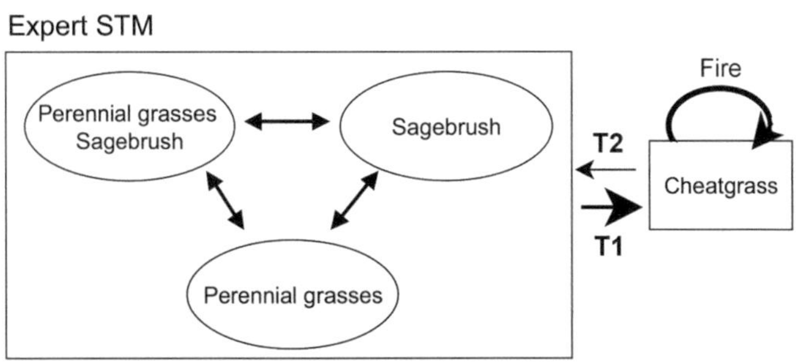

Fig. 11.6 An "expert" states and transition model for sagebrush steppe vegetation dynamics (adapted from Bagchi et al. 2013). The three sagebrush and perennial grass-dominated states are often referred to as "phases" that do not have strong thresholds between them, while the transition to cheatgrass (*T1*) is often argued to be irreversible, because of the difficulty of crossing the strong threshold leading back to perennial grass and sagebrush types (*T2*). While researchers generally agree that *T1* is strongly facilitated by fire (Young and Clements 2009), the role of grazing today is less clear. Two studies using long term data have found that *T2* does sometimes occur over time in the absence of fire (Allen-Diaz and Bartolome 1998; Bagchi et al. 2013). Fires are one way the cheatgrass state is maintained.

The role of grazing in this scenario is one of many interacting dramatic changes, and changes without directionality. Removal of grazing may have little or no impact, and threshold dynamics may result in multiple stable states that have little to no relation to a "climax" or previously identified "potential" vegetation. There is no simple story here.

11.12 Holistic Resource Management: A New Story

The states and transition format, based on non-equilibrium theory, lacks the directionality and purpose of a good story— lacks a moral compass, as it were. Instead, a states and transition model is a set of boxes representing vegetation states and a "catalogue of transitions," a simple listing of what drives change from one box to another. While it may be possible to see a moral message in some transitions, it is not as simple or compelling as the linear, Clementian succession story. Perhaps an analogy to the succession story is a story with stereotypical characters that "make sense" to the reader, like an old style western pulp novel, or a bodice-ripping romance: In a states and transition model, on the other hand, things just happen. Within this narrative vacuum, the concept of Holistic Resource Management has flourished.

"Holistic Resource Management" is a package of rangeland management concepts and practices with origins in the early 1980s, oriented around a central, powerful, story. As the most well known promoter of the programme puts it, livestock grazing should mimic the herds of wild grazers that once wandered the grassland (Savory and Butterfield 1999). The wildlife mimicry carries a positive moral loading: this must be a better, "natural" way of doing things. Livestock are moved regularly from one pasture to another, in what has been termed a "rotational" grazing scheme, attempting to emulate, though at a much smaller scale, the way a wild herd moves from one grazing ground to another, allowing plants that have been grazing to "recover" before the next visit. Part of the HRM canon is that the soil surface should be broken by hooves, allowing seeds to be worked into the soil. In fact, the first range research carried by the USFS was a test of Sampson's theory that you could rest or sow seed, then run sheep through to plant seeds. It failed when he tried it in Oregon in 1908 and has never been shown to work since (Sampson 1909; Skovlin et al. 2000).

Typical of ecological or environmental stories, the wild herd narrative that is used to support HRM has if nothing else been over-extended to ecosystems that not only have a different climate, but that never had such animals. Perhaps more important, scientific review has shown a lack of ecological benefits from rotational grazing in arid systems (Heady 1961, Bartolome 1993; Briske et al. 2008). However, the herd management aspect comes with a variety of other interventions that may or may not be effective in improving management. For one example, encouraging the manager to spend more time observing and working with the stock is probably beneficial (Briske et al. 2011), and ranchers mention other benefits, including that it makes stock more gentle because of frequent human contact.

Implementing the required grazing system can be costly to ranching enterprises, while fencing and rotational grazing may be detrimental to some ecosystem services. But the story is so compelling that it continues to attract adherents in the livestock and management communities. Holistic resource management might even be seen as a story that helps a producer communicate good intentions, and a commitment to protecting the environment, to agency regulators and managers.

Both Clementian succession models and HRM are based on compelling stories that overshadow the need for evidence, particularly on arid lands.

11.13 Conclusions

Desertification is an ill-defined term that has been used sporadically to discuss grazing issues in the United States. Unlike in many parts of the world, it is not a part of everyday discourse about grazing. The terminology of rangeland condition used widely on public rangelands has seemingly supplanted the language of desertification. A review of journal articles reporting research in the United States with desertification in the title according to the Web of Science shows only about a dozen published in the last 10 years, and they use differing definitions of the term. It has mostly been applied in the Southwest, in research on the causes of shrub encroachment, usually mesquite (*Propsopis* sp.), into warm desert grasslands, and the possible roles and inter-relationships of grazing, fire, warming temperatures, and rainfall in that encroachment. There are more papers with "desertification" as a keyword term, but a cursory review reveals that many of these do not actually mention desertification within the paper.

Despite the difficulties of identifying and defining desertification, estimates of the amount of "desertification" for North America have been published a number of times. In 1976 a symposium on the topic was held at the University of Arizona in Tucson (Paylore and Haney 1976). In the introduction, the definitional problems are discussed, and the following definition adopted, stating that desertification is a process that:

> …deals with the extension of typical desert landscapes and landforms to areas where they did not occur in the recent past, one taking place for our purposes in marginal arid zones bordering deserts under average annual rainfalls of approximately 50–300 mm, areas characterized by increasing aridity and intensification of distinct geomorphological processes, desiccation and increasing salinity of soils, and a manifest degradation of vegetative cover. The term implies a change, whether by the long term (in the geologic sense) climatic change or by short term climatic fluctuation, or by man's intervention through careless extension of agricultural developments, burning, overgrazing, urbanization, and increasing population pressures.

The purpose of the symposium is stated to be to contribute to the development of measures for reversing desertification. It is implied that the most important measure is population control, and the document calls for the establishment of reserves by the government, with compensation for those excluded. The author of the lead article provides a map of desertification in North America (Fig. 11.7; Dregne 1976). Desertification is attributed to the combined impacts of "man's activities and drought." The symposium includes a paper on the Papago Indians of the southwest,

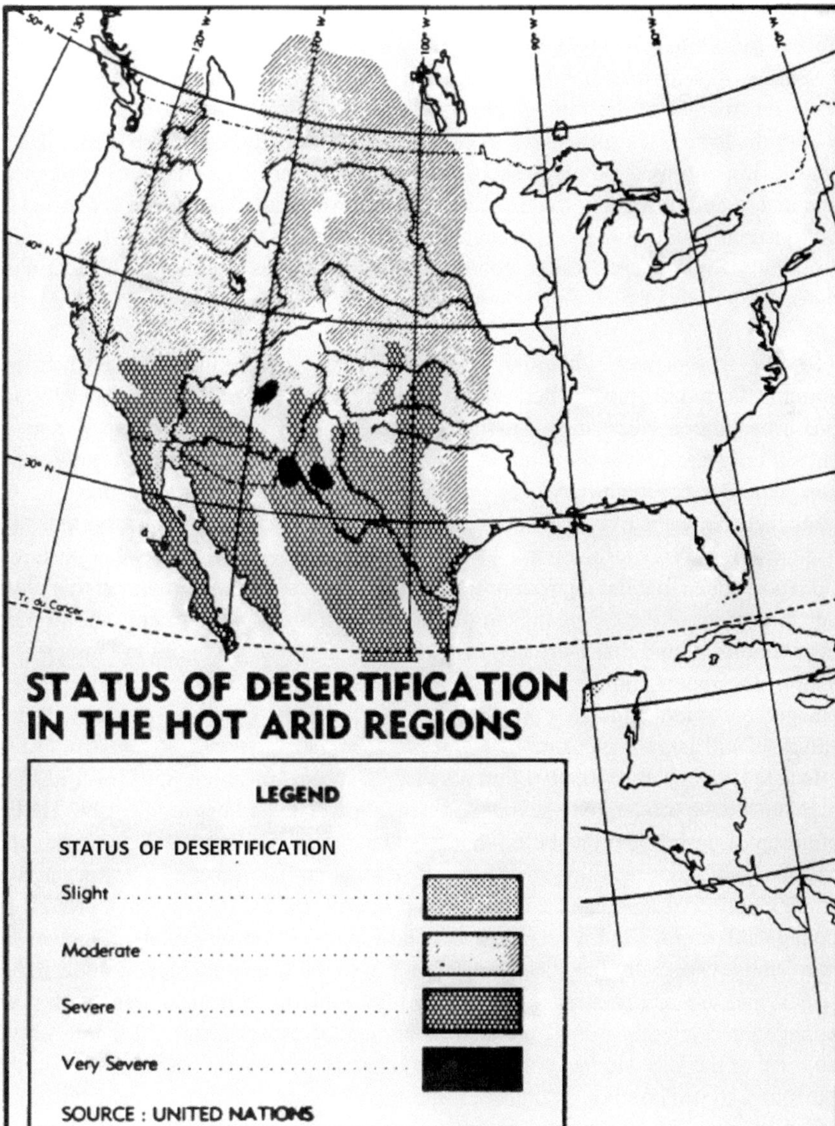

Fig. 11.7 Map of desertification in the United States from the United Nations published in a 1982 BLM report on desertification (Sabadell et al. 1982). Aside from a few spots, degree of desertification seems cartographically strongly linked to prevailing temperatures and aridity.

but only to discuss their livestock grazing practices. Their use of fire, once a feature of the hunting economy (Stewart et al. 2002), is not mentioned.

In 1982 the federal agency managing the largest area of U.S. arid rangelands, the Bureau of Land Management, published a document titled "Desertification in the United States: Status and issues" (Sabadell et al. 1982). The document states that "an assessment of the problem at the national and regional levels has been recognized by world and national organizations as a first priority in combating and preventing desertification." Desertification is defined as "the sustained decline and/or destruction of the biological productivity of arid and semiarid lands caused by man-made stresses sometimes in conjunction with natural extreme events. Such stresses, if continued or unchecked, over the long term may lead to ecological degradation and ultimately to desert-like conditions." Range condition is defined as "The present productivity on a range site as related to the potential natural plant community for that site. Range condition is expressed as excellent, good, fair, or poor, on the basis of how much the present plant community has departed from the potential."

Several factors were identified in the report as contributing to desertification, including farming, grazing, herbivore population growth, mining, energy production, urbanization, recreation—particularly the use of off road vehicles, and in general, competition for the land base among diverse uses. The loss of indigenous management is not mentioned. For grazing, the report argues that a balance must be maintained between livestock numbers and the land's carrying capacity, and that during drought years, grazing must be reduced and shifted to feed or other pastures. The report states that the Agricultural Research Service in 1974 estimated that more than 71 percent of the rangelands in the 17 Western States were in only fair to poor range condition, and that there is a high correlation of range condition to degree of erosion. The report comments that although conditions are improving under federal management, "the damage done by excessive grazing prior to the turn of the century is still largely present."

In 2004, Lal et al. estimated that about 2.75 m km^2 or about 85 % of arid U.S. rangelands have moderate to severe desertification. The authors used a 1992 UNEP definition of desertification stating that it is "land degradation in arid, semiarid, and dry subhumid areas resulting mainly from adverse human impact." They argue that the degradation of rangeland vegetation is primarily caused by excessive prolonged grazing and removal of vegetation by anthropogenic perturbations. Changes in biodiversity, especially the species mix leading to the predominance of undesirable species, and use of irrigation water with a high mineral content in areas with poor drainage, among other things, are also mentioned as causal factors. It is interesting that three of the U.S. studies published in the last 10 years with desertification in the title found by this author evaluated some type of faunal diversity and two found higher or equivalent species diversity on the "desertified" sites (Kerley and Whitford 2000; Bestelmeyer 2005; Klass et al. 2012).

None of these papers and documents considers the role of the suppression of indigenous practice in desertification. Aggressive, nearly paramilitary fire suppression, accompanied by the criminalization of indigenous burning and the destruction of native cultures, has a complex role in processes of desertification that have vast implications for western rangelands and forests. Rather than leading to the "recovery" of a "climax" state, the removal of grazing and fire suppression can mutually reinforce the frequency and intensity of conflagrations. In some climate zones shrublands and woodlands may become more fire prone in the absence of grazing. Dried grasses on ungrazed grasslands make fire starts more common. On rangelands, some invasive species were introduced and spread as part of livestock grazing and cultivation. Livestock dispersed seed, and disrupted native vegetation, creating opportunities for invasives. How much livestock continue to influence erosion, species introductions, and species change is the subject of debate, and no doubt varies place to place and time to time, but in many cases they remain branded as the bad actors behind it all.

The changes in ecosystems that are occurring as a result of all these factors are often irreversible and widespread. The science of "novel ecosystems" has developed to study such unprecedented assemblages of vegetation (Seastedt et al. 2008). In some cases livestock grazing is a tool used to control non-native species that are the new actors in desertification, prevent fires like those in cheatgrass, or to control the encroachment of shrubs and trees that once were constricted by indigenous burning. However, the "deviation from pristine" model for evaluating rangeland conditions, and the mythology of the "untouched" American wilderness, has fed into a still powerful ideology of avoiding management interventions and relying on natural "recovery" that challenges rangeland managers today as they attempt to cope with changing species and climatic conditions. Doing nothing often becomes the most viable alternative for government land managers faced with lawsuits and controversy over management practices. There is little institutional accountability for the results of "hands off" management.

Against the background of vast changes on rangelands, climate change is a factor today. In fact, in rangeland management discourse, climate change or the specter of climate change has begun to overshadow the recognition of a legacy of mismanagement of western ecosystems. The story is perhaps shifting away from correcting past management to laying most of the blame at the foot of a situation beyond the manager's control. Yet the lessons learned from the past should not be forgotten as we move forward. In particular here we find a warning against charismatic stories—somewhat as Elinor Ostrom warns against policies rooted in compelling metaphors (1990). Stories encourage and conceal deep-rooted, untested assumptions, simplify complex relationships, and universalize truths that may hold true only in a single time and place.

Aknowledgements Thank you to Paul Starrs, Roy Behnke, and Sarah Robinson for their reviews of this manuscript.

References

Allen-Diaz, B., & Bartolome, J. W. (1998). Sagebrush-grass vegetation dynamics: Comparing classical and state- transition models. *Ecological Applications, 8*, 795–804.

Anderson, K. (2005). *Tending the wild: Native American knowledge and the management of California's natural resources*. Berkeley, CA: University of California Press.

Bagchi, S., Briske, D. D., Bestelmeyer, B. T., & Wu, X. B. (2013). Assessing resilience and state-transition models with historical records of cheatgrass *Bromus tectorum* invasion in North American sagebrush-steppe. *J Appl Ecol, 50*, 1131–1141.

Bailey, L. R. (1980). *If you take my sheep: The evolution and conflicts of Navajo pastoralism*, 1630–1868. Pasadena, CA: Westernlore Publications.

Baker, W. L. (2011). *Pre-Euro-American and recent fire in sagebrush ecosystems*. Berkeley, CA: University of California Press.

Balch, J. K., Bradley, B. A., D'Antonio, C. M., & Gomez-Dans, J. (2013). Introduced annual grass increases regional fire activity across the arid western USA (1980-2009). *Glob Change Biol, 19*, 173–183.

Bartolome, J. W. (1993). Application of herbivore optimization theory to rangelands of the western United States. *Ecological Applications, 3*, 27–29.

Behnke, R. H., Scoones, I., & Kerven, C. (Eds.). (1993). *Range ecology at disequilibrium: New models of natural variability and pastoral adaptation of African savannas*. London: Overseas Development Institute/International Institute for Environment and Development/ Commonwealth Secretariat.

Bestelmeyer, B. T. (2005). Does desertification diminish biodiversity? Enhancement of ant diversity by shrub invasion in south-western USA. *Diversity and Distributions, 11*, 45–55.

Blackburn, T. C., & Anderson, K. (Eds.). (1993). *Before the wilderness: Environmental management by native Californians*. Norman, OK: University of Oklahoma Press.

BLM [Bureau of Land Management]. (1991). *Final environmental impact statement for vegetation treatment on BLM lands in thirteen western states*. Washington, D.C: Bureau of Land Management.

BLM [Bureau of Land Management]. (2015a). *Public Lands Statistics*. Washington, D.C: United States Department of the Interior, http://www.blm.gov/public_land_statistics/pls14/pls2014.pdf. Accessed Jan 10, 2015.

BLM [Bureau of Land Management]. (2015b). *Fact Sheet on the BLM's management of livestock grazing*. http://www.blm.gov/wo/st/en/prog/grazing.html. Accessed Jan 10, 2015.

Bowers, E. A., Fernow, B. E., Olmstead, F. L., Rothrock, J.F., Colvin, V., Roosevelt, T., Pinchot, G., Shaler, N. S., Riordan, D. M., Muir, J., Abbe, C., Palmer, W. J., Anderson, G. S. (1895). Comments on professor Charles S. Sargent's scheme of forest preservation by military control. *Century Magazine [Cornell University Library, Making of America Archive] 49*, 626–634.

Brandt, C. A., & Rickard, W. H. (1994). Alien taxa in the North-American shrub-steppe 4 decades after cessation of livestock grazing and cultivation agriculture. *Biological Conservation, 68*, 95–105.

Briske, D. D., Derner, J. D., Brown, J. R., Fuhlendorf, S. D., Teague, W. R., Havstad, K. M., et al. (2008). Rotational grazing on rangelands: Reconciliation of perception and experimental evidence. *Journal of Range Management, 61*, 3–17.

Briske, D. D., Sayre, N. F., Huntsinger, L., Fernandez-Gimenez, M., Budd, B., & Derner, J. D. (2011). Origin, persistence, and resolution of the rotational grazing debate: Integrating human dimensions into rangeland research. *Rangeland Ecology & Management, 64*, 325–334.

Brooks, M. L., D'Antonio, C. M., Richardson, D. M., Grace, J. B., Keeley, J. E., DiTomaso, J. M., et al. (2004). Effects of invasive alien plants on fire regimes. *Bioscience, 54*, 677–688.

Climate Central. (2012). *The age of western wildfires*. Princeton, N.J.: Climate Central.org, http://www.climatecentral.org/wgts/wildfires/Wildfires2012.pdf. Accessed Jan 1, 2015.

Chambers, J. C., Roundy, B. A., Blank, R. R., Meyer, S. E., & Whittaker, A. (2007). What makes Great Basin ecosystems invisible by *Bromus tectorum*? *Ecological Monographs, 77*, 117–145.

Clements, F. (1916). Plant succession, an analysis of the development of vegetation. *Carnegie Institution of Washington Publication, 242*, 1–512.

Cole, K. L., Henderson, N., & Shafer, D. S. (1997). Holocene vegetation and historic grazing impacts at Capitol Reef National Park reconstructed using packrat middens. *Great Basin Naturalist, 57*, 315–326.

Cronon, W. (1992). A place for stories: Nature, history, and narrative. *The Journal of American History, 78*, 1347–1376.

Daubenmire, R. F. (1940). Plant succession due to overgrazing in the *Agropyron* bunchgrass prairie of southeastern Washington. *Ecology, 21*, 55–64.

Davis, D. (2007). *Resurrecting the granary of Rome: Environmental history and French colonial expansion in North Africa*. Athens, OH: Ohio University Press.

de Buys, W. (1985). *Enchantment and exploitation: The life and hard times of a New Mexico mountain range*. New Mexico: University of New Mexico Press.

Dregne, H. E., (1976). Desertification: Symptom of a crisis. In P. Paylore & R. A. Haney (Eds.), Desertification: Process, problems, perspectives (pp. 11–24). Tucson, AZ: Office of Arid Lands Studies, University of Arizona.

Dykesterhuis, E. J. (1949). Condition and management of range land based on quantitative ecology. *Journal of Range Management, 2*, 104–115.

Eckert, R. E, Jr, & Spencer, J. S. (1987). Growth and reproduction of grasses heavily grazed under rest-rotation management. *J Range Manag, 40*, 156–159.

Ellis, J. E., & Swift, D. M. (1988). Stability of African pastoral ecosystems—Alternate paradigms and implications for development. *Journal of Range Management, 41*, 450–459.

Fairfax, S. K., Huntsinger, L., & Adelburg. C. (1999). Lessons from the past. *Forum for Applied Research and Public Policy* (Summer), 85–89.

Fleischner, T. L. (2002). *Land held hostage: A history of livestock and politics*. Washington, D.C.: Island Press.

Farquhar, F. P (1930). *Up and Down California in 1860–1864: the Journal of William H. Brewer*. New Haven, CN: Yale University Press.

Forero, L. C. (2002). *Grass, graziers and tenure: A case study on the Shasta-Trinity National Forest*. Ph.D. dissertation. Berkeley, CA: University of California, Berkeley. http://oskicat.berkeley.edu/record=b21322700~S1 Accessed January 1, 2015.

Fortmann, L. P., & Fairfax, S. K. (1989). American forestry professionalism in the third world: Some preliminary observations. *Economic and Political Weekly 12*, 18–24.

Gallant, A. L., Hansen, A. J., Councilman, J. S., Monte, D. K., & Betz, D. W. (2003). Vegetation dynamics under fire exclusion and logging in a Rocky Mountain watershed, 1856–1996. *Ecol Appl, 13*, 385–403.

Gleason, H. A. (1917). The structure and development of the plant association. *Bulletin of the Torrey Botanical Club, 43*, 463–481.

Goforth, B. R, & Minnich, R. A. (2008). Densification, stand-replacement wildfire, and extirpation of mixed conifer forest in Cuyamaca Rancho State Park, southern California. *Forest Ecology and Management*, 36–45.

Gorte, R. W., Vincent, C. H., Hanson, L.A., & Rosenblum, M. R. (2012). *Federal Land Ownership: Overview and Data. Report for Congress 7-5700; R42346*. Washington D.C.: Congressional Research Service.

Hagmann, R. K., Franklin, J. F., & Johnson, K. N. (2013). Historical structure and composition of ponderosa pine and mixed conifer forests in south-central Oregon. *Forest Ecology and Management*, 492–504.

Hardin, G. (1968). The tragedy of the commons. *Science, 162*, 1243–1248.

Heady, H. F. (1961). Continuous vs. specialized grazing systems: A. review and application to the California annual type. *Journal of Range Management, 14*, 182–193.

Holechek, J. L.,Pieper, R. D., & Herbel, C. H. (2010). *Range management: Principles and practices* (6th Edn.). New Jersey: Prentice Hall.

Humphrey, R. R. (1947). Range forage evaluation by the range condition method. *J Forest, 45*, 10–16.

Huntsinger, L., Forero, L. C., & Sulak, A. (2010). Transhumance and pastoralist resilience in the western United States. *Pastoralism: Research, Policy, and Practice, 1*, 1–15.
Huntsinger, L., & Starrs, P. F. (2006). Grazing in arid North America: A biogeographical approach. *Secheresse, 17*, 219–233.
Hutchinson, C. F. (2000). *John Wesley Powell and the New West. Cosmos 2000.* Washington D. C.: Cosmos Club. http://www.cosmosclub.org/web/journals/2000/hutchinson.html. Accessed May 2014.
Keeley, J. E. (2006). Fire management impacts on invasive plants in the western United States. *Conserv Biol, 20*, 375–384.
Kerley, G. I. H., & Whitford, W. G. (2000). Impact of grazing and desertification in the Chihuahuan Desert: Plant communities, granivores and granivory. *American Midland Naturalist, 144*, 78–91.
Klass, J. R., Peters, D. P. C., Trojan, J. M., & Thomas, S. H. (2012). Nematodes as an indicator of plant-soil interactions associated with desertification. *Applied Soil Ecology, 58*, 66–77.
Lal, R., Sobecki, T. M., Livari, T., & Kimbell, J. M. (2004). *Chapter 10: Desertification In Soil Degradation in the United States: Extent, Severity, and Trends,* 150–154. Boca Raton, FL: Lewis Publishers-CRC Press.
Laycock, W. A. (1987). Grazing management systems and tactics in the sagebrush ecosystem. In J. A. Onsager (Ed.), *Integrated pest management: State of the art in the sagebrush ecosystem*. Washington, D.C.: US Agricultural Research Service ARS-50.
Mack, R. N., & Thompson, J. N. (1982). Evolution in steppe with few large hooved animals. *The American Naturalist, 119*, 757–773.
Mack, R. N., & Pyke, D. A. (1983). The demography of *Bromus-tectorum*—variation in time and space. *J Ecol, 71*, 69–93.
McFerrin, R., & Wills, D. T. (2013). Searching for the big die-off: An event study of the 19th century cattle markets. *Economics and Business History Journal, 31*, 33–52.
Menakis, J. P., Keane, R. E., & Long, D. G. (2000). Mapping ecological attributes using an integrated vegetation classification system approach. *Journal of Sustainable Forestry, 11*, 245–263.
Merrill, K. R. (2002). *Public lands and political meaning: Ranchers, the government, and the property between them*. Berkeley, CA: University of California Press.
Miller, J, D., & Safford, H. D. (2012). Trends in wildfire severity: 1984 to 2010 in the Sierra Nevada, Modoc Plateau, and southern Cascades, California, USA. *Fire Ecology*, 41–57.
Miller, J. D., Safford, H. D., Crimmins, M., & Thode, A. E. (2009). Quantitative evidence for increasing forest fire severity in the Sierra Nevada and southern Cascade Mountains, California and Nevada, USA. *Ecosystems*, 16–32.
Morrow, J. O. (1918). Letter to Pete Hufford, Whitmore, CA. from the Forest Supervisor's office, United States Forest Service, Shasta-Trinity National Forest, April 12. Hufford Family Archives, Red Bluff, CA.
Muir, J. (1911). *My first summer in the Sierra*. Boston and New York: Houghton Mifflin.
Nelson, R. H. (1995). *Public lands and private rights*. Lanham, MD: Rowman and Littlefield.
NIFC [US National Interagency Fire Center]. (2015). *Wildfire statistics*. Boise, ID:National Interagency Fire Center. www.nifc.gov/fireInfo/fireInfo_stats_totalFires.html. Accessed January 1, 2015.
Ostrom, E. (1990). *Governing the commons: The evolution of institutions for collective action*. Cambridge; New York: Cambridge University Press.
Paylore, P., & Haney, R. A. (1976). *Desertification: Process, problems, perspectives*. Tucson, AZ: Office of Arid Lands Studies, University of Arizona.
Perevolotsky, A., & Seligman, N. G. (1998). Role of grazing in Mediterranean rangeland ecosystems—Inversion of a paradigm. *Bioscience, 48*, 1007–1017.
Powell, J. W. (1879). Report on the lands of the arid region of the United States, with a more detailed account of the lands of Utah. With maps. Washington, D.C.: U.S. Government Printing Office.

Powell, J. W. (1890). The non-irrigable lands of the arid region. *Century Magazine [Cornell University Library, Making of America Archive] 39*, 915–922.

Pyne, S. J. (2008). *Year of the fires: The story of the great fires of 1910*. Missoula, Montana: Mountain Press Publishing Company.

Raban, J. (1996). *Bad land: An American romance*. New York: Random House.

Raumann, C. G., & Cabik, M. E. (2008). Change in the forested and developed landscape of the Lake Tahoe basin, California and Nevada, USA, 1940–2002. *For Ecol Manage, 255*, 3424–3439.

Roth, F. (1901). *Grazing in the forest reserves [Reprint from the Yearbook of Department of Agriculture for 1901]*. Washington, D.C.: Government Printing Office.

Sabadell, J. E., Risley, E. M., Jorgensen, H. T., & Thornton. B. S. (1982). Desertification in the United States. Status and issues. Final report. Washington, D.C.: Bureau of Land Management.

Sampson, A. W. (1909). Natural revegetation of depleted mountain grazing lands. *Circular*, 169, 28. Washington, D.C.: U.S. Department of Agriculture, Forest Service.

Savory, A., & Butterfield, J. (1999). *Holistic management: A new framework for decision making*. Covelo, CA: Island Press.

Sayre, N. F. (2002). *Ranching, endangered species, and urbanization in the Southwest*. Tucson, AZ: University of Arizona Press.

Sayre, N. F., & Fernandez-Gimenez, M. (2003). The genesis of range science, with implications for current development policies. In *Proceedings of the 7th International Rangelands Congress*, 26 July-1 Aug Durban, South Africa:1976–1985.

Schama, S. (1996). *Landscape and memory*. New York NY: Vintage.

Seastedt, T. R., Hobbs, R. J., & Suding, K. N. (2008). Management of novel ecosystems: Are novel approaches required? *Frontiers in Ecology and the Environment, 6*, 547–553.

Skovlin, J. M., Strickler, G. S., Peterson, J. L., & Sampson. A. W. (2000). *Interpreting landscape change in high mountains of northeastern Oregon from long-term repeat photography*. F.S. U. S. Department of Agriculture, Pacific Northwest Research Station. Albany, CA: PNW-GTR-505.

Spiegal, S., Huntsinger, L., Hopkinson, P., & Bartolome. J. W. (2015). Chapter 37, Range ecosystems. In Zavaleta, E., & H. Mooney (Eds.) *Ecosystems of California*, 835–864. Berkeley, CA: University of California Press.

Starrs, P. F. (1998). *Let the cowboy ride: Cattle ranching in the American West*. Baltimore, MD: Johns Hopkins University Press.

Stewart, O. C., Lewis, H. T., & Anderson, M. K. (2002). *Forgotten fires: Native Americans and the transient wilderness*. Norman, OK: University of Oklahoma Press.

Tansley, A. G. (1935). The use and abuse of vegetation concepts and terms. *Ecology, 16*, 284–307.

Taylor, A. H. (2000). Fire regimes and forest changes in mid and upper montane forests of the southern Cascades, Lassen Volcanic National Park, California. *USA. Journal of Biogeography, 27*, 87–104.

USFS [United States Forest Service]. (2015). *Grazing statistical summary*. Washington (DC). http://www.fs.fed.us/rangelands/ftp/docs/GrazingStatisticalSummaryFY2014.pdf. Accessed Jan 10, 2015

van der Valk, A. G. (2013). From formation to ecosystem: Tansley's response to Clements' climax. *Journal of the History of Biology, 47*, 293–321. doi:10.1007/s10739-013-9363-y).

Westoby, M., Walker, B., & Noymeir, I. (1989). Opportunistic management for rangelands not at equilibrium. *Journal of Range Management, 42*, 266–274.

Worster, D. (1979). *Dust Bowl: The southern plains in the 1930s*. Oxford: Oxford University Press.

Worster, D. (1994). *Nature's economy: A history of ecological ideas*. Cambridge: Cambridge University Press.

Young, J. A., & Clements, C. D. (2007). Cheatgrass and grazing rangelands. *Rangelands, 29*, 15–20.

Young, J. A., & Clements, C. D. (2009). *Cheatgrass: Fire and forage on the range*. Reno, NV: University of Nevada Press.

Part III
Regional and Country Case Studies

Chapter 12
Pastoral System Dynamics and Environmental Change on Ethiopia's North-Central Borana Plateau—Influences of Livestock Development and Policy

D. Layne Coppock

Abstract The Borana Plateau is a heavily studied pastoral ecosystem. Knowledge generated with respect to the history and politics of the region, pastoral system dynamics, and patterns of environmental change is instructive in understanding how environmental degradation is linked to pastoral development priorities and policies. Despite decades of research and development efforts, the Borana pastoral system is characterized by increasing poverty, wealth stratification, and food insecurity. The condition and trend for rangelands in the north-central portion of the plateau is deemed generally poor due to chronically intense pressure from large numbers of livestock in support of a growing human population. Particularly noteworthy is the emergence of extensive gullying and bush encroachment in several locations; in both cases the productivity of the pastoral system has been reduced. This trend is attributed to decades of development priorities and policies that have failed to provide incentives for surplus pastoralists to discover alternative livelihoods, or for herd owners to adopt resource management tactics that could promote environmental stewardship. The way forward is to dispense with laissez faire attitudes regarding condition and trend of natural resources and forge new collaborative partnerships among pastoral communities, researchers, and development agents to reverse environmental degradation where possible. This requires a transformation of local governance so that stocking rates can be better managed and innovative land-management systems can be adopted to better accommodate the increasing population pressure and more frequent droughts. Over the longer term, assisting those who wish to leave pastoralism can reduce demands on natural resources and promote system sustainability.

D.L. Coppock (✉)
Department of Environment and Society, Utah State University,
Logan, UT 84322-5215, USA
e-mail: Layne.Coppock@usu.edu

R.H. Behnke and M. Mortimore (eds.), *The End of Desertification?*
Springer Earth System Sciences, DOI 10.1007/978-3-642-16014-1_12

Keywords Pastoral population · Pastoral governance · Natural resource management · Food security · Pastoral development · Pastoral land use · Land degradation · Bush encroachment

12.1 Introduction

Background. As will be described, the north-central Borana Plateau (henceforth called the "plateau") is now one of the most studied pastoral regions in Sub-Saharan Africa. The Borana pastoral system has undergone at least three distinct periods of development, socioeconomic change, and research observation. The first period occurred before 1970 when the pastoralists were relatively isolated from the outside world. Development investment was limited and research documentation was rare. The second period covered the 1970s and 1980s. This was a time when large, rural infrastructure-development projects began, a few non-governmental organizations (NGOs) arrived—largely to distribute food aid—outside interest grew concerning the marketing of highly-valued Boran cattle, and research became more common—although most studies were conducted by investigators representing only a few institutions. The third period covers the 1990s–2014. Particularly since 1995, this has been a time of rapid change for much of Ethiopia as well as for the plateau.[1] At the national level there have been expanded investments in rural infrastructure as well as devolution of development resources from the federal capital of Addis Ababa to the Oromia Regional State and local districts. Locally many rangeland towns and settlements have grown. All-weather roads are being upgraded and transportation has improved. International and local NGOs have proliferated and their mandates now include more pastoral development activities. The pastoralists have become more connected to the outside world via enhanced livestock marketing and improved access to mobile phones, formal education, and other services. In some cases pastoral livelihoods have begun to diversify away from a sole reliance on livestock to include more small-business activities. Research activity continues to expand and the diversity of investigators and research institutions has markedly increased. Research in many cases has shifted from purely observational to more participatory in nature. Major demographic, ecological, and political factors underpin this backdrop of change. Major droughts have occurred in the north-central region of the Borana plateau about once every five to six years since 1985, a time when researchers began to keep records. Livestock mortality rates during droughts are high—especially for cattle—the key livestock species in the system. The cattle population therefore regularly undergoes a "boom-and-bust" dynamic where periods of rapid

[1]Economic growth in Ethiopia has averaged 10.6 percent per annum from 2004 to 2012 http://www.worldbank.org/en/country/ethiopia/overview. Change is most evident in large cities as evident from urban construction activities. There has also been considerable infrastructure development (primarily roads) nationwide.

herd growth are followed by precipitous herd crashes. The human population has steadily grown over the past 30 years; this is due to internal reproduction as well as immigration. Despite improvement in marketing and service provision, marketed offtake rates for livestock remain modest. Food insecurity and the incidence of poverty appear to be increasing. While certain rangeland landscapes exhibit a high degree of resilience in response to chronically heavy use by livestock, there are also areas suffering from pronounced environmental degradation.

The Borana pastoral system has existed under several national governments in the modern era.[2] Prior to 1974, it was the Imperial regime of Emperor Haile Selassie. From 1977 to 1991 it was the military socialist Derg regime under Colonel Mengistu Haile Mariam. From 1991 to 1994 it was the Transitional Government of Ethiopia. From 1994 to the present it has been the Federal Democratic Republic of Ethiopia (FDRE) which has embraced a mixed (i.e., capitalist and socialist) economic-development model. This has been largely led by Mr. Meles Zenawi (from 1995 to 2012) and subsequently by Mr. Hailemariam Desalegn. Shifts in development priorities and approaches with each regime have dramatically affected citizens in both the urban and rural sectors. The current regime created a federalized system of nine regional states, each largely based on distinctions of ethnicity and language. The Oromia Regional State—home to over 27 million people living in urban and rural settings in both the moist highlands and dry lowlands—is where the plateau is located.

Objectives. The first objective of this chapter is to provide a concise overview of pastoral system dynamics so readers can grasp the context of the plateau. Details will provide evidence for points already made. The second objective is to describe patterns of environmental degradation, altered land use, and environmental resilience in the north-central region with a focus on how degradation has been caused by chronic grazing pressure; mechanisms of change are illuminated by an understanding of pastoral system dynamics over several decades. The third objective is to propose the extent that environmental degradation is linked to pastoral development priorities and policy frameworks. The fourth objective is to describe a way forward for development agents, policy makers, and local communities so they can make better decisions that promote the sustainability of the pastoral system.

12.2 Overview of Pastoral System Dynamics

The Borana Plateau (aka the plateau) is located in southern Ethiopia and it is roughly 95,000 km^2 in size. It is bounded by the Kenyan border (to the South), the Somali Ogaden (East), the lower Omo Valley (West), and the southern highlands

[2]Details concerning Ethiopian governments can be found at http://www.ethioembassy.org.uk/fact%20file/a-z/history.htm.

(North). The region has been occupied by the Borana pastoralists for centuries. It has been the subject of several rangeland development projects and numerous scientific investigations over the past 50 years. The history of pastoral development, research, and social change in the region up through 1991 is summarized in Coppock (1994).

While there is a significant level of continuity in some aspects of the research conducted on the plateau, most studies only contribute "snapshots" at certain points in time or space. For topics related to rangeland ecology and livestock dynamics, high-quality data at varied spatial and temporal scales remain rare. For rangeland ecology in particular, differences in where and how studies have been conducted—as well as a lack of long-term protected sites that could serve as benchmarks to assess vegetation change—make rigorous assessments problematic. Despite such challenges, however, it is argued that the overall trends for vegetation, livestock, and people in this system are clear.

12.2.1 Environment and Pastoral Society

Climate. This review is largely limited to a discussion of the north-central plateau, a 16,000-km^2 area bisected from North to South by the main tarmac road that runs past the outskirts of Yabelo town to Moyale town on the Kenya border (Fig. 12.1). Climate patterns for the 1980s are described in Coppock (1994). The elevation varies from 1200 to 1500 m. The climate is semi-arid. Precipitation on the north-central plateau has traditionally ranged from 400 to 850 mm per year and increases from South to North as well as at higher elevations. The coefficient of variation for annual precipitation is about 12 % in warmer, drier locations versus 6 % in cooler, wetter locations. The rainfall distribution is bimodal, with about 60 % received between March and May and 30 % received between September and November. Annual air temperatures are relatively cool, with daily maxima and minima on the order of 28 and 15 °C, respectively. The interaction of rainfall and temperature results in four distinct seasons (Coppock 1994). The long rainy season (March to May) is followed by a cool dry season (June to August), a short rainy season (September to November), and a warm dry season (December to February). Climate change has been recently implicated for Ethiopia overall as well as for the plateau (Funk et al. 2012). The prognosis for the nation is that it is gradually becoming warmer. The prognosis for the plateau is that it is gradually becoming drier from East to West.[3]

[3]Climate change has been strongly implicated for sub-Saharan Africa (Thornton et al. 2011) and Ethiopia (Funk et al. 2012; Evangelista et al. 2013). Pastoralists on the Borana Plateau report that their environment is becoming warmer and drier with shifts in precipitation (Coppock, unpublished observations). Funk et al. (2012) documents that Ethiopia is exhibiting a warming trend overall, with growing-season rainfall declining nationally by 15–20 %. The 250-mm contours for the long rainy season of March to June in the southern rangelands have shifted westward between

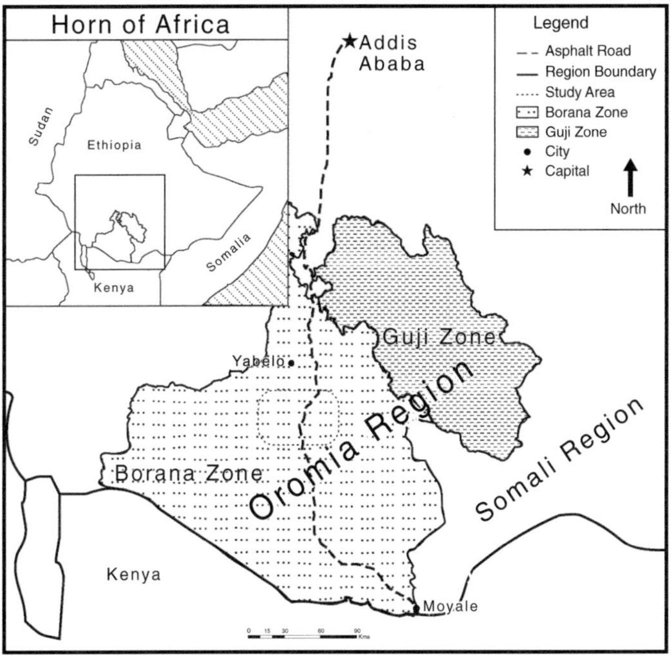

Fig. 12.1 Map showing the southern Ethiopian rangelands as demarcated by regional state and zones within the Oromia Regional State. The study area germane to this chapter is located in the north-central Borana Plateau. It is where much of the research related to pastoral households and environment has been conducted by the International Livestock Center for Africa (ILCA) in the 1980s as well as by USAID-supported projects from 1997 to the present. The land-cover change work by Mesele (2006) also occurred here. The four pastoral associations referred to in recent investigations are roughly coincident with four quadrants of the study area; namely Harweyu (to the northeast), Dikale (northwest), Medecho (southeast), and Denbala Badana (southwest) (*Source* D. Layne Coppock).

Vegetation and soils. Vegetation and soils are described in Coppock (1994). The vegetation is typified by mixed savanna dominated by perennial grasses, shrubs, and trees. Net primary productivity for herbaceous plants in a near-average rainfall year has been estimated to vary from 1.5 to 2.7 tons of dry matter per hectare per year for drier or wetter locations, respectively. Well-drained upland soils dominate 88 % of the landscape in the north-central region. They vary in color from red to brown or gray and are composed of sand (53 %), clay (30 %), and silt (17 %). Upland soils are often relatively shallow, erodible, and low in nutrient content. Common plants on upland soils include valuable perennial grasses such as *Cenchrus* sp. and *Chrysopogon* sp., while woody plants include *Acacia* and

(Footnote 3 continued)

1960 and 2009, a pattern that will continue for the next 25 years. In sum, the plateau is becoming drier from East to West (Funk et al. 2012).

Fig. 12.2 Trough area for a deep tula well on the central Borana Plateau. These wells have provided water for pastoral livestock over hundreds of years (*Photo credit* D. Layne Coppock).

Commiphora spp. Valley bottoms, in contrast, occur across 12 % of the landscape. Soils in these sites typically consist of poorly drained black or brown Vertisols with a high clay-content (>60 %). Common plants on Vertisols include perennial grasses such as *Pennisetum* spp. and small trees (*A. drepanolobium*). Traditionally, valley bottoms have been important dry-season grazing areas because deep-rooted *Pennisetum* plants have persistently green leaves. In recent decades, however—with the advent of maize cultivation—many valley bottoms have been cultivated, especially to the North (see below).

Water resources, pastoral mobility, and governance. In sum, the area is a high-elevation rangeland environment where livestock production has been primarily limited by a scarcity of surface water (Coppock 1994). During the warm dry season water is lifted from deep (tula[4]) wells to livestock troughs by people (Fig. 12.2). Small, leather buckets of water are passed along chains of laborers who stand on narrow ledges in the well shafts. Tula wells are hundreds of years old. They have had profound effects on the social organization of the Borana pastoralists because hundreds of people must be coordinated to provide water on a regular schedule for thousands of livestock. Clusters of tula wells also help define 29 units of traditional resource allocation called madda. Madda contain villages or encampments (olla) and grazing locations for resident (e.g., milking) herds (warra) that forage locally and satellite herds (forra) that forage at distant locations. The olla supply the labor to operate the tula wells in dry periods. Ponds provide water during and soon after rainy periods. Ponds vary widely in capacity but most are ephemeral. Some were hand-dug by pastoralists, while others have been excavated with heavy machinery during infrastructure-development projects in the 1970s and 1980s (Coppock 1994). In the 1960s a few very large ponds were excavated with heavy machinery near the northern edge of the plateau. These became perennial water

[4]Henceforth in the text local terms for various entities on the Borana Plateau—such as wells, villages, types of cattle herds, political leadership structures, spatial units of territoriy, etc., are referred to in the regional language called Oromifa.

sources, allowing pastoralists who had left the well-based madda system to the South to permanently settle. This matters because this northern area was a traditional grazing reserve for cattle prior to the 1960s. People did not live there then, so forage supplies could be conserved for future use during a drought, for example. Once people began to settle, the area could no longer serve as a grazing reserve (Coppock 1994).

In general, the traditional pastoral system is best described as semi-nomadic (Coppock 1994). The relatively high and consistent levels of forage productivity here have reduced the need for a truly nomadic lifestyle where pastoral households pick up and move on a monthly (or even weekly) basis as they search for adequate supplies of forage and water. On the plateau—while forra herds can be highly mobile—the associated warra herds and pastoral households can be relatively sedentary. In the past an olla could re-locate perhaps a couple of times per year. As will be shown, however, this is rarely the case today. Pastoral mobility has been declining for decades in response to many factors.

The customary (informal) system of pastoral governance [i.e., the hierarchical and generational Gumi Gaayo of the Gaada system (Legesse 1973)] that is connected to the traditional social organization of madda has been augmented over the past 40 years by a formal governance structure called Pastoral Associations (PAs).[5] Pastoral Associations provide an interface between local people and the regional and national governments. The PA had a strong socialistic or collective character during the Derg regime, but with subsequent political change after 1991 it has become more market-driven and diverse in terms of socioeconomic functions. In some cases the area governed by a PA is coincident with boundaries of a madda, but in others one large madda may contain several PAs.

The PA has reportedly gradually eroded Gumi Gaayo authority on the plateau (Homann et al. 2008; Mellisse 2014; Wassie 2014). The increased authority of PAs has allowed for recent trends that include less control over general access to pond water, de facto privatization of plots by wealthy elites for use in cultivation and as fodder banks, and less regulatory oversight for grazing management (reviewed below).

The Gumi Gaayo has recently been assuming more of an advisory role to the state and oversees cultural ritual ceremonies. On the positive side, PAs have improved access of their communities to services such as education, health care, and markets (Coppock, unpublished observations).

[5]The PA administers the local community on behalf of the national and regional governments. This includes taxation, providing community political leadership and policing, arbitrating local conflicts, controlling resource use and trespass, assisting in the targeting of food aid, and serving as the interface between the community and development agents. Each PA has a chairperson, manager, secretary, a governing council of advisors, and one or two government extension agents. There are also subcommittees that deal with specific issues such as water, grazing, and social safety nets. In some cases subcommittees invoke hybrid governance (Coree) between PA and Gumi Gaayo elements. The PA manager is a literate, government employee who oversees political matters. The other positions are unpaid. Power relations between chairpersons and managers are idiosyncratic across PAs depending on personalities involved.

Human population, livestock marketing, and the Boran cattle breed. During the late 1980s the pastoral population of the north-central region was estimated at around 325,000 with an annual net growth rate of 2.5 % in non-drought years (Coppock 1994). In subsequent decades census taking has become more systematic. In the early 1990s the total human population of the region (including the pastoral and non-pastoral sectors) was estimated at about 1.4 million by CSA (1994). By the early 2000s the total human population was estimated at over 2 million with 12 % residing in towns and settlements, representing an increase of 44 % within a decade (CSA 2007). Although the CSA figures are only rough estimates, it is obvious that the number of people has increased at a high rate. This is due to internal recruitment as well as migration originating from elsewhere in Ethiopia as well as Kenya and Somalia (Coppock, unpublished observations).[6]

Local expansion of livestock marketing opportunities began around 2005. This was largely a result of privatization of Ethiopia's agro-industries, growth in urban demand, and opening of export markets to the Gulf States (Desta et al. 2006). The Boran have also needed to sell more animals (e.g., males) to buy food in recent years, a consequence of a declining ratio of cattle to people (Desta and Coppock 2004).

The Borana pastoralists have largely depended on the Boran cattle breed for milk and asset accumulation (Coppock 1994, Fig. 12.3). The Boran breed is highly regarded for its productivity and tolerance to drought and local diseases. Boran cattle have been in high demand—both domestically and abroad—and thus the breed has provided a key incentive for range-livestock market development beginning in the 1970s. As will be shown, small ruminants, equines, and camels have tended to play much smaller roles in the traditional livestock economy of the Boran when compared to that for Boran cattle, but recently the species diversification of herds in some locations has increased.

Ratios of cattle to people. During the 1980s the peak cattle population across the entire Borana Plateau was estimated to be on the order of 1 million head. When combined with the 350,000 figure for pastoralists (above) this gives a ratio of cattle to people of 2.85 on average (Coppock 1994). Such a low ratio requires households to secure food via non-pastoral options such as maize cultivation, food relief, and selling animals or animal products to purchase grain. This is because the traditional dietary mainstay—milk—has been declining on a per capita basis, and the ratio of cattle to people is an indicator of this process. Desta and Coppock (2004) found that numbers of cattle per person were dropping for all pastoral wealth classes in the north-central region. The declining number of cattle per capita is a fundamental driver of change in this system. It pushes pastoralists to become agro-pastoralists and it is an indicator of increasing poverty and famine risk. The ratio slowly ratchets downward over time as a function of a growing human population in combination

[6]The population is also increasingly spatially confined. Population sizes of neighboring groups are growing (CSA 2007) and the landscape is becoming more crowded. Rigorous demarcation of the border between the Oromia and Somali Regional States has resulted in losses of key ecological resources for the Boran (Coppock et al. 2011).

Fig. 12.3 Boran cattle. This breed is known for productivity and adaptability to dry conditions and is in high demand for export (*Photo credit* Brien E. Norton).

with a boom-and-bust pattern for cattle numbers. If it assumed that: (1) At least 10 cattle per person would be required to sustain one person on a traditional milk-based diet (Coppock 1994); and (2) the carrying capacity in an average rainfall year is about 30 head of cattle per square kilometer (Coppock 1994), this implies that about 3 people per square kilometer could be reliably supported. Scenarios described by Coppock (1994) indicated that the number of people actually living on these rangelands in the 1980s could have been as high as 11 per square kilometer. The population was then projected to double by 2018 given the 2.5 % rate of net annual growth (Coppock 1994). Hence the growing need on the plateau to find other food sources besides livestock.

Spatial variation in human population density. Table 12.1 displays information collected from four Pastoral Associations (PAs) in the north-central region of the plateau during 2013. These data illustrate high spatial variation in human population density and cattle per person even across a relatively small area—the maximum distance between any two PAs is less than 30 km. The data provide an inadequate test of predictions in Coppock (1994) because the land area represented is only 13 or 2 %, respectively, of the north-central region or the entire plateau. Despite this, it is notable that the data do not support extreme scenarios of a human population on pace to exceed 22 persons per square kilometer by 2018 or a ratio of cattle to people

Table 12.1 Population and land statistics covering four pastoral associations on the north-central Borana Plateau 2013 (*Source* Coppock et al. (2014a)[a])

Pastoral association	Area (km^2)	Human population	Cattle		Camels		Sheep and Goats		Cattle: person		Humans per km^2	Cattle per km^2
			No.	TLU[b]	No.	TLU[b]	No.	TLU[b]	No.	TLU[b]		
Dikale	315	4409	25,420	17,794	671	671	22,460	2246	5.8	4.0	14.0	81
Harweyu	567	3067	8700	6090	900	900	11,400	1140	2.8	2.0	5.4	15
Medecho	392	3763	12,484	8739	594	594	16,480	1648	3.3	2.3	9.6	32
Denbala Badana	767	4229	6245	4372	671	671	23,740	2374	1.5	1.0	5.5	8
Average	510	3867	13,212	9249	709	709	18,520	1852	3.3	2.3	8.6	34

[a]Variation in human and livestock numbers largely reflect local variation in ecological site potential. The maximum distance separating these Pastoral Associations is 90 km^2 (between Dikale and Medecho). Higher or lower cattle numbers for Dikale and Harweyu, for example, reflect (in part) a higher degree of bush encroachment and less herbaceous forage production in the latter. Each Pastoral Association provides most of the water and forage for resident herds in most years; during drought years all livestock species may seek forage and water away from the home pastoral association

[b]Tropical Livestock Units (TLU) based on relative variation in the live weights of adult animals, where cattle = 0.70 TLU, camels = 1.0 TLU, and sheep or goats = 0.10 TLU (Jahnke 1982)

appreciably different from the 2.85 value observed in the late 1980s. This could reflect data inadequacy. Or, if we assume the numbers are accurate, they could indicate that the pastoral system has "bottomed-out," with either surplus people finding the means to secure non-pastoral livelihoods, or that fertility rates have been in decline. Other interview data in the same PAs, however, support the idea of a persistent and large food deficit even in near-average rainfall years. For example, households designated as poverty-stricken by relief agencies in 2013 are eligible to receive up to 100 kg of free food grain per month (Bedasa Eba, OARI, personal communication). Food aid is reportedly pervasive in the system today (Coppock et al. 2014a)

Boom-and-bust dynamics of the regional cattle herd. Desta and Coppock (2002) focused on cattle population dynamics and concluded that major crashes of the regional cattle herd occurred at an interval of 5 to 6 years between 1980 and 2001. These crashes killed from 37 to 62 % of all cattle. This analysis indicated that mortality spikes were caused by high stocking rates combined with drought conditions. The pattern was thus a dynamic, oscillating equilibrium where cattle density and precipitation level jointly affected forage supplies. The key was the regular interval in the cycle; the 5 to 6 year period is the time required for the cattle population to recover from the previous drought and achieve a density of about 30 head per square kilometer. Attaining such a density then makes the regional herd vulnerable to even modest dips in seasonal precipitation. When combined with a slowly degrading natural resource base (see below) and the loss of traditional, drought-grazing reserves due to overpopulation (in the northern portion of the plateau, as above), the cattle production system essentially has become a vicious cycle characterized by a boom-and-bust.

Desta and Coppock (2002) predicted that the next herd crash would occur by 2005. This was verified by subsequent observations (Coppock et al. 2008) and the overall time series is shown in Fig. 12.4. A fifth major herd crash occurred in 2011–12, still roughly fitting the predicted pattern (Solomon Desta and Getachew Gebru, MARIL,

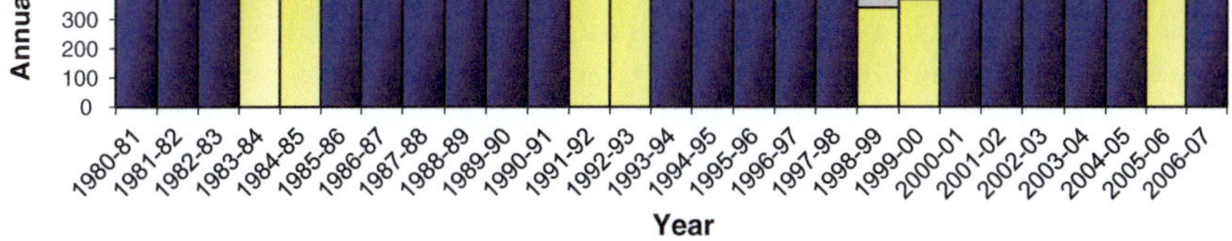

Fig. 12.4 Periodicity of crashes in the cattle population in relation to annual rainfall totals on the central Borana Plateau, 1980–2007. Average rainfall was calculated across 4 meteorological stations at Yabelo, Negele, Mega, and Arero. *Yellow bars* denote years when a major crash occurred (*Source* D. Layne Coppock).

personal communications). This most recent crash has not been included in the time series, however, because the rainfall data collected by government at the four representative stations has been very spotty—and thus unanalyzable—since 2008.

Considered over several sources, the database is inadequate to assess whether the regional cattle population is growing or shrinking over time, despite the oscillation. It is reasonable to speculate, however, that further growth in cattle numbers is unlikely. This is because the grazing resources for cattle have dramatically declined due to land cover changes (below).

Spatial variation in cattle density. Data in Table 12.1 illustrate that the cattle density per unit area varied widely among four PAs in 2013, from 8 to 81 head per km^2. Each PA experienced the drought-related herd crash during 2011–12, so 2013 was early in the subsequent recovery process (Coppock et al. 2014a). If it is assumed that the relative impact of drought was similar for each PA—given their close proximity—this information again illustrates the large effect of local ecological variation on base livestock numbers. The rule of thumb that considers 30 cattle per km^2 as a threshold for herd vulnerability to variation in precipitation is thus only useful as a crude generalization at very large spatial scales. Sites like Dikale PA are relatively favorable overall given its position at the northern edge of the plateau, hence the higher densities of cattle and people there.

High variation in cattle density among the four PAs probably reflects local variability in grass production, the preferred forage for cattle. A critical factor is the steep rainfall gradient from North to South in this area (Coppock 1994). For example, Dikale PA to the North (with 81 head per km^2) has the highest annual rainfall, and hence higher grass productivity. Similarly, Denbala Bedana to the South (with 8 head per km^2) is more arid with less grass productivity. Despite being near Dikale in a relatively mesic location, Harweyu PA (with 15 head per km^2) suffers from a very high degree of bush encroachment, with concomitant reductions in grass production (Coppock 1994; Forrest et al. 2014).

Shifts in livestock composition. The database is also inadequate to assess whether there have been significant alterations in the sex composition of the regional cattle herd. Estimates for the 1980s and 1990s are similar at 71 % females and 29 % males (Desta and Coppock 2004). We would expect the proportion of cows to increase over time as the marketing of bulls increases and pressures on grazing resources escalate, but there is no evidence of such a trend.

Recent observations suggest that livestock species composition in the central plateau has undergone a shift over the past 30 years. In the 1980s, for example, cattle comprised about 90 % of the total livestock biomass, while camels comprised only 3 % (Coppock 1994). Ethiopian government data sampled for 10 districts across the plateau in 2011 indicate that cattle now comprise 77 % of livestock biomass, while camels comprise 16 %. Small ruminants and donkeys make up the remainder—there are no discernible shifts for these species (Coppock, unpublished observations). Boru et al. (2014) have detected similar trends. The reasons for the relative increase in camels are now under study. There are several possible explanations, one being that an increased emphasis on camels reflects a need to

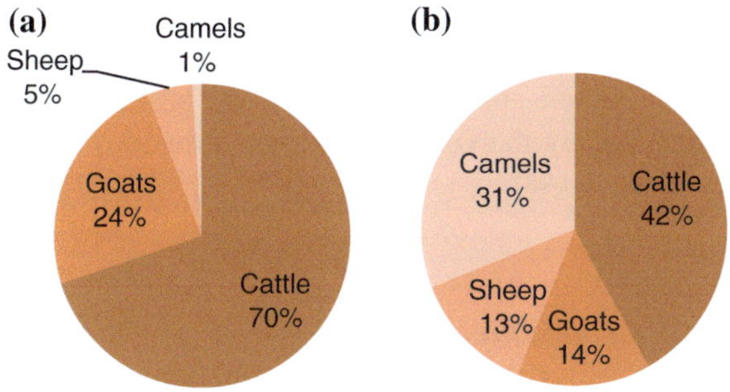

Fig. 12.5 (**a**, **b**) Actual versus desired composition of livestock holdings as expressed by 24 Boran herdowners during 1988 in the north-central region of the Borana Plateau. (**a**) Actual herd composition; (**b**) Desired herd composition (*Source* D. Layne Coppock).

make better use of increasing browse resources relative to grass (below). Another is that camels can provide more milk than cattle during dry periods and droughts. And yet another is that camels can now bring exceptionally high prices in the marketplace compared to cattle; camels have also been exported from Ethiopia over the past decade (Desta et al. 2006). Multiple explanations for increasing adoption of camels may thus be anticipated.

The cattle-keeping Borana have wanted more camels in their livestock holdings for many years. Figure 12.5 illustrates interview data showing actual versus desired percentages of livestock species in the holdings of Borana pastoralists in 1988. Why has it taken them so long to acquire more camels? Camels reproduce much more slowly than cattle and offer new management challenges to producers less accustomed to camel husbandry (Coppock 1994; Megersa et al. 2008). Camels have also traditionally been held by Muslim groups in this region. The Borana are largely non-Muslim. This difference in religion may have contributed to cultural or market-related barriers in camel acquisition (Desta and Coppock 2004).

12.3 Environmental Degradation, Altered Land Use, and Environmental Resilience

The focus in this section is on patterns of environmental degradation and resilience in the rangelands. Degradation is defined here as a reduced capacity of the system to support livestock (e.g., cattle) and people in a sustainable and equitable fashion; reduction in capacity is judged relative to a previous time when human and livestock populations were lower. The system components subject to degradation include soils, preferred forage plants, and water (both quantity and quality). Resilience here refers to the ability of the system to endure heavy use and then recover from at least a superficially degraded state. Although long-term benchmarks are lacking, it is very reasonable to believe—based on 30 years of observations—that the north-central plateau has become markedly degraded in terms of its ability to sustain cattle and people. In some respects, the pastoral system has already collapsed—as evidenced by the volume of food aid that now routinely enters the region (Coppock et al. 2014a).[7]

12.3.1 Bush Encroachment

The central and western portions of the plateau have experienced large-scale changes in the dominant vegetation communities. By the mid-1980s, 40 % of this region was determined to be affected by bush encroachment, with 19 % affected by topsoil erosion. These trends were coincident and most apparent in semi-arid sites having hilly relief in the vicinity of permanent water development. Both bush encroachment and topsoil erosion have long-been postulated to occur because of over-grazing by pastoral livestock (J.C. Billé et al. cited in Coppock 1994). This has also been repeatedly confirmed when pastoralists are interviewed about their perceptions concerning the cause and effects of ecological change (Oba et al. 2000; Angassa and Beyene 2003; Solomon et al. 2007; Coppock et al. 2014a).

In an earlier era of fewer pastoralists and livestock, the grazing system is postulated to have resembled a patch dynamic (J.C. Billé et al. cited in Coppock 1994). Namely, people and animals regularly moved around the landscape and intensively grazed the same parcels of land for relatively short periods of time. Then they moved on—not to return for several years—thus allowing local forage resources to recover. The herbaceous standing crop also provided sufficient fuel that allowed the managed use of fire by the pastoralists to treat the landscape. These fires were tolerated by the perennial grasses that can re-sprout from roots and rhizomes in the soil, but fires inhibited the establishment and growth of woody plants, hence maintaining a mixed-savanna mosaic. Growth in human and livestock populations—accompanied by the development of permanent water—gradually promoted longer-term pastoral settlement. This, in turn, has contributed to chronically heavy grazing of local landscape parcels. This pressure has not allowed herbaceous forages to recover from previous use. The reduced standing crop also markedly

[7]A series of Participatory Rural Appraisals (PRAs) in 2013 clearly and independently revealed the existence of a significantly large, "very poor" class of pastoralists having few livestock who routinely rely on food aid for survival. Food aid is distributed by the leadership of the Pastoral Associations (PAs) where these people reside. Food aid has been part of local development programmes here for a number of years. Calculations of food energy budgets based on population sizes for people and livestock, and productivity of livestock, for Harweyu PA also indicate a food energy gap for the people that is not being met by either livestock production or purchased food; food relief is thus a likely contributor (Forrest 2014). This situation may be unlike what has been previously reported where food aid only makes a small or irregular contribution to the diets of pastoral households in southern Ethiopia or northern Kenya, as evidenced by survey research conducted during 2000–2002 (McPeak et al. 2012).

lowered fuel loads, which has undermined the use of fire and thus allowed woody plants to establish and proliferate.

About 16 woody species have been implicated as encroachers here. They have varied effects on the environment; some are positive, but most are negative. Some *Acacia* bush species suppress herbaceous production by 50 % or more in their understories by out-competing grasses or forbs for light, water, and nutrients (Coppock 1994; Forrest et al. 2014). Figure 12.6 illustrates the dramatic effects of bush removal on herbaceous forage yield following the long rainy season of 2013 in Harweyu PA. A few encroaching species (e.g., *A. brevispica, Commiphora* spp.) can offer nutritious browse for small ruminants or camels. Typically, however, these forages are either physically inaccessible due to tree height or thorniness of stems, or their leaves and stems have low nutritive value because of high concentrations of toxic secondary compounds (Coppock 1994).

A more recent study of land-cover change for a 400-km^2 section in the north-central region used remotely sensed images over a 30-year period, from 1973 to 2003; associated work was conducted on soil properties of sites representing the

(a) (b) (c)

Fig. 12.6 (**a**, **b**, **c**) Comparison of rangeland sites at Harweyu Pastoral Association during May, 2013: (**a**) Site occupied by mature specimens of the invasive bush species *Acacia reficiens*—note the scanty herbaceous understory; (**b**) an adjacent site cleared by pastoralists using hand tools a few months previously—note the recovery of herbaceous plants; and (**c**) close up of native grasses in bush-cleared sites that included high-value *Chloris, Cenchrus,* and *Pennisetum* species (*Photo credits* Brien E. Norton).

Table 12.2 Percent change in land cover for Yabelo District between 1973 and 2003 as interpreted from LandSat images

Land cover class	Area in 1973		Percent change in land cover		
	(km^2)	%	1973–1986	1986–2003	1973–2003
Bushlands	80	20.0	+25.0	+15.0	+43.8
Bushed-grasslands	134	33.5	+20.1	+23.0	+47.8
Grasslands	173	43.3	−38.7	−77.4	−86.1
Croplands	13	3.3	+153.8	+84.8	+384.6

Source Mesele (2006)

main land-cover types (Mesele 2006). Dramatic shifts among grassland, bushland, and cultivated land are evident. Overall, the grasslands declined from 173 to 24 km^2 while the bushed sites increased from 214 to 326 km^2 and cultivation expanded from 13 to 50 km^2 (Table 12.2; Fig. 12.7). Data for soil texture, bulk density, and soil compaction across site types indicated that the bushed sites had soils that were the least permeable to water infiltration when compared to soils at grassland sites. There was also a sequential decline in soil organic-matter content from grassland (3.44 %) to bushed grassland (2.16 %) and bushland (1.36 %) sites. In sum, the soil changes associated with bush encroachment (a higher degree of soil compaction and less herbaceous biomass) were determined to be detrimental to sustained herbaceous productivity (Mesele 2006).

Bush encroachment has been recently shown to have important implications for food security. In an integrated linear programming analysis for Harweyu PA—a site where 95 % of the land suffered from bush encroachment—Forrest et al. (2014) found that a 30 % reduction in bush cover could lead to a similar-sized reduction in food aid requirements for the resident population of over 3000 people. Given large reductions in forage yields caused by invading bush, the consequences of bush encroachment include reductions in milk production from cattle—the traditional dietary staple. Food aid, in part, occurs to help make up deficits in traditional food sources such as cattle milk.

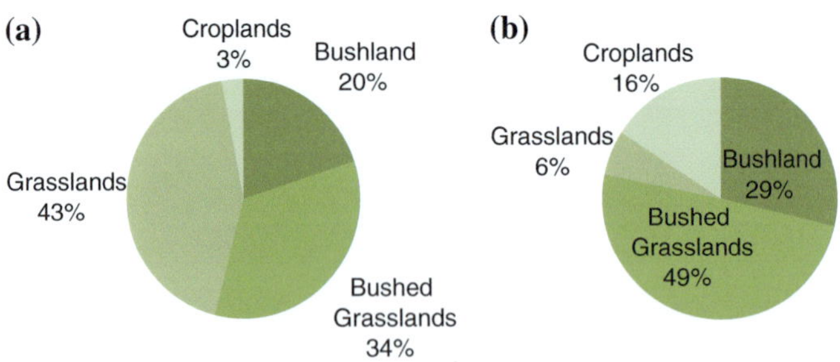

Fig. 12.7 (**a**, **b**) Land-cover composition based on LandSat imagery for 400-km^2 in the Yabelo District in the north-central region of the Borana Plateau in (**a**) 1973 and (**b**) 2003 (*Source* Mesele 2006).

Forrest et al. (2014) noted that the costs of bush clearing are high and unlikely to be borne by pastoral communities themselves. Rather, it is recommended that development agents target donor resources to create endowments of cleared land that are subsequently maintained by the local people. In such cases bush clearing could be combined with fenced fodder banks to assist in a process of grazing management that hinders the rapid return of bush seedlings. Re-introduction of prescribed fire along with bush clearing via human labor with axes and machetes would provide a more cost-effective management strategy (Pratt and Gywnne 1977), but the ever-present challenge is the lack of sufficient fine fuels in dry seasons to carry fires hot enough to kill troublesome bush species (LaMalfa and Coppock 2005).

It has long been known that some encroaching bush species on the plateau can yield valuable charcoal (Coppock 1994). Although charcoal use is pervasive in Ethiopia, charcoal making has been illegal for decades. Officially sanctioned charcoal production has reportedly begun in north-eastern Ethiopia to help control the invasive woody species *Prosopis juliflora* (Kebede 2009), and it can be considered as an option in Borana. The challenge is to adequately regulate charcoal production once it is underway at a suitable scale. Regulations should promote the use of noxious woody species for charcoal production and prohibit the use of valuable species, but enforcement could be a serious problem. In 2014 a pilot project began in Harweyu PA whereby a private firm from Addis Ababa was granted a license to make charcoal. However, this operation closed in less than a year, and the reason is unclear. Combining revenue generation from charcoal production with manual bush clearing and use of fire could be an effective development package for managing the landscape in places like Harweyu PA.

12.3.2 Soil Erosion

Gullying. Currently, there is extensive surface erosion and gullying for the dominant red upland soils in Dikale PA (Fig. 12.8). Gullying has become a serious problem that restricts livestock mobility. It can also lead to a permanent reduction of herbaceous forage production by drying out the upper layers of the soil profile. Gullying reportedly began in Dikale in the late 1990s (Coppock et al. 2014a). While gullying has not yet been carefully studied here, it is postulated that heavy continuous grazing at the local level—combined with severe rainfall events—have been important contributing factors.

Dikale is one of several PAs located in the northern zone of the Borana Plateau. As previously noted, pastoralists first began to reside in the north after large ponds were constructed using heavy machinery in the 1960s. In part, this initiative was intended to expand the settlement area because the central region had become over-populated. The northern zone had previously been a drought fallback area; hence by allowing settlement the utility of the zone as a grazing reserve was lost (Coppock 1994). Dikale PA is thus atypical in that the dry-season water resources are founded on permanent ponds rather than deep wells (described above). It is

(a)

(b)

Fig. 12.8 (**a**, **b**) Gully erosion in the Dikale Pastoral Association in May 2013, with examples of (**a**) early stages of gully formation and (**b**) a deep gully with cattle foraging (*Photo credits* Brien E. Norton).

postulated that the availability of pond water keeps more livestock in the local vicinity year round compared to the situation for the deep wells. Grazing pressure is therefore more chronically intense, contributing to land denudation and the vulnerability of top soil to wind and water erosion.

Pond siltation. Use of Participatory Rural Appraisal (PRA) among the residents of four PAs in 2013 revealed that the most critical need of communities was improved access to drinking water for both people and livestock (Coppock et al. 2014a, b). One intervention priority was to improve the water quantity and quality held in small- to medium-sized ponds that have been traditionally maintained by human labor. The means to achieve this was to reduce excessive pond siltation by protecting catchments from indiscriminate livestock use with bush fencing. The fencing would exclude livestock from all parts of the catchment except for certain entry points to the pond edge where trampling effects on soil would be minimized (Coppock et al. 2014b). Recovered vegetation in protected catchments can help filter sediment in runoff; this should extend pond life by reducing siltation and improve water quality for human consumption (Coppock et al. 2014b). An example of a typical denuded pond catchment is shown in Fig. 12.9, and illustrates another facet of the challenges that arise from excessive soil erosion. Denuded pond catchments and heavily silted ponds are the rule rather than the exception throughout the north-central plateau.

Fig. 12.9 Denuded pond catchment at Dikale Pastoral Association (*Photo credit* Brien E. Norton).

12.4 Altered Use of Communal Grazing Land

Cultivation. The emergence of small-scale cultivation and chronic, localized grazing pressure on the plateau in the 1970s and 1980s has led to important, contemporary challenges for land management. These challenges concern a trend towards increasingly restricted access of many pastoralists to what were once common-pool resources. Cultivated fields—largely planted to maize—have been expanding to provide more human food given the growing imbalance in the ratio between cattle and people (previously noted above). Cultivation can also help reduce the need for the pastoralists to sell animal assets to buy food. Given that crop production here is non-irrigated and dependent on rainfall, it is a high-risk endeavor. For example, the probability of crop failure for a typical plot is about 40 % in non-drought years (Desta and Coppock 2004).

Most cultivated fields occur in valley bottoms and other moist locations. The fields are typically bush-fenced. While definitive research remains to be done concerning who has access to cultivated fields, it has been speculated that prime farming sites tend to be controlled by wealthier pastoralists (Solomon Desta, MARIL, personal communication). It is clear—regardless of ownership—that once cultivation is established it precludes the use of valley bottoms for dry-season grazing by livestock (Coppock 1994). Although crop residues can be consumed by animals, residues have a lower nutritive value than perennial grasses such as *Pennisetum mezianum,* a valuable species typically found on valley bottoms. In addition, crop residues tend to be sparse in quantity and are often depleted before the height of the warm dry season (i.e., January and February), the bottleneck when forage is most needed (Coppock 1994).

Cultivated fields now obstruct livestock travel routes in some cases, hindering herd mobility and contributing to landscape fragmentation (Desta and Coppock 2004). This situation is highly variable throughout the area, however; some sites such as Dikale PA have a sizable area (556 ha) devoted to cropland, while in three others it varies from virtually nil to 143 ha (Coppock, unpublished observations). On a relative basis, the land area covered for each PA is small (less than 2 %), but the location may be important in relation to livestock travel routes or key resources. Communities in general are concerned about the proliferation of cultivation (Desta and Coppock 2004; Coppock et al. 2014a).

Dry-land cultivation has long-been known as a major source of soil degradation —whether via soil nutrient depletion or soil erosion (Holecheck et al. 2001). Importantly, this has never been carefully studied on the plateau. Because many cropping plots are confined to valley bottoms, negative effects on soil properties are assumed to be modest. If plots occur on hillsides or other highly erodible sites this could be problematic, but evidence for this is rare.

How best to manage cultivation has been debated by pastoralists and policy makers here for at least 15 years; a laissez faire approach has prevailed with respect to the control of cultivation given the pastoral system is under pressure and local maize production has been perceived as important for food security and herd asset preservation. Recently, however, a comprehensive land use plan has been approved by government and this may lead to land use regulations (see below). Besides policy, another major factor that could control cultivation is climate change. A warmer, drier climate on the plateau (Funk et al. 2012) could increase rates of crop failure and hence gradually eliminate cultivation as a livelihood option in many locations.

Fodder banks (kalo). In addition to cultivation, there has been an emergence of protected forage sites on the plateau called kalo (Coppock 1994). Atsedu (1990) studied kalo and found them to be very common across six madda that he surveyed. Averaging 12 hectares in size—with a range of 1–80 ha—kalo were typically located on prime grazing sites near olla where they could be supervised. Smaller ones were bush-fenced and larger ones were simply protected by decree. The purpose of kalo is to provide a fodder bank for local calves and sickly cows during the hot dry season. This is a system of deferred grazing, as the plants are allowed to grow unhindered during growing periods. Use of kalo is determined by local leaders in conjunction with community members. Kalo, however, represent a change in traditional land tenure in that the forage resources are annexed from the common property system.

At the time of Atsedu's research, kalo averaged 6.4 years of age. Six years of seasonal protection at that time was enough to cause a shift in composition of the herbaceous layer from forb to grass dominance, which can improve cattle diets given their preference for grasses (Atsedu 1990). The occurrence of kalo has probably expanded here over the past 20 years. Some are now hundreds of hectares in size, and in a few cases managed fire has been implemented to improve forage quality (Fig. 12.10).

Although kalo have important implications for improved land management—and some communities that have developed kalo have been recognized with governmental land-stewardship awards—there is concern about who ultimately controls the largest kalo, and what this portends for broader community access to land. For example, there is evidence that a few wealthier, more powerful pastoral leaders may use kalo to augment forage supplies for commercial livestock production (Napier and Desta 2011). The specter of poorer people being gradually displaced is a concern. The poor now comprise about 56 % of the pastoral population in four recently studied PAs, and yet they control only 31 % of the livestock. The very wealthiest households, in contrast, comprise about 7 % of the population and control at least 20 % of the livestock (Coppock et al. 2014a). The trend is for more households to slide into poverty and food insecurity (Desta and Coppock 2004; Coppock et al. 2014a). At present, however, the numbers and cumulative acreages for kalo across the north-central plateau—while reportedly growing—is anticipated to still be modest (Coppock et al. 2014a).

(a)

(b)

Fig. 12.10 (**a**, **b**) Adjacent range sites in Dikale Pastoral Association, May 2013, showing (**a**) a large, decades-old kalo where restricted and deferred grazing has been practiced and (**b**) a typical location that has been continuously grazed (*Photo credits* Brien E. Norton).

12.4.1 Indicators of Environmental Resilience

We have been unable to conduct a systematic assessment of rangeland condition and trend throughout the north-central Borana Plateau over the past 30 years. Our recent observations in four PAs, however, have revealed high spatial variation in ecological conditions across a relatively small study area (2041 km^2). Some grassland sites—especially in Medecho PA—appear relatively stable in response to intensive grazing by pastoral herds, so it is incorrect to assume that the study area is uniformly degraded. The Dikale PA is one that appears to be the most prone to soil erosion, while the Harweyu PA has been overwhelmed by bush encroachment. In addition to supporting its own local livestock herds, Denbala Bedana PA endures very heavy livestock traffic from across the region during the warm dry season because this PA contains a large cluster of tula wells. All four PAs suffer from denuded pond catchments and high rates of pond siltation. In other words, each PA has unique attributes with respect to environmental condition and trend.

This diversity of observations supports studies by other investigators that have primarily sought to characterize rangeland condition and trends for certain sites and better understand how the Boran perceive environmental change (Dalle et al. 2006; Angassa and Oba 2008). In our case, results from community participatory activities clearly show that the four communities recognize major problems in their production system and are discouraged by the high human and livestock populations in their respective PAs (Coppock et al. 2014a, b). They readily understand that population pressures restrict options to manage natural resources and thus undermine the scope for improvements in their livelihoods (Coppock et al. 2014a). This general picture is bleak.

Despite such findings we have also noticed some surprising elements of rapid ecological recovery and resilience in the plant communities. Research concerning site restoration is in progress (Brien Norton et al., unpublished observations), but initial observations are enlightening. For example, heavily bush-encroached sites at Harweyu—once cleared and protected from continuous grazing—yield impressive stands of perennial grasses (Fig. 12.6). Badly degraded, hard-pan sites north of Dikale—once protected from grazing, lightly tilled with hand tools, and planted with native and appropriate exotic species—show remarkable levels of forage production after just one rainy season (Fig. 12.11). Gullies—once protected from grazing and subject to physical interventions such as check dams and sieve dams—begin to heal (Fig. 12.12). Pond catchments—once protected from grazing—exhibit remarkable recovery patterns for native herbaceous material, especially in the lower portions near the water's edge where soil moisture is readily available (Fig. 12.13). The north-central plateau has outstanding perennial forage grasses, and it seemingly does not require too much intervention to facilitate their re-establishment.

Another local grass species has recently caught our attention, *Cynodon dactylon* (Brien Norton et al., personal observations). This species has a prostate, creeping growth form and is reliably green in a wide variety of field situations (Fig. 12.14). Originally from the Middle East, the species is widespread across the plateau

Fig. 12.11 Adjacent sites at a hard-pan site north of our study area at Jegesa as studied by OARI staff showing (to the *right*) a denuded site prior to intervention, and (to the *left*) effects of site protection from grazing for one year with plantings of local and introduced forage species (*Photo credit* Brien E. Norton).

(Coppock 1994). It is known worldwide as being tolerant to drought, trampling, and grazing. Perhaps most impressive in our case is that we have seen *C. dactylon* proliferating at high-traffic locations in Borana such as livestock holding grounds, entry points for tula wells, and in the immediate vicinity of olla. *Cynodon dactylon* could prove to be an effective, indigenous technology for ecological site restoration here.

12.5 Development Priorities, Policy Frameworks, and Environmental Degradation

12.5.1 Sequence Leading to Environmental Degradation and Land Use Change

It is useful to remind ourselves that the north-central plateau has been grazed by pastoral livestock for centuries. That the system still shows signs of ecological resilience today in several respects is nothing less than remarkable. The key is how

(a)

(b)

Fig. 12.12 (**a**, **b**) Pond catchment sites in Dikale PA where vegetation is recovering following protection from grazing and installation of (**a**) sieve dams on smooth overflow areas or (**b**) check dams in gullies. (*Photo credits* Brien E. Norton).

Fig. 12.13 Recovery of the lower reaches of a pond catchment at Dikale Pastoral Association after only six weeks of protection from grazing during the long rainy season of May 2014 (*Photo credit* Brien E. Norton).

to alter resource use patterns to capture opportunities to enhance system productivity, and this will be described in the next section.

While there may have been times in history when this region was over-populated with people or animals, the classical response would have been for the Boran to expand their resource base and make war on neighbors, if necessary (Legesse 1973). Alternatively, food shortages or disease epidemics could also have helped limit population size (Coppock 1994). However, it seems reasonable to expect that population densities of humans and livestock have never been as high as they are today. Ultimately, it is unsustainable populations of people and livestock that are the drivers for environmental degradation here. The steps leading to the current situation are briefly recapped as follows:

1. Although poorly documented, it seems plausible there was a population surge for people and livestock on the central plateau in the 1950s and 1960s (Coppock 1994). Given no other opportunities to support non-pastoral livelihoods during this time, human population growth had to be internally absorbed. As previously noted, surplus pastoralists expanded into the previously "vacant" region along the northern edge of the plateau—permanent water development was then needed in this area because of the lack of tula wells (Coppock 1994). The USAID effort

(a)

(b)

Fig. 12.14 (**a**, **b**) The low-stature and creeping habit of *Cynodon dactylon*—along with its resistance to grazing, drought, and trampling—makes it an ideal species for restoring cover for exposed soil on the Borana Plateau: (**a**) Close up and (**b**) situation at a high-traffic location (*Photo credits* Brien E. Norton).

to construct Beke Pond and other large water sources in this area during the 1960s was the critical step to address acute humanitarian needs, but it was also the trigger for widespread settlement that eventually led to the pronounced environmental degradation evident today in places like Dikale. Settlement of the northern region also removed a vital drought fallback resource, undermining herd stability in response to future droughts and setting the stage for the boom-and-bust dynamics of the regional cattle herd after 1984. In the 1960s there were also regional vaccination campaigns—it is likely that such campaigns had a positive effect on encouraging growth in animal numbers here (Coppock 1994);

2. By the 1970s and 1980s, bilateral projects like the Third Livestock Development Project (TLDP) were involved in rangeland development that focused on building roads, expanding access to veterinary inputs, and creating the initial elements of cattle markets (Coppock 1994). A key intervention at this point was the use of heavy machinery that was primarily tasked to build roads. This machinery was informally enlisted, however, to assist the Boran to create more medium- and large-sized ponds (Coppock, unpublished observations). This further alleviated seasonal water constraints for livestock throughout the region. Herds could continue to grow, and incentives to market animals were few because the ratio of cattle to people was higher than it is today and there was less need for regular cash income. More milk per capita, for example, translated into less demand to purchase non-pastoral foods. Indiscriminate water development was deemed to be the fundamental cause of increased pastoral settlement and bush encroachment (J.C. Billé et al. cited in Coppock 1994). Bush encroachment has also been encouraged by national policies that were interpreted by local administrators to prohibit the use of managed fire on the rangelands (LaMalfa and Coppock 2005);
3. Food relief initially became common in Borana during the mid-1980s, coincident with the widespread drought in Ethiopia that received worldwide attention. Unlike the famine-stricken farming regions in the northern highlands, however, the pastoral regions to the south suffered much less. One relief camp was located in the north-central plateau. Human mortality from starvation was almost completely mitigated by the timely arrival of food aid (Coppock 1994). Food relief is reportedly pervasive today in our four PAs under study (Coppock et al. 2014a). Food relief is obviously a vital safety net given the surplus pastoralists who have few options to find employment outside of the system. However, another view is that food relief provides disincentives for other forms of problem solving;
4. Cultivation began to proliferate after the 1984-1985 drought when pastoralists had lost most of their animals and needed another source of food; it was common to see maize planted in vacated livestock corrals (Coppock 1994). A laissez faire attitude of government towards cultivation then encouraged cultivation to expand; the driver for this trend has been the declining ratio of

cattle to people (i.e., increasing poverty and food insecurity.) Although it is common that the Boran complain about the negative implications of farming for pastoralism, PAs differ greatly in the extent of cultivation and it's impacts (Coppock et al. 2014a);

5. Kalo have proliferated in response to a growing perception that adequate access to local forage resources in dry periods is uncertain. This is also a symptom of increasing population pressure from livestock and people; and
6. In the last decade more resources have been directed towards expanding some of the tula-well complexes (Coppock, unpublished observations). While development agents have always attempted to help the Boran rehabilitate existing wells, digging new wells has been a more recent activity. This may lead to additional growth in livestock numbers.

12.5.2 Development Priorities and Policy Frameworks

Discerning development priorities and policy frameworks that affect pastoralists in Ethiopia is challenging (Desta and Coppock 2002). This is because there is a lack of transparency in official documentation, as well as bureaucratic inertia, spatial and temporal variation in policy application and implementation, and shifts in national political strategies. And pastoralists have never been a priority for any Ethiopian government.

Empirical observations over the past 30 years (Coppock, unpublished observations) suggest several themes that constitute the broadest and most consistent development priorities and policies for the Borana Plateau: (1) Maintaining physical security and political conformity among the citizenry; (2) promotion of commercial cattle trading, especially trade that generates hard currency from livestock exports; and (3) protection of natural resources, especially with regards to the conservation of remnant wildlife populations and residual patches of woodlands. These themes markedly vary in terms of the level of investment and implementation—the first is by far the most important. The themes also have tended to be top-down in origin.

The fundamental development and policy errors that have undermined the integrity of the natural-resource base—and eventually damaged livelihoods—on the plateau are neither unique to Ethiopia nor to pastoralists. They relate to the government's inability to manage the human population in positive ways.

Lack of economic development. The historically low levels of economic development across Ethiopia have greatly hindered the provision of public services and job creation in all sectors. A generally low level of human-capital development is attributable to poor educational systems, especially prevalent in rural areas (Tekeste 1996). Confronted with the decision of whether or not to leave the Boran pastoral system because traditional livelihoods are largely inaccessible, most pastoralists have little choice but to stay amid poverty and despair. This is due to lack of other options, and this in turn is caused by a lack of employment and little exposure to formal education. Human overpopulation has been the fundamental driver of unsustainable patterns of natural-resource use on the plateau, and this is ultimately linked to a lack of personal choice.

Proliferation of livestock. A second error has been to allow livestock numbers to proliferate. There have been no systematic programmes to provide creative incentives for herd owners to periodically destock by marketing animals and diversifying their income sources and capital assets. The situation has led to unsustainable levels of grazing that gradually undermine the productive basis of the pastoral system. This has resulted from laissez-faire attitudes towards the condition and trend of forage, soil, and water resources as well as an inability to regulate resource use. The growth in livestock numbers has been driven by human population growth in tandem with unfettered water development over the past 40 years. Larger herds may promote prosperity for a minority of herd owners when buffeted by drought and other calamities, but for most households the outcome is a downward spiral because the regional forage base is simply insufficient to support the current densities of people.

Importantly, government per se is not proposed here as an agent that has attempted to phase out traditional pastoralism via aggressive settlement plans or land-annexation policies, as has been observed in other settings (FAO 2001). Perspectives from district-level government[8] indicate that population and social factors are the major reasons as to why pastoralists have settled. However, despite that more pastoral households are settling, it is notable that in most cases their livestock can still be highly mobile. Part of a household, for example, can reside in town and take advantage of public services while other family members herd animals to remote locations to find feed and water (Coppock and Desta 2013). The role of government has been to assist pastoralists to adapt to their settled circumstances, as possible. Government does not have a sedentarization policy on the plateau.

In recent years—the district government in collaboration with NGOs—has re-configured or re-sited some olla to achieve more efficient use of local grazing and water resources. For example, some olla that are too close to key water points have been moved further away. Paddocks have been established to support community concepts for grazing rotations. Such settlement patterns are thus being "rationalized" in 14 of 23 PAs in Yabelo District (Coppock, unpublished observations.)

After many years of debate, a national land-use plan was finalized in 2011. Now a land-use plan for Yabelo District has also been finalized in 2013, but it is not yet released. The plan for Yabelo District is focused on natural resource conservation and settlement patterns (Coppock, unpublished observations). The land-use plan will be implemented by the district office, and it will provide a basis for the regulation of cultivation in collaboration with PA leaders, as one example. In places

[8]Mr. Indaalee Haptee, Yabelo District Office for Land Use Planning and Environmental Protection, personal communication 2014.

where cultivation is a sound and sustainable option, it will be strengthened. Where cultivation should be discontinued, it will be stopped; if a warming and drying of the plateau continues (Funk et al. 2012) rain-fed cultivation may slowly vanish, regardless. The land-use plan also can help in conflict management among neighbors based on competition for dwindling natural resources. This should reduce ambiguities in the rights and responsibilities among adjacent PAs. Essentially, the planning tools are an effort to improve pastoral governance.

12.6 Is Environmental Degradation Reversible?

The question of whether patterns of rangeland degradation can be slowed or reversed is important, but any positive effects of technical intervention will ultimately be constrained by continued population growth and persistent poverty. This becomes a vicious cycle. Attention therefore needs to be given to providing education and opportunities for pastoralists to leave the pastoral sector, if given a choice. This is an important means to increase the ratio of cattle to people given the limits on cattle numbers, and release more food to those who remain in the system.

The other means to increase the ratio of cattle to people is to boost herbaceous forage productivity. Evidence previously reviewed suggests that much of the rangelands in the north-central region are ecologically resilient—grass forage for cattle can be recovered in some cases via deferred grazing (kalo), bush clearing, and renovation of eroded locales. In tandem, this could be all regarded as a form of sustainable intensification for increased productivity (Pretty et al. 2011). While it is promising that we have observed a positive and rapid response of some forage grasses to grazing protection, equally important is the question as to whether innovative management systems could be adopted by the pastoralists.

While the gradual decline of herd mobility has been well documented for over 30 years and it is an important concern, it remains unclear what can be done about it. Herd mobility and access to key resources are affected by population growth as well as local, regional, and international politics. For example, herd mobility can be ultimately constrained by enforcement of national borders (i.e., between Ethiopia and Kenya) as well as enforcement of state boundaries (i.e., between the Oromia Regional State and the Somali Regional State in southern Ethiopia.) Such factors are typically beyond the control of the Ethiopian Boran or their advocates. There are, however, other efforts underway to restore traditional rights of grazing access on the plateau at spatial scales of resolution that greatly exceed those of madda or the PAs. These units are referred to in the Oromifa vernacular as deedha and reera. How the rights and responsibilities of users have become eroded for deedha and reera is unclear, but the value of re-establishing traditional patterns of grazing access at such scales is a key priority of the recently initiated PRIME project (Solomon Desta, MARIL, personal observation).

12.7 The Way Forward: How Should This Region Be Managed?

The Borana Plateau has a storied history of being the homeland for a highly organized pastoral society that has exhibited wise stewardship of the natural resources over many generations. Recent evidence, however, indicates that portions of the rangeland environment have changed for the worse and that a growing human population is the ultimate driver. The fundamental challenge for the proper management of common property regimes is over-population (Hardin 1968), an old lesson that still holds here.

Management of this system should therefore embrace a holistic perspective that fundamentally places people and their needs at the forefront. In other words, change agents need to help empower pastoralists to better control their own destiny. This was recognized over 20 years ago by Coppock (1994) but has yet to take hold in a widespread fashion. Similar proposals have been advanced under the banner of "endogenous livestock development" whereby indigenous knowledge is combined with modern insights and technologies as a basis for progress (Homann et al. 2008). The steps of such an approach would include the following:

1. Assist pastoralists to leave the rangelands where possible. Some people will prefer to stay in the traditional system, but some will rather depart. Development is about providing choices for people. There is a fundamental need to provide appropriate education, skills training, and job opportunities that allow some people to leave the system and thus reduce the demand on natural resources. In recent years we have increasingly seen pastoral youths enroll in school and even attend university. Until such processes are aggressively underway, attempts to improve natural resource management will eventually be negated by the ever-growing pressure to overharvest the rangelands;
2. Similarly, promote diversified livelihoods among pastoralists where possible. There are exciting examples where pastoralists have cleverly diversified their livelihood activities to include pastoral and non-pastoral income streams and assets (Coppock et al. 2011; Coppock and Desta 2013). Often all that is needed is for people to receive some careful mentoring, role modeling, and seed capital to begin new endeavors;
3. One part of diversified livelihoods also involves improving the access of pastoralists to livestock markets. More options to buy and sell stock is always a good thing—it can help promote wise destocking when the environment is less able to support grazing livestock;
4. Assist pastoral communities in diagnosing their natural-resource management problems at a local and regional scale, and empower them to create (or recreate) enforceable rules and regulations that underpin sustainable resource use. Today we see Borana society in flux—ultimately as a result of population pressure. There are emerging trends that appear to pit wealthier herd-owners versus poorer herd-owners, or commercially-oriented producers versus traditional producers.

This primarily involves aspects of land privatization and control, much of which would never have been sanctioned by traditional leaders decades ago. The pastoral society may benefit from the wise input of third parties that can help articulate problems and codify solutions that promote equity and wise resource use. Traditional leaders and policy makers can be engaged in this process. Oba and Kotile (2001) contend that indigenous ecological knowledge among the Boran gives highly correlated results with range condition and trend assessments from rangeland professionals, thus indigenous knowledge can be a useful input when devising landscape-level rehabilitation plans;

5. Actively strengthen local governance. This can include revitalization of customary institutions (e.g., Gumi Gaayo) along with better integrating these with PA structures to better solve problems. Both forms of governance must co-exist in the modern world. This can include facilitating group access to key technologies and training, developing group constitutions and bylaws, and even business plans. We and others have been engaged in such efforts on a pilot basis where community groups are being empowered to deal with bush clearing, rangeland restoration, gully management, and pond de-siltation; and
6. Promote restoration of herd mobility where possible, primarily by relaxing or removing barriers for livestock use of key water points and neighboring territories having strategic grazing resources. This will continue to be a formidable challenge, however, for reasons previously mentioned.

In sum, the current need to improve natural resource management in pastoral areas will not be markedly affected by more conventional research—or conventional development tactics, for that matter. Rather, the need is for change agents to form complementary partnerships and help empower pastoralists so they can better fend for themselves. However, research can still document empowerment efforts, success stories, and reflect on lessons learned. This research needed is best described as participatory action research (Whyte 1989), "research for development" (Ashby 2003), or innovation systems research (Coppock et al. 2009) where researchers and development agents collaborate closely with pastoralists in a process of stepwise problem-solving and mutual learning.

Acknowledgements The author appreciates support for field research provided from the following sources: (1) The International Livestock Center for Africa (1980-91); (2) Utah State University (1991-2014 on a discontinuous basis); (3) The Rockefeller Foundation (1995-1997); (4) the Global Livestock Collaborative Research Program (GL-CRSP) of the US Agency for International Development (USAID) Global Bureau (Grant no. PCE-G—00-98-00036-00; 1997-2009); and (5) the Feed the Future Innovation Laboratory: Adapting Livestock Systems to Climate Change of USAID Global Bureau (Grant no. EEM-A-00-10-00001; 2012-2015). Professional colleagues and pastoral community members who contributed to the work described in this chapter are warmly acknowledged. Dr. Brien E. "Ben" Norton is especially thanked for providing his excellent photographs and his important insights regarding rangeland degradation and resilience. This publication was made possible through support provided by the Bureau for Economic Growth, Agriculture, and Trade, USAID, under the terms of the two grants listed above. The opinions expressed herein are those of the author and do not necessarily reflect the views of USAID or the US government.

References

Angassa, A., & Beyene, F. (2003). Current range condition in southern Ethiopia in relation to traditional management strategies: The perceptions of Borana pastoralists. *Tropical Grasslands, 37*, 53–59.

Angassa, A., & Oba, G. (2008). Herder perceptions on impacts of range exclosures, crop farming, fire ban, and bush encroachment on the rangelands of Borana, southern Ethiopia. *Human Ecology, 36*, 201–215.

Ashby, J. (2003). Introduction: Uniting science and participation in the process of innovation—research for development. In B. Pound, S. Snapp, C. McDougall, & A. Braun (Eds.), *Managing natural resources for sustainable livelihoods—uniting science and participation* (pp. 1–19). London: Earthscan Publications Ltd.

Atsedu, M. (1990). *Ecology of calf pastures and supplementary feeding by Borana pastoralists of southern Ethiopia*. Master's thesis. Fort Collins: Department of Range Science, Colorado State University.

Boru, D., Schwartz, M., Kam, M., & Degen, A. (2014). Cattle reduction and livestock diversification among Borana pastoralists in southern Ethiopia. *Nomadic Peoples, 18*(1), 115–145.

Coppock, D. L. (1994). The Borana plateau of southern Ethiopia: Synthesis of pastoral research, development and change, 1980–91. Systems Study No. 5. Addis Ababa: International Livestock Centre for Africa. https://books.google.com/books?hl=en&lr=&id=TeLRgwQU1HEC&oi=fnd&pg=PR10&dq=Coppock+Borana+Plateau+&ots=0r1Gs3pWdW&sig=flHKtzvRUe9KwG9rhpAjGt3s6F8#v=onepage&q=Coppock%20Borana%20Plateau&f=false

Coppock, D. L., & Desta, S. (2013). Collective action, innovation, and wealth generation among settled pastoral women in northern Kenya. *Rangeland Ecology and Management, 66*, 95–105.

Coppock, D. L. Gebru, G., Mesele, S., & Desta, S. (2008). Are drought-related crashes in pastoral cattle herds predictable? More evidence of equilibrium dynamics from the southern Ethiopian rangelands. In *Multifunctional grasslands in a changing world* (Vol. II). Organizing Committee of the 2008 International Grasslands Congress/International Rangelands Congress (IGC/IRC), 1090. Beijing: Guagdong People's Publishing House.

Coppock, D. L., Desta, S., Tezera, S., & Gebru, G. (2009). An innovation system in the rangelands: using collective action to diversify livelihoods among settled pastoralists in Ethiopia. In A. Waters-Bayer, C. Wettasinha, J. Njuki, P. Sanginga, & S. Kaaria (Eds.), *Innovation Africa: Enriching farmer's livelihoods* (pp. 104–119). London: Earthscan Publications.

Coppock, D. L., Desta, S., Tezera, S., & Gebru, G. (2011). Capacity building helps pastoral women transform impoverished communities in Ethiopia. *Science, 334*(6061), 1394–1398.

Coppock, D. L., Tezera, S., Eba, B., Doyo, J., Tadele, D., Teshome, D., Husein, N., & Guru, M. (2014a). *Preliminary results from participatory rural appraisals (PRAs) and follow-up investigations held at four pastoral associations on the north-central Borana Plateau, Ethiopia*. Logan: Department of Environment and Society, Utah State University. http://digitalcommons.usu.edu/envs_facpub/900

Coppock, D. L., Tezera, S., Eba, B., Doyo, J., Tadele, D., Teshome, D., Husein, N., & Guru, M. (2014b). *Sustainable pastoralism in Ethiopia: Preliminary results from participatory community assessments on the north-central Borana Plateau*. Research Brief-16-2014, Feed the Future—Adapting Livestock Systems to Climate Change, Colorado State University, Fort Collins, CO, USA. 4 pp. http://digitalcommons.usu.edu/envs_facpub/902

CSA (Central Statistical Agency of Ethiopia). (1994). *National Census*. www.csa.gov.et

CSA (Central Statistical Agency of Ethiopia). (2007). *National Census*. www.csa.gov.et

Dalle, G., Maass, B., & Isselstein, J. (2006). Rangeland condition and trend in the semi-arid Borana lowlands, southern Oromia, Ethiopia. *African Journal of Range & Forage Science, 23*(1), 49–58.

Desta, S., & Coppock, D. L. (2002). Cattle population dynamics in the southern Ethiopian rangelands, 1980–97. *Journal of Range Management, 55*, 439–451.

Desta, S., & Coppock, D. L. (2004). Pastoralism under pressure: Tracking system change in southern Ethiopia. *Human Ecology, 32*(4), 465–486.

Desta, S., Gebru, G., Tezera, S., & Coppock, D. L. (2006). Linking pastoralists and exporters in a livestock marketing chain: Recent experiences from Ethiopia. In J. McPeak & P. Little (Eds.), *Pastoral livestock marketing in eastern Africa: research and policy challenges* (pp. 109–127). Rugby: ITDG Publishing.

Evangelista, P., Young, N., & Burnett, J. (2013). How will climate change spatially affect agricultural production in Ethiopia? Case studies of important cereal crops. *Climatic Change*. doi:10.1007/s10584-013-0776-6

FAO (Food and Agricultural Organization of the United Nations). (2001). Pastoralism in the New Millennium. *Animal Production and Health Paper No. 150*. Rome: UNFAO. http://www.fao.org/docrep/005/y2647e/y2647e00.HTM.

Forrest, B. (2014). *Analysis of building resiliency in an Ethiopian pastoral system: Mitigating the effects of population and climate change on food insecurity*. MS thesis. Logan: Department of Applied Economics, Utah State University.

Forrest, B., Bailey, D., Ward, R., & Coppock, D. L. (2014). *Can bush-clearing, deferred grazing, or camels help mitigate climate-change and population effects for Borana pastoralists? An Economic Analysis of Potential Interventions*. Research Brief-19-2014, Feed the Future—Adapting Livestock Systems to Climate Change, Colorado State University, Fort Collins, CO, USA. 4 pp. http://digitalcommons.usu.edu/envs_facpub/901/

Funk, C., Rowland, J., Eilerts, G., Kebede, E., Biru, N., White, L., & Galu, G. (2012). *A climate trend analysis of Ethiopia. Fact Sheet 2012–3053*. Published by the United States Geological Survey in collaboration with the Rolla Publishing Service Center.

Hardin, G. (1968). The tragedy of the commons. *Science, 162*, 1243–1248.

Holecheck, J., Pieper, R., & Herbel, C. (2001). *Range management—principles & practices* (4th ed.). Upper Saddle River: Prentice Hall.

Homann, S., Rischkowsky, B., Steinbach, J., Kirk, M., & Mathias, E. (2008). Towards endogenous livestock development: Borana pastoralists' responses to environmental and institutional changes. *Human Ecology, 36*, 503–520.

Jahnke, H. (1982). *Livestock production systems and livestock development in tropical Africa*. Kiel: Kieler Wissenschaftsverlag Vauk.

Kebede, A. (2009). *Sustaining the Allideghi grassland of Ethiopia: Influences of pastoralism and vegetation change*. Ph.D. Dissertation. Logan: Department of Wildland Resources, Utah State University.

LaMalfa, E., & Coppock, D. L. (2005). Rebuilding the capacity to burn: Community development and rangeland productivity. Pages 8–11 in the Spring Issue of *Ruminations—Newsletter of the Global Livestock Collaborative Research Support Program*. University of California, Davis.

Legesse, A. (1973). *Gada—three approaches to the study of African society*. New York: The Free Press.

McPeak, J., Little, P., & Doss, C. (2012). *Risk and social change in an African rural economy: livelihoods in pastoralist communities*. New York: Routledge.

Megersa, B., Regassa, A., Kumsa, B., & Abunna, F. (2008). Performance of camels *(Camelus dromedarius)* kept by pastoralists with different degrees of experience in camel-keeping in Borana, southern Ethiopia. *Animal Science Journal, 79*(4), 534–541.

Mellisse, B. (2014). *Implications of constrained mobility on livestock production and pastoral livelihoods of the Borana Plateau, southern Ethiopia*. TIRI Research Brief-02-2014, Feed the Future—Adapting Livestock Systems to Climate Change, Colorado State University, Fort Collins, CO, USA. http://www.vivo.colostate.edu/lccrsp/briefs/TIRI%20RB-02-2014.pdf

Mesele, S. (2006). *Land use dynamics and impact on selected physiochemical properties of soils in Yabelo woreda of the Borana lowlands, southern Ethiopia*. Master's thesis. Harer: Faculty of Rangeland Management, Alemaya University.

Napier, A., & Desta, S. (2011). *Review of pastoral rangeland enclosures in Ethiopia*. Boston: Feinstein International Center Publications. http://fic.tufts.edu/publication-item/review-of-pastoral-rangeland-enclosures-in-ethiopia/

Oba, G., & Kotile, D. (2001). Assessments of landscape level degradation in southern Ethiopia: Pastoralists versus ecologists. *Land Degradation and Development, 12*, 461–475.

Oba, G., Post, E., Syvertsen, P., & Stenseth, N. (2000). Bush cover and range condition assessments in relation to landscape and grazing in southern Ethiopia. *Landscape Ecology, 15*, 535–546.

Pratt, D., & Gywnne, M. (1977). *Rangeland management and ecology in east Africa*. London: Hodder and Stoughton.

Pretty, J., Toulmin, C., & Williams, S. (2011). Sustainable intensification in African agriculture. *International Journal of Agricultural Sustainability, 9*(1), 5–24.

Solomon, T., Snyman, H., & Smit, G. (2007). Cattle-rangeland management practices and perceptions of pastoralists towards rangeland degradation in the Borana zone of southern Ethiopia. *Journal of Environmental Management, 82*(4), 481–494.

Tekeste, N. (1996). *Rethinking education in Ethiopia*. Uppsala: Nordiksa Afrikainstitet.

Thornton, P., Jones, P., Ericksen, P., & Challinor, A. (2011). Agriculture and food systems in sub-Saharan Africa in a 4 C+ World. *Philosophical Transactions of the Royal Society A,*. doi:10.1098/rsta2010.0246.

Wassie, Y. (2014). *Impacts of change in customary rangeland governance institutions on pastoral livelihoods of the Borana Plateau, southern Ethiopia*. TIRI Research Brief-05-2014, Feed the Future—Adapting Livestock Systems to Climate Change, Colorado State University, Fort Collins, CO, USA. http://www.vivo.colostate.edu/lccrsp/briefs/TIRI%20RB-05-2014.pdf

Whyte, W. (1989). Advancing scientific knowledge through participatory action research. *Sociol Forum, 4*, 367–385.

Chapter 13
Humans Cause Deserts: Evidence of Irreversible Changes in Argentinian Patagonia Rangelands

Gabriel Oliva, Juan Gaitan and Daniela Ferrante

Abstract Argentinian Patagonia is a 750,000 km^2 semi-arid region. Sheep were introduced there in the late 19th century, peaked at about 20 million head in 1960 and fell sharply thereafter reaching 8 million at present. Decline in productive capacity was blamed on overgrazing or climate change alternatively, but maps in 1998 indicated that 34 % of the region was severely desertified. We analysed the regional FPAR in the GIMMS NDVI dataset from NOAA satellite imagery in the last 30 years, a period in which the land has been destocked and under generally improved management. The trend of FPAR does not show a generalized "greening", or recovery process, even when the rainfall trend was removed. A case study in Los Pozos Farm in the south shows that total stock has fallen monotonically since 1930 under a rainfall regime that remained constant. Variable stocking rates that tracked the productivity introduced in 1990 stabilized grazing capacity but did not induce a clear recovery in forage production. The erosion of unstable soils and damage to perennial vegetation probably explain the farm production losses, and management rather than climate change is the most probable cause. Production analysis shows that stocking rates that maximize short-term income in wool are too high and will probably damage the land. Good management has the potential to stabilize and maybe slowly revert long-term mismanagement effects on these cold desert ecosystems, but under the present market economy and land tenure system it seems unlikely that producers will give up profits to avoid possible long-term degradation.

Keywords Desertification · South America · Patagonia · Semiarid · Grazing · Livestock · NDVI · AVHRR

G. Oliva (✉) · D. Ferrante
INTA EEA Santa Cruz, Mahatma Gandhi 1322, 9400, Río Gallegos Santa Cruz, Argentina
e-mail: geoliva@speedy.com.ar

G. Oliva · D. Ferrante
Universidad Nacional de la Patagonia Austral, Unidad Académica Río Gallegos. Av. Gregores y Piloto "Lero" Rivera—CP (Z9400KVD), Santa Cruz, Argentina

J. Gaitan
INTA EEA Bariloche, Modesta Victoria 4450, 8400, Río Negro Bariloche, Argentina

R.H. Behnke and M. Mortimore (eds.), *The End of Desertification?*
Springer Earth System Sciences, DOI 10.1007/978-3-642-16014-1_13

13.1 Introduction

Argentinian Patagonia is a 750,000 km^2 region in south Argentina (Fig. 13.1). It is mostly semi-arid and covered by shrub or grass steppes. Pastoral use of these rangelands is relatively recent and even today there are less than 2.3 inhabitants per square kilometer (INDEC 2001).

Colonization of this vast land took place in the late 1880s, after military campaigns and mainly with Argentine and Chilean 'criollos', locally born people of Spanish ancestry but also to settlers of European or Middle-Eastern descent. Native peoples were severely reduced in number after colonization and nowadays constitute about 1.4–3.5 % of the residents. Early colonization (1880–1900) gave access to pastoral leases in large areas of the most productive or readily accessible land. At the beginning of the 20th century, poorer settlers took part in the colonization of arid or inaccessible areas. The land was divided geometrically into parcels of about 10–20 thousand hectares (Barbería 1995), and freehold rights consolidated land tenure of most of the big stations, locally called 'estancias'. Poorer settlers, including some indigenous people, occupied small areas that remained mostly under informal or traditional land use, a great number of them

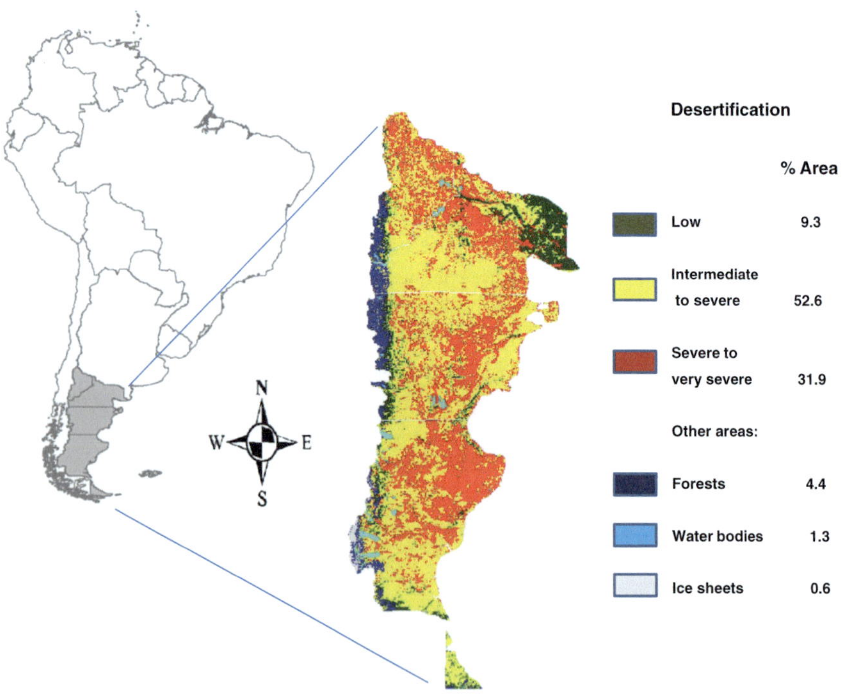

Fig. 13.1 Location of the Argentinian Patagonia in South America (*left*) and desertification degrees based on Del Valle et al. (1998) map.

unfenced, and mostly located in Northern Patagonia. A good description of the area and the sheep industry may be found in (Cibils and Borrelli 2005) (Photo 13.1).

Concurrently with the well-known land degradation crisis in Africa's Sahel, in the late 1970s Patagonia experienced a crisis that reduced sheep numbers drastically affecting the base of the economy of rural areas, and impacting the livelihoods of people. The causes of the loss of production capacity and degradation of rangelands have been debated (DHV-SWEDFOREST 1998). As in the Sahel case, some stakeholders emphasized that overgrazing was to blame, and others pointed out that climate had triggered the process. In 1990, Argentina joined the international programmes to evaluate desertification, and focused mainly on the Patagonian region (Goergen 1995). A desertification map of Patagonia shown in Fig. 13.1 was produced following the FAO guidelines. Desertification is defined in this work as the degradation of land due to several factors including climate and management. It was assessed by means of indicators of water and wind erosion, salinization of soils and modification of vegetation (Photo 13.2). An extrapolation of detailed 1:250.000 scale maps that were made throughout the region reached the conclusion that 34 % of the region was severely affected (Photo 13.3) (Del Valle et al. 1998).

Sheep numbers, that had reached a maximum of about 20 million head around 1950 (Mendez Casariego 2002) fell steadily starting in 1978, and dropped by half in 20 years (Fig. 13.2). A small increase in cattle numbers was registered at that time, but it did not compensate the losses in sheep.

Photo 13.1 Sheep are tended in the Magellan Steppes, south Patagonia (Horacio Cordoba).

Photo 13.2 Sand dunes associated with coarse textured quaternary deposits and clay dunes in the Central Plateaus are common and poor management often makes them active.

Photo 13.3 Sand dunes affecting Magellanic steppe.

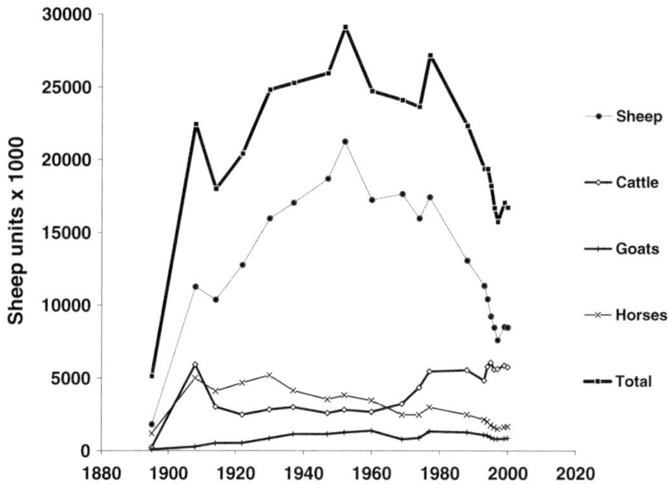

Fig. 13.2 Evolution of livestock in Patagonia in the 1890–2002 period in Sheep Units (SU) Herbivore equivalences were 1 cattle unit = 6 SU, 1 goat = 1 SU, 1 horse = 6 SU. The full, *bold line* indicates the cumulative total. Drawn from data of Mendez Casariego (2002).

The decreasing number of sheep at regional scale was mostly the consequence of low weaning rates that prevented farmers from replacing ageing animals. Losses were heavier in some areas such as Santa Cruz province, that at the turn of the 20th century had a very low herbivore stocking rate that was equivalent to 0.16 sheep $ha^{-1}year^{-1}$ (one sheep per 6 ha), while Neuquén in the north and Tierra del Fuego in the south had 0.64 and 0.83 sheep $ha^{-1}year^{-1}$ respectively (one sheep per 1.5–1.2 ha) (Fig. 13.3). The grazing pressure of guanacos, the only large native herbivore was negligible at that time (Fig. 13.3). Differences in remaining stock in Fig. 13.3 reflect the effects of the crisis on different types of producers: while those in Santa Cruz are mostly commercial with a mean flock size of about 4000 sheep and freehold rights to the land, typical small rural producers of Neuquén are subsistence-type, with a flock of 200–300 sheep and goats. When faced with dwindling animal production and natural disasters such as volcanic ashes or very severe snowfalls, producers of big stations simply walked off the land, which still remains ungrazed by domestic herbivores. In contrast, those in small allotments responded to hardships with other types of strategies and remained on the land, tending their small flocks. Big stations in Tierra del Fuego with a mean of 10,000 sheep and moister climate did not suffer comparable land degradation at this time and retained their stock.

Concerns about overgrazing and degradation led government agencies to develop range evaluation methods to estimate forage biomass and adjust stocking rates (Borrelli and Oliva 1999; Siffredi and Becker 1999; Borrelli 2001; Borrelli and Oliva 2001; Elissalde et al. 2002). They were applied widely in desertification programmes in the early 90s and have become mandatory for stations that requested

Photo 13.4 Sheep grazing during a long trip from summer camp to wintering areas (Francisco Milicevic).

financial help from the government. Nowadays about 6 million hectares, close to 10 % of the region have been surveyed, mapped, and has a formal grazing management plan in place. Grazing plans are designed to adjust animal numbers to forage availability, keeping the traditional, continuous year-round grazing system (Borrelli 2001).

Given that management is supposed to be the main cause of degradation, the drastic reduction in total regional inventories since the 1980s and the application of

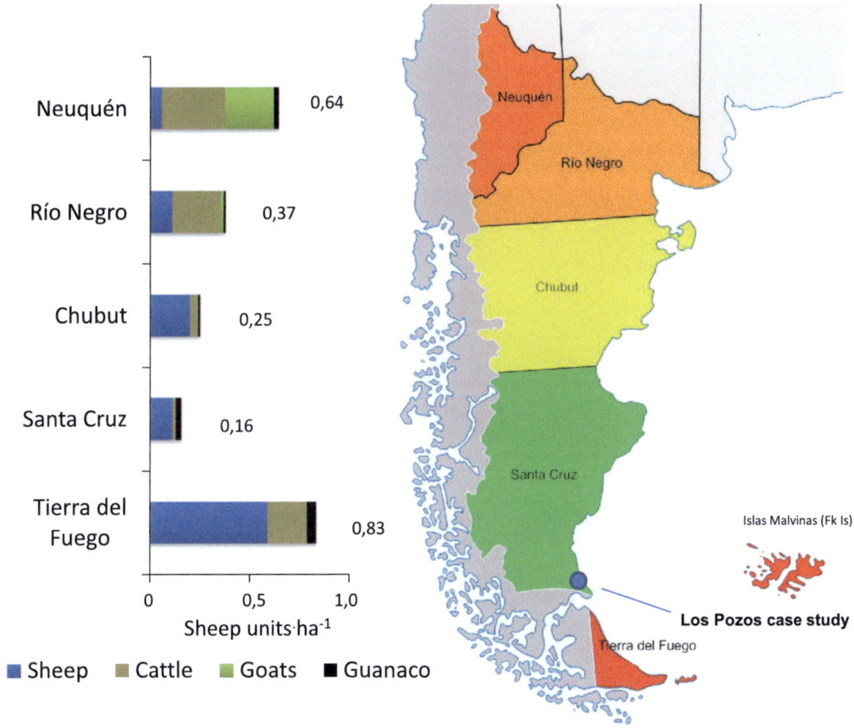

Fig. 13.3 Total estimated herbivore stocking rates in the year 2000 across the provinces of the Region in sheep units (SU) ha^{-1}. Herbivore equivalences as in Fig. 13.2, guanaco numbers from Amaya et al. (2001). The *circle* indicates the location of Los Pozos station, our study case at a station-scale.

better management plans should have stopped the process and possibly triggered recovery of the land, somewhat like the "greening of the Sahel" that has been reported after the desertification events in the 1960s (Herrmann et al. 2005). An important difference in this case is that climate changes for Patagonia in the last three decades have been variable throughout the region: According to the Global Precipitation Climatology Project (GPCP) data-set (Adler et al. 2003), south Patagonia showed positive rainfall trends in the 1980–2010 period, no change was registered in the center and a reduction was experienced in the Andean NW (Fig. 13.4).

The objective of this paper is to review rangeland trends in the last three decades at two distinct scales: regional and local. At the regional scale, satellite data were used to detect "greening" or "browning" processes. Due to the different trends in rainfall over the period and considering its impact on vegetation growth, an attempt was made to remove the precipitation signal from the satellite data time series. This type of analysis highlights areas that displayed a negative or positive trend independently of the local climate (Evans and Geerken 2004). At the local scale, a sheep

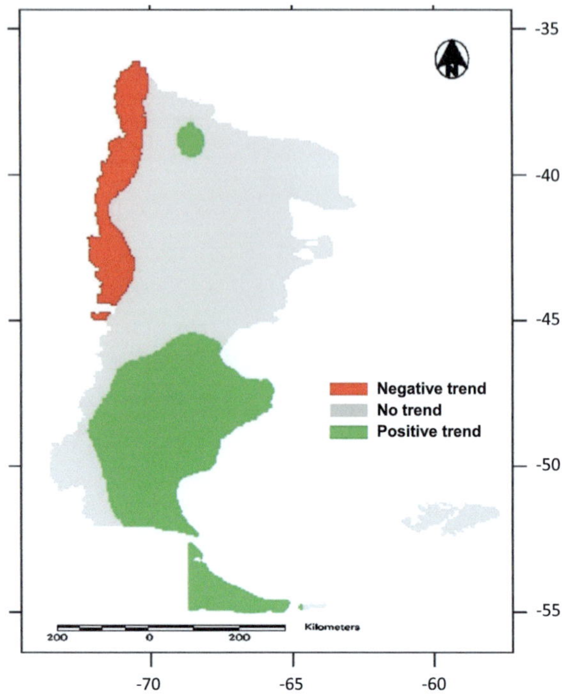

Fig. 13.4 General trends in rainfall in Patagonia. Data from the Global Precipitation Climatology Project (*GPCP*) data-set (Adler et al. 2003).

station case study was analysed. The station had been gradually reducing animal numbers since 1930 and the managers decided to adjust management drastically in the late 1980s, lowering stocking rates and adjusting them annually thereafter, tracking the grazing capacity of the rangelands. Given that the station kept unusually detailed productive and climatic records since 1930, a description of the long-term trends and the response to management was possible.

The study area was the Argentine portion of Patagonia, South America between 39 and 55°S (Fig. 13.1).

13.1.1 Regional Analysis from Remotely-Sensed Data

Regional trends were studied using vegetation indices (VIs) that relate to ecosystem structure and function, particularly biomass (e.g. Tucker et al. 1985), leaf area index (e.g. Steltzer and Welker 2006), net primary productivity (e.g. Paruelo et al. 1997) and vegetation cover (e.g. Gaitán et al. 2013). In particular, the Normalized Difference Vegetation Index NDVI (Tucker 1979), based on red and near-infrared reflectance was used. The trends in NDVI have been used as a proxy for greening or browning (de Jong et al. 2011), and Gaitán et al. (2013) found that they are a suitable tool to estimate functional and structural ecosystem attributes in Patagonia.

Remotely sensed data were obtained from "Advanced Very High Resolution Radiometer" (AVHRR) sensors, on board of the "National Oceanic and Atmospheric Administration" (NOAA) satellites that provide one of the most extended time series of remotely-sensed data for long-term observations. The main difficulties of the NOAA-AVHRR series stem from navigation uncertainties, sensor degradation and calibration, atmospheric contamination, orbital drift and limited spectral and directional sampling. In order to improve data quality, NDVI data series that follow different processing streams are available. One of the most commonly used long-term NDVI data sets was processed by the Global Inventory Modeling and Mapping Studies (GIMMS). The latest version of GIMMS, termed the Third Generation NDVI Data Set (NDVI3g) has been recently produced with AVHRR sensor data from NOAA's 7–18 satellites, and is available at 8 km horizontal resolution every 15 days from July 1981 to December 2011. GIMMS processing includes corrections for sensor calibration, volcanic aerosols, and other sensor degradation and contamination issues. It is the only continuous and up-to-date global NDVI dataset that is continually assessed and validated (Fensholt et al. 2012). GIMMS is generally considered the best dataset to use for the analysis of long-term trends (Beck and Goetz 2011). The GIMMS NDVI product has been used previously to diagnose and detect land degradation and greening trends (e.g. Herrmann et al. 2005).

A product derived from NDVI3g: the Fraction of Photosynthetically Active Radiation absorbed by vegetation (fPAR) data set (Zhu et al. 2013) was used. The fPAR is a relative measure of the vegetation-absorbed radiation in the 0.4–0.7 μm spectral region of solar radiation, and hence, characterizes the energy that is potentially used in the process of photosynthesis. The derived fPAR product is generated based on a neural network algorithm that uses the NDVI3g and an improved version of Collection 5 Terra MODIS fPAR, the MODIS BNU product (Beijing Normal University version). The fPAR can be readily interpreted within an ecological context, while NDVI itself is a ratio of spectral reflectances that is a proxy for vegetation activity.

The growing season (October to March) mean fPAR (fPAR mean) was analysed in order to describe temporal trends in green biomass. A linear regression between time (years) and fPAR mean was estimated for each pixel per decade (1981–82 to 1990–91, 1991–92 to 2000–01 and 2001–02 to 2010–11) and for the entire period (1981–82 to 2010–2011) (Fig. 13.5). The positive or negative slope of this line is considered to reflect an increase or decrease in green biomass.

Rainfall measurements from rain gauge stations are conventionally considered the most accurate and reliable source of rainfall data, but in Patagonia rain gauges are sparse and of varying reliability. For climate data we therefore used gridded satellite precipitation estimates instead. They combine satellite observations from different sources and ground measurements: the Global Precipitation Climatology

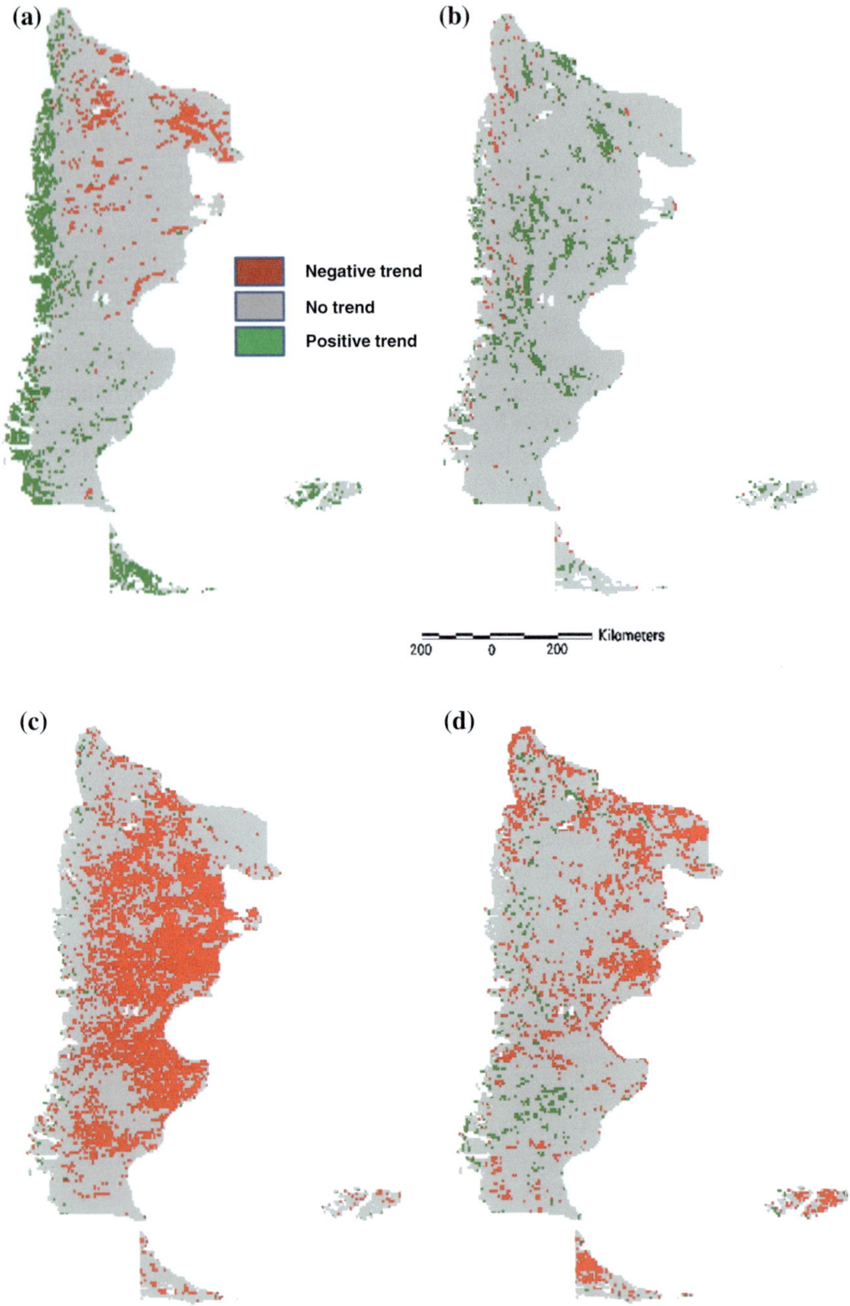

Fig. 13.5 Overall trends in vegetation greenness throughout the decades 1981–82 to 1990–91 (**a**), 1991–92 to 2000–01 (**b**), 2001–02 to 2010–11 (**c**) and the whole period (1981–82 to 2010–2011) (**d**) based on growing season mean fPAR (October to march) of AVHRR GIMMS fPAR3g time series. Significant trends: $p < 0.05$.

Project (GPCP) data-set (Adler et al. 2003). The GPCP data-set consists of global monthly precipitation estimates produced in a 2.5° latitude-longitude grid for the period 1979–present. We obtained the cumulative rainfall (from July to June) for each growing season. These data were interpolated using the Kriging method in order to match the 8 km resolution of fPAR data-set.

A linear regression analysis through the time series for each pixel was carried out with fPARmean as the dependent variable and cumulative rainfall (from July to June) as the independent variable. The regression equations allow the calculation of predicted values of fPARmean for each growing season and each pixel from the estimated precipitation values. The fPARmean residuals (difference between observed and predicted fPARmean) were then computed for each growing season. The residuals represent the portion of the observed fPARmean value which is not explained by precipitation. Any significant temporal trends in the residuals could be interpreted as long-term human-induced effects on vegetation: greenness or regeneration (positive trend) or land degradation (negative trend). Residual trends for each pixel were computed on the entire time series (1981–82 to 2010–11) and significant trend slopes were mapped in order to assess spatial patterns (Fig. 13.6).

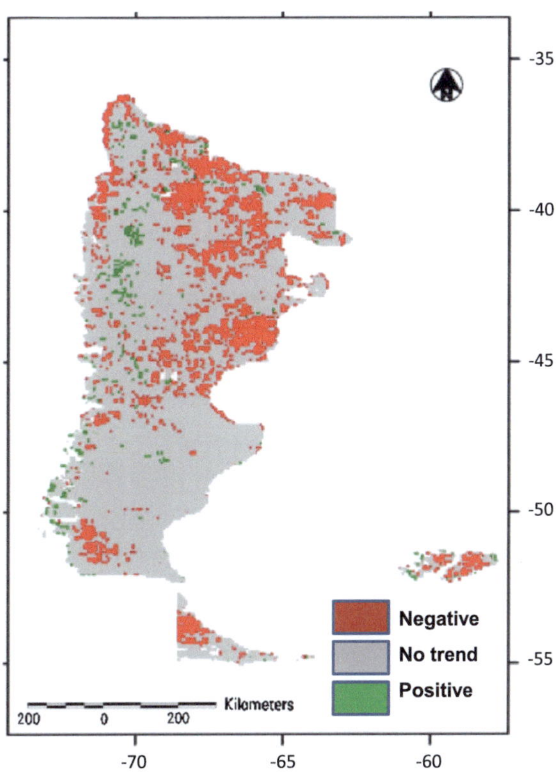

Fig. 13.6 Overall trends in the residual fPARmean throughout the period 1981–82 to 2010–11 based on regression of vegetation greenness (AVHRR GIMMS fPAR3g) on cumulative rainfall (July–June) (GPCP estimate). Significant trends: $p < 0.05$.

13.1.2 Station Scale Case Study

Los Pozos station, described in detail in Oliva et al. (2012) was used as a case study. The station is located in the south of the region, 30 km north of Río Gallegos, Santa Cruz, Argentina (69° 19′W, 51° 9′S) and close to the Magellan Strait. The climate is maritime with 239 mm mean annual rainfall and 12.7 °C mean temperature in summer and 1.4 °C in winter. The station occupies 27,200 ha of terraces of the Gallegos River and sedimentary plateaus, 150 m above sea level. Vegetation consists of perennial tussock steppes of *Festuca gracillima,* with an array of short high-quality grasses (Faggi 1985). The station has ten large management units and a number of smaller paddocks and produces lambs and wool mostly from Corriedale ewes.

The Santa Cruz range assessment method (Borrelli and Oliva 1999; Borrelli and Oliva 2001) was used to estimate available forage. Herbage availability was estimated by clipping short grasses, graminoids and forbs in 0.2 m^2 frames at peak herbage mass (January) at predetermined sites distributed across a paddock. These plants have the highest nutritive value and constitute most of the sheep diet (Posse et al. 1996). Sampling of herbage was performed annually at three permanent sampling sites in each paddock, with three quadrats harvested from each site. The samples were oven dried for 48 h at 40 °C and weighed. Grazing capacity was estimated from total herbage mass of short grasses, graminoids and forbs and adjusted yearly based on the estimated grazing capacity of each paddock using an allowance of 513 kg dry matter of forage for each sheep equivalent per year (Borrelli 2001). The sheep equivalent used covers the energy requirements of a 49-kg ewe shorn in spring (September) that weans a 20-kg lamb after 100 days of lactation.

Sheep production data were obtained from the historic records of the station. Flock numbers included all categories of sheep remaining on the station after lambing and selling. Lambing rate (%) was estimated from lambs marked (January) and ewes mated (May of the previous year). Total clean wool weight sold was estimated from greasy wool weight multiplied by the percentage of clean wool. The relationships between stocking rate, lamb marking rate and clean wool production were explored using linear regressions included in SAS/Stat v9.0 statistical package.

13.2 Results

The fPAR trends for pixels in the northern extra-Andean Patagonia were mostly negative in the first decade 1981–82 to 1990–91 (Fig. 13.5a). South of the region, a small number of pixels showed a positive trend. The Andes, west of the region and mostly covered with Nothofagus forests and high altitude grasslands, showed a clear positive trend. In the 1991–92 to 2000–01 decade (Fig. 13.5b), few pixels showed a clear trend, and in the extra-Andean area they were mostly positive. In the 2001–02 to 2010–11 decade (Fig. 13.5c), a negative trend was evident in about 40 % of the region, mostly in the North-East. The trend for the entire 30-year period, form 1981–82 to 2010–2011 (Fig. 13.5d) shows red spots in the Rio Negro province in the north, and in Tierra del Fuego at the south.

13.2.1 Station Case Study

Mean annual rainfall in Rio Gallegos airport, close to Los Pozos was 235 mm. No significant trend in the mean annual rainfall was observed throughout the 80-year record, although a dry period was evident during the 1960s, when the mean annual rainfall was only 160 mm.

High stocking rates with about 18,000 sheep (0.70 sheep $ha^{-1}year^{-1}$) were used in the 1930s, but a steep downward trend was observed from 1945 until 1989. In that year, the first range evaluation was performed and the stock further reduced to 12,000. During this 60-year period the flock was maintained above the grazing capacity of 10,600 that was estimated through yearly forage evaluations from 1990 to 2010.

From 1990 to 2010, the adaptive management period, animal numbers were adjusted to forage availability, following a tracking strategy (Campbell et al. 2006), and the flock fluctuated around 8000 sheep (0.35 sheep $ha^{-1}year^{-1}$), with half the initial stocking rates (Photo 13.5).

Mean annual herbage mass in the adaptive management period was 194 kg Dry matter ha^{-1} that corresponds to a mean grazing capacity of 0.38 sheep $ha^{-1}year^{-1}$.

Photo 13.5 The Magellanic tussock steppe in a well-conserved plot close to the study case area.

Photo 13.6 Degraded Magellanic Steppe with dominance of subshrubs after years of overgrazing.

The annual herbage masses were generally well above the value of 100 kg Dry Matter ha^{-1} that was suggested by Borrelli and Oliva (1999) to prevent restriction in intake by the sheep (Photo 13.6). Exceptions were the dry years of 1998 and 1999 when some of the paddocks only reached 70 kg Dry Matter ha^{-1} (data not shown).

The slope of the regression of Grazing Capacity in relation to time was not significantly different from zero, indicating that the station has not lost productivity in this period (Fig. 13.10).

Lamb marking rates increased from 72 to 87.3 % with the adaptive management. Although the total stock numbers on the station were halved, a similar number (about 4000) sheep were sold. Wool production on a per animal basis also increased from 2.21 to 2.82 kg wool $sheep^{-1}year^{-1}$, but this increase did not compensate for the lower number of animals at shearing, and total wool production of the station fell from 36,102 to 26,771 kg wool $year^{-1}$ (Oliva et al. 2012).

Lamb marking rate (%) showed a linear negative relationship with stocking rate. When the stocking rates used were close to the estimated grazing capacity of the station, lamb marking rate was above 80 % (Fig. 13.11). Clean wool production on a per area basis showed a quadratic relationship with SR and the inflexion point of this curve is 0.65 sheep $ha^{-1}year^{-1}$, where the station obtained about 2.5 kg wool $ha^{-1}year^{-1}$. With stocking rates close to the grazing capacity, wool production ranged between 1 and 2 kg wool $ha^{-1}year^{-1}$ (Fig. 13.11).

13.3 Discussion

The overall trend in annual mean fPAR was zero or negative over a large portion of the Patagonian region (Fig. 13.5d). This trend may be partly explained by climate change, as rainfall is closely correlated to fPAR. The long- term patterns in rainfall for this period vary across Patagonia, with no change in the central areas, a small increase in the south and a decrease in the north-west (Fig. 13.5). This pattern cannot be observed clearly in the 30-year fPAR trends (Fig. 13.5d). Some "greening" is observed in southern Santa Cruz province, but south of this, the Tierra del Fuego steppes have a negative trend that appears unrelated to rainfall data. The central-north area of Rio Negro and Chubut provinces show a negative trend despite rainfall remaining stable in this area. At a decadal scale, localized droughts may have generated the clear negative trend in fPAR in central-north Chubut and Rio Negro provinces during the 2001–02 to 2010–11 period (Fig. 13.5c). Drought in the last decade and the high stocking rates that still applied in these northern Patagonian provinces may have interacted to show the negative trends in the north-center area of this region.

Trends in the fPAR residuals were analysed in an attempt to separate the climate signal from other factors, as in Archer (2004). In these maps, large areas mostly in the south show no significant trends (Fig. 13.6) indicating that vegetation greenness corresponds closely to what is expected from rainfall dynamics. There is still a considerable proportion of areas of negative residual trends, i.e. areas in which the vegetation has been losing vigor at higher rates than would be expected by rainfall alone, and they are mostly concentrated in the north-east of Patagonia and in the southern Magellanic steppes of Tierra del Fuego (Fig. 13.6). Both areas are still grazed at comparatively high stocking rates (Fig. 13.2).

The widespread "browning" trends indicate that land degradation is still active in some areas, and the prevalence of no-trend pixels shows that land recovery has not happened. This could point to ongoing unsustainable management even when these three decades have seen a significant reduction in domestic animal numbers in some areas of Patagonia. Explanations other than overgrazing may be a mosaic of local factors that have influenced different parts of the region in this 30 year-period: Volcanic eruptions in vast areas of Santa Cruz in 1992 and central-north provinces of Chubut in 2008 and Rio Negro and Neuquén in 2011 that may have reduced fPAR signal through ash deposition. Relocation of stock has increased grazing pressure in some areas because crops are now grown on former animal production areas of the Pampas and cattle have been moved to marginal areas, affecting especially the northern Rio Negro province. Unsustainable dry land farming for annual crops followed by intense erosion also affected this north-east area. The central plateau of Santa Cruz, where sheep have been excluded from an area of about 10 million ha is probably too degraded to recover, and native guanacos have probably managed to compensate sheep losses, reestablishing the total grazing stock and slowing down the recovery of rangelands. Overall, the areas that were desertified in the 1998 map of Fig. 13.1 show no signs of recovery or a negative trend in our 1980–2010 analysis of Fig. 13.6, and those that were classified as intermediate or low, such as NE Rio Negro and the Magellan Steppes of SW Santa Cruz and N of Tierra del Fuego appear to have a negative trend in the residual fPAR.

The remotely sensed data provide valuable insights into the last 30 years, but sheep were introduced into the region over a century ago and presumably most of the degradation of rangelands would have taken place in the early stages of land settlement. Our southern Patagonia case study allows better analysis of the effects of management by focusing both on a smaller area and a longer time scale using more precise rain-gauge records and production data for the 1930–2010 period. Additional information from yearly forage availability was also available for the 1990–2010 period. Mean rainfall in the last 80 years showed no clear trend in this southern location, although the variability increased in the last 20 years (Fig. 13.7). Regardless of rainfall, the station has been reducing animal numbers at a close-to-linear rate since the introduction of sheep, when over 20 thousand animals were raised, falling to 12 thousand head by the and the end of the 1980s (Fig. 13.8). This pronounced negative trend during the 20th century is typical of the Patagonian stations, and agrees with Golluscio et al. (1998b) who showed five other long-term examples in the region.

In 1989, a range survey of Los Pozos indicated that a 30 % stock reduction was necessary to match the forage on offer with animal demand (Fig. 13.8). Careful yearly adjustments in stocking rates thereafter tracked the forage availability resulting in about half the maximum historical numbers (Fig. 13.9). Grazing capacity of the rangeland varied in relation to rainfall and did not show an overall negative trend in these twenty years (Fig. 13.10). Adaptive management was

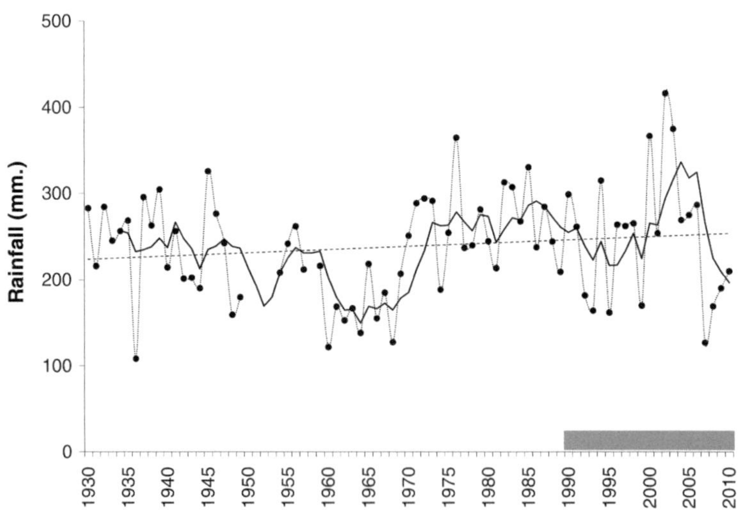

Fig. 13.7 Yearly rainfall in the Rio Gallegos airport, close to the Los Pozos station from 1930 to 2010. *Solid line* is a 5-year moving mean and *dashed line* a linear regression y = 0.377x + 223.3 R^2 = 0.0202 P = 0.220. The *shaded bar* indicates the adjusted management period.

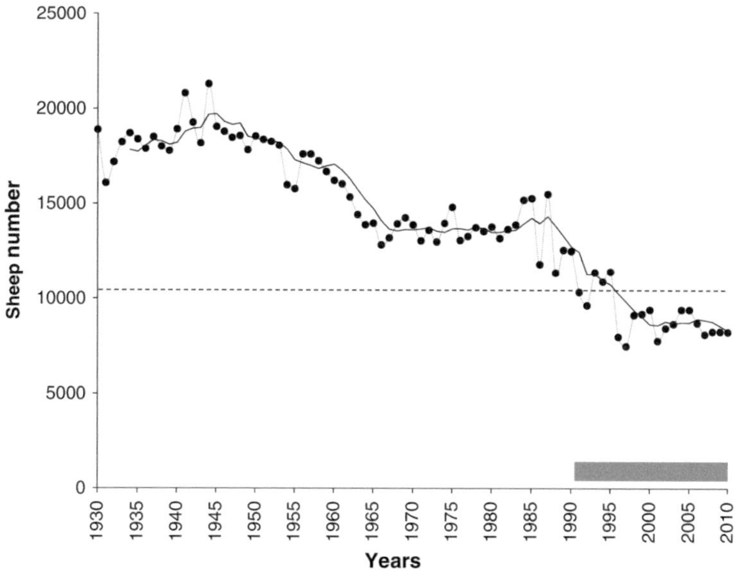

Fig. 13.8 Number of animals (sheep units) in the Los Pozos station from 1930 to 2010. The *gray bar* indicates the 1990–2010 adaptive grazing period and the *dashed line* the mean estimated grazing capacity of the station in the same period.

therefore able to halt the negative historical trend and preserve the capacity of the land to convert rainfall into forage over two decades, and is sustainable by definition (Pickup and Stafford Smith 1993). The downside is that no clear recovery of the grazing capacity was evident during this period. Similar results have been described in continuous grazing experiments with flexible stocking rates close to Los Pozos (Oliva et al. 1998) and in different locations in Patagonia (Escobar et al. 2011). Some researchers regard the lack of recovery as a consequence of continuous grazing schemes that do not enable regeneration even at low stocking rates, and recommend short duration grazing (Savory and Parson 1980), but well-documented case studies of short duration, rotational schemes, also fail to clearly regenerate the land (Golluscio et al. 1998a).

The reasons for a persistent, irreversible change in Patagonia's rangelands are unclear. The classical studies of Milchunas et al. (1988) and Milchunas and Lauenroth (1993) suggested that the sensitivity of the vegetation communities in Patagonia may be explained by a short evolutionary history of grazing, but later papers acknowledged that the vegetation shows herbivory defenses and that guanacos, that have been present and abundant since the Miocene (Raedeke 1979), may have exerted a high herbivory pressure in evolutionary times (Lauenroth 1998). Given these evidences, Patagonia was even used as an example of the opposite: a long history of grazing in Adler et al. (2005), so that the issue remains unresolved.

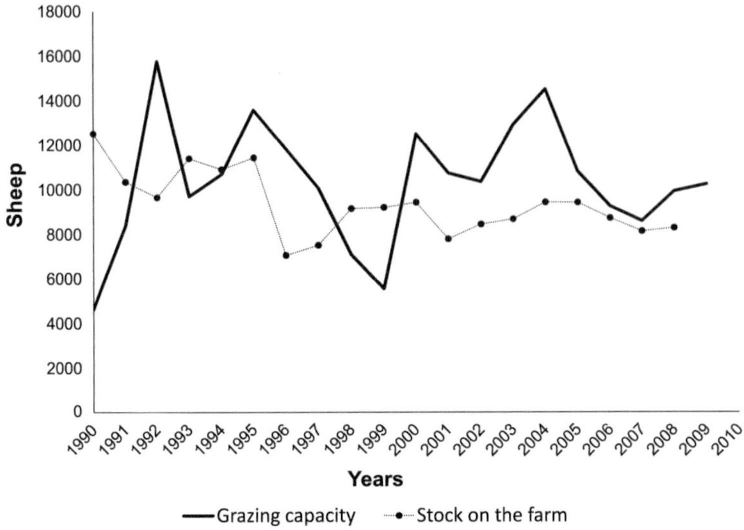

Fig. 13.9 Grazing capacity (sheep $ha^{-1}year^{-1}$) estimated by range evaluation (*solid line*) and actual stock in the station (*dashed line*) from 1990 to 2010. From Oliva et al. (2012).

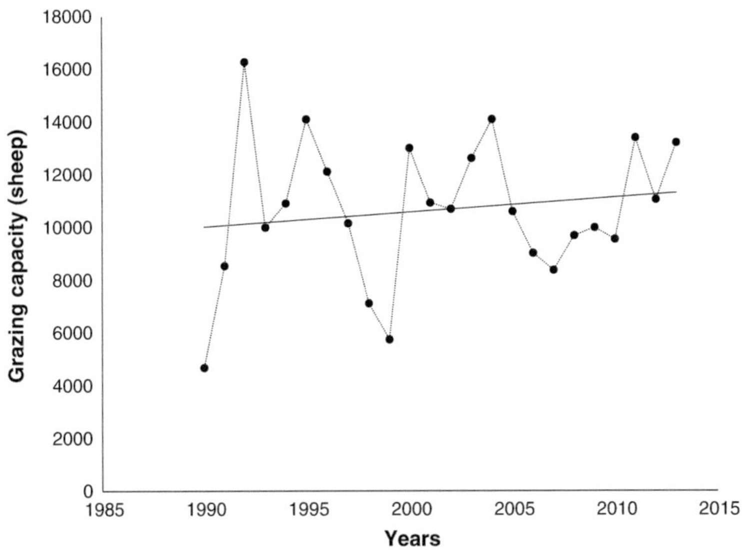

Fig. 13.10 Trend of grazing capacity (*GC*) estimated by range evaluation (*solid line*) from 1990 to 2013. Regression is GC = 56.0 year – 101437 (R^2 = 0.0213; P = 0.497).

Climate variability is lower in Patagonia than in other deserts and this may have actually favored the degradation: in our Los Pozos example the mean rainfall is 239 (63 SD) mm with a coefficient of variation of only 26 %, that is clearly in the lower range of deserts of the world (von Wehrden and Wesche 2007). The non-equilibrium concept suggests that rangeland degradation is more likely to occur where rainfall variability is low, because highly variable climates regulate herbivore numbers through drought-related deaths (Illius and O'Connor 1999). A threshold value of the coefficient of variation of rainfall of 33 % was suggested by (Ellis and Swift 1988), and the non-equilibrium concept states that below this value, degradation can occur if animal numbers are high enough, as in the case of Patagonia.

A number of state and transition models (Westoby et al. 1989), that catalog the existing vegetation communities and hypothesize the mechanisms that are involved in the shifts, have been compiled for Patagonia (Paruelo et al. 1993), and most of them include non-reversible transitions (Bertiller and Bisigato 1998). They are associated to the dominance of long-lived, perennial grasses of *Festuca, Poa* and *Stipa* (Leon et al. 1998), the lack of annuals and colonizer plants (Soriano 1990), and the structure of soil that is particularly prone to wind erosion (Chartier and Rostagno 2006). High stocking rates also interfere with the regeneration of dominant cool-season grasses that grow slowly and mostly vegetatively in these steppes (Oliva et al. 2005). Low variability rainfall and the fact that plants can depend on a reliable water pulse originated in winter rains or snow that thaws in springtime enable perennial life forms to dominate. Long-lived grasses usually devote only a moderate amount of resources to seed forming (Fenner 1985; Folland et al. 1986; Bertiller and Coronato 1994) as their main reproduction is through tillering. Their propagules also lack specific dormancy mechanisms that would enable long term soil seed banks (Soriano 1960; Bertiller and Coronato 1994). Aridisol-type soils dominate the region, with most of the organic matter and nutrients in superficial layers, and concentrated in patches associated with shrubs or tussocks (Rostagno and Del Valle 1988). When the dominant tussock grasses and big shrubs are lost to grazing, they are easily eroded exposing abundantly pebbled clay layers that are partially occupied by dwarf shrubs of genus *Nassauvia* or *Larrea* (Bisigato and Bertiller 1997a). Regeneration of degraded rangelands in this context has to rely on the establishment of grasses through seed of the few remaining perennial grasses that colonize cracks and crevices in desert pavements, which is likely to be a slow process regardless of the grazing system used.

The most important management challenge in Patagonia seems clear in Fig. 13.11. The vertical line indicates the sustainable stocking rate as estimated with 20-years of range surveys, and the points are yearly stocking rates used during the 80-year period. A smaller number of better-fed animals showed weaning rates over 80 % and overall the 8000-sheep station sold as many animals (lambs, rams and wethers) as the historical flock of 20 thousand (Oliva et al. 2012). Wool production, on the other hand was not closely related to animal condition and its decrease was proportional to the reduction of animal numbers. The optimal stocking rate for wool production appears to be 0.65 sheep $ha^{-1}year^{-1}$, close to the historical rates that were used in the early years of land settlement. This level of grazing would surely damage the rangeland in the long term (Oliva et al. 2005), a conflict that is evident in other systems where economic optimal stocking rates are higher than grazing carrying capacity (Cingolani et al. 2008).

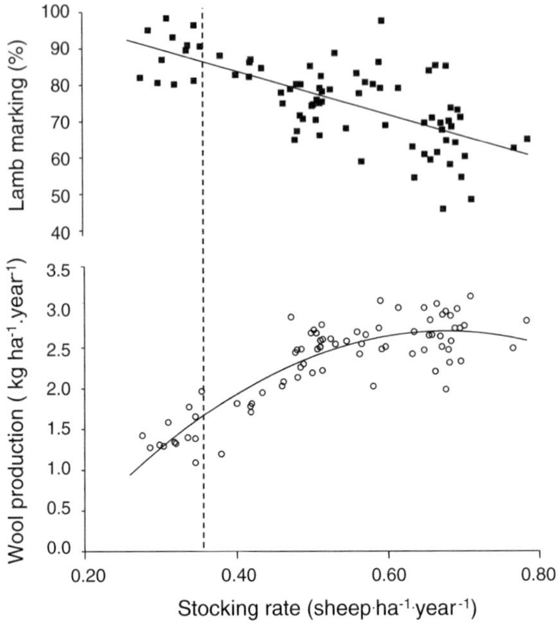

Fig. 13.11 Stocking rate (*SR*) in sheep ha^{-1} year^{-1} in relation to wool production (*WP*) in kg ha^{-1} year^{-1} and lamb marking (*LM*) rate (number of lambs obtained from 100 mated ewes in %). *Regression lines* are linear for LM = 1.08−0.60 SR ($R^2 = 0.47$; $P < 0.01$), and quadratic for WP = −2 + 14 SR−10 SR^2 ($R^2 = 0.76$; $P < 0.01$). *Vertical line* indicates the mean grazing capacity of the station between 1990 and 2010. From Oliva et al. (2012).

13.4 Conclusions

Do humans cause deserts? Patagonia gives a different insight into the dilemma that Reynolds and Stafford Smith (2002) pointed out. In the Sahel example, climate appears to have been the main cause both of degradation and recovery (Evans and Geerken 2004), with little effect of management (Seaquist et al. 2009). In Patagonia, land use has triggered a general degradation effect that is evident in sheep production records and in remotely sensed data. In other examples of non-sustainable use of grazing lands in the world, as in the Western Division of New South Wales in Australia, the loss of grazing capacity seems to have been faster, and in less than two decades the wool industry collapsed (Condon 1976). There are also examples of remarkable recoveries elsewhere such as the Sahel case that show that the land has resilience (Herrmann et al. 2005). In comparison, Patagonian systems endured over eight decades of high stocking rates until the generalized losses of the early 1980s. Some vegetation changes documented in Patagonia have resulted in desert-like conditions (Bertiller and Bisigato 1998), and they have not shown clear signs of recovery thereafter, even with careful, conservative management of the land and in a unchanging or slightly positive rainfall scenario. This delay in revealing signs of degradation and the fact that recovery seems to be difficult, slow, or absent, point to the existence of rangelands with stable states, that change slowly and more or less permanently at a time-scale that is difficult to perceive (Oliva et al. 1998). Well-documented conservative management examples such as the Los Pozos station indicate that it is possible to stop the downward trend but at the cost of foregoing short-term income. It is unlikely that producers will decide to give up profits in order to avoid possible future degradation, as this requires an understanding of the problems that overstocking causes that are slow and not evident in the manager's experience (Stafford Smith et al. 2007). It also requires a sound financial status that is not the case in most of the smaller Patgonian stations that struggle to make ends meet or have subsistence-type economies. Promises of short-term transformation via grazing pulses seem unlikely, and will at any case take a long time to regenerate some of the lost grazing capacity. Overall, the prospects under good management seem at best to be those of stabilization. The old-timer's description of rich Patagonian meadows with grasses that reached the belly of the horses may not be easily seen again, after over a century of unsustainable pressure on the natural resources of these unique cold deserts.

Acknowledgments The authors wish to recognize Edward Halliday and Silvina Puig, who gave access to all the historic information of the Los Pozos Station. INTA (Instituto Nacional de Tecnología Agropecuaria/Argentina) financed the farm study. David Preston contributed reviewing the manuscript.

References

Adler, R. F., Huffman, G. J., Chang, A., Ferraro, R., Xie, P., Janowiak, J., et al. (2003). The version-2 global precipitation climatology project (GPCP) monthly precipitation analysis (1979–present). *Journal of Hydrometeorology, 4*, 1147–1167.

Adler, P. B., Milchunas, D. G., Sala, O. E., Burke, I. C., & Lauenroth, W. K. (2005). Plant traits and ecosystem grazing effects: Comparison of US sagebrush steppe and Patagonian steppe. *Ecological Applications, 15*, 774–792.

Amaya, J., von Tüngen, J., & de Lamo, D. (2001). Informe preliminar del Proyecto de Relevamiento y Distribución de guanacos en la Patagonia. Resultados sobre la densidad de guanacos en las diferentes provincias patagónicas. INTA-GTZ-TÖB. Area de Recursos Naturales. fauna. INTA EEA Bariloche., Bariloche.

Archer, E. R. (2004). Beyond the "climate versus grazing" impasse: Using remote sensing to investigate the effects of grazing system choice on vegetation cover in the eastern Karoo. *Journal of Arid Environments, 57*, 381–408.

Barbería, E. (1995). Los dueños de la tierra en la Patagonia Austral. Universidad Federal de la Patagonia Austral, Buenos Aires.

Beck, P. S., & Goetz, S. J. (2011). Satellite observations of high northern latitude vegetation productivity changes between 1982 and 2008: Ecological variability and regional differences. *Environmental Research Letters, 6*, 045501.

Bertiller, M., & Coronato, F. (1994). Seed bank patterns of Festuca pallescens in semiarid Patagonia: A possible limit to bunch reestablishment. *Biodiversity and Conservation, 3*, 57–67.

Bertiller, M. B., & Bisigato, A. (1998). Vegetation dynamics under grazing disturbance. The state-and-transition model for the Patagonian steppes. *Ecología Austral, 8*, 191–199.

Bisigato, A. J., & Bertiller, M. (1997). Grazing effects on patchy dryland vegetation in northern Patagonia. *Journal of Arid Environments, 36*, 639–653.
Borrelli, P. (2001). Planificación del pastoreo. In P. Borrelli & G. Oliva (Eds.), *Ganadería ovina sustentable en la Patagonia Austral* (pp. 183–196). Buenos Aires: INTA.
Borrelli, P., & Oliva, G. (1999). *Managing grazing: experiences from Patagonia* (pp. 441–447). In Vth International Rangeland Congress. Townsville: Australia.
Borrelli, P., & Oliva, G. (2001). Evaluación de Pastizales. In P. Borrelli & G. Oliva (Eds.), *Ganadería ovina sustentable en la Patagonia Austral: Tecnología de manejo extensivo* (pp. 163–168). Buenos Aires: INTA.
Campbell, B., Gordon, I. J., Luckert, M., & Pehterama, L. (2006). In search of optimal stocking regimes in semi-arid grazing lands: One size does not fit all. *Ecological Economics, 60*, 75–85.
Cibils, A., & Borrelli, P. (2005). Grasslands of Patagonia. In J. M. Suttie, S. G. Reynolds, & C. Batello (Eds.), *Grasslands of the world* (pp. 121–170). Rome: Food and Agriculture Organization of the United Nations.
Cingolani, A., Imanual, N.-M., Renison, D., & Cabido, M. (2008). La ganadería extensiva, ¿es compatible con la conservación de la biodiversidad y de los suelos? *Ecología Austral, 18*, 253–271.
Condon, R. W. (1976). History and administration of the Western Division. *Agricultural Gazzette, 87*, 2–5.
Chartier, M. P., & Rostagno, C. M. (2006). Soil erosion thresholds and alternative states in northeastern Patagonian rangelands. *Rangeland Ecology and Management, 59*, 616–624.
de Jong, R., de Bruin, S., de Wit, A., Schaepman, M. E., & Dent, D. L. (2011). Analysis of monotonic greening and browning trends from global NDVI time-series. *Remote Sensing of Environment, 115*, 692–702.
Del Valle, H., Elissalde, N., Gagliardini, D., & Milovich, J. (1998). Status of desertification in the Patagonian Region: Assessment and mapping from satellite imagery. *Arid Soil Research and Rehabilitation, 12*, 95–122.
DHV-SWEDFOREST. (1998). Diagnóstico, estrategias y acciones propuestas para el uso sostenible de los recursos naturales en la Patagonia. Secretaria de agricultura, ganadería, pesca y alimentación. Buenos Aires: República Argentina.
Elissalde, N., Escobar, J., & Nakamatsu, V. (2002). *Inventario y evaluación de pastizales naturales de la zona arida y semiarida de la Patagonia*. Trelew: Instituto Nacional de Tecnología Agropecuaria. Centro Regional Patagonia sur. EEA Chubut.
Ellis, J. E., & Swift, D. M. (1988). Stability of African pastoral ecosystems: Alternate paradigms and implications for development. *Journal of Range Management, 41*, 6.
Escobar, J. M., Beloqui, A., Buono, G., Massara Paletto, V., La Torraca, A., & Nakamatsu, V. B. (2011). Effects of flexible stocking rates strategies on rangeland dynamics and sheep production performance in Patagonian steppe. In S. RO, G. Feldman, & M. Sacido (Eds.), IX International Rangeland Congress (p. 384). INTA- AAMPN-IRC: Rosario, Argentina.
Evans, J., & Geerken, R. (2004). Discrimination between climate and human-induced dryland degradation. *Journal of Arid Environments, 57*, 535–554.
Faggi, A. (1985). Las comunidades vegetales de Rio Gallegos, Santa Cruz. In O. Boelcke, D. Moore, & F. Roig (Eds.), Transecta Botánica de la Patagonia Austral (pp. 592–633). CONICET Instituto de la Patagonia (Chile) Royal Society (Gran Bretaña): Buenos Aires.
Fenner, M. (1985). *Seed ecology*. London, New York: Chapman and Hall.
Fensholt, R., Langanke, T., Rasmussen, K., Reenberg, A., Prince, S. D., Tucker, C., Scholes, R. J., Le, Q. B., Bondeau, A., & Eastman, R. (2012). Greenness in semi-arid areas across the globe 1981–2007—An Earth Observing Satellite based analysis of trends and drivers. *Remote Sensing of Environment, 121*, 144–158.
Folland, C., Palmer, T., & Parker, D. (1986). Sahel rainfall and worldwide sea temperatures, 1901–85. *Nature, 320*, 602–607.
Gaitán, J. J., Bran, D., Oliva, G., Ciari, G., Nakamatsu, V., Salomone, J., et al. (2013). Evaluating the performance of multiple remote sensing indices to predict the spatial variability of ecosystem structure and functioning in Patagonian steppes. *Ecological Indicators, 34*, 181–191.

Goergen, J. (1995). Lucha contra la desertificación en Patagonia con ayuda de un sistema de monitoreo ecológico (SME). In H. Del Valle, G. Eiden, H. Mensching, & J. Goergen (Eds.), Lucha contra la desertificación en la Patagonia. Cooperación técnica argentino - alemana. Instituto Nacional de Tecnología Agropecuaria y GTZ: Bariloche.

Golluscio, R., Deregibus, A., & Paruelo, J. (1998a). Sustainability and range management in the Patagonian Steppes. *Ecología Austral 8*, 265–284.

Golluscio, R., Giraudo, C., Borrelli, P., Montes, L., Siffredi, G., Cechi, G., et al. (1998b). Utilización de los recursos naturales en la Patagonia. In C. DHV/SWEDFOREST (Eds.), ECOSISTEMAS DESÉRTICOS PATAGÓNICOS (p. 80). Secretaría de Agricultura, Ganadería, Pesca y Alimentación. República Argentina: Buenos Aires.

Herrmann, S. M., Anyamba, A., & Tucker, C. J. (2005). Recent trends in vegetation dynamics in the African Sahel and their relationship to climate. *Global Environmental Change, 15*, 394–404.

Illius, A., & O'Connor, T. (1999). On the relevance of nonequilibrium concepts to arid and semiarid grazing systems. *Ecological Applications, 9*, 798–813.

INDEC. (2001). Censo Nacional de Población, hogares y viviendas 2001. Serie 5: Grupos poblacionales N°2: Jóvenes, hogares y población. In I. N. d. E. y. Censos (Ed.), INDEC, Buenos Aires.

Lauenroth, W. (1998). Guanacos, spiny shrubs and the evolutionary history of grazing in the Patagonian steppe. *Ecología Austral 8.*

Leon, R., Bran, D., Collants, M., Paruelo, J., & Soriano, A. (1998). Grandes unidades de vegetación de la Patagonia extra andina. *Ecología Austral, 8*, 125–144.

Mendez Casariego, H. (2002). Sistema de soporte de decisiones para la producción ganadera y forestal de la provincia de Neuquén. INTA EEA Bariloche: Bariloche.

Milchunas, D. G., & Lauenroth, W. K. (1993). Quantitative effects of grazing on vegetation and soils over a global source. *Ecological Monographs, 63*(4), 328–366.

Milchunas, D. G., Sala, O. E., & Lauenroth, W. K. (1988). A generalized model of the effects of grazing by large herbivores on grassland community structure. *The American Naturalist, 132*, 87–106.

Oliva, G., Cibils, A., Borrelli, P., & Humano, G. (1998). Stable states in relation to grazing in Patagonia: A 10-year experimental trial. *Journal of Arid Environments, 40*, 113–131.

Oliva, G., Collantes, M., & Humano, G. (2005). Demography of grazed tussock populations in Patagonia. *Rangeland Ecology and Management, 58*, 466–473.

Oliva, G., Ferrante, D., Puig, S., & Williams, M. (2012). Sustainable sheep management using continuous grazing and variable stocking rates in Patagonia: A case study. *The Rangeland Journal, 34*, 285–295.

Paruelo, J. M., Bertiller, M. B., Schlichter, T. M., & Coronato, F. R. (1993) Secuencias de deterioro en distintos ambientes patagónicos (p. 109). Su caracterización mediante el modelo de estados y transiciones.

Paruelo, J., H. E. Epstein, W. K. Lauenroth, & I. C. Burke. (1997). ANPP estimates from NDVI for the central grassland region of the US. *Ecology.*

Pickup, G., & Stafford Smith, D. (1993). Problems, prospects and procedures for assessing the sustainability of pastoral land management in arid Australia. *Journal of Biogeography, 20*, 471–487.

Posse, G., Anchorena, J., & Collantes, M. (1996). Seasonal diets of sheep in the steppe region of Tierra del Fuego, Argentina. *Journal of Range Management, 49*, 24–30.

Reynolds, J. F., & Stafford Smith, D. (2002). *Do humans cause deserts. Global desertification: Do humans cause deserts* (pp. 1–21).

Rostagno, C. M., & Del Valle, H. F. (1988). Mounds associated with shrubs in aridic soils of northeastern Patagonia: Characteristics and probable genesis. *Catena, 15*, 347–359.

Savory, A., & Parson, S. (1980). The Savory grazing method. *Rangelands, 2*, 234–237.

Seaquist, J., Hickler, T., Eklundh, L., Ardö, J., & Heumann, B. (2009). Disentangling the effects of climate and people on Sahel vegetation dynamics. *Biogeosciences 6.*

Siffredi, G. L., & Becker G. F. (1999). Pastoreo de veranadas. Cordillera del Neuquén. Guía de recomendación de carga animal para pampas de coirón blanco.

Soriano, A. (1960). *Germination of twenty dominant plants in patagonia in relation to the regeneration of vegetation* (pp. 154–158). In 8th International Grassland Congress.

Soriano, A. (1990). Missing strategies for water capture in the patagonian semidesert. *Academia Nacional de Ciencias Exactas Fisicas y Naturales, 5*, 135–139.

Stafford Smith, D. M., McKeon, G. M., Watson, I. W., Henry, B. K., Stone, G. S., Hall, W. B., & Howden, M. S. (2007). Learning from episodes of degradation and recovery in variable Australian rangelands. *PNAS, 104*, 20690–20695.

Steltzer, H., & Welker, J. M. (2006). Modeling the effect of photosynthetic vegetation properties on the NDVI-LAI relationship. *Ecology, 87*, 2765–2772.

Tucker, C. J. (1979). Red and photographic infrared linear combinations for monitoring vegetation. *Remote Sensing of Environment, 8*, 127–150.

Tucker, C. J., Townshend, J. R., & Goff, T. E. (1985). African land-cover classification using satellite data. *Science, 227*, 369–375.

von Wehrden, H., & Wesche, K. (2007). Relationships between climate, productivity and vegetation in southern Mongolian drylands. *Basic and applied dryland research, 1*, 100.

Westoby, M., Walker, B., & Noy-Meir, I. (1989). Opportunistic management of rangelands not at equilibrium. *Journal of Range Management, 42*, 266–274.

Zhu, Z., Bi, J., Pan, Y., Ganguly, S., Anav, A., Xu, L., et al. (2013). Global data sets of vegetation leaf area index (LAI) 3g and Fraction of Photosynthetically Active Radiation (FPAR) 3g derived from Global Inventory Modeling and Mapping Studies (GIMMS) Normalized Difference Vegetation Index (NDVI3g) for the period 1981 to 2011. *Remote Sensing, 5*, 927–948.

Chapter 14
Interpreting Environmental Changes in the Southern Bolivian Andes: Rural Responses and Political Actions

David Preston

Abstract Rural people in the southern Andes mountains of Bolivia may choose to plant different crops, to keep livestock under different conditions and to diversify their livelihood strategies in ways that involve less investment of time in farming and more involvement in commercial activities. These changes are in response to change in both physical and cultural environments and are examined here. Policies and activities by local and national government that are intended to reduce the impact of a range of environmental changes on people and their physical environments are considered and evaluated. We conclude that there is limited evidence that changes in the natural environment such as soil erosion and degeneration of vegetation in the Altiplano and adjacent valleys in western Tarija are more rapid than in the recent past. Changes in the use of different areas for livestock and crops reflect accommodation to culturally-determined household preferences for cattle over other ruminants. Farmers are used to change and readily modify their use of different areas of land to ensure that livestock can survive and necessary crops can be harvested. With respect to local political responses to environmental issues, the history of action that started in 1930 is notable. We conclude that such action had only limited environmental consequences in the Tarija valleys. There is no evidence to suggest that recent regional governments have been adequately informed or politically motivated to take remedial action to help those suffering from extreme climatic events or other environmental stresses.

Keywords Environmental change · Household livelihoods · Landscape evolution

D. Preston (✉)
Latin American Centre, University of Oxford, Oxford, UK
e-mail: d.a.preston@leeds.ac.uk

R.H. Behnke and M. Mortimore (eds.), *The End of Desertification?*
Springer Earth System Sciences, DOI 10.1007/978-3-642-16014-1_14

14.1 Introduction

Environmental changes have affected people in Andean America since long before the arrival of Europeans in the continent. Such changes occur as a consequence of many factors; changes in climate resulting in a gradual change in vegetation, in stream flow in the rivers as well as the ways in which native and introduced livestock consume a range of resources. Many other changes took place in southern Bolivia following the development of mining during the colonial period, as we shall show.

At the present time, rural people in southern Andean Bolivia may choose to plant different crops, to keep livestock under different conditions and to diversify their livelihood strategies in ways that involve less investment of time in farming and more involvement in commercial activities to best ensure the wellbeing of household members. These changes are in response to change in both physical and economic environments. We shall also seek to identify what sorts of changes can be observed in the physical landscapes in three distinct parts of southern Bolivia—high plateaux, a deep valley in the mountains and the temperate valleys on the eastern margins of the Andes. Actions by local and national government that are intended to reduce the impact of a range of environmental changes on people and their physical environments will also be considered and evaluated. This will enable an assessment of the extent to which natural environmental changes are an important part of the broad range of changes affecting people, plants and animals.

This chapter will review the range of environmental changes in one part of the southern Andes of Bolivia on the basis of our research in the Department of Tarija since 1993 and in the broader context of relevant research in the Central Andes of Peru and Bolivia as well as north-western Argentina. Many of our conclusions have been strongly influenced by what we have learnt from women and men, children and old people in formal interviews, discussions in community meetings and informal conversation in a wide range of rural communities. We have also benefitted from participation in meetings with scientists and policy makers at the administrative Departmental level (Fig. 14.1).

Changes in the weather, such as rainfall, frost, floods and storms, inevitably affect farmers although they have many customary ways of coping with such situations. Changes in the vegetation in some areas have occurred, through clearance following burning, or elimination of some plants by selective grazing by livestock but also as a result of plant invasion, and the introduction of new species to serve particular purposes—for example eucalypts or pines for timber.

Characterizing environmental change by a single term such as desertification is simplistic, because from a land user's point of view, it is imperative to consider the whole range of medium to long-term changes in both human and physical environments. Rural people are very conscious that weather is constantly variable and also that free-ranging livestock roam according to their perception of where good grazing may be found. For example introducing new types of livestock, often raised elsewhere, presents new challenges for pastoralists who may have to introduce such

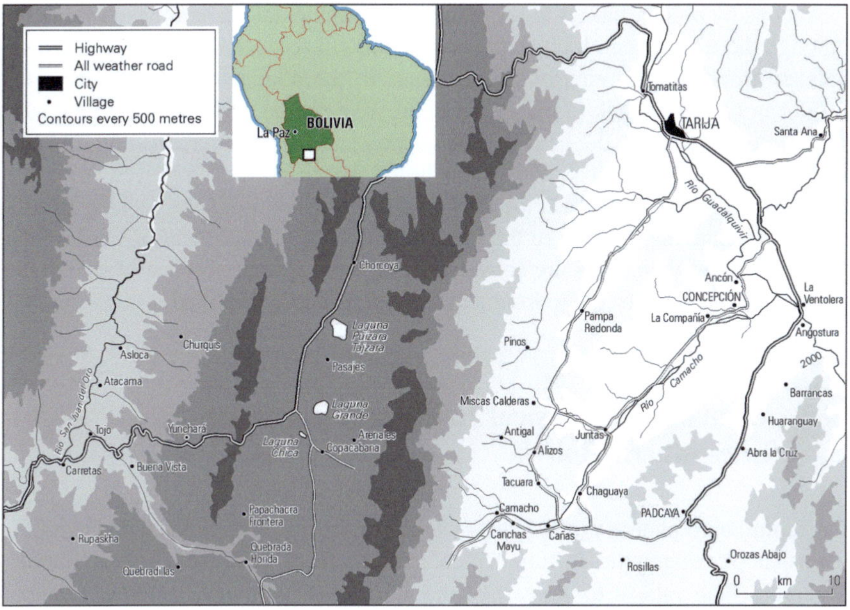

Fig. 14.1 Western Tarija.

animals to the particular grazing behaviour of long-established sheep, cattle or goats. In the course of our field work our team has helped people in several communities to introduce new breeds of goats that provide more milk and are larger and worth more when sold; however their inability to search for and find palatable grazing initially proved challenging.

When asked about changes in the weather, rural people cite extreme events—such as floods, hail and frost—as most worrying. This has emerged from interviews conducted during our research in rural communities in southern Bolivia over the past twenty years and is also reported in the northern Bolivian Altiplano (Chaplin 2007; Valdivia et al. 2013). However it is essential to place contemporary changes in the context of the 4000–5000 years of change in the use which humans have made of the Andean environments; this starts from their first use and subsequent domestication of native animals and plants and their adaptation to long-term climate changes which are well-documented in scientific literature, discussed later.

Some recent changes in Andean environments have occurred as a consequence of where people now prefer to live, particularly the increasing proportion of regional populations living or working in urban areas. Labour scarcity was frequently mentioned in interviews in rural communities, reflecting periodic migration, mainly to nearby Argentina, but also to nearby urban areas. However, changes in the rhythms of the long-established climatic variations—most notably the El Niño-Southern Oscillation (ENSO)—also affect human populations. Such changes are increasingly commented on in local and national media.

Actions by policy-makers in response to environmental changes need evaluation, although there have been limited actions in southern Andean Bolivia. The origins of activism in relation to environmental issues in the administrative Department of Tarija can be traced back to the 1930s; some 40 years later a well-funded technical organization was created—PERTT (Programa ejecutivo de rehabilitación de tierras en el departamento de Tarija)—which lasted 16 years and employed a team of natural scientists from Bolivia together with overseas consultants, seeking ways of combatting land degradation in the western part of the Department of Tarija. Some of the results of their work are considered and the apparent absence of importance given by policymakers to environmental conservation is outlined.

14.2 Human and Physical Landscape Evolution in Southern Bolivia

The central high plateaux (altiplano) of south-western Bolivia are just one part of an extensive physiographic feature which extends from southern Peru to northern Argentina. The valleys to the east of the altiplano have a very different but complementary environment and people living in the valleys and the high plateaux have traded and moved between highlands and lowlands since long before the arrival of Europeans (Lathrap 1973; Saignes 1995; Murra 1972). Inter-montane valleys, such as that of the Río San Juan del Oro, which forms the boundary between the Departments of Tarija and Potosí offer a third environment influenced by river water which offers both opportunities for crop farming and threats from seasonal flooding.

In order to place the discussion in a clear geographical context it should be noted that the western part of Tarija department comprises three distinct environment zones, in which the research that provides much of the basis of this review of natural and human environmental changes has taken place.

The Tarija Altiplano lies at about 3500 m mean altitude and is bordered on both west and east by rolling mountainsides whose slopes rise to more than 4200 m before dropping firstly to a gently-sloping shelf at a similar altitude to the altiplano and then fall sharply to the north-south valley of the Río San Juan del Oro at around 2600 m above sea level. The slopes on the sides of the Rio San Juan include both extensive near-level areas and steep, and eroded valley sides but livestock can readily reach areas where grazing is attractive. To the east of the Altiplano, mountainsides drop steeply to the network of temperate valleys at about 1900 m. in the northern part of which lies the largest urban centre and political capital, the city of Tarija. The passage of animals and people between the altiplano, the mountains and the temperate valleys presents few difficulties and there is ample evidence of this movement having occurred for over 600 years.

Moist air brings rain and cloud formations from the east and, accordingly, precipitation varies from around 600 mm in the temperate eastern valley to just

Fig. 14.2 Cross-section of western Tarija Department.

Fig. 14.3 Landscapes in western Tarija Department. (**a**) Río San Juan del Oro valley. (**b**) Altiplano. (**c**) Temperate valleys.

300 mm in the more level land of the altiplano and also in the deep valley of the Río San Juan. Figure 14.2 presents a cross-section of the three regions in which our research has been centred (Fig. 14.3).

14.2.1 Natural Environmental Histories Related to Present Landscapes

Natural environmental history in the Andes has long been influenced by the activity of the human population, in particular during the past 5000 years during which the people, domesticated plants and animals have transformed livelihoods and landscapes. Alongside the changes in human land use have been variations in climate, so that it is crucial to understand that the natural environment is ever-changing. Early human settlement was associated with the cultivation of different varieties of potatoes and chenopods (such as quinoa) as well as the domestication of wild llamas and alpacas. One result was that already by 4000 years ago, much forest cover in many parts of the Andes had been cleared and both agriculture and livestock became important for human livelihoods (Chepstow-Lusty et al. 1998). At

Fig. 14.4 Llamas from the highlands carrying produce for exchange in the Tarija valleys, Alisos, 1992.

lower altitudes, maize was also introduced, probably from Central America, and there seems to have been continuous contact between the altiplano and valley areas. This contact, associated with trading produce from contrasting ecological zones, has continued until quite recently. During fieldwork in the valleys of Tarija in the early 1990s, columns of llamas still descended from the altiplano, many days' travel to the west. They were loaded with products such as salt (from the salt lakes) and dried meat that were traded for maize and other goods produced in the lower altitude valleys (Fig. 14.4).

With the arrival of Europeans some 400 years ago, new types of livestock, particularly sheep, cattle and goats were introduced as well as European cereals, particularly wheat. The appropriation of areas of land from the native population by the invading Europeans transformed the landscape in temperate valleys and the altiplano. The founder of the town of Tarija in 1574 described the surrounding area as fertile, in a broad valley with plenty of water, with many fish in the rivers, abundant wild animals and well capable of growing grape vines, olives and crops to which the Spanish were accustomed (Corrado and Comajuncosa 1990, p. 11).

Very soon after the establishment of the Spanish in the Andes, mining for a range of minerals, highly prized in Europe, became an important activity in many parts of the Andes and the mines of Potosí, to the north of Tarija, grew until the town was one of the most important economic centres in the continent in the 17th and 18th centuries. The urban and industrial population created a demand for a range of produce for both food and clothing. Agriculture in the temperate valleys flourished to supply such markets and trade routes were established along which mining goods and labour moved. The historical changes in what is now Tarija and resulting changes in the natural landscape have been documented by Presta (2001). When Luiz de Fuentes, the founder of the town of Tarija, died in 1598 just 38 years after his arrival, his will lists, among his possessions, 2000 cattle in the nearby highlands and 4000 sheep in what is now San Lorenzo in the valleys adjacent to Tarija (Preston 1998).

The extension of great estates during the colonial period undoubtedly affected the vegetation of the grazed areas but such changes can only be inferred. Using data from the early 20th century, some changes in farming may be detected but it is important to note that there were still areas of the highlands and hillsides in the temperate valleys that were only occasionally used (Preston 1998). Wild vicuña (camelids) can still be observed in isolated parts of the Tarija altiplano and old people remember people going hunting wild animals on the altiplano in the early 20th century. Small areas in the valleys and highlands of Tarija Department still have woodlands with native tree species dominant and to the east of the Tarija valleys where there is higher rainfall, the more luxuriant vegetation is seasonally grazed by transhumant livestock from the valleys of Tarija.

Large estates dominated many areas of Tarija Department until the land reform of 1953. After agrarian reform, most of the former estates became autonomous communities and former serfs had better access to larger areas of land and therefore a wider range of resources. Changes noted during the past 30 years in these areas include a continued loss of population in rural areas, both numerically and proportionately as more people move to urban centres or to other rural areas, particularly in nearby Argentina where wages were much higher. Excellent data from two communities in the Camacho valley using household registers for school attendance and interviews with older people showed that over the 23 year period 1975–98 in Alizos, 29 % of those living in the community had left to work in Argentina. In Juntas 43 % had gone to Argentina over the period 1969–93 (Preston and Punch 2001).

14.3 Environmental Change in the Bolivian Andes

The current preoccupation with environmental change (or global warming) in the media and reflected in popular perceptions of weather patterns, makes it necessary to review scientific evidence of changes in the long-term patterns of weather. In addition, changes in the environment, in particular vegetation, erosion and deposition in hillsides and fields, should be understood in order to assess the influence of human causation of environmental change through modification of vegetation by expansion of crop production areas as well as the introduction of more grazing livestock. This assessment is based on our recent work in Tarija which focussed on the temperate valleys around and to the south of the city of Tarija, the highland (altiplano) areas to the west and the valley of the Río San Juan del Oro to the west of the Altiplano. We have conducted many interviews with farmers in each of these areas in order to identify their perceptions of weather/climate changes that they felt affected livelihoods. In conclusion we review the existing evidence of climatic changes in this part of highland Bolivia.

14.3.1 Change in the Tarija Valleys

Recent work in catchments in Tarija Department using detailed field mapping, lichenometric and ^{14}C dating and clast size measurement concluded that alluviation was concentrated during wet La Niña phases of the Southern Oscillation. Flood magnitude may have been greater during the Little Ice Age (c.1450 to c.1850) when a reduced vegetation cover would have increased weathering and sediment supply and decreased after about AD1880 (Maas et al. 2001). It is suggested that contemporary degradation is probably less than it was a century ago, although the recent greater frequency of dry El Niño-related periods will have contributed to a decrease in vegetation cover and the likelihood of erosion acceleration during the following wetter years.

Striking areas of badlands occur in some parts of the central valleys of Tarija. They result from the erosion by wind and water-borne sediments. It is generally believed that erosion is widespread, despite the absence of any thorough scientific examination of the extent of erosion in the Tarija area. The national government established an agency (PERTT) specifically to combat erosion in Tarija. The belief that there is an environmental crisis in Tarija is related to exaggerated estimates of the areas subject to erosion. A report from the German government agency GTZ in 1988 asserted that 75 % of the Central Valley was eroded and that 800 ha. of farmland in the valleys was being lost annually (GTZ 1988). It was apparently based on a study [uncited] by W D Beattie although no definitions of 'eroded' or of the methods by which such data were calculated are indicated. These statistics are widely and uncritically quoted in a range of Bolivian and other publications (e.g. Freeman 1980) and form part of the image of what Zimmerer has claimed to be 'the country's catastrophic erosion' (Zimmerer 1996, p. 111).

The extent of active erosion seems to us to be much less than the statistic quoted by GTZ would suggest. Our work in the Alizos catchment in the Río Camacho valley suggests that probably no more than 10 % of the surface is subject to active erosion and visits to other parts of the central valleys of Tarija have suggested that this lower figure may be true of the region as a whole. Further detailed field studies are needed to determine more exactly the extent of active erosion. We found no evidence to support the view that erosion has increased during the present century and, on the contrary, it may be neither rapid nor as widespread as previously asserted.

One of the possible causes of erosion could be livestock grazing, which remains an important component of household livelihoods. Research in a number of communities in the valleys to the south of Tarija, and including one above the valley floor level on the mountain slopes 1000 m higher at 2900 m found that the flexible use of common areas on nearby hills as well as transhumance was a critical part of the community's grazing management. This management minimizes the likelihood that vegetation would become permanently degraded, which would reduce its capacity for feeding grazing animals.

Before the agrarian reform of the 1950s, some farmers in the Tarija valleys used to send sheep and some goats during the wet season to graze some little-used areas in the adjacent altiplano, with the acquiescence of the estate owners, and also some small stock were sent across the altiplano and down to the higher grazing areas above the Río San Juan valley. Following the formation of new community self-government organizations (*sindicatos*) after the land reform, highland people wished to exclude other livestock and the practice of such transhumance ceased (Preston et al. 2003, p. 146). The increasing numbers of cattle and declining sheep numbers may have been partly in response to this change, as cattle were easily managed with minimum labour input when left to graze freely on the higher mountainsides above the Tarija valleys.

Our research on changes in livestock numbers during the 20th century in three communities in the Tarija altiplano suggests that numbers may have increased in the first half of the century (although data for the beginning of the century are unreliable) but, following the land reform, livestock densities may have fallen by as much as 30 % (Preston 2000). A comparison of the species structure of livestock populations for the Camacho valley to the east of the mountain slopes suggests that during the 20th century the proportion of sheep decreased from around 62 % to just 33 % while the importance of cattle increased from 18 % to almost 40 % (Preston 1998). This may be related to the fact that sheep demand more supervision while cattle ownership confers a degree of social status. Changes in livestock numbers during the decade of the 1990s, as reported by farmers in a series of communities in which we have recently worked, are discussed later.

14.3.2 Change in the Highlands

On the Tarija altiplano, farming depends on the cultivation of suitable areas for potatoes and broad beans as well as raising sheep, llamas, donkeys and cattle. The effects of grazing are seen in the composition of the flora. The consequences of trampling and compaction of the soil, as well as close grazing of selected plants has led to the elimination of some palatable species from certain areas, particularly close to settlements where livestock pass daily. For instance, the swards of annual gramineae of *Poa* and *Agrostis*, and perennials such as *Deyeuxia* and *Festuca* seem to have been replaced in some areas by hardier and less palatable species such as *Stipa leptostachya, Festuca orthophylla,* and *Tetraglochin cristatum* (Beck et al. 2001).

Conclusive evidence is lacking that the rangelands of the Tarija altiplano are chronically overgrazed as a result of current management practices. The use of multiple ecological zones and practices of transhumance and zone management has to be included in any assessment of the sustainability of current practices (Fairbairn et al. 2000).

Altiplano households depend on a combination of livestock and arable farming as well as work outside the community. They engage in activities such as spinning, weaving and basket-making as well as the collection and sale of fuel wood and sheep manure. Informants repeatedly commented that they depended on all their activities to get by; difficulties resulting from drought or the loss of sheep through disease are readily compensated by migration to work in nearby Argentina—only two hours by vehicle from the Tarija altiplano—in order to earn enough to get by during a period when household resources are reduced. Because households are not totally dependent on their livestock, they are not tempted to put pressure on grazing resources by keeping more sheep so as to insure against the effect of drought or high animal mortality. This resilience allows other resources to be used until the local environment has recuperated (Preston el al. 2003).

14.3.3 Change in the Río San Juan del Oro Valley

The Río San Juan valley is characterized by fertile valley bottom land, often irrigated, producing a range of fruit, maize, beans and vegetables. The valley sides are often steep with limited grazing potential but a few goats, donkeys and sheep are kept there by some families. Farming is sensitive to regular droughts but in bad years people go to work in nearby Argentina for short periods until better weather allows a resumption of farming to satisfy their daily needs. People in the communities mentioned that the greatest environmental stresses that they faced were from the occasional droughts and hail storms. Surpluses of maize, fruit and vegetables are traded in regional fairs and informally with people from the altiplano. Livestock grazing is of secondary importance but some households with larger numbers of sheep send them to graze high up on the nearby altiplano in the care of individuals in altiplano communities who have access to ample pastures, in return for payment of fruit, sugar and a bit of cash (Fairbairn 1999). Erosion is limited and most noticeable on the steepest slopes but there is little evidence that it is human-induced (Coppus et al. 2003).

14.3.4 Summary of Environmental Changes

This review of evidence of environmental changes in one area of southern highland Bolivia has indicated that the evolution of the physical landscapes has been influenced by a number of factors. While changes in vegetation have undoubtedly occurred, it is over-simplistic to conclude that such changes could reasonably be said to signify land degradation. Grazing by both domesticated and wild animals inevitably modifies the vegetation in comparison with its pristine state. A widespread assumption exists in parts of the scientific community and in Bolivian local government that a form of grazing exists that damages the environment and can be labelled 'overgrazing'. Such a term is misleading because any grazing removes plants that would otherwise be undisturbed and thus might constitute overgrazing. A number of papers have questioned the validity of the term whose use is rarely supported by evidence. Human-induced factors clearly play a part in reducing and/or changing vegetation cover in rangelands; however few studies quantify the influence of grazing on erosion and loss of productivity. Those that have made such an attempt highlight the complexity of vegetation cover changes after intensive grazing. Wilson and MacLeod (1991) for example, comment on the lack of concrete evidence to back up broad claims that the rangelands of the world, (particularly in Australia, USA and Africa) are 'overgrazed'.

14.4 Old and New Vegetation Components of Tarija Vegetation

Grazing and trampling may have led to the loss of palatable components of the vegetation as had the collection of some bushes for fuel. Rotation of grazing livestock seasonally has reduced the loss of some plants and the growing of fodder crops for livestock has had positive effects in some areas. In a study of livelihood strategies in the Altiplano and the Rio San Juan del Oro, Fairbairn (1999) emphasized the wide range of livestock movements that are undertaken to ensure that livestock feed adequately and grazing areas maintain their capacity to feed grazing animals. However it is important to note that informants repeatedly commented that vegetation cover is less than one or two generations ago, in particular nearer the villages, because of clearance for fuel and also on account of the frequent droughts.

Very considerable variations in vegetation occur over the Altiplano, according to patterns of human use, but the most isolated areas still have comparatively abundant natural vegetation cover, including parts of the altiplano above the east bank of the Río San Juan.

It should also be noted here that many small areas of woodland with native trees, such as alders (*Alnus acuminata*) survive in some hillsides in the temperate valleys and Pino del cerro (*Podocarpus parlatorei*) and Keñua (*Polylepis crista-galli*) also occur on some higher mountain slopes. These species are particularly prized since they can grow on steep slopes not necessarily suitable for grazing and serve to prevent soil erosion (Fig. 14.5).

The leguminous Tipa (*Tipuana tipu*) survives well in poorer valleys and is used for fuel, for fencing and its leaves are grazed by more agile livestock. It is believed to have been much more widespread before the increase in human population and associated livestock. It has in part been replaced by the Churqui *Acacia caven,* but

Fig. 14.5 Surviving alder woodlands near Alisos.

Fig. 14.6 Tipa (*Tipuana tipu*) tree surviving erosion.

Fig. 14.7 Churqui (*Acacia caven*), a recent colonist.

its ability to survive in areas subject to extensive erosion indicates its importance (Fig. 14.6).

Churqui (*Acacia caven*) has colonized the temperate valleys over more than 400 years, with its foliage consumed by cattle and its seed pods consumed by goats which assists its advance in the valleys. It is notable that a good grass cover grows beneath the trees as a result of the shade provided; this is favoured by small stock. Its wood is also valued for construction and for fuel, while its rapid spread is the result of deliberate propagation (Fig. 14.7).

Detailed studies of erosion features, soil surface characteristics and biomass of palatable species in 36 sites in 15 representative physiographic units in the Rio San Juan, the altiplano and the Tarija valleys concluded that current soil and vegetation degradation processes observed are largely associated with climatic and geomorphological factors (Coppus et al. 2003).

14.5 Climatic Changes in Bolivia

A number of papers have reviewed the accumulating evidence of recent climatic changes taking place in highland Bolivia but much less have been published on changes in areas to the east of the Andes. Such changes need to be taken account of and placed in the context of rural peoples' perceptions of and response to change. The timing of rainfall is now later in the year as well as more frequent and more intense throughout the central Andes. The impact of decreasing springtime soil moisture on crops will present challenges to farmers (Thibeault 2010; Seiler et al. 2013; Valdivia et al. 2013). Research with farmers in the northern Bolivian Altiplano near Lake Titicaca and in the highlands east of the altiplano in Oruro

Department confirm these changes, while farmers report unreliable crop yields and smaller numbers of sheep being kept (Chaplin 2007). There is therefore little doubt that climate changes are occurring and being detected by farmers.

Relatively little has been published seeking to identify climatic changes in the valleys east of the Andes and the eastern lowlands, although Nordgren (2011) has carried out a simple study using local meteorological data and matching these with rural people's perceptions of changes. In this work, studies were carried out in the altiplano, the upper valleys of Cochabamba, the Chaco and Amazonia. He found varying degrees of accord between rural people's perceptions of changes in climatic events and local meteorological data, but it may reasonably concluded that there is widespread perception of a range of climatic changes that rural people have noticed in the past 20 years. Our research in Tarija communities has shown that people perceive a range of challenges presented by weather extremes, that are sometimes linked to their perceptions of climate change.

14.6 Contemporary Rural Household Livelihoods in Relation to Environmental Resources

Although changes in farming systems have occurred in response to the opportunities available in different locations, such as water for irrigation, ample areas for grazing animals and fields with fertile soils for growing potatoes, maize or beans, other changes also occur as a result of labour shortage as more children attend school for longer, people go to work in Argentina and others combine a job in a nearby town with farming in their spare time. Such changes occur widely in many parts of the world and are placed in a convincing theoretical framework by van der Ploeg (2008).

Arable farming is for both subsistence and also for sale but growing crops for sale is more common in the temperate valleys where there is better access to urban markets. Livestock are kept for subsistence use but also for sale of wool, milk and/or cheese when there is access to markets: in many ways cattle serve as a savings bank and sudden need for an important ceremony or to survive an economic crisis can be satisfied by the sale of an animal. Cattle grazing on mountains demand less care and taking cattle to the humid valleys to the east and leaving them to be cared for by local farmers also reduces demands on the owners. On the Altiplano, sheep provide meat, wool, dung/fertilizer, but need daily care. Access to grazing land is variable but includes common land and use of tenuous rights to hillsides in all three areas studied in Tarija. Being able to leave cattle for weeks or months unvisited on mountainsides is an attractive option for households short of labour power.

Detailed studies of vegetation and pasture use in Tarija suggest that households manage some degree of control of the pastures but their limited understanding of carrying capacity of different vegetation types has led to a sub-optimal use and occasional gradual degradation of some vegetation types. The use of different micro-ecological areas for grazing according to the season (or the actual weather) is important for reducing the impact of grazing by livestock (Paniagua and Yevara 2000).

Our most recent research in Tarija included the collection of data from a sample of 40 households in the temperate valleys and on the eastern mountainsides (Preston et al. 2011). This enabled an assessment of changes in the livelihoods of households in the 10 year period 2000–2010. Lack of access to grazing areas does not seem to be a common problem. Environmental changes were seldom mentioned and only 10 % of households mentioned weather changes such as increased frost and hail losses and also lower crop yields.

These findings can usefully be compared with those of the work by Fairbairn (1999) ten years earlier, on households living on the altiplano and in the Rio San Juan valley. Households were asked about the relative importance of the various activities in which household members were engaged and also about how they responded to challenges presented by extreme weather. In addition they were asked about how grazing was managed to ensure livestock well-being. It is clear from the responses from many households that a wide range of grazing patterns were employed to ensure that pastures in different locations are used in order to provide adequate grazing for the animals, whether sheep, goats, cattle or camelids, and to allow time for plants in each grazing location to recover (Fairbairn 1999). Livestock and agriculture were clearly the most important activities but, in both areas, work away from home was also important. The people are for ever travelling to one place or another, or to distant workplaces or with their livestock seeking areas of good grazing. Improvements in transport over the past 40 years make seasonal migration easier while a series of factors encourage migration—climatic hazards such as droughts or floods, the increasing need to afford secondary education for the children and, among young people, the perception that farming is difficult and unproductive.

Research in neighbouring northern Argentina in association with an earlier EU-funded international programme suggested that, in the altiplano, many similar changes have occurred with no inference of associated environmental degradation. People moving away from the area have led to flocks becoming smaller and there seemed a greater reliance on subsistence production. Some areas that used to be cultivated are now grazed but grazing has intensified nearer to settlements and there is a balancing improvement of vegetation in the more distant areas that are now less used (Arzeno 2000).

14.7 Political Interpretations of and Responses to Environmental Situations

We have indicated that a range of environmental changes are occurring in the Bolivian Andes and that these are viewed in different ways by farmers and scientists. Change is ever-present and a feature of natural and human environments. For example, foods that are widely consumed in Europe and North America, such as the potato, maize and, most recently, quinoa, are known to have originated in this part of South America.

Concern with the negative effects of environmental change on the human population in Bolivia is relatively recent. This concern has not often been translated into political actions to try to identify those environmental changes that affect large populations negatively and to develop policies that minimize negative environmental change. Such actions would enable local populations to benefit from human-induced change. It was notable that, on first planning research on environmental change in southern Bolivia, we realized that a well-funded government agency existed in Tarija specifically charged with 'rehabilitating land in the Department of Tarija (Programa ejecutivo de rehabilitación de tierras en el departamento de Tarija—PERTT). The history of this organization indicates how such an organization can be created, the nature of the funding that it received and the limitations on what it has achieved (Ruíz Martínez 2001).

Action relating to environmental issues is likely to originate in the main city and political heart of effective activism. The first organization created in Tarija that was concerned with the conservation of natural resources was the Defenders of Tarija Natural Resources (Los defensores de los recursos naturales de Tarija) founded in 1930 by Jorge Paz Rojas. He believed that what he saw as the problem of erosion in the badlands close to some parts of the city should be tackled by the planting of trees, largely of foreign origin, such as the eucalyptus, and by controlling grazing. There was no prior scientific study of soil erosion and gullying, or of the role of existing native trees in controlling run-off in the area around the city. A further organization to 'save' the Tarija valleys was started in 1977; this encouraged the creation of a new organization, with the cooperation of the national Ministry of Rural Affairs, the Prefecture of Tarija Department and the Centre for Forestry as well as the main Tarija university. The result was the creation of PERTT in 1978. This was the first organization in Bolivia specifically charged with erosion control. At the time there was no national government agency charged with soil conservation or river basin management. PERTT obtained financial support from United States aid agencies (USAID), World Food Programme and the German government development aid.These agencies were concerned with creating local level support for remedial work in the countryside, irrigation programmes, river basin management and agricultural development.

A report in 1993 indicated that the main achievements of PERTT included tree planting—eucalypts, cypresses and pines—in a series of small catchments, all probably within an hour's drive of the city. In addition some fruit trees, soft fruit and fruit-bearing cacti were also planted and 14 irrigation systems benefitting 450 households were established (PERTT 1993). There was only limited follow-up of the work done and no post-project evaluation of any sort. It was concluded by a well-qualified person who had worked with the organizations mentioned that the results of the work by PERTT over a decade and a half between 1978–94 were very limited, considering the US$9.7 million budget covering this period (Ruíz Martínez 2001).

The benefits of a keen interest in environmental issues by a politically well-placed person such as Paz Rojas in the 1930s were limited by the absence of a scientific study that could identify the main environmental and social issues in the Tarija valleys and ways in which these issues might be overcome. The need for such a study was not appreciated at that time. Political authorities viewed it necessary to plan remedial action that could benefit households living in areas whose progress was impeded by environmental problems. At the same time, taking remedial actions such as irrigation and flood control could readily be seen by urban people as worthwhile, as produce from these rural areas was reaching city markets.

Based on a limited number of meetings with PERTT staff in the early 2000s, it appeared that their work with some communities was part of a logical approach to help make farmers aware of the opportunity to change some of their existing practices in order to better maintain the potential of farming areas. However PERTT seemed to be undertaking little community-focussed action which might have helped diffuse broader knowledge of alternative resource conservation practices.

During the time our team was working in Tarija there was little evidence of public debate about environmental issues. It was obvious that the Guadalquivir River which passes through the city of Tarija received water contaminated in various ways but public opinion on such matters seemed rarely expressed in the media. One environmental activist with whom we have maintained contact now has a television programme and will shortly be publishing a magazine dedicated to environmental issues. The existence of such opportunities for environmental debate indicates at least some degree of awareness that such issues are relevant to contemporary life in urban as well as rural areas.

14.8 Conclusions

There is limited evidence that changes in the natural environment, soil loss or erosion in the altiplano and adjacent valleys in the western part of Tarija, are more rapid than in the recent past, or that vegetation is more sparse. Flexibility in the use of different areas in rural communities for livestock and crops facilitates accommodation to preferences for cattle over other ruminants. Farmers are used to change and, to a certain extent, can modify their use of different areas of land to ensure that livestock survive and some crops can still be harvested. The research conducted in three very different parts of western Tarija Department over more than a decade offers an insight into how rural people have managed their environmental resources in such a way as to permit the maintenance of a satisfactory quality of life for households, despite both cultural and environmental changes.

In the valleys not far from the city of Tarija, in spite of dramatic evidence of extensive erosion in certain areas, we concluded that degradation is probably less than a century ago. The areas of active erosion seemed to be much less than was reported 20 years ago. Livestock remain important for rural households and the flexible use of nearby hills and of adjacent high mountainsides allows satisfactory rearing of cattle and sheep. Data on livestock holdings over a number of decades suggest that overall livestock numbers have decreased but that cattle have become more numerous.

On the altiplano to the west of the Tarija valleys livestock are important but they graze a variety of micro-ecological locations according to the season. A range of complementary activities such as weaving, using wool from their own sheep, collecting livestock manure for sale to farmers in the valleys as well as the sale of fuel wood allow people to get by. Particularly importance is attached also to the proximity of Argentina where seasonal earnings can help households overcome shortages caused by weather extremes.

In the Río San Juan valley, to the west of the altiplano, farmers have even better access to Argentina for occasional work when necessary but also their agricultural produce from land in the well-watered valley fields can readily be traded at regional fairs and purposive visits to Tarija or even Tupiza on the border with Argentina. Relatively few livestock use the higher hillsides and this allows some households to make seasonal use of such areas.

We have suggested, on the basis of many discussions with rural people in the areas of Tarija where we have worked over two decades, that rural people are conscious of noteworthy changes in weather patterns. However, as we have indicated, it is expected that farming needs to take account of irregular patterns of weather as well as animal disease and fluctuating prices of farm produce. The existence of areas where some degree of communal access to grazing is possible allows flexibility on use of grazing resources.

Particular extreme weather events such as floods, droughts and unusual seasonal weather are widely commented upon, perhaps as popular awareness of climate change increases. Clearly visible evidence of change such as the remarkable decrease in snow accumulation on Mount Chacaltaya in La Paz, where once tourists were taken to watch skiing, is an indication of different weather patterns. In Tarija, occasional extensive flooding affecting major highways is treated in the media as evidence of climate change and there is now widespread coverage in the Bolivian news media of such events throughout the country. This results in a relatively widespread awareness of climate change and studies in different parts of highland Bolivia demonstrate some degree of recognition of this in rural communities.

Many changes in land use are the result of the ways households manage their lives rather than the deterioration of environmental resources. For example, greater mobility and the widespread use of modern communication technology improve the awareness of prices paid for some rural products in urban markets.

In respect of a local political response to environmental issues, the history of action is notable that was started in 1930 and was partly instrumental in the foundation of an environmentally aware regional agency in the 1970s. We have concluded that this had only a limited impact on the environment in the Tarija valleys. There is no evidence to suggest that recent regional governments have been adequately informed or politically motivated to take remedial action to help those suffering from extreme climatic events. National government policies in respect of environmental protection for those affected by extreme events (storms, droughts etc.) seem to have had little effect in Tarija. However, it is remarkable that recent legislation offering social protection through state pensions and financial support for parents of schoolchildren, although minimal, is appreciated and may contribute to some rural households believing that their quality of living is maintained, even if at a lower level than that of urban people.

References

Arzeno, M. (2000). Desestructuración de la economía campesina y problemáticas ambientales en un área de la Quebrada de Humahuaca. Informe Técnico, Proyecto Ambiente y Sociedad en los Andes: Estrategias y Políticas, Instituto de Geografía, Universidad de Buenos Aires. https://sites.google.com/site/dpresearch36/home/bolivia/tarija.

Beck, S., Paniagua, N., Yevara, M., & Liberman, M. (2001). La vegetación y uso de la tierra del altiplano y de los valles en el oeste del departamento de Tarija. In S. Beck, N. Paniagua & D. Preston (Eds.), *Historia, ambiente y sociedad en Tarija, Bolivia* (pp. 47–93). La Paz: Instituto de Ecología, Universidad Mayor de San Andrés.

Chaplin, A. (2007). *Percepciones de comunarios y comunarias del altiplano Boliviano sobre los cambios en el clima y sus efectos*. La Paz: CIPE, CIPCA and Christian Aid.

Chepstow-Lusty, A. J., Bennett, K. D., Fjeldsa, J., Kendall, A., Gallano, W., & Tupyachi Herrera, A. (1998). Tracing 4000 years of environmental history in the Cuzco area, Peru, from the pollen record. *Mountain Research and Development, 18*, 159–172.

Coppus, R., Imeson, A., & Sevink, J. (2003). Identification, distribution and characteristics of erosion sensitive areas in three different Central Andean ecosystems. *Catena, 513*, 315–328.

Corrado, A., & Comajuncosa, A. (1990). *El colegio franciscano de Tarija y sus misiones* (Vol. 1). Tarija: Editorial Franciscana.

Fairbairn, J. (1999). Estrategias de sobrevivencia de hogares en el altiplano tarijeño y el valle del rio San Juan del Oro, Bolivia, Documento de Trabajo 99/01 School of Geography, University of Leeds. https://sites.google.com/site/dpresearch36/home/bolivia/tarija.

Fairbairn, J., Preston, D., Paniagua, N., & Maas, G. (2000). *Grazing and environmental change on the Tarija altiplano*, Working Paper 00/04, School of Geography, University of Leeds. https://sites.google.com/site/dpresearch36/home/bolivia/tarija.

Freeman, P. H. (1980). Bolivia. State of the environment and natural resources. A field study. McLean, Va.: JRB Associates.

Gesellschaft für Technische Zusammenarbeit (GTZ). (1988). *Bolivia, desarrollo agroforestal en el valle de Tarija. Relato Final*. La Paz: GTZ.

Lathrap, D. W. (1973). The antiquity and importance of long-distance trade relationships in the moist tropics of pre-Columbian South America. *World Archeology, 5*, 170–186.

Maas, G., Macklin, M., Warburton, J., Woodward, J., & Meldrum, E. (2001). A 300-year history of flooding in an Andean mountain river system: the Río Alizos, southern Bolivia. In D. Maddy, M. Macklin, & J. Woodward (Eds.), *River basin sediment systems: Archives of environmental change* (pp. 297–323). Rotterdam: Balkema.

Murra, J. V. (1972). El control vertical de un máximo de pisos ecológicos en la economía de las sociedades andinas, In I Ortiz de Zúñiga, Visita a la provincia de León de Huánuco [1562] (Vol. 2, pp. 429–476). Huánuco: Universidad Nacional Hermilio Valdizan.

Nordgren, M. (2011). Cambios climáticos. Percepciones, efectos y respuestas en cuatro regiones de Bolivia. La Paz: CIPCA.

Paniagua, N., & Yevara, M. (2000). Caracterización de las pastures y el pastoreo en las comunidades de Tojo (Cuenca del río San Juan del Oro) y Juntas (Cuenca del Río Camacho), Tarija, Bolivia. Informe Técnico 2000/02. https://sites.google.com/site/dpresearch36/home/bolivia/tarija.

van der Ploeg, J. D. (2008). The new peasantries. Struggles for autonomy and sustainability in an era of empire and globalization. London: Earthscan.

Presta, A. M. (2001). 'Hermosos, fértiles y abundantes'. Los valles centrales de Tarija y su población en el siglo XVI. In S. Beck, N. Paniagua & D. Preston (Eds.), *Historia, ambiente y sociedad en Tarija, Bolivia* (pp. 25–39). La Paz: Instituto de Ecología, Universidad Mayor de San Andrés.

Preston, D. (1998). Postpeasant capitalist graziers: The 21st century in southern Bolivia. *Mountain Research and Development, 18*, 93–122.

Preston, D. (2000). Pastoreo en el altiplano Tarijeño: cambios 1906–1999, EU Project Working Paper 00/02, School of Geography, University of Leeds. https://sites.google.com/site/dpresearch36/home/bolivia/tarija.

Preston, D., Montaño, R., & Cano, M. I. (2011). Trajectories of change: household livelihood dynamics in rural communities in southern Bolivia, Research Report, Proyecto HEDECOM, Tarija. https://sites.google.com/site/dpresearch36/home/bolivia/tarija.

Preston, D., & Punch, S. (2001). Estrategias de vida del hogar, sistemas de producción y medio ambiente: el valle Camacho, Bolivia. In S. Beck, N. Paniagua & D. Preston (Eds.), *Historia, ambiente y sociedad en Tarija, Bolivia* (pp. 95–122). La Paz: Instituto de Ecología, Universidad Mayor de San Andrés.

Ruíz Martínez, J. (2001). Lucha contra la erosión en el Valle de Tarija, un acercamiento a la visión del trabajo institucional. In S. Beck, N. Paniagua & D. Preston (Eds.), *Historia, ambiente y sociedad en Tarija, Bolivia* (pp. 217–230). La Paz: Instituto de Ecología, Universidad Mayor de San Andrés.

Saignes, T. (1995). Indian migration and social change in 17th century Charcas. In B. Larson & O. Harris (Eds.), *Ethnicity, markets and migration in the Andes* (pp. 167–195). Durham: Duke University Press.

Seiler, C., Hutjes, R. W. A., & Kabat, P. (2013). Likely ranges of climate change in Bolivia. *Journal of Applied Meteorology and Climatology, 52*, 1303–1317.

Thibeault, J. -M. (2010). Changing climate in the Bolivian Altiplano, PhD, University of Connecticut. http://digitalcommons.uconn.edu/dissertations/AAI3415565/.

Valdivia, C., Thibeault, J., Gilles, L., Garcia, M., & Seth, A. (2013). Climate change trends and projections for the Andean Altiplano and strategies for adaptation. *Advances in Geosciences, 33*, 69–77.

Wilson, A. D., & MacLeod, N. D. (1991). Overgrazing: Present or absent? *Journal of Range Management, 44*, 475–482.

Zimmerer, K. S. (1996). Discourses on soil loss in Bolivia. Sustainability and the socioenvironmental 'middle ground'. In R. Peet & M. Watts (Eds.), *Liberation ecologies. Environment, development and social movements* (pp. 110–124). London: Routledge.

Chapter 15
Media Perceptions and Portrayals of Pastoralists in Kenya, India and China

Mike Shanahan

Abstract Resilient food systems require policies that let people use their own adaptive capacity. Pastoralists use mobility to take advantage of pasture and water that are patchily distributed in space and time. Pastoralism can therefore strengthen food security, livelihoods and economic prosperity. However, policy makers, donors and the public rarely acknowledge these benefits. This is partly because development and media narratives paint pastoralism as something bad that must change. This paper explores how the media portrays pastoralism in Kenya, China and India. In Kenya, pastoralists feature mostly in 'bad news' stories of conflict and drought. They appear vulnerable and lacking in agency. Stories rarely mention the benefits that pastoralists bring. In China, the media presented pastoralists as the cause of environmental degradation and as (generally happy) beneficiaries of government investment and settlement projects. In India, newspapers portrayed pastoralists with more pity, as vulnerable people whose livelihoods were at risk. Overall coverage of pastoralism in India was rare, however. Journalists surveyed there said pastoralists are "invisible" to editors. In each country, important topics such as climate change, the economic value of pastoralism and the links between mobility and resilience, were under-reported. Most articles about pastoralists failed to quote these people. Improved media coverage of pastoralism is part of the institutional capacity that is needed to support resilient food systems. Improved eco-literacy among journalists and editors can help strengthen the resilience of vulnerable communities and national food systems alike, and will become more important as climate change takes hold.

Keywords Media narratives · Development narratives · Kenya · India · China · Pastoralism · Drylands · Climate change

---

M. Shanahan (✉)
International Institute for Environment, 80-86 Gray's Inn Road, London WC1X 8NH, UK
e-mail: mikeshanahan@yahoo.com

R.H. Behnke and M. Mortimore (eds.), *The End of Desertification?*
Springer Earth System Sciences, DOI 10.1007/978-3-642-16014-1_15

15.1 Introduction

Mobile pastoralism—in which herders move livestock across landscapes so they can exploit resources such as pasture and water that are patchy in space and time—is an ancient livelihood that contributes to food security, plays a vital role in the ecology of drylands, and provides pastoralists with flexible strategies for dealing with uncertainties, such as a variable climate. But development narratives have tended to disagree. Development narratives are strategic simplifications that help in the face of situations whose complexity can paralyse policy making (Roe 1991). They generate consensus around major policies and make political action possible. But they can also be problematic. As simplifications, narratives are fundamentally different from scientific theories. While scientific facts are falsifiable, narratives are not. They escape the checks and balances of science—such as publication of evidence, peer review and replication. Narratives need the support of scientific authority but at the same time they need to avoid the complexity and conditional nature of scientific knowledge and this is why they exist. Narratives can be fairly relevant representations of the situation they are designed to address. But like wide-angle camera lenses that capture a huge range of variety, the scenarios they produce are increasingly distorted at the edges. One area of public policy where narratives have been contentious is in relation to pastoralism and other forms of food production in the world's drylands. The nature of such narratives—and whether they can be modified or improved—will only grow in importance with climate change. This is not only because climate models predict more extreme and more variable climatic patterns, but also because knee-jerk policy responses to the threats posed by climate change can create new problems for food production in the drylands, such as investments in large-scale irrigated agriculture that are not well suited to a more variable climate.

Krätli and Enson (2013) have reviewed current and recent public policy narratives on the drylands, promoted by various global institutions. Their review provided the basis for discussion among researchers working in Kenya, China and India who met in September 2012 at a workshop in Kenya (Shanahan 2012). Participants at the workshop agreed that dominant policy narratives cast pastoralism as a backward, wasteful and irrational livelihood that takes place in fragile, degraded and unproductive ecosystems and creates a catalogue of problems for non-pastoralists. The narratives frame pastoralism as something that should be replaced, because it is uneconomic, archaic and ungovernable. They frame pastoralists as lazy, poor and at times criminal and dangerous. And they portray the mobility that makes pastoralism possible as problematic, random, unproductive and a cause of conflict and disease. There is more nuance to these narratives at a national and subnational level. In China, for instance, the dominant policy narrative frames nomadic herding as a livelihood that damages grasslands, and says that when herders settle in towns they will have a better, more economically productive life.

Pastoralists themselves might of course disagree. And a growing body of recent research shows that the dominant narratives are far from accurate, that mobility is an asset (see de Jode 2009) and that pastoralism is an economic powerhouse. In the Horn of Africa alone, the informal livestock trade is estimated to be worth more than US$1 billion each year (Catley et al. 2012). A modified narrative could show that pastoralism has inbuilt adaptability and can harness environmental variability in a positive way—something that will be critical as our climate changes. It could enable pastoralism to meet its potential to increase equity, environmental sustainability and economic output in the drylands. As part of a larger project that aims to identify ways to influence policy narratives around pastoralism (Hesse 2013), I examined the role of the media in reinforcing dominant narratives and asked how journalistic coverage of the sector could improve. To assess media perspectives on pastoralism I analysed media reports from Kenya, China and India—and asked dozens of journalists in those countries to complete an online survey.

15.2 Media Content Analysis

The project aimed to understand how journalists portray pastoralists and pastoralism, who speaks for and about pastoralists in the media, and in what contexts the media reports on pastoralism. I used the LexisNexis newspaper database (LexisNexis 2006) and the websites of individual newspapers to find articles that mentioned any of the following terms: pastoralist, pastoralists, pastoralism, herding, herder, herders. For China, I searched the China Daily and People's Daily websites. For India, I searched the Times of India, The Hindu and Hindustan Times websites. I scored each article for the presence or absence of 106 content types (e.g. "Article refers to meat or milk"; "Article refers to drought"; "Article quotes government official"). For this, I used a binary coding system that I based on the one used by Billett (2010) in his study of Indian newspaper coverage of climate change. I had already tested the coding system on a sample of articles from The Guardian (UK) newspaper and refined it accordingly before using it in this study.

Readers should note that the study covered only English language media. In the case of India, this meant missing vernacular language press in pastoral regions in favour of articles in the national media. In the case of China, this meant that the study largely focused on articles in state-owned media that are aimed at English-speaking (i.e. foreign and urban elite) audiences. I analysed 100 media articles from Kenya, 50 from China and 20 from India (Tables 15.1 and 15.2). These numbers reflect the relative abundance of stories about pastoralism in the media sources searched in each country. Tables 15.3, 15.4, 15.5, 15.6, 15.7, 15.8 and 15.9 present some of the main trends and inter-country differences in media coverage that our content analysis revealed. This showed how the media in each country portrayed pastoralism in a very different way, and that in each case, the portrayal was close to that of the dominant policy narratives.

Table 15.1 Time distribution of media articles on pastoralism analysed

	1982	1985	1998	1999	2000	2001	2002	2003	2004	2005	2006	2007	2008	2009	2010	2011	2012	Total
Kenya					5	2		4	6	7	17	3	5	15	8	24	4	100
China	3	1	2					1			1	1	7	6	10	5	13	50
India									1			5	1	1	4	2	6	20

Table 15.2 Sources of media articles from each country

Kenya		China		India	
The Nation	65	People's Daily	21	The Times of India	7
Nairobi Star	15	Xinhua News Agency	12	The Hindu	6
The East African Standard	13	China Daily	11	DNA (Daily News + Analysis)	2
The Star	3	Global Times	5	The Indian Express	2
The East African	2	Shanghai Daily	1	Indo-Asian News Service	1
East African Business Week	1			Press Trust of India	1
Business Daily	1			Tehelka	1
				Hindustan Times	1
Total	100		50		20

15.2.1 What's the Story in Kenya?

In Kenya, pastoralists tend to star only in bad-news stories—93 % of those analysed here referred to conflict or drought. Otherwise, the media tends to ignore pastoralists. This sentence, from a 2006 article in The Nation, encapsulates the overall narrative: "Banditry, robberies, infiltration of small arms, poaching in the region's game reserves and national parks and frequent outbreak of livestock diseases are now being attributed to the uncontrolled movement of pastoralists and their animals." While 51 % of stories that mention conflict presented pastoralists as a cause of problems, only 5.7 % suggested that pastoralists might be the victims of the actions (or inactions) of others (e.g. farmers or government policies). While 28 % of articles reported efforts to evict or move pastoralists, in only one-fifth of them did the journalist describe where the pastoralists might go—and in every case it was back to where they had come from, back to the problems they left behind. An astonishing 22 % of all articles referred to pastoralists as "invaders" or as having "invaded" land. Pastoralists clearly have an image problem in the Kenyan media. Stories of pastoralists achieving, contributing or leading are extremely rare. For example:

> Isiolo North MP Dr Mohamed Kuti yesterday called on the security agents to mobilize all its resources in its disposal to ensure that the raiders were arrested and prosecuted for the offence. He regretted that pastoralists have continued to embrace outfashioned culture of cattle rustling and banditry and reminded that cattle raids are a thing of the past and that they must grow and change with the changing world. (Salesa 2011)
>
> Mr Warfa urged pastoralists to discard retrogressive cultural practices like cattle raids. (Kipsang 2012)
>
> "It's very hard to convince uneducated person to stop cattle rustling. To them, it is like a hobby. They participate in the raid to achieve respect and dignity in their communities," explains Mr Joseph Lekolua, a local politician. (Letiwa 2008).

Table 15.3 The most common themes in each country, by percentage of articles in which they appear

Kenya	Percentage (%)
Conflict/disputes over resources	70
Drought (in general)	51
Specific drought	47
Violence (or threat of violence)	43
Portrays pastoralists as vulnerable/needing help	43
Quotes pastoralist	41
Conflict between pastoralists and non-pastoralists	38
Portrays pastoralists as source of problems	37
Describes pastoralists as trespassing/encroaching	35
Dead livestock linked to climatic extreme (drought)	35
China	
Government acting to help pastoralists	86
Quotes government official	82
Refers to resettlement/sedentarization	52
Refers to land/soil degradation or to desertification	44
Implies pastoralism contributes to degradation	36
Refers to grassland restoration/conservation	36
Quotes pastoralist	36
Describes government investment in grassland areas	32
Puts currency value on government investment	30
Describes houses built for pastoralists to live in	30
India	
Portrays pastoralists as victims of external problems	60
Quotes Indian civil society organization	50
Refers to pastoralist rights or empowerment	45
Describes threats to the survival of livestock breeds	35
Refers to the mobility of pastoralists	35
Refers to scientific assistance to pastoralists (e.g. veterinary/improved grass or livestock)	30
Pastoralists portrayed as vulnerable/needing help	30
Quotes Indian scientists	30
Portrays pastoralism as having been marginalized by government	25
Refers to dwindling pasture	25
Quotes pastoralist	25

Half of all stories depicted pastoralists as poor and vulnerable. For Kenyan newspaper readers, this persistent narrative must seem depressingly familiar. And while it illustrates a failure of government to tackle the causes of conflict, it also reveals a failure of journalism to explore why this is the case. Kenyan media stories

Table 15.4 Comparison among countries by percentage of articles including each theme

	Kenya	China	India
Conflict	70	4	2
Climatic extremes	51	20	5
Climate change	3	8	10
Overgrazing/degradation	16	36	0
Pastoralists cause problems	37	12	5
Pastoralists are victims of problems	23	10	60
Food security	1	4	10
Pastoralist rights	2	6	45

Table 15.5 Sources quoted (percentage of articles)

	Kenya	China	India
Government/officials	70	82	15
Pastoralists	41	36	25
Scientists	7	26	30
National NGO/CSO staff	13	2	50
International NGO/CSO	9	0	10
Private sector	4	0	0
UN agency	5	0	15

Table 15.6 Percentage of articles with special reference to women and children in pastoralist communities (%)

	Kenya	China	India
Women	6	0	15
Children	16	8	5

Table 15.7 Different portrayals of mobility (% of articles)

Content type	Kenya	China	India
Refers to mobility of pastoralists	67	12	35
States that problems arose or are anticipated after movement of pastoralists	47	0	5
Includes statement that explicitly supports (or calls for support to) mobility as a way pastoralists can overcome resource scarcity	6	0	0

Table 15.8 Percentage of articles that mention climate change

Content type	Kenya	China	India
Refers to extreme climatic event (drought or flood)	51	20	5
Refers to climate change	3	8	15

Table 15.9 Percentage of articles that mention aspects of food security

Content type	Kenya	China	India
Mentions meat/milk	17	14	15
Refers to ways pastoralism contributes to food security beyond pastoralists	1	4	10

make virtually no mention at all of specific government policies and only a small proportion report on initiatives that could improve the lives of pastoralists, reduce conflict and promote sustainable development.

15.2.2 *What's the Story in China?*

In China, by contrast, pastoralists tend to feature in 'good news' stories in the English language publications. The media narrative is made up largely of stories about herders who have settled in towns and are largely happy with the change. These stories highlight government investments in housing and infrastructure to improve the wellbeing of poor communities. They often quote pastoralists who tell how they have gained materially since abandoning their nomadic lifestyle.

> "Practices have shown that settlement of local herders helps develop animal husbandry in a large scale and promote cultural, technological and educational undertakings in the pastoral areas," Qi Jingfa said. The way of settling down is also the best option for herders in need to become better off or become affluent, he noted. (Xinhua News Agency 1998)

> "I have never dreamed of living in such a nice place. The water and electricity are so convenient. I even can watch television," Nyima, a 70-year-old herder in Yushu prefecture of northwest China's Qinghai province, said Thursday. (Xinhua News Agency 2012a)

> In his cosy, furnished home, Dorjie recalled the nomadic lifestyle he lived just two years ago. At that time, he lived with his family in a shabby adobe structure on the pasture about 9 km away from his new home. "Raising 100 sheep and 30 heads of cattle, I earned only half of what I do now," Dorjie said. (Xinhua News Agency 2012b)

Although some articles describe support for pastoralism, they don't explain much about why nomads move in the first place. Many (36 %) articles blamed pastoralists for degrading grasslands.

> Over the years, nearly 1 million herders across the Qinghai-Tibet Plateau have settled or relocated to prevent the ecological degradation of the grassland. (Xinhua News Agency 2012c)

> Decades of global warming combined with over-grazing have degraded 90 percent of the grassland, forcing the government to push forward a series of environmental protection measures, including a massive human migration to preserve the region's delicate ecological balance. (Zou 2010)

> Long-term overgrazing has caused severe degradation of the grassland and a marked decline in its herd-carrying capacity. (Wei 2011)

15.2.3 What's the Story in India?

In India, the media narrative is quite different. It tends to present the pastoralist communities as victims (60 % of articles) who have lost access to grazing land because of the growth of industrial agriculture, the dominance of more powerful social groups, and limits to grazing in forested land, among others. Examples include a 2007 story in The Hindu, which stated that:

> The changing pattern of land use, rapid expansion of the irrigation area and privatisation of tenancy in the rain-fed areas are some of the factors responsible for erosion of livelihood security of pastoral people, whose way of life has come under threat from the mainstream development paradigm. Experts called upon the policy planners to recognize the potential of pastoralists to contribute to the growth process and look beyond the "rigid development model" which they said was only promoting the sedentary life. (Anon 2007).

…and a 2010 article in Tehelka, a weekly political magazine, which included this quotation from a herder:

> "In Mehsana district our grazing lands were encroached by upper castes. When we migrate, we are forced to live in cremation grounds outside village boundaries. Schools do not want to enrol our children. They think if we move, it will reflect in their school's dropout rates," says Hirabhai Bharwad, a Bharwad community leader. (Yadav 2010).

The concept of pastoralist rights appears often in the Indian articles (45 % of those analysed), as in this Indian Express story about the pastoral Gujjar people in Jammu and Kashmir, which included this quotation from Dr Javaid Rahi, Secretary of the Tribal Foundation:

> We have already written to the Prime Minister to intervene into the matter as forest rights were available to Gujjars even before independence. In erstwhile Dogra rule, Gujjars were enjoying forest rights, which were later snatched from them through legislations after the establishment of forest department in 1950s, (Anon 2010).

A relatively common theme in the Indian coverage, featuring in 35 % of articles, was about threats to local breeds of livestock and efforts to conserve genetic diversity. An example is an article published in 2012 in The Hindu, which included the following paragraph:

> Globalisation has led to a situation where the traditional role of pastoralists as custodians of animal genetic resources is on the wane. These indigenous breeds, which were maintained after a meticulous process of selection and breeding, could withstand local environment conditions. They are disease-resistant and culturally and religiously are part of our social imagination as property resource. The traditional herdsmen followed this process over centuries but they are all fading into memory, says Mr. Vivekanandan. (Karthikeyan 2012)

Indian articles were more likely (compared to Kenya and China) to describe how pastoralism can be a source of resilience to environmental change, and said more than those from the other countries about the value of pastoralism—to both the environment and the economy. As Sudha Passi wrote in a story for the Press Trust of India:

> Pastoralists or herders have traditionally never owned land, but have utilized forest resources judiciously and have significantly contributed to economy, ecology and preserving biodiversity. (Passi 2004).

But such framing was still relatively rare overall (see below).

15.2.4 Missing Voices

The voices of pastoralists feature in less than half of the articles about them (41 % of articles in Kenya, 36 % in China and 25 % in India; Table 15.3). If pastoralists as a whole are missing, the perspectives of pastoralist women and children are even more so (Table 15.6). Government representatives dominate the articles in China (quoted in 82 % of articles) and Kenya (71 %), but in India only 15 % of the articles included a quotation from an official. There, civil society organizations had the biggest say (quoted in half of the stories, compared to 21 % in Kenya and just 2 % in China). Scientists had a quote in few Kenyan stories (7 %) compared to China (26 %) and India (30 %). While there is no ideal mix of voices in a story, there are marked differences between each country and this will influence the overall narrative that emerges from media coverage.

15.2.5 Missing Money

In both Kenya and India, the reports made rare mention of what government investment in pastoralist communities could mean. By contrast, one-third of the Chinese articles mentioned investment and in 94 % of these, there was a hard currency value attached. Very few articles in any of the three countries referred to the economic importance of pastoralism (4 % in Kenya, 12 % in China and 15 % in India).

15.2.6 Missing Mobility

Mobility is the key that pastoralists use to unlock the scattered riches of arid lands. The landscape may appear barren, extreme and risky to city-based journalists but the pastoralists have the knowledge and skills to take advantage of the land's variability and diversity. Stories that presented mobility in a positive light were rare. Just 6 % of Kenyan stories included a statement that explicitly supported mobility as a way pastoralists can access resources that vary in space in time. None of the articles in either China or India did. This is despite mobility itself being a common theme in the articles (Table 15.7). Indeed, in Kenya, nearly half of all the stories linked mobility to problems. This contributes to a false narrative, one that is blind to the true nature of the lands the pastoralists move across, and to the knowledge they draw upon to take advantage of resources that are distributed there in an unpredictable way.

15.2.7 *Missing Climate Change*

The media also fail to cover climate change in the context of pastoralism and the extreme climatic conditions that pastoralists face, and which their mobility can help overcome. In Kenya, although 51 % of stories mentioned drought, only 3 % referred to climate change. The topic got slightly more coverage in China (8 %) and India (15 %) (Table 15.8). When the media did mention climate change, it was to highlight the vulnerability of pastoralists, as in this example.

> The pastoralists are running out of ideas. They have exhausted every known coping mechanism. [...] The current situation gives urgency to the question of whether nomadic pastoralism is viable in an overpopulated environment and with worsening climate change. (The East African 2009).

15.2.8 *Missing Meat and Milk*

Very few articles mentioned how pastoralism contributes to food security outside of pastoralist communities (Kenya: 1 %; China 4 %; India: 10 %, see Table 15.9).

15.3 Surveys of Journalists

To complement the content analysis, I invited several hundred of the International Institute for Environment and Development's media contacts in Kenya, India and China to complete a short survey. The questions asked what journalists think and know about pastoralists and pastoralism, and about how the media covers this subject.

In Kenya, 42 out of 250 invited journalists responded (response rate 17 %). They work for media outlets that include print (e.g. The Nation, The East African, The Standard, The Star), broadcast (Baraka FM radio, Kenya Broadcasting Corporation, BBC World Service) and both domestic and international news agencies (Kenya News Agency and China's Xinhua news agency). Three of the journalists who responded were themselves from pastoralist communities (including Turkana and Borana) and two more had married pastoralists. These journalists had a combined 477 years of experience of journalism (average 11.4).

In India, 61 out of 207 invited journalists responded (response rate 29 %). Their media outlets include print (e.g. The Economic Times, Times of India, The Hindu, Hindustan Times, India Today, Deccan Chronicle), broadcast (UTV Bloomberg, ABP Majha) and both domestic and international news agencies (Press Trust of India, Inter Press Service). These journalists had a combined 790 years of experience of journalism (average 12.9).

In China, 16 out of 130 invited journalists responded (response rate 12 %). Their media outlets include print (e.g. South China Morning Post, NewsChina Magazine, Environmental Protection magazine, Private Economy News, China Daily, Southern Weekly), broadcast (China Central Television) and both domestic and international websites (Caixin Online, chinadialogue.net, SciDev.Net). These journalists had a combined 173 years of experience of journalism (average 10.8).

Figures 15.1, 15.2 and 15.3 show word-clouds created from what the journalists wrote when asked which five words or phrases they associated with nomadic herders or herding in general. In each figure, the most commonly mentioned words appear largest. The commonest words described environmental and social dangers in Kenya, poverty in India and a somewhat more romantic vision of nature in China.

In the online survey, journalists were given an opportunity to say anything they would like to about media coverage of pastoralism. In India, 33 (67 %) of the 49 journalists who answered this question chose to say that the media has "neglected", "ignored", "forgotten" or otherwise under-reported the issues that affect pastoralist communities. They explained that this was because urban media consumers (and editors) are not interested in rural affairs. "They do not get the attention they deserve," said one journalist. "The media highlights only negative news like nomads selling girl children". Another commented: "rather little coverage and mostly as 'the other' or 'the untrustworthy'".

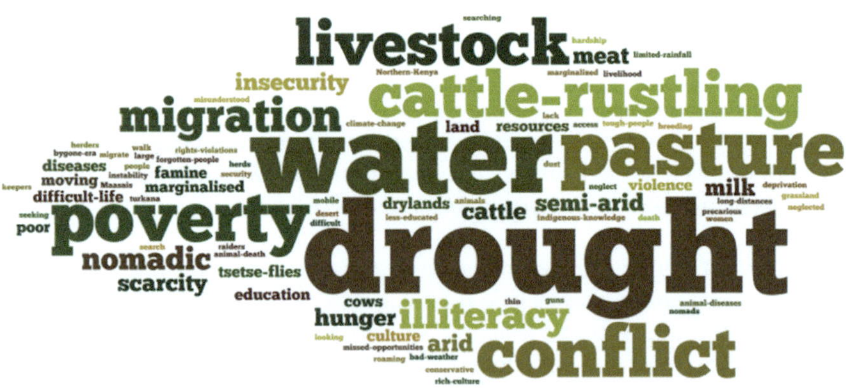

Fig. 15.1 Words Kenyan journalists associate with pastoralists; *commonly mentioned words appear largest.*

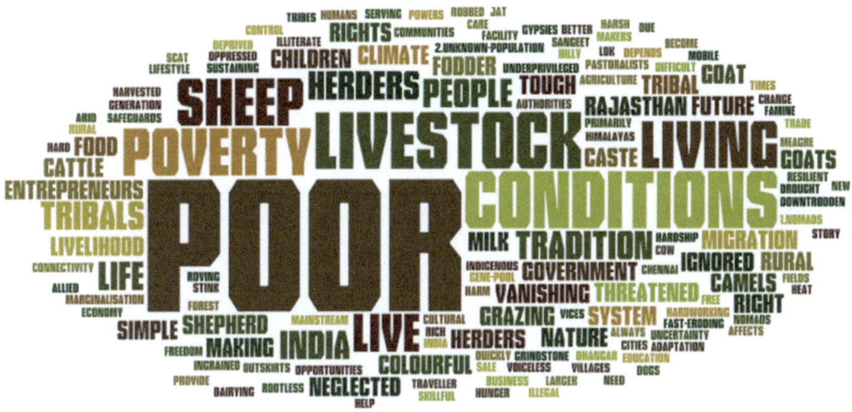

Fig. 15.2 Words Indian journalists associate with pastoralists; *commonly mentioned words appear largest.*

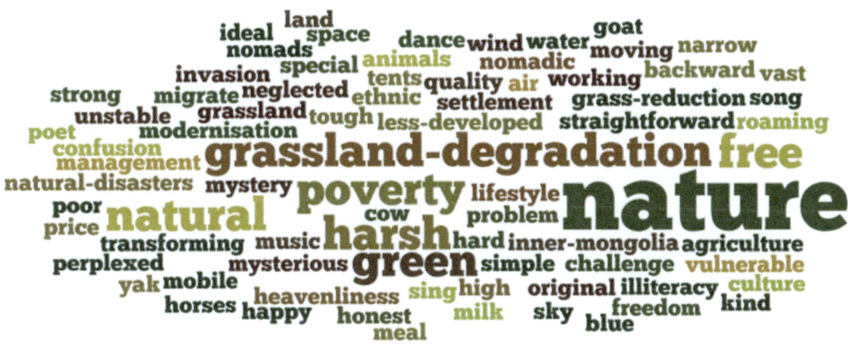

Fig. 15.3 Words Chinese journalists associate with pastoralists; *commonly mentioned words appear largest.*

Likewise in China, 55 % of the journalists who answered this question said that the media should report more on pastoralism. One noted that they could hardly recall a single Chinese media article on the subject. Another commented that "This is not a quite hot topic in media coverage in China. However, sporadic reports on climate change or nature reserves have indicated that desertification invades many parts of the grassland regions—including in Qinghai Tibetan Plateau and the Inner Mongolia regions—which affected the lives of herders in those regions."

Journalists in Kenya confirmed the way their media outlets connect pastoralism with conflict and woe. "The media only gives special attention to pastoralists or pastoralism when there is a crisis, like a major drought or famine where large numbers of people and animals have died," commented one. Another said: "Pastoralism is generally ignored. It only makes headlines when there is cattle rustling and scores of people are killed." Journalists pointed to neglect as a critical factor. "Pastoralism is misunderstood. Government and media have neglected pastoralist communities over the years," said one, while another stated: "The media has neglected pastoralism since its takes place in far flung areas of northern Kenya which the government has neglected for years."

Another Kenyan journalist put it bluntly: "Pastoralism is seen as a less glamorous beat. Very few journalists cover it."

When asked more specific questions, the journalists in all three countries revealed knowledge and opinions (Tables 15.10 and 15.11) that seem to contradict the dominant narrative presented in the national media.

Most (91 %) Kenyan journalists acknowledged, for instance, the importance of pastoralism to the national economy, with more than half of them stating that this is major. This is surprising given that this theme was invisible in the stories analysed. Only 4 % the Kenyan articles mentioned it, and not one published a figure such as a shilling, dollar or GDP value. Other things the journalists said suggest that there is

Table 15.10 Percentage of journalists in online survey who agreed with each statement

		Disagree strongly	Disagree somewhat	Don't know	Agree somewhat	Agree strongly
Pastoralists are to blame for conflict over resources such as land and water	Kenya	41.5	22.0	0.0	24.4	12.2
	China	20.0	40.0	6.7	26.7	6.7
	India	54.1	24.6	4.9	6.6	9.8
Pastoralists are poor, vulnerable and need help	Kenya	12.5	30.0	0.0	17.5	40.0
	China	0.0	6.7	20.0	46.7	26.7
	India	6.6	4.9	1.6	34.4	52.5
Pastoralism is backward and not suited to the modern world	Kenya	57.1	21.4	0.0	11.9	9.5
	China	40.0	40.0	0.0	20.0	0.0
	India	39.3	39.3	4.9	11.5	4.9
The government has neglected pastoralists	Kenya	10.3	15.4	0.0	23.1	51.3
	China	7.1	35.7	21.4	35.7	0.0
	India	4.9	8.2	4.9	37.7	44.3
Pastoralists need to settle and farm land instead of herding livestock	Kenya	46.3	22.0	2.4	22.0	7.3
	China	14.3	57.1	14.3	14.3	0.0
	India	16.7	16.7	8.3	35.0	23.3
Pastoralism helps to maintain a healthy environment	Kenya	19.5	31.7	4.9	31.7	12.2
	China	7.1	21.4	14.3	42.9	14.3
	India	6.7	6.7	26.7	43.3	16.7
Pastoralists cause environmental harm, e.g. overgrazing, land degradation, threats to species	Kenya	15.0	22.5	7.5	42.5	12.5
	China	28.6	42.9	7.1	14.3	7.1
	India	41.7	31.7	5.0	16.7	5.0

Table 15.11 Percentage of journalists in online survey who agreed with each statement

	Kenya	China	India
Pastoralism is highly vulnerable to climate change	81.0	64.3	44.1
Pastoralism is somewhat vulnerable to climate change	9.5	14.3	28.8
Pastoralism is no more or less vulnerable to climate change than other sectors	4.8	14.3	18.6
Pastoralism is somewhat resilient to climate change	4.8	7.1	3.4
Pastoralism is highly resilient to climate change	0.0	0.0	5.1
Pastoralism creates a major barrier to food security	4.8	0.0	3.4
Pastoralism creates a partial barrier to food security	4.8	7.1	10.2
Pastoralism has no overall effect on food security	7.1	57.1	39.0
Pastoralism makes a partial contribution to food security	45.2	35.7	35.6
Pastoralism makes a major contribution to food security	38.1	0.0	11.9
Pastoralism is a major burden to the economy	2.4	0.0	5.1
Pastoralism is a minor burden to the economy	0.0	21.4	1.7
Pastoralism is neither a burden nor a contributor to the economy	7.1	50.0	33.9
Pastoralism is a minor contributor to the economy	38.1	28.6	40.7
Pastoralism is a major contributor to the economy	52.4	0.0	20.3

an opportunity for a new narrative to emerge in the Kenyan media, one that does not ignore the social, economic and environmental benefits pastoralists provide:

- "There's a lot that the media can do to better the lives of the pastoral communities and integrating them in the modern economy without losing the essence of pastoral lives."
- "Pastoralism has a chance to become a key growth sector for Kenya's economy if supported by media and policy makers alike."
- "Pastoralists, if well harnessed, can play a bigger role in Kenya."
- "Livestock can contribute in a big way to the economy if properly nurtured. Kenyan media are not giving adequate coverage."

In China, most journalists (71.5 %) felt that herding did not cause damage to the environment, and more than two-thirds (67.8 %) even felt that herding had a positive effect on the environment. Among these journalists 71.4 % disagreed that herders need to settle instead of herding livestock, and 42.8 felt that the government had neglected herders. As one journalist commented: "Their lives are strongly impacted by the policy to make them settle down for reason of keeping stability, [and this] damages already-vulnerable ecology (herders could no longer graze in areas rich in grass)." Another said "In my heart, I know nomadic herders are good for the environment." These views contrast the dominant narrative in the English-language stories analysed here.

The answers from Indian journalists diverged from the media narrative most in the cases of a minority of respondents who expressed doubts about the value of a nomadic lifestyle.

Table 15.12 Where journalists get their information on pastoralism (%)

	Kenya	India	China
Media	47.5	36.8	69.2
Pastoralists	35	29.8	53.8
NGOs	20	7.0	15.4
Internet	15	14.0	7.7
Government	17.5	7.0	0
Journals	10	7.0	0
Researchers	7.5	1.8	0
Books	2.5	1.8	0
UN reports	2.5	0	0
Policy brief	2.5	0	0
Aid agencies	2.5	0	0

- "Media often looks at them as a public nuisance. The idea that people have to migrate because of poverty is not something that often gets attention. On pastoralism itself, I think most people who are nomads here are not doing it by choice."
- "Media give no attention to these people because they are lesser and don't contribute to society."
- "They need to be made part of a respectable living system."
- "Pastoralism cannot go on forever. It's simply too archaic to make economic sense."

Such strong viewpoints were absent in the Indian articles analysed, which in general portrayed pastoralists in a more sympathetic light.

The survey also asked journalists to identify their main sources of information about nomadic pastoralists and their lifestyle. In each country, the media was the most frequent answer and few journalists counted researchers among their sources (Table 15.12).

15.4 Toward New Narratives

A modified policy narrative around pastoralism might show how governments can make sensible decisions in the face of climate change and population growth by investing in pastoralism and, critically, in pastoralists on their own terms. The analysis of media articles presented here suggests that a modified media narrative would have a role to play, while the comments from surveyed journalists suggest that great potential for change exists. But there is still much work to be done.

In 1999, Saverio Krätli and Jeremy Swift wrote a report on pastoralism in Kenya in which they said:

> "The way pastoral conflict is reported [by the media…]—as a relatively unimportant, backward, tribal activity—is part of the problem. There is a need to improve press reporting […]. This should include working with the editors of major newspapers in order to promote better coverage and more accurate and up-to date reporting about the logic of pastoral system. Positive images […] must be circulated to combat the widespread view that pastoralism is backward and must change into sedentary, more agriculture-based, activities. Journalists who understand about pastoral districts must be identified and supported." (Krätli and Swift 1999).

A decade and a half later these recommendations still apply, not only in Kenya, but also in India and China. In each country, the media present pastoralism through a very narrow lens, one that is likely to create barriers to sustainable development. Opportunities to reframe pastoralism abound. In Kenya, for instance, an alternative narrative could show how the new constitution could work best for the drylands and their communities. In India, an alternative narrative could show how herding is part of the wider dryland agriculture system that can increase food security in the context of climate change. In China, an alternative narrative can relate how support for pastoralism can increase food security and better manage rangelands for economic benefits.

This analysis highlights what is missing. It points to areas that journalists and editors can pursue in creating a more balanced, more nuance and more accurate narrative around pastoralism. That will involve reporting on the economics of pastoralism, as well as on the other values of pastoralism that are harder to price. It will involve a better understanding of mobility and markets, of resilience and vulnerability. It will require journalists and researchers to communicate better together and it will require the media to give more voice to the pastoralists themselves. Donors and development agencies can act to encourage more accurate, relevant and useful media coverage of pastoralism by supporting training programmes, opportunities for journalists to travel to areas where pastoralists live, and initiatives that bring together journalists, pastoralists, dryland researchers and policymakers. Ultimately, though, it is editors not reporters who decide how a media outlet will cover an issue. Any effort to improve the media narratives around pastoralism will need to engage these gatekeepers as well as the journalists with stories to tell. If media narratives fail to improve, pastoralists and their advocates will need to take advantage of new media tools and other communication tactics to bypass these intermediaries and speak more directly with policymakers, fellow citizens and other audiences.

Acknowledgements This research was undertaken when the author was employed at the International Institute for Environment and Development (IIED). The chapter here is an adaptation of Shanahan (2013). As such, the chapter benefits from discussions between the author and the following participants in a workshop in Kenya in 2012: Jeremy Swift, Saverio Krätli, Ced Hesse, Gongbuzeren, Hu Jingping, Li Wenjun, Srijit Mishra, Ravindra Adusumilli Wassan, Kamal Kishore, Ilse Koehler Rollefson, Lila Buckley, Victor Orindi, Izzy Birch and Michael O. Odhiambo. Ria Sen assisted with the initial testing of coding system used in the article content analysis. The research presented here, and the writing that explains it, were funded by the Ford Foundation within the context of the IIED project 'New perspectives on climate resilient drylands development'. The views expressed do not necessarily reflect the views of the Ford Foundation.

References

Anon. (2007). Experts call for recognition of Rajasthan as a pastoral region. The Hindu (India). November 25, 2007. www.thehindu.com/todays-paper/tp-national/tp-otherstates/article1955447.ece. Accessed April 22, 2013.

Anon. (2010). J-K: Muslim Gujjars demand extension of Forest Rights Act 2006. Indian Express. September 12, 2010. www.indianexpress.com/news/jk-muslim-gujjars-demand-extension-of-forest-rights-act-2006/680679/0. Accessed April 22, 2013.

Billett, S. (2010). Dividing climate change: Global warming in the Indian mass media. *Climatic Change, 99*, 1–2.

Catley, A., Lind, J., & Scoones, I. (Eds.). (2012). *Pastoralism and development in Africa: Dynamic change at the margins*. London: Routledge.

de Jode, H. (Ed.). (2009). *Modern and mobile. The future of livestock production in Africa's drylands*. London: IIED and SoS Sahel.

Hesse, C. (2013). *New perspectives on climate resilient drylands development*. London: IIED.

Karthikeyan, D. (2012). Pulikulam cattle approved as indigenous breed. The Hindu. July 18, 2012. www.thehindu.com/news/cities/Madurai/article3652470.ece. Accessed April 22, 2013.

Kipsang, W. (2012). Give up arms, PC orders herders. The Nation (Kenya). February 10, 2012. LexisNexis newspaper database. Accessed April 22, 2013.

Krätli, S., & Enson, S. J. (2013). *New perspectives on climate resilient drylands development: Refining the arguments*. London: IIED.

Krätli, S., & Swift, J. (1999). *Understanding and managing pastoral conflict in Kenya. A literature review. Institute of Development Studies*. Brighton: University of Sussex.

Letiwa, P. (2008). Why herders won't surrender their firearms just yet. The Nation (Kenya). April 30, 2008. LexisNexis newspaper database. Accessed April 22, 2013.

LexisNexis. (2006). LexisNexis newspaper database. http://web.lexis-nexis.com. Accessed April 22, 2013.

Passi, S. (2004). Cattle farmers seek rights to graze on forest land. Press Trust of India. August 20, 2004. LexisNexis newspaper database. Accessed April 22, 2013.

Roe, E. (1991). Development narratives, or making the best of blueprint development. *World Development, 19*, 287–300.

Salesa, H. (2011). Ten people shot dead in Isiolo as raiders strike. Nairobi Star (Kenya). June 27, 2011. LexisNexis newspaper database. Accessed April 22, 2013.

Shanahan, M. (2012). What's in a narrative? In policy, everything or nothing. IIED blog. September 28, 2012. www.iied.org/what-s-narrative-policy-everything-or-nothing. Accessed April 22, 2013.

Shanahan, M. (2013). Media perceptions and portrayals of pastoralists in Kenya, India and China. *Gatekeeper, 154*, 1–25.

The East African. (2009). Taking stock of drought. The East African. October 12, 2009. LexisNexis newspaper database. Accessed April 22, 2013.

Wei, S. (2011). Herders face five year ban on grazing. China Daily. August 1, 2011. LexisNexis newspaper database. Accessed April 22, 2013.

Xinhua News Agency. (1998). Vice-minister says all nomads expected to settle by 2000. March 18, 1998. LexisNexis newspaper database. Accessed April 22, 2013.

Xinhua News Agency. (2012a). Tibetan nomads settle down in residential communities. March 1, 2012. LexisNexis newspaper database. Accessed April 22, 2013.

Xinhua News Agency. (2012b). China exclusive: nomads on Qinghai-Tibet plateau bid farewell to life on plains. March 16, 2012. LexisNexis newspaper database. Accessed April 22, 2013.

Xinhua News Agency. (2012c). Xinhua Insight: Settled Tibetan nomads pin hopes on children's Mandarin studies. June 28, 2012. LexisNexis newspaper database. Accessed April 22, 2013.

Yadav, A. (2010). Migrant communities. The Hindu (India). July 24, 2010. http://www.thehindu.com/todays-paper/tp-national/tp-otherstates/article1955447.ece Accessed April 22, 2013.

Zou, L. (2010). Fenced out. Global Times (China). August 9, 2010. LexisNexis newspaper database. Accessed April 22, 2013.

Chapter 16
Much More than Simply "Desertification": Understanding Agricultural Sustainability and Change in the Mediterranean

Mark Mulligan, Sophia Burke and Andrew Ogilvie

Abstract In the socio-economic and climatic complexity of a Mediterranean environment, desertification involves many processes and is much more spatio-temporally sophisticated than is often suggested by both the policy and academic communities. Here we examine the spatial variability of the Mediterranean environment and its agriculture—a supposed key driver of desertification in the region. We examine trends in agricultural production and climate in the region and the differences between north and south. We review the recent history of Mediterranean desertification research and its changing policy context. In particular, we examine the evidence for Mediterranean desertification, and what we now understand of it, with the hindsight of three decades of research and having 'come out of the other side' of decadal drought events that prompted large-scale concern over the desertification of Mediterranean Europe in the 1990 s. We focus particularly on the spatial heterogeneity of the Mediterranean and the implications of this heterogeneity for understanding perceived desertification in the region, and explore implications for land use and management policy-making, indicating the need for a much more locally nuanced approach.

Keywords Desertification · Land degradation · Europe · Mediterranean · Heterogeneity · Policy

M. Mulligan (✉) · A. Ogilvie
Department of Geography, King's College London, Strand, London WC2R 2LS, UK
e-mail: mark.mulligan@kcl.ac.uk

S. Burke
AmbioTEK Community Interest Company, London, UK

A. Ogilvie
Institut de Recherche Pour Le Développement (IRD), UMR G-eau, Montpellier, France

R.H. Behnke and M. Mortimore (eds.), *The End of Desertification?*
Springer Earth System Sciences, DOI 10.1007/978-3-642-16014-1_16

16.1 Introduction

Desertification is an umbrella term that is used to represent a range of biophysical and socio-economic processes that may combine to reduce the biological potential of the land in arid and semi-arid agricultural areas. In contrast, at the edges of true deserts, desertization can be a simple process of desert encroachment by migrating dunes. In the socio-economic and climatic complexity of a Mediterranean environment, desertification involves many more processes than desertization and is much more spatio-temporally sophisticated than is often suggested by both the policy and academic communities. Here we examine the Mediterranean environment and its agriculture—a supposed key driver desertification in the region—and look into the recent history of Mediterranean desertification research. In particular, we examine the evidence for Mediterranean desertification, and what we now understand of it, with the hindsight of three decades of research and, moreover having 'come out of the other side' of decadal drought events that prompted large-scale concern over the desertification of Mediterranean Europe in the 1990s. We focus particularly on the spatial heterogeneity of the Mediterranean and the implications of this heterogeneity for understanding perceived desertification in the region in the 1990s and now, having access to much more detailed spatial data.

16.2 The European Mediterranean: Environments and Agriculture

The Mediterranean is characterized by a semi-arid to sub-humid climate that is highly seasonal for both solar radiation (and thus temperature) and for cloud generation (and thus rainfall). The long, hot and dry summer 'drought' followed by heavy October rainfall provides both a constraint to vegetation production and conditions of low cover and intense rainfall which favour soil erosion and degradation. Degradation is variously attributed to overgrazing, poor soil management, poor drainage on irrigated land, and the over-abstraction of woodland for fuel wood.

Over the long-term the Mediterranean climate has combined with the region's geology to create rugged, highly dissected landscapes with sparse vegetation and poor, stony soils. Traditional agriculture is well adapted to this environment through sparse cropping of hardy species (Almond, *Prunus dulcis;* Olive, *Olea europea;* Fig, *Ficus carica*) combined with careful management of water through: runoff capture; mulching of soils; terracing and occasional irrigation. However, recent trends include larger fields, mechanized ploughing, significant soil movements (Ramos and Mulligan 2005; van Wesemael et al. 2005), and intensively irrigated crops including greenhouse vegetables, rice, wheat and citrus. These may not be so well adjusted to both the seasonal water shortages and inter-annual and longer-term climate variability that characterize the region. Agricultural change in the Mediterranean is driven by: complex changes in population distribution (especially urbanization); demography and socio-economic change (including the development of the service sector and of tourism); changes to transportation costs; subsidies and markets in Europe and beyond; and technological change.

We examine the countries of the northern (largely European) Mediterranean with a coastline on the Mediterranean sea. The Mediterranean countries have strong latitudinal precipitation (P) and temperature (T) gradients. Precipitation decreases from north to south and from west to east and temperature increases along the same gradients. There are, however, significant montane and coastal effects with mountain and coastal areas generally being cooler and wetter.

Rainfed agriculture dominates the northern and western Mediterranean countries (Fig. 16.1) and is intensive in northern France and Spain. If >50 % cover signifies 'intensive' then some 16.6 % of the study area can be considered intensive rainfed cropping. Irrigated cropland (Fig. 16.2), on the other hand, is concentrated in a few areas (NE Italy, and parts of Spain, France and Greece) at high intensity but is common over small areas throughout the region. Irrigation is a significant drain on groundwater resources in some of the drier regions and is estimated to use some 16 % of groundwater production in the Mediterranean (Siebert et al. 2010). If >50 % cover signifies intensive then some 2.1 % of the study area can be considered intensive irrigated cropping.

Figure 16.3 shows grazing is much more extensive than intensive in the Mediterranean and occurs throughout the Northern and Eastern Mediterranean,

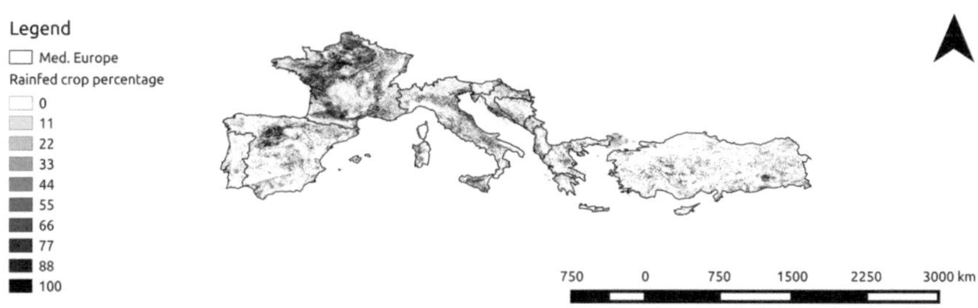

Fig. 16.1 Rainfed crops in Mediterranean countries according to GlobCover continuous fields (Mulligan 2009a).

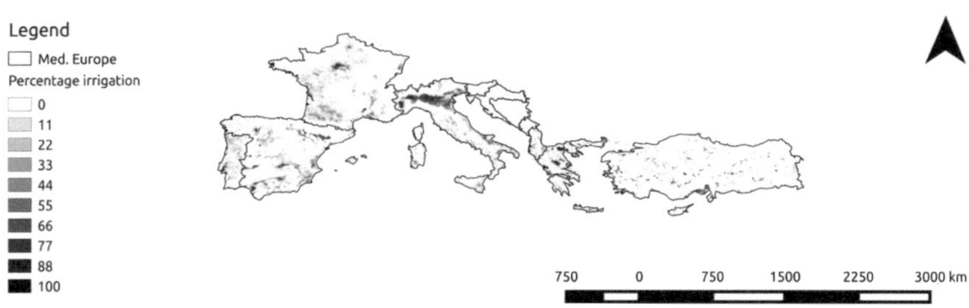

Fig. 16.2 Percentage of land irrigated. *Source* Siebert et al. (2007).

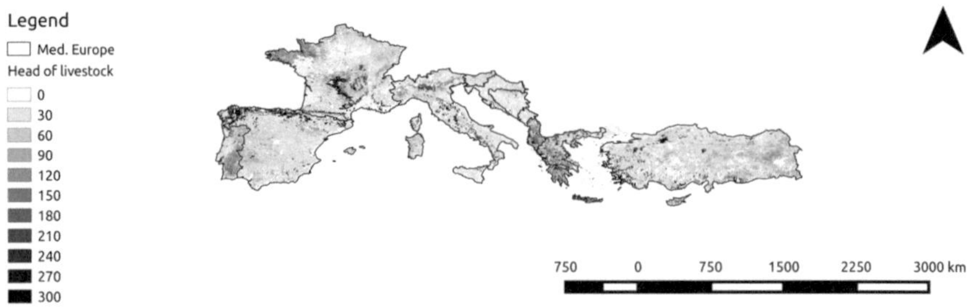

Fig. 16.3 Total grazers (headcount/km^2). *Source* Wint and Robinson (2007).

especially in mountains, coastal areas and river margins. If >200 head per square km signifies intensive then some 4.9 % of the study area can be considered intensive grazing land.

The greatest cereals producers in the EU and southern Mediterranean countries in 2007 (an average year for the decade) according to EUROSTAT were France (59.4 million tonnes), Germany (40.6 million tonnes) and Poland (27.1 million tonnes). These are followed by Spain (23.8 million tonnes), UK (19.4 million tonnes) and Italy (18.8 million tonnes). Non-EU southern Mediterranean countries, on the other hand, have lower cereal production, most countries producing 5 million tonnes of cereal annually or less, with the exception of Egypt which produced as much as 15.3 million tonnes of cereal in 2007 (Petit 2007). Though statistics for the southern Mediterranean countries are not as comprehensive, productivity remains lower than for the EU. Fruit and vegetable production is also very variable across the Mediterranean, with Spain, Italy and Egypt having the highest production in the last decade according to EUROSTAT. The European Union is by far the largest producer of olives, producing about 75–80 % of the world's supply (Muller 2004). Greece, Spain and Italy are the top olive producers; followed by Portugal, Algeria, Egypt, Argentina, Lebanon, Libya, Morocco, Palestine, Tunisia and the USA. Production of olives has greatly benefited from EU subsidies. Italy, Spain, and France are the top grape and wine producers in the world, followed by the USA and China. Of the top twelve global producers, the EU producers (Italy, Spain, France and Germany) account for about 67 % of wine production (Patterson and Josling 2005, using FAOSTAT data).

Population is distributed throughout the north and east Mediterranean landmasses with some significant urbanizations also present in those areas (Fig. 16.4). In the southern Mediterranean there are large areas with very low or no populations, people being highly concentrated in the coastal fringe and the major river valleys. This reflects the distribution of climatic and agricultural suitability and also determines the demand for water and the human pressure (directly and through agriculture) on the environment.

The northern Mediterranean is highly urbanized (Fig. 16.5) with these urbanizations being dense (in population terms), extensive and highly concentrated around coasts (e.g. Spain, Italy) and major rivers (e.g. Egypt).

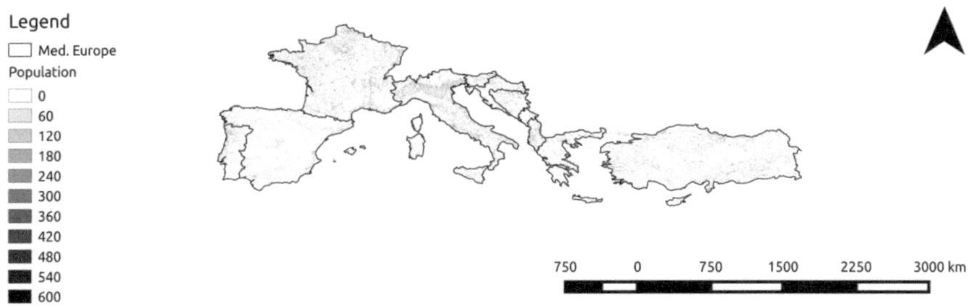

Fig. 16.4 Population (persons/km^2). *Source* Landscan (2007).

Fig. 16.5 Urban areas. *Source* CIESIN (2004).

Figures 16.1, 16.2 and 16.3 indicate that the Mediterranean landscape is both extensively and intensively agriculturalized, whilst Fig. 16.6 shows dry matter productivity for the Mediterranean. This measure is for all plant systems, not only for crops and shows strong gradients with the highest observed values in the northern areas (of Italy, Sardinia, northern Spain, Greece) and a strong decline towards the southern areas (especially of Spain, Turkey). Towards the South and East, water becomes limiting because of the gradient of decreasing precipitation in this direction.

Overall, the north and west Mediterranean is productive and extensively used for dryland or irrigated agriculture or grazing. Despite this intensive, extensive and productive agricultural situation, the Mediterranean has long been considered an

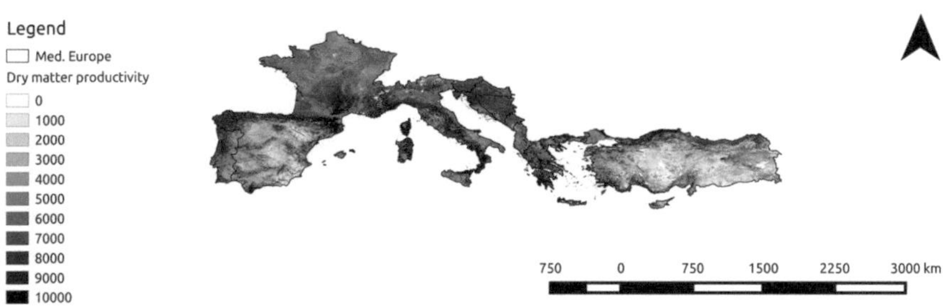

Fig. 16.6 Dry matter productivity (Dg/ha/day) for the Mediterranean. *Source* Mulligan (2009b).

area under threat of desertification and land degradation driven by unsustainable exploitation of water resources, abandonment and subsequent degradation of managed farmland and significant changes to farming practices over the last decades (Brandt and Thornes 1996; Puigdefábregas and Mendizabal 1998; Geeson et al. 2003; Briassoulis 2004; Rubio and Recatalá 2006; Safriel 2009).

16.3 Recent Drivers for "Mediterranean Desertification"

Here we review some of the reasons that Mediterranean desertification has received a great deal of scientific, political and media attention over the last three decades. Some of the reasons are biophysical whilst others are socio-economic in origin. These drivers include agricultural land use change, change to grazing lands, changes in agricultural policy, climate variability and change and, finally, changes to human exposure to land degradation.

16.3.1 Agricultural Land Use Change in the Mediterranean

Agricultural production in the Mediterranean has significantly increased over the past fifty years. Since 1961, production increased 3-fold for cereals, 2.5-fold for vegetables and 5-fold for citrus fruits and these crops account for over 85 % of the total agricultural production in the Mediterranean (Plan Bleu 2009). The growth rate in agricultural production has started to fall in recent years in all Mediterranean countries with the exception of Egypt, which showed remarkable increases in the period 1961–1983 and still grew (but at a much slower growth rate) in the period 1984–2007 (Plan Bleu 2009). This has affected trade balances and Mediterranean countries are becoming increasingly dependent on foreign food imports. Nevertheless, southern Europe exports much of its produce.

Although rain-fed agriculture predominates in the Mediterranean, it is on irrigated land that the greatest productivity and productivity gains have been achieved in recent decades (Montgomery 2009), as is also the case outside the Mediterranean (Siebert and Döll 2010). The total irrigated area in the Mediterranean countries doubled in the previous 40 years to exceed 26 million hectares in 2005, possibly representing over 20 % of the land under cultivation (Plan Bleu 2009). The area under arable and permanent cropping has not increased alongside annual agricultural production, indicating a trend for intensification rather than extensification, again replicating the global trend. In fact, since 1961, arable land per capita in both the north and south of the Mediterranean has halved (Plan Bleu 2009) as land that is suitable for cultivation is lost to urbanization and abandonment every year, largely because of migration to coastal urban areas but also in response to EU set-aside policy.

16.3.2 Recent Changes to Grazing Lands

In the European Mediterranean—there has been a major shift in land use from livestock that grazed on extensive rangelands to intensively farmed cereals and intensive grazing systems. This is partly driven by markets and regional agricultural policies, in particular subsidies designed to increase biodiversity but also as a more sustainable method of management (Roeder et al. 2008). Taking animals to graze is a hard and time-demanding job (Ruiz et al. 2001) with low economic rewards compared with many others. Intensification necessitates the utilization of supplementary feedstuffs, either on-farm (cultivated pastures and crops) or most frequently off-farm through concentrates and forages (Riedel 2007).

In the Southern Mediterranean, cultivation has encroached onto rangelands, driven by population growth and national policies, often resulting in land degradation (Safriel 2009). In the south and the east (from Morocco to Turkey), despite rural to urban migration, the agricultural population increased by 10 million (16 %) within 40 years, reaching 71 million in 2000 (Thivet 2007). This can be contrasted with the Northern Mediterranean where, from Spain to Greece, the agricultural population decreased from 46 million to 12 million over the same period (Thivet 2007).

16.3.3 Recent Changes in Agricultural Policy

Much of the rise in productivity in the EU has been encouraged by the Common Agricultural Policy (CAP), which was set up initially with the treaty of Rome (1957) to control food production across Europe and latterly focuses on environmental concerns. Through the provision of incentives and a number of accompanying policies, the CAP is a key driver of agricultural change in the EU. The impact of the CAP on Mediterranean land use has been well-documented (Shucksmith et al. 2005; van Meijl et al. 2006; Verburg et al. 2008; Serra et al. 2008). The effect of the CAP also needs to be seen alongside other drivers, such as the substantial level of national funding for agriculture, the fluctuations on world markets for food products and the negotiations taking place within the World Trade Organization (Gay et al. 2005). In particular the CAP has provided support (via intervention buying and storage) for a few Mediterranean products such as olive oil, tobacco and cotton, although this no longer takes place. Increased interest in energy efficiency and climate change is increasingly raising questions about the relevance of current CAP objectives (Cooper and Arblaster 2007). The likely continued future enlargement of the Union will also necessitate significant change to future CAP policy. The Mediterranean will also be affected by the establishment of a Mediterranean Free Trade Zone (MFTZ), which will encompass 12 non-EU nations of the Mediterranean region as well as the entire European Union. This was expected to be completed by now but has been delayed by: foreseen negative social and environmental consequences; the Sarkozy initiative to set up a Union for the countries that border the Mediterranean sea; and questions as to the relationship between this and the EU-GCC (Gulf Cooperation Council) free trade initiative.

16.3.4 Recent Climate Variability, the Prolonged 1990s Drought and Climate Change Scenarios

The climate of the Mediterranean region is characterized by hot, dry summers and cool, wet winters (Conacher and Conacher 1998) with December, January and February precipitation (DJF) a major contributor to the annual total. It is a region of significant climate variability, both spatially and temporally. This climate variability is due to the Mediterranean's location in a transitional zone between the competing "marine west coast climate" and the "subtropical desert climate" (Lionello et al. 2006). The Mediterranean climate is exposed to the South Asian Monsoon in summer and the Siberian high pressure system in winter. Variability is expressed as changes in annual total rainfall (both inter-annually and on decadal time scales) and changes in rainfall seasonality and the frequency and magnitude of extreme rainfall events and seasonal drought episodes. Rainfall records since 1900 indicate strong multi-decadal variations in regional precipitation (meteorological drought) coupled with a long term trend in the Palmer Drought Severity Index (PDSI) suggestive of increased land surface aridity associated with observed long term temperature increase since the 1960s (Mariotti 2010). DJF precipitation in the Mediterranean showed a striking decrease during the 1960–1990s which may result from circulation changes in the North Atlantic climate Oscillation (NAO). This meteorological drought has been realized as hydrological drought with river discharge for major Mediterranean rivers such as the Ebro, Jucar and Arno showing a considerable decline over the 20th Century. Current rainfall trends in the region continue the clear downward trend during the second half of the 20th Century (de Luis et al. 2009; García-Ruiz et al. 2011) but much less dramatically than during the 1990s. This drying is experienced in the form of more frequent and intense drought, especially in the winter with 10 of the 12 driest winters since 1902 occurring from 1992–2012 (Hoerling et al. 2012).

Future climate change scenarios for the Mediterranean are also complex. Generally, a change in circulation patterns is expected for the Mediterranean region as a result of increases in sea level pressure, a northward shift of the Atlantic storm track and a deflection of storms north of the Mediterranean into higher latitude areas (Giorgi and Lionello 2008). These changes in circulation are expected to lead to a general reduction in precipitation for the Mediterranean regions and an increase for the northern European regions. Changes in variability of precipitation and temperature are also expected. Schar et al. (2004) concluded that as a result of an increase in inter-annual variability of temperature along with mean long-term warming, the Mediterranean region might experience a much more frequent occurrence of extremely high temperature events and heat waves. As a region with already marginal climate for some types of agriculture, the Mediterranean is expected to be affected significantly by climate change.

Giorgi (2006) analysed IPCC SRES AR4 simulations to calculate a climate change index based on precipitation and temperature means and variation. The Mediterranean was amongst a small number of climate change "hot-spots" globally. Weib et al. (2007) considered socio-economic and climate changes as indicated by the IPCC SRES A2 and B2 scenario forcing the ECHAM4 GCM. Under this scenario a current 100-year drought would likely occur 10 times more frequently in the Northern Mediterranean while in North Africa, a current 100-year drought will occur less frequently. There is thus the expectation of significant negative effects of climate change in the Mediterranean. According to Giorgi and Lionello (2008), climate change scenarios from general circulation models (GCMs) indicate a "*pronounced decrease in precipitation, especially in the warm season, except for the northern Mediterranean areas (e.g. the Alps) in winter*". These effects are projected to lead to a progressive decline in streamflow (García-Ruiz et al. 2011).

Using the more recent Coupled Model Intercomparison Project Phase 5 (CMIP5) data, Peleg et al. (2014) compared current climate with projected mid-21st century climate based on Representative Concentration Pathway (RCP) RCP4.5 and RCP8.5 scenarios using four of the available CMIP5 models. The models projected a decreasing frequency of rainfall in the region with a 10–22 % reduction in wet events. Also using CMIP data, Diffenbaugh and Giorgi (2012) indicate the Mediterranean as one of a few hotspot areas for prominent climate change using a multi-variable metric including rainfall and temperature.

However, a large source of uncertainty in projections for the Mediterranean region derives from unrealistic modelling of the climate influence of the Mediterranean Sea (Somot et al. 2008; Lionello et al. 2008). Furthermore, particularly for regional models, uncertainties in future land use and land cover change adds considerably to uncertainty in climate projections. Uncertainties in climate simulations with RCMs for Europe were addressed in the PRUDENCE project (http://prudence.dmi.dk/) which found that, particularly for temperature, uncertainty in boundary forcing plays a greater role than the RCM used (Déqué et al. 2007). In addition to these climate changes, the future of the Mediterranean environment will also be determined by changes in population, technology, agriculture and thus land use and cover.

16.3.5 Changing Exposure: Population Growth and Economic Development

Spatial projections for population and GDP are publicly available in map form only for the Special Report on Emissions Scenarios (SRES) B2 scenario (Fig. 16.7).

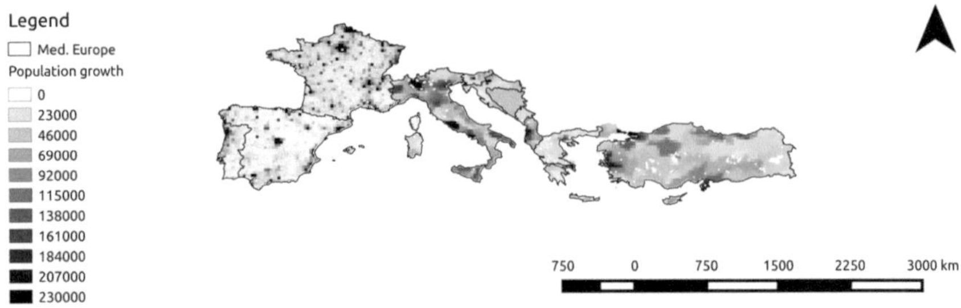

Fig. 16.7 Population projection for 2025 (persons/km^2, SRES B2). *Source* CIESIN (2002).

Under SRES B2, population increases by 2025 are expected to be greatest in the south and east of basin, concentrated in coastal areas. De-population is expected to occur in parts of the north Mediterranean (especially Spain, Italy and Greece).

16.3.6 The Water Resources Situation

In addition to the observed low productivity in some areas of the Mediterranean (and expectation that this will reduce further with land degradation and climate change), water resources are also a key constraint on productivity and sustainable land management in this environment. We have calculated water balance as rainfall (Hijmans et al. 2005) minus actual evapotranspiration (Mu et al. 2011) on a 1 km resolution basis with negative local water balances (representing areas where evapotranspiration is supported by lateral surface or subsurface inflows) set to zero. The results of this analysis are shown in Fig. 16.8a. Dark areas have plenty of water, whereas those in lighter shades show progressively lower water balance. The most water stressed areas are in the southern Mediterranean, in the eastern Mediterranean and in coastal northern Mediterranean. Water stress can lead to over-cultivation and to the negative effects of irrigation in drylands. Note that this water balance does not include inflows of water from rivers nor unsustainable use of fossil water but focuses on the locally produced renewable water resources (the water balance) since this will be critical for sustainable agricultural production. Some 8 % of the area has a local water balance of zero, indicating all local rainfall consumed by local evapotranspiration. Figure 16.8b shows the same data averaged over watersheds, indicating clearly the more water stressed catchments. Only 0.3 % of basin area experiences basin closure (all rainfall evaporated within the basin) on an annual basis under current conditions. Though a positive water balance will lead to the generation of runoff and thus river flow, this analysis does not explicitly account for available water in rivers and groundwater that is used to supplement the local water balance at the pixel scale in some regions, though these are determined entirely by water balance at the basin scale.

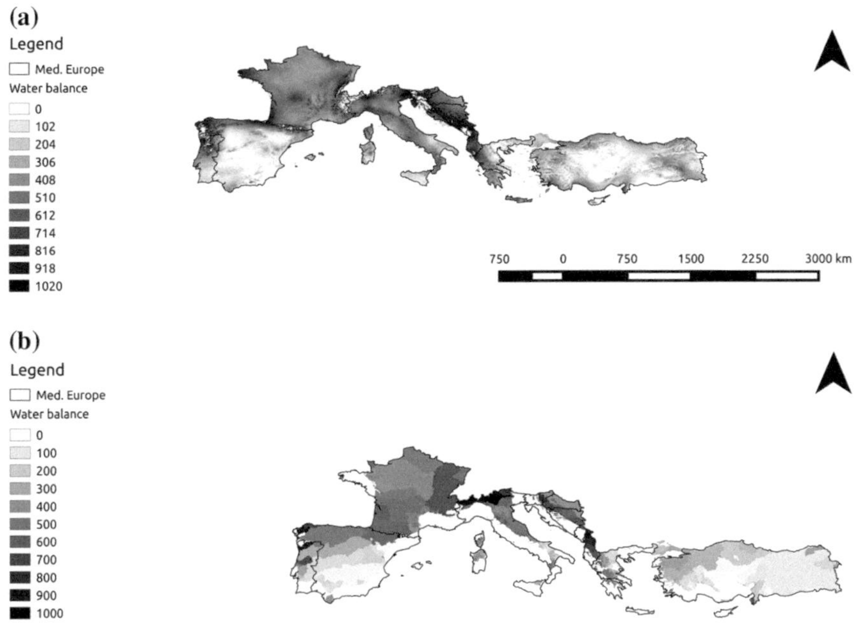

Fig. 16.8 **a** Water balance (rainfall-ET) (mm/yr). **b** Water balance (rainfall-ET) (mm/yr) by catchment.

Table 16.1 Evapotranspiration by different land covers and as a percentage of rainfall as calculated by WaterWorld

All vegetation	Irrigated croplands	Croplands	Pastures	Natural vegetation
Actual evapotranspiration (percent of total)	0.41	26.8	16.1	56.1
Actual evapotranspiration (percentage of rainfall)	0.2	15.3	9.2	31.9

We carry out an analysis of evapotranspiration though the different types of vegetation cover in the region (Table 16.1), calculated using WaterWorld[1] and based on the actual evapotranspiration (AET) dataset of Mu et al. (2011), the irrigated cropland dataset of Siebert et al. (2007) and the cropland and pasture dataset of Ramankutty (2008). Results indicate that, some 0.41 % of actual evapotranspiration in the region occurs through irrigated croplands, 26.8 % through all croplands, 16.1 % through pastures and 56.1 % through natural vegetation, with the

[1]WaterWorld is a modelling platform containing a range of global databases and the capacity to run detailed algorithms at scales from local through to continental. It is available free for non-commercial use from www.policysupport.org.

Fig. 16.9 Watersheds of dams (Mulligan et al. 2011).

remainder over bare soil and water bodies. The majority of AET thus occurs through natural vegetation, though croplands also evaporate a significant volume. In total AET consumes 56 % of rainfall in the region, some 32 % of which is evaporated through natural vegetation though only 15 % is evaporated through croplands and 0.2 % through irrigated croplands. The remaining 44 % is available to fill surface stores and generate runoff. On this basis irrigation consumes relatively little of the available rainfall, most of which evaporates through natural vegetation. The pressure points on water are thus highly localized to low rainfall and high AET areas, including those under intensive irrigation and pressure is not extensive throughout the region. The French and Italian Alps in particular will experience very different conditions that lower elevation areas with warmer, drier climates.

The demand for water storage (in reservoirs) during the dry season in the Mediterranean is evidenced by the distribution of dammed watersheds in the Mediterranean (Fig. 16.9), with the exception of certain dams in central France whose primary output is hydropower. The black areas indicate those watersheds that outflow at a dam according to the global census of Mulligan et al. (2011). Some 45 % of the European Mediterranean area, 41 % of its rainfall according to WorldClim (Hijmans et al. 2005) and 33 % of its water balance (according to WorldClim rainfall minus Mu et al. (2011) AET) drains into a dam. This represents a very significant human appropriation of the hydrological fluxes of this region, supporting irrigation, industrial and domestic use.

As well as the annual total, the seasonality of rainfall and evaporation is a key component of water stress in the Mediterranean. Figure 16.10 shows the Walsh and Lawler (1981) seasonality index calculated using WorldClim monthly rainfall (mean 1950–2000) and indicates that seasonality is greatest in the SW of the Iberian peninsula, in southern Italy, Greece and Turkey. Some 8.2 % of the area is considered seasonal and some 0.7 % is markedly seasonal using this metric. In these areas a high winter rainfall elevates the annual total but the summer rainfall is a key limit to crop growth.

Fig. 16.10 The Walsh and Lawler (1981) seasonality index for the Mediterranean.

16.4 The Scientific Response to Perceived Mediterranean Desertification

Concern over desertification in the European Mediterranean has led to the funding of a range of field and modelling based assessments in the region. The field programmes aimed to understand the processes operating and assess desertification at various sites whereas the modelling projects aimed to provide a detailed analysis of causes and consequences in a spatially explicit manner, at the landscape scale. We will focus here on the modelling projects since they operate at more policy relevant scales than the field programmes, though they are—of course—informed by data and knowledge from the field. A series of models have been developed at scales from the hillslope to the continent. The MEDALUS I project developed and applied models for soil erosion at the hillslope scale (Thornes et al. 1996), MEDALUS II combined these soil erosion models with models for vegetation growth and development as well as soil hydrology and MEDALUS III culminated in the development of the MEDRUSH model for forecasting hydrology, soil erosion and sediment yield in catchments of up to 2000 km^2 in a spatially explicit fashion using the GRASS GIS (Kirkby et al. 2002). The MODMED project (1995–1999, Mazzoleni and Legg 1998) built predictive models for understanding natural vegetation dynamics and their response to changes in grazing and burning pressure resulting from developments in agricultural policy. The MODULUS project (1998–1999) led by the Research Institute for Knowledge Systems (RIKS) took models and datasets developed in a range of previous EC projects (in particular EFEDA 1988–1991 (Bolle 1995), MEDALUS and MODMED 1995–1999) and combined them in an integrated model for climate, hydrology, plant growth and vegetation development, soil erosion and socio-economic processes as an operational decision support system (DSS) aimed at regional policy-makers and applied in the Marina Baixa, SE Spain and the Argolid of Greece (Oxley et al. 2004). The hydrology, crop growth, climate and erosion models were based on the PATTERN model (Mulligan 1996, 1998) which was developed alongside the EFEDA project.

MODULUS was a technical experiment in that it took existing models that had been developed in previous projects, left them in their native code and wrapped them with a software 'layer' to permit their integration as a single model capable of running simulations for scenarios of land use and climate change and provided results as graphics and maps. The project showed that models developed for scientific purposes could be integrated together to tackle policy-relevant questions and then supported by a sophisticated but user-friendly interface and thus approach the requirements for policy application. MedAction (2001–2004) further developed the MODULUS DSS into a Policy Support System (PSS) capable of advising policy. This required the engagement of end users at the start of the project, the definition of clear policy goals with them and the re-engineering of models from the bottom up for careful integration into an integrated model. The PSS was applied to land use and water management issues in the Guadalentin basin (SE, Spain) (van Delden et al. 2007). In the DESURVEY project (2005–2010), the MedAction PSS was further developed for application across Europe and North Africa as the DESURVEY surveillance system. It was enhanced through the incorporation of multi-scale modelling with a simpler model, based on PESERA (Kirkby et al. 2003) being applied for application at the pan-European scale (1 km resolution) and a more complex and data demanding model based on PATTERN (Mulligan 1998)—as implemented in MedAction—used for application for local scale analyses (at 1 hectare resolution). New process descriptions, policy options and datasets necessary for application to North Africa were also developed alongside much more extensive training and application throughout Europe.

Throughout this period of model building, the models have become more and more sophisticated both in terms of the processes represented, the degree of feedback between processes, the degree of integration and inclusion of biophysical and socio-economic processes, and the degree of spatial detail and spatial heterogeneity incorporated. This evolution reflects our improved understanding of the complexities of processes that are grouped together under the umbrella of desertification. Later projects built upon these efforts to interface and adapt models for use in advising policy—driven by increasing pressure from the EC to connect its science with its policy agendas, though there is little evidence that such tools are used in supporting policy at the EC.

16.5 Revisiting Mediterranean "Desertification"

With 20 years of hindsight we now revisit the science on Mediterranean desertification with a view to understanding better the current thinking around this issue.

16.5.1 *Where Actually Desertified?*

According to the literature, land degradation is said to affect 80 % of the arid and dry areas of the Mediterranean, especially pasturelands and rain-fed croplands, but even irrigated land is considered under threat (Thivet 2007). Degradation is variously attributed to overgrazing, poor soil management, poor drainage on irrigated land, and the over-abstraction of woodland for fuel wood. This has been linked to population migration in the past, for example the Thessaly Plain was grazed for centuries by transhumance flocks and herds but after significant immigration in the early 1920s these areas were converted for wheat cultivation (Kosmas et al. 1999). For sure groundwater resources in the region have declined significantly, largely reflecting increasing agricultural abstraction (Iglesias et al. 2007).

Yet, agricultural production has continued to increase in the European Mediterranean (Leetmaa et al. 2004; Montgomery 2009) at the same time that much land has been returned to nature. Moreover, the rainfall-use-efficiency corrected negative NDVI (normalized difference vegetation index) analysis of Bai et al. (2008) indicate negative trends in vegetation cover (possibly reflecting degradation) over only 1 % of the study region, largely in the intensive agricultural areas of SW Spain, NE Italy and Montenegro—which are likely the result of major land use changes rather than actual land degradation. Coastal parts of North Africa and of the eastern Mediterranean are also considered to be degrading (García-Ruiz et al. 2011). Therefore developments so far in agricultural practices and technology seem to be keeping pace with any degradation so that 'desertification' appears not to be having large-scale societal impacts in the region. Increasingly fewer families depend on agricultural livelihoods, especially in the EU; agriculture is generally larger scale and more 'commercialized'; and much of the population lives and works in an urban setting. The land degradation picture is thus more heterogeneous and complex than the large scale 'desertification' of the region that was anticipated in the 1990s.

16.5.2 *What Did We Really Mean by Mediterranean "Desertification"?*

Desertification is a term used to cover a 'multitude of evils' and can fairly well incorporate any socio-economic or biophysically-driven problem in a dryland context from the multiple impacts of climate change, through land degradation, through application of over-intensive methods to overgrazing, water resource stress and other issues associated with poor management of common pool resources. So what did (do) we really mean by Mediterranean desertification? We can identify three different lenses that have been used.

16.5.2.1 The Land Degradation Lens (Pressure-Response)

In the land degradation phase, desertification is considered to be the response of the landscape to the application of pressure (drought, overcropping, overgrazing, over-extraction of groundwater). As the pressure increases, the environment reacts to create a problem (agricultural drought, soil exhaustion and erosion, grazing-land degradation, salinization of groundwater), see for example Catalinas and Cantón 2012). The system is seen as a pressure response system and the kinds of pressure-response tools (models) used suggested continued progressive degradation into the future (see Mulligan 2009a, b, c). Each element of the system is considered independently as an independent pressure response system and research is usually focused on one set of processes e.g. soil erosion, ignoring the links between different pressure-response systems. The focus is on land degradation as the outcome (to be avoided) rather than as we might see it today: as a driver itself that will act to reduce economic and agricultural pressure in a negative feedback cycle (where alternatives to agricultural livelihoods exist).

16.5.2.2 The Sustainability Lens (Pressure-Response-Feedback)

With this approach the focus is on understanding the limits of environmental systems and the likely impacts once these limits are reached. This approach examines desertification in a more integrated manner looking into a range of processes and systems and focusing on the changes to socio-economic systems that would be required to achieve sustainability. The focus is thus on sustainability as the outcome (to be achieved), see Laraus (2004), Perry (2006) and Turner et al. (2007).

16.5.2.3 The Ecosystem Services (ES) Phase (Tipping Points and Complexity)

With this approach the focus is on maintaining the ES that are critical to local populations. This involves understanding how and where ES are delivered, the role of natural environments and of agriculture in delivery, and who benefits from them. The science concentrates on highlighting the pressures on ecosystems that are likely to result in degradation of the ecosystem functions that provide these ES and seeks to maintain these pressures at levels below those tipping points that would lead to significant impact on humanity. Maintenance of ecosystem service provision whilst maintaining and increasing agricultural production is the outcome sought (see for example Schröter et al. 2005; Benayas et al. 2007; Bugalho et al. 2011; Willaarts et al. 2012).

16.6 The Importance of Policy, Place, Space and Variability in Assessing and Managing Desertification

As Figs. 16.1, 16.2, 16.3, 16.4, 16.5, 16.6, 16.7, 16.8, 16.9 and 16.10 highlight, the European Mediterranean shows significant spatial variability in the biophysical and socio-economic factors that might influence land degradation. This variability combines with the policy context to determine the impact of policies on desertification. Farming policy in Mediterranean Europe is dominated by the CAP. Current trends in the CAP could lead to further abandonment of production in zones of high nature conservation value (Caraveli 2000) and this may lead to reduced land degradation (Ruecker et al. 1998) or increased land degradation especially on steep slopes (e.g. Koulouri and Giourga 2007) in the areas abandoned. Mediterranean water policy is largely focused on drought management. According to Iglesias et al. (2007) water management in the region works best when carried out at the basin scale and when institutional responsibilities are clear but they argue that all countries in the region show weak inter-institutional cooperation and gaps the roles of the state, the region and the river basin authorities. This situation impairs the implementation of legislation. The European Water Framework Directive (2000/60/EC) does attempt to develop legislation and institutions for water management and planning at the basin level. However, key issues in implementation of the WFD include the difficulties of transboundary cooperation (across regions as well as countries) and lack of provision for water quality management during drought conditions (Iglesias et al. 2007). Indeed monitoring, early warning and management of drought, including the provision of contingency sources needs much further development throughout the Mediterranean region. Legislation around the management of groundwater pumping for irrigation also requires a much more adaptive and locally nuanced approach that takes better account of varying climatic and hydro-geological conditions over time and space.

According to WDPA (2013) some 18 % of the area studied here is at least nominally protected with 0.06 % of this with RAMSAR designation as wetlands of international importance. Protected areas cover 13.75 % of the rainfall of the region calculated with WorldClim data (Hijmans et al. 2005) and 17.25 % of the water bodies calculated with SRTM SWBD (lakes, rivers, USGS 2006) data. To the extent that protected area designation is effective at reducing pressure on the ground, these represent an important mechanism to manage land degradation.

South of the Mediterranean Sea, a large divide exists between national desertification policy objectives and their local implementation, notably as a result of institutional complexity and inadequacies. Institutions suffer from weak resources and capacities, as well as the legal pluralism generated by the overlap of traditional, state and project-led institutions which can undermine the effective regulation of practices (Briassoulis 2004). Though the success of these policies hinge largely on changing end user practices, consultation of these stakeholders in the design and implementation of policies remains low, leading to weak participation and concern for desertification issues. Degradation of resources (overgrazing, over-abstraction etc.) often ensues with officials and traditional authorities turning a blind eye to support smallholder's agricultural practices in the absence of alternative livelihoods (Le Goulven et al. 2009).

Past policies also suffered from ministries addressing complex socio-environmental problems from an isolated, single perspective, leading to unexpected results such as reducing aquifer recharge as a result of local irrigation efficiency improvements (Molle 2009). Concepts such as green, blue and virtual water accounting (Allan 1998) or payment for ecosystem services have started to filter into policies (Besbes et al. 2010) and further holistic methods are required for researchers and stakeholders to understand the mutual interactions between hydrological and agricultural processes as well as wider socio-economic influences (subsidies, land tenure, water pricing, etc.) on land degradation. Data scarcity (both absolute and as a result of deficient data sharing) on the spatial and temporal gravity of desertification also currently impedes the targeting of resources towards local, site specific strategies.

16.7 Conclusions: If the Problem Is Complex, the Solution Is Complex

Change in agricultural policy is often not tested for long-term sustainability in the face of soil degradation and long-term climate variability and this has the potential to lead to severe degradation that may render short term economic gains as long term losses. To be of real benefit, land use strategies and incentives for the Mediterranean must be sustainable and profitable in the long term and land management options must also be effective for the long-term. The impacts of both land use and land management strategies must also be understood for the range of socio-economic and biophysical conditions that exist in the region of influence of the proposed policy, since response will not be uniform across variable landscapes and those applying policy should not assume it will. The legislative and institutional framework must recognize this need for spatio-temporal sophistication to build robust and sustainable agriculture.

Policies that need to be tested in this way include:

(a) investment or incentivization of particular crop choices or land management techniques including irrigation, terracing, contour ploughing, ploughing and slope reforming,
(b) infrastructural investments such as dams, water transfers, desalinization facilities,
(c) conservation and protection schemes such as designation of protected areas, deforestation, buffer strips, check dams.

All of these have complex interactions with biophysical processes and thus stores and fluxes of water, sediment, soil and plant productivity and on human behaviour such as crop choice and land management.

It is clear that the desertification narrative of the 1990s was not representative of the spatial complexity and subtlety of land degradation processes in the Mediterranean (see Geeson et al. 2003; Mulligan et al. 2004) and the spatial heterogeneity that is clear in Figs. 16.1, 16.2, 16.3, 16.4, 16.5, 16.6, 16.7, 16.8, 16.9 and 16.10 here. The significant declines in rainfall throughout the region in this decade provoked concerns about large-scale desertification that then never materialized. Even under current intensive agriculture, much of the rainfall in the region still evaporates through natural vegetation rather than agriculture and there is little evidence (for example in agricultural or economic output) of large scale degradation in the region. In fact the abandonment of agricultural land is generally reducing soil and water degradation (Ruecker et al. 1998), except in areas of very steep slopes where terraces are now breaking down (Koulouri and Giourga 2007) but will revegetate and recover in time. Over spatially and temporally explicit landscapes the kinds of large scale desertification mapping efforts of the 1990s (e.g. UNEP 1997) do not capture this spatial heterogeneity at all. An effective way to manage such complexities and test policies in silico before they are applied in vivo is through the application of data and process-based, spatially detailed modelling of the type used here but also to ensure that such modelling is accessible in a policy-support context (*cf.* Mulligan 2009a, b, c) and is well supported by extensive field and remote sensing datasets.

References

Allan, J. A. (1998). Virtual water: A strategic resource. *Ground Water, 36*(4), 545–546.

Bai, Z. G., Dent, D. L., Olsson, L., & Schaepman, M. E. (2008). *Global assessment of land degradation and improvement 1. Identification by remote sensing*. GLADA Report 5. Version 5.

Benayas, J. R., Martins, A., Nicolau, J. M., & Schulz, J. J. (2007). Abandonment of agricultural land: An overview of drivers and consequences. *CAB Reviews: Perspectives in Agriculture, Veterinary Science, Nutrition and Natural Resources*, 2(57), 1–14.

Besbes, M., Chahed, J., Hamdane, A., & de Marsily, G. (2010). Changing water resources and food supply in arid zones: Tunisia. *Water and Sustainability in Arid Regions* (pp. 103–121). Netherlands: Springer.

Bolle, H. J. (1995). Identification and observation of desertification processes with the aid of measurements from space: Results from the European Field Experiment in Desertification-threatened Areas (EFEDA). *Environmental Monitoring and Assessment, 37* (1–3), 93–101.
Brandt, C. J., & Thornes, J. B. (1996). *Mediterranean desertification and land use*. Hoboken: Wiley.
Briassoulis, H. (2004). The institutional complexity of environmental policy and planning problems: The example of Mediterranean desertification. *Journal of Environmental Planning and Management, 47*(1), 115–135.
Bugalho, M. N., Caldeira, M. C., Pereira, J. S., Aronson, J., & Pausas, J. G. (2011). Mediterranean cork oak savannas require human use to sustain biodiversity and ecosystem services. *Frontiers in Ecology and the Environment, 9*(5), 278–286.
Caraveli, H. (2000). A comparative analysis on intensification and extensification in Mediterranean agriculture: Dilemmas for LFAs policy. *Journal of Rural Studies, 16*(2), 231–242.
Catalinas, M. A. R. T. A., & Cantón, Á. G. (2012). Relationships between anthropogenic pressures and ecological status in a Mediterranean river basin district in Spain. In *9th International Symposium on Ecohydraulics 2012 Proceedings*. ISBN: 978-3-200-02862-3.
Center for International Earth Science Information Network (CIESIN). (2002). *Country-level Population and Downscaled Projections based on the B2 Scenario 1990–2100 [digital version]*. Palisades NY: CIESIN Columbia University. Available online at http://www.ciesin.columbia.edu/datasets/downscaled
Center for International Earth Science Information Network (CIESIN). (2004). International Food Policy Research Institute (IFPRI), and Centro Internacional de Agricultura Tropical (CIAT): Global Rural-Urban Mapping Project (GRUMP). Socioeconomic Data and Applications Center (SEDAC) *Alpha Version: Urban Extents* Palisades, NY: CIESIN, Columbia University.
Conacher, A. J., & Conacher, J. (1998). Introduction. In A. J. Conacher & M. Sala (Eds.), *Land degradation in Mediterranean environments of the World*. Chichester, UK: Wiley.
Cooper, T., & Arblaster, K (2007). *Climate change and the rural environment in a European context: Implications for land use and land use policy*. Background Paper for the LUPG/BfN Conference on 'Future Policies for Rural Europe. 2013 and Beyond', held in Brussels, 19–20 September 2007. Available online at http://www.ieep.eu/publications/publications.php?pub=68412
de Luis, M., González-Hidalgo, J. C., Longares, L. A., & Štepánek, P. (2009). Seasonal precipitation trends in the Mediterranean Iberian Peninsula in second half of 20th century. *International Journal of Climatology, 29*(9), 1312–1323.
Déqué, M., Rowell, D. P., Lüthi, D., Giorgi, F., Christensen, J. H., Rockel, B., et al. (2007). An intercomparison of regional climate simulations for Europe: Assessing uncertainties in model projections. *Climatic Change, 81*(1), 53–70.
Diffenbaugh, N. S., & Giorgi, F. (2012). Climate change hotspots in the CMIP5 global climate model ensemble. *Climatic Change, 114*(3–4), 813–822.
Florke, Weib M., Menzel, L., & Alcamo, J. (2007). Model-based scenarios of Mediterranean droughts. *Advances in Geosciences, 12*, 145–151.
García-Ruiz, J. M., López-Moreno, J. I., Vicente-Serrano, S. M., Lasanta–Martínez, T., & Beguería, S. (2011). Mediterranean water resources in a global change scenario. *Earth-Science Reviews, 105*(3–4), 121–139.
Gay, S. H., Osterburg, B., Baldock, D., & Zdanowicz, A. (2005). *Recent evolution of the EU Common Agricultural Policy (CAP): State of play and environmental potential.* Report from MEACAP project n°SSPE-CT-2004-503604, Document number: MEACAP WP6 D4b, Institute of European Environmental Policy. Available onlione at: http://www.ieep.eu/publications/pdfs/meacap/WP6/WP6D4B_CAP.pdf
Geeson, N. A., Brandt, C. J., & Thornes, J. B. (Eds.). (2003). *Mediterranean desertification: A mosaic of processes and responses*. Hoboken: Wiley.
Giorgi, F. (2006). Climate change hot-spots. *Geophysical Research Letters, 33*(8).
Giorgi, F., & Lionello, P. (2008). Climate change projections for the Mediterranean region. *Global and Planetary Change, 63*(2), 90–104.

Hijmans, R. J., Cameron, S. E., Parra, J. L., Jones, P. G., & Jarvis, A. (2005). Very high resolution interpolated climate surfaces for global land areas. *International Journal of Climatology, 25*, 1965–1978.

Hoerling, M., Eischeid, J., Perlwitz, J., Quan, X., Zhang, T., & Pegion, P. (2012). On the increased frequency of Mediterranean drought. *Journal of Climate, 25*(6), 2146–2161.

Iglesias, A., Garrote, L., Flores, F., & Moneo, M. (2007). Challenges to manage the risk of water scarcity and climate change in the Mediterranean. *Water Resources Management, 21*(5), 775–788.

Kirkby, M. J., Abrahart, R. J., Bathurst, J. C., Kilsby, C. G., McMahon, M. L., Osborne, C. P., et al. (2002). MEDRUSH: A basin-scale, physically based model for forecasting runoff and sediment yield. In N. A. Geeson, C. J. Brandt, & J. B. Thornes (Eds.), *Mediterranean Desertification: A mosaic of processes and responses* (pp. 203–229). Chichester: Wiley.

Kirkby, M. J., Jones, R. J. A., Irvine, B., Gobin, A, Govers, G., Cerdan, O., …. & Van Lynden, G. (2003). *Pan-European Soil Erosion Risk Assessment: The PESERA Map*, Version 1 October (2003). Explanation of Special Publication Ispra 2004 No.73 (S.P.I.04.73). European Soil Bureau Research Report No. 16, EUR 21176, 18 pp. Office for Official Publications of the European Communities, Luxembourg.

Kosmas, C., Kirkby, M. J., & Geeson N. (1999). *Manual on key indicators of desertification and mapping environmentally sensitive areas to desertification.* European Commission Publication EUR 18882, 87 pp

Koulouri, M., & Giourga, C. (2007). Land abandonment and slope gradient as key factors of soil erosion in Mediterranean terraced lands. *Catena, 69*(3), 274–281.

LandScanTM Global Population Database. (2007). Oak Ridge, TN: Oak Ridge National Laboratory. Available online at http://www.ornl.gov/landscan/

Laraus, J. (2004). The problems of sustainable water use in the Mediterranean and research requirements for agriculture. *Annals of Applied Biology, 144*(3), 259–272.

Le Goulven, P., Leduc, C., Bachta, M. S., & Poussin, J.-C., (2009). Sharing scarce resources in Mediterranean river rasin: Wadi Merguellil in Central Tunisia. In F. Molle & F. Wester (Eds.), *River Basin Trajectories: Societies, environments and development* (pp. 147–170). CAB International.

Leetmaa, S. E., Arnade, C., & Kelch, D. (2004). *A comparison of US and EU agricultural productivity with implications for EU enlargement.* Leetmaa: US-EU Food and Agriculture Comparisons. Mary Anne Normile and Susan E. 33.

Lionello, P., Malanotte-Rizzoli, P., & Boscolo, R. (Eds.). (2006). *Mediterranean climate variability.* Amsterdam: Elsevier.

Lionello, P., Platon, S., & Rodo, X. (2008). Trends and climate change in the Mediterranean region. *Global and Planetary Change, 63*, 87–89.

Mariotti, A., (2010). Recent changes in Mediterranean water cycle: A pathway toward long-term regional hydroclimatic change? *Journal of Climate, 23*(6), 1513–1525.

Mazzoleni, S., & Legg, C. (1998). Modmed: Modelling vegetation dynamics and degradation in Mediterranean ecosystems. In P. Mairota, J. B. Thornes, & N. Geeson (Eds.), *Atlas of Mediterranean environments in Europe—the desertification context* (pp. 14–18). Chichester: Wiley.

Molle, F. (2009). Water scarcity, prices and quotas: A review of evidence on irrigation volumetric pricing. *Irrigation and Drainage Systems, 23*(1), 43–58.

Montgomery, R. (2009). *Crop production in the Mediterranean Partner Countries continues to increase.* Statistics in Focus, Eurostat, Office for Official Publications of the European Communities. Available online at http://www.pcbs.gov.ps/Portals/_PCBS/Documents/KS-SF-09-012-EN.pdf

Mu, Q., Zhao, M., & Running, S. W. (2011). Improvements to a MODIS global terrestrial evapotranspiration algorithm. *Remote Sensing of Environment, 115*, 1781–1800. doi:10.1016/j.rse.2011.02.019.

Muller, K. (2004). California and the Global Olive Oil Industry.

Mulligan, M. (1996). *Modelling the hydrology of vegetation change in a degraded semi-arid environment*. Unpublished Ph.D. Thesis, King's College London, University of London.

Mulligan, M. (1998). Modelling the geomorphological impact of climatic variability and extreme events in a semi-arid environment. *Geomorphology, 24*(1), 59–89.

Mulligan, M. (2009a). *Vegetation continuous fields by land cover class based on GLOBCOVER*. Available online at http://www.ambiotek.com/vcfclasses. Source Data: © ESA/ESA Globcover Project, led by MEDIAS-France/Postel

Mulligan, M. (2009b). *Global mean dry matter productivity based on SPOT-VGT (1998–2008)*. http://www.ambiotek.com/dmp

Mulligan, M. (2009c). Integrated environmental modelling to characterise processes of land degradation and desertification for policy support. In J. Hill & A. Roeder (Eds.), *Remote sensing and geo information processing in the assessment and monitoring of land degradation and desertification*. UK: Taylor and Francis.

Mulligan, M., Burke, S. M., & Ramos, M. C. (2004). Climate change, land use change and the 'desertification' of Mediterranean Europe. In S. Mazzoleni, G. di Pasquale, M. Mulligan, P. di Martino, & F. Rego (Eds.), *Recent dynamics of the Mediterranean vegetation and landscape* (pp. 259–279). Chichester: Wiley.

Mulligan, M., Saenz, L., & van Soesbergen, A. (2011). Development and validation of a georeferenced global database of dams. Available online: http://www.ambiotek.com/dams

Oxley, T., Winder, N., McIntosh, B. S., Mulligan, M., & Engelen, G. (2004). Integrated catchment modelling & decision support tools: A Mediterranean example. *Environmental Modelling and Software, 19*(11), 999–1010.

Patterson, L. A., & Josling, T. E. (2005). *Mediterranean agriculture in the global marketplace: A project comparing policy approaches in California and the Southern EU States*. Report on Stage 1. European Forum Institute for International Studies, Stanford University. Available online at http://aic.ucdavis.edu/research1/patt_jos_med_ag_glob_mkt.pdf

Peleg, N., Bartov, M., & Morin, E. (2014). CMIP5 predicted climate shifts over the East Mediterranean: Implications for the transition region between Mediterranean and semi arid climates. *International Journal of Climatology*. Early View.

Perry, A. (2006). Will predicted climate change compromise the sustainability of Mediterranean tourism? *Journal of Sustainable Tourism, 14*(4), 367–375.

Petit, M. (2007) *Interactions between the Euro/Med liberalization process: World markets and International negotiations*. Deliverable D25 of Impacts of agricultural trade liberalization between the EU and Mediterranean countries (EU-MED AGPOL), Project no. SSPE-CT-2004-502457. Available online at: http://eumed-agpol.iamm.fr/html/publications/prj_report/d25.pdf

Plan Bleu. (2009). State of the Environment and Development in the Mediterranean. *Athens: UNEP/MAP-Plan Bleu*.

Puigdefábregas, J., & Mendizabal, T. (1998). Perspectives on desertification: Western Mediterranean. *Journal of Arid Environments, 39*(2), 209–224.

Ramankutty, N., et al. (2008). Farming the planet: 1. Geographic distribution of global agricultural lands in the year 2000. *Global Biogeochemical Cycles*, *22*, GB1003. doi:10.1029/2007GB002952.

Ramos, M. C., & Mulligan, M. (2005). Spatial modelling of the impact of climate variability on the annual soil moisture regime in a mechanized Mediterranean vineyard. *Journal of Hydrology, 306*(1–4), 287–301.

Riedel. (2007) Dairy goat grazing systems in Mediterranean regions: A comparative analysis in Spain, France and Italy. *Small Ruminant Research, 85*(1), 42–49.

Roeder, A., Udelhoven, Th, Hill, J., et al. (2008). Trend analysis of Landsat-TM and -ETM+ imagery to monitor grazing impact in a rangeland ecosystem in Northern Greece. *Remote Sensing of Environment, 112*(6), 2863–2875.

Rubio, J. L., & Recatalá, L. (2006). The relevance and consequences of Mediterranean desertification including security aspects. In *Desertification in the Mediterranean region. A security issue* (pp. 133–165). Netherlands: Springer.

Ruecker, G., Schad, P., Alcubilla, M. M., & Ferrer, C. (1998). Natural regeneration of degraded soils and site changes on abandoned agricultural terraces in Mediterranean Spain. *Land Degradation and Development, 9*(2), 179–188.

Ruiz, F. A., Mena, Y., Castel, J. M., Guinamard, C., Bossis, N., Caramelle-Holtz, E., et al. (2001). Typology of cereal-sheep farming systems in Castile-La Mancha (South-Central Spain). *Agr. Syst., 68*, 215–232.

Safriel, U. N. (2009). Status of desertification in the mediterranean region. In J. L. Rubio, U. Safriel, R. Daussa, W. Blum & F. Pedrazzini (Eds.), *Water scarcity, land degradation and desertification in the mediterranean region* (pp 33–73). Proceedings of the NATO Advanced Research Workshop on Water Scarcity, Land Degradation and Desertification in the Mediterranean Region DEC 10-11, 2007 Valencia, SPAIN. Book Series: Nato Science for Peace and Security Series C—Environmental Security.

Schar, C., Vidale, P. L., Lüthi, D., Frei, C., Häberli, C., Liniger, M. A., & Appenzeller, C. (2004). The role of increasing temperature variability in European summer heatwaves. *Nature, 427*, 332–336.

Schröter, D., Cramer, W., Rik Leemans, I., Prentice, C., Araújo, M. B., Arnell, N. W., et al. (2005). Ecosystem service supply and vulnerability to global change in Europe. *Science, 310* (5752), 1333–1337.

Serra, P., Pons, X., & Saurí, D. (2008). Land-cover and land-use change in a Mediterranean landscape: A spatial analysis of driving forces integrating biophysical and human factors. *Applied Geography, 28*(3), 189–209.

Shucksmith, M., Thomson, K. J., & Roberts, D. (Eds.). (2005). *The CAP and the regions: The territorial impact of the common agricultural policy*. CABI.

Siebert, S., Burke, J., Faures, J. M., Frenken, K., Hoogeveen, J., Döll, P., & Portmann, F. T. (2010). Groundwater use for irrigation–a global inventory. *Hydrology and Earth System Sciences Discussions, 7*(3), 3977–4021.

Siebert, S., & Döll, P. (2010). Quantifying blue and green virtual water contents in global crop production as well as potential production losses without irrigation. *Journal of Hydrology, 384* (3), 198–217.

Siebert, S., Döll, P., Feick, S., Frenken, K., & Hoogeveen, J. (2007). *Global map of irrigation areas version 4.0.1*. Germany, and FAO, Rome, Italy: University of Frankfurt (Main).

Somot, S., Sevault, F., Déqué, M., & Crépon, M. (2008). 21st century climate change scenario for the Mediterranean using a coupled atmosphere-ocean regional climate model. *Global and Planetary Change, 63*, 112–126.

Thivet, G. (2007). *Towards rural revival in the Mediterranean*. Blue Plan Notes No. 5. United Nations Environment Programme—Mediterranean Action Plan, Plan Bleu.

Thornes, J. B., Shao, J. X., Diaz, E., Roldan, A., Hawkes, C., & McMahon, M. (1996). Testing the MEDALUS hillslope model. *Catena, 26*, 137–160.

Turner, B. L., Lambin, E. F., & Reenberg, A. (2007). The emergence of land change science for global environmental change and sustainability. *Proceedings of the National Academy of Sciences, 104*(52), 20666–20671.

UNEP (United Nations Environment Programme). (1997). *World atlas of desertification* (2nd ed.). London: UNEP.

USGS. (2006). *Shuttle radar topography mission water body dataset*. Available online at: http://edc.usgs.gov/products/elevation/swbd.html

van Delden, H., Luja, P., & Engelen, G. (2007). Integration of multi-scale dynamic spatial models of socio-economic and physical processes for river basin management. *Environmental Modelling and Software, 22*(2), 223–238.

van Meijl, H., Van Rheenen, T., Tabeau, A., & Eickhout, B. (2006). The impact of different policy environments on agricultural land use in Europe. *Agriculture, Ecosystems & Environment, 114* (1), 21–38.

van Wesemael, B., Rambaud, Xavier, Poesen, Jean, Mulligan, Mark, Cammeraat, Erik, & Stevens, Antoine. (2005). Spatial patterns of land degradation and their impacts on the water balance of rainfed treecrops: A case study in South East Spain. *Geoderma, 133*, 43–56.

Verburg, P. H., Eickhout, B., & van Meijl, H. (2008). A multi-scale, multi-model approach for analyzing the future dynamics of European land use. *The Annals of Regional Science, 42*(1), 57–77.

Walsh, P. D., & Lawler, D. M. (1981). Rainfall seasonality: Description, spatial patterns and change through time. *Weather, 36*, 201–208.

Weib, M., Florke, M., Menzel, L., & Alcamo J. (2007). Model-based scenarios of Mediterranean droughts.*Advances in Geosciences, 12*, 145–151.

Willaarts, B. A., Volk, M., & Aguilera, P. A. (2012). Assessing the ecosystem services supplied by freshwater flows in Mediterranean agroecosystems. *Agricultural Water Management, 105*, 21–31.

World Database on Protected Areas. (WDPA). Annual Release. (2013). (web download version), November 2013. Available online at: http://www.protectedplanet.net

Wint, G. R. W., & Robinson, T. P. (2007). Gridded livestock of the world 2007. Rome: Food and Agricultural Organization of the United Nations, 131pp.

Chapter 17
Land Degradation in Central Asia: Evidence, Perception and Policy

Sarah Robinson

Abstract The introduction of communism into Central Asia brought agricultural transformation on a massive scale. Irrigation projects, expansion of livestock numbers and ploughing of the northern steppe modified vegetation and soils. Despite initial censorship, in the late Soviet period resources available for the study of land degradation processes were substantial and large scale mapping projects defined uniform criteria for degradation type and severity. Scientists found that degradation of vegetation cover from grazing was widespread, although productivity losses were slight in many areas. Tighter regulation led to stabilization of forest cover. Perhaps the most acute form of degradation was soil salinization, and the related Aral Sea disaster. Independence brought economic crisis: privatization turned salaried workers into subsistence farmers, dependant on local resources for survival. The early years were characterized by ploughing of marginal land in the mountains; abandonment of steppe fields for want of machinery; collapse in livestock inventories; and increasing reliance on wood for fuel. These changes led to a new mixture of degradation and recovery processes. Yet these were poorly documented, as funding for science collapsed and trained personnel migrated or retired. Institutes came to depend on environment-focussed development projects, so incentives to keep degradation on the agenda became strong. Such projects fund little basic science—so most statistics used to justify them were based on data from the 1980s or on more recent national data unaccompanied by documentation of methodology. Some research funding became available through international scientific collaborations, which have improved our understanding of specific processes such as grazing, soil erosion and deforestation. But much of this research is case-study based and cannot be scaled up. Studies at the national or regional scale often involve time-series analysis of remotely sensed vegetation indices. These have revealed responses to climatic factors, but so far have provided only speculative documentation of anthropogenic degradation processes over large areas.

S. Robinson (✉)
Division of Ecology and Evolution, Imperial College London, London, UK
e-mail: s.robinson@imperial.ac.uk; sarah.robinson09@gmail.com

R.H. Behnke and M. Mortimore (eds.), *The End of Desertification?*
Springer Earth System Sciences, DOI 10.1007/978-3-642-16014-1_17

Keywords Central asia · Land degradation · Science policy · Soviet union

This chapter reviews research on land degradation in the former Soviet republics of Kazakhstan, Kyrgyzstan, Tajikistan, Turkmenistan and Uzbekistan. The chapter examines the degradation processes which have been described in the region, and how both the scientific evidence for these processes and related public policies have changed since the countries became independent in 1991.

A set of physical characteristics specific to Central Asia largely determines the types of degradation processes that are important there (Fig. 17.1). Firstly, the climate is both arid and continental, having extremely cold winters and hot dry summers. Precipitation averages around 150 mm or less in the Karakum and Kyzylkum deserts of Turkmenistan and Uzbekistan respectively, increasing northwards to around 400 mm in the steppes of northern Kazakhstan. Desert-like conditions prevail in the Pamir Mountains of eastern Tajikistan, whilst the Tien Shan Mountains of Kazakhstan may receive over 800 mm per year, most of it as snow. The bulk of precipitation falls in winter and early spring, followed by a virtual absence during the rest of the growing season and for this reason, in the semi-arid southern areas, irrigated agriculture is dependent on snow and glacier melt from the mountains. It is this locally abundant and predictable water supply, combined with very high temperatures in the plains, which favours soil salinization.

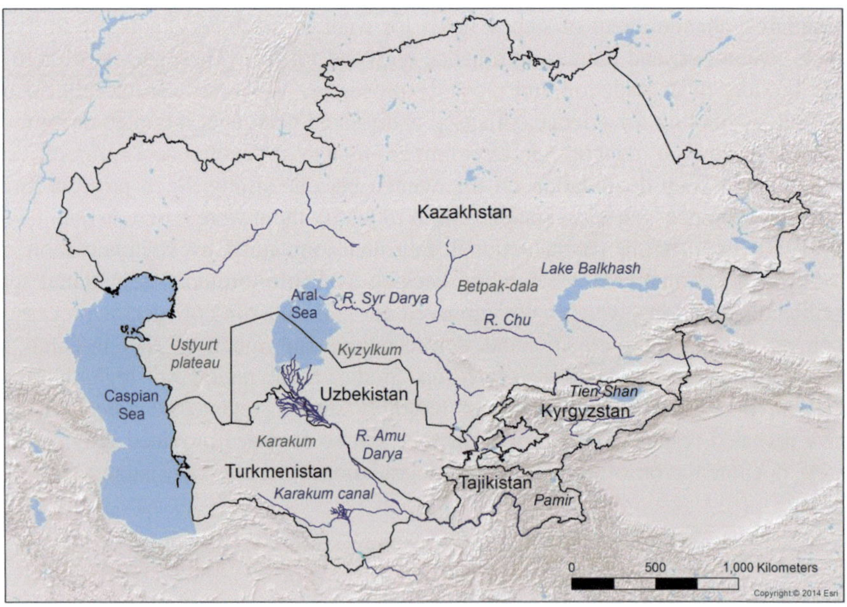

Fig. 17.1 Key physical features of the Central Asian Region (shaded relief—ESRI 2014. *Notes* names in black italics are mountain ranges, names in grey italics refer to deserts.

The second important characteristic of the Central Asian climate is its stability. Inter-annual variation in precipitation is low; for example the coefficient of variation (CV) varies between 34 and 25 % in desert and semi-desert areas of Kazakhstan and Uzbekistan (Robinson 2000; Gintzburger et al. 2005).[1] Due to this stability, numbers of livestock and wild ungulates have historically been little affected by catastrophic droughts [with the exception of Turkmenistan (Fedorovich 1973; Babaev 1996)]. Instead, mass mortality events were caused by severe winters and extreme snowfall events known as *dzhut*, occurring approximately every 10–12 years (Alekseevna 2004; Sludskii 1963). It is possible that such events have prevented the type of density-dependant feedback effects which might be expected of herbivore-plant relationships in equilibrium environments (Robinson and Milner-Gulland 2003).

High levels of aridity mean that crop agriculture is limited mostly to irrigable areas; thus arable land covers a small percentage of the territory of each republic (see Table 17.1). The exception is Kazakhstan, where arable land is concentrated in the wetter north; only 10 % of the cultivated area of the country is irrigated. Elsewhere, this percentage ranges from 68 % (Tajikistan) to 99 % (Turkmenistan) (Gintzburger et al. 2005). For the same reason, forest cover is also low (see Table 17.1), consisting of thin strips of Tugai forest along rivers or in deltas, areas of *Haloxylon* spp. and other large desert shrubs, or open walnut or juniper forests. Only in the wettest areas of the Tien Shan and Altai do closed canopy pine forests exist (Plate 17.1 shows photos of a number of the ecosystems described here). Overwhelmingly the region is dominated by vast rangelands, which cover over 60 % of its total area, or 90 % of land classified usable for agricultural purposes.

The introduction of communism by the Russian Bolshevik state began in the 1920s, when Central Asian society was still pre-industrial. The transformations in land use which were to follow were extraordinary both in rapidity and magnitude. Three in particular stand out. Firstly, vast areas of northern Kazakhstan were ploughed up during the 'Virgin Lands' campaign of 1954–1960. Secondly, across the remaining pastoral rangeland regions, livestock numbers rose to historical highs as supplies of winter feed provided by the state removed the limits to growth previously imposed by winter conditions, and well construction opened areas of previously unused desert pastures. Lastly, the area of irrigated land was expanded—gigantic canal construction projects resulted in the almost total diversion of the waters of the Pamir and Tien Shan mountains to irrigated agriculture in the Amu Darya and Syr Darya basins of Turkmenistan and Uzbekistan.

A survey of research from the Soviet period shows how different processes of land degradation were categorized and measured. The collapse of communism in the 1990s led to a second rapid transformation of land use patterns and to the partial loss of the scientific establishment in the region. The question is raised whether

[1]Note that the CV of primary productivity may be higher than that of rainfall (Ash and Melvor 2005).

Table 17.1 Statistics on land degradation in Central Asian Republics (and 2011 statistics on land use type as percentage of total area in each republic from FAOSTAT)

Total area of each republic km^2	% arable	% pasture	% Forest	Source of degradation statistic	Total % area degraded (and total area affected in km^2)	Degradation of vegetation cover	Salinization caused by sea level lowering (%)	Salinization of arable land	'Technogenic degradation' (see p. 458) (%)	Wind and water erosion of soils
Kazakhstan 2,724,900	9	68	1[1]	1985 degradation map[a]	59.9 % of arid area (632,931 km^2)	45 %	6.9	0.6 %	5.6	–
				Kazakhstan NAP[b]	66 % of national territory	14 % of pastures (severely)	–	20 % of irrigated land		11 % of national territory
Kyrgyzstan 199,949	7	46	5	Kyrgyzstan NAP[c]	40 % of agricultural land (pasture and arable)	74 % of pastures (30 % severely)	–	–	–	60 % of ploughed land
Tajikistan 142,550	7	27	3	Tajikistan NAP[d]	–	90–95 % of pastures	–	15 % of irrigated land	–	–
Turkmenistan 488,100	4	63	8	1985 degradation map[a]	66.5 % of arid area (277,220 km^2)	40.4 %	2.8	5.4 %	3.1	11.5 % wind erosion 2.8 % water erosion
				Turkmenistan NEAP[e]	100 % of all usable land (446,287 km^2)	76.8 %[2]	3.0	96 % of irrigated land[3]	0.2	1.8 % wind erosion 1.4 % water

(continued)

Table 17.1 (continued)

Total area of each republic km^2	% arable	% pasture	% Forest	Source of degradation statistic	Total % area degraded (and total area affected in km^2)	Degradation of vegetation cover	Salinization caused by sea level lowering (%)	Salinization of arable land	'Technogenic degradation' (see p. 458) (%)	Wind and water erosion of soils
Uzbekistan 447,400	10	49	7	1985 degradation map[a]	59.2 % of arid area (163,277 km^2)	35.7 %[4]	2.9	1.4 %	15.2	4 % wind erosion
				Uzbekistan NAP[f]	–	73 % of pastures[5]	–	53 % of irrigated land	–	73 % wind erosion; 18 % water erosion (of agricultural land)

Note In the 'total' column, those land categories included in the calculation of degraded areas are given (e.g. national territory, usable land, agricultural land etc.). Percentages given in subsequent columns are all calculated out of those land categories specified in the 'total' column, unless otherwise specified. Where land categories used to calculate percentages of land affected by each degradation type differ across columns, the sum will not add up to 100 %. In the case of row 1, figures do not quite add up due to small errors in the original cited dataset

[a]Kharin and Kiril'tseva 1988; Babaeva 1999; Babaev 1985. These figures are for arid lands only and exclude mountain areas. The category concerning degradation of vegetation cover includes both areas exhibiting general degradation and areas with local degradation around wells

[b]Ministry of Environmental Protection of the Republic of Kazakhstan 2005 (Similar figures with additional information are given in Kulov and Zhooshov (2007) and appear also in the Kyrgyz 2007 Natural Resource Sector Study (Asian Development Bank 2007a)

[c]Ministry of Agriculture and Water Resources of the Republic of Kyrgyzstan 2000

[d]Ministry of Nature Conservation of the Republic of Tajikistan 2001

[e]National Environmental Action Plan (Government of Turkmenistan 2002) [The 1997 NAP currently on the UNCCD website was not used as it does not include data on the status of land degradation. The 2002 National Environmental Action Plan of Turkmenistan (NEAP), includes a review of land degradation based on data sourced from the maps of Human Induced Degradation in Turkmenistan (unavailable but described in Babaev 1996) and of Soils in the Aral Sea Basin (Kharin et al. 1993). The figures are based on a three class scale of desertification (slight, moderate severe) with no base year. All usable lands of the republic (91 % of total area) are classed as degraded in some way. Data collection periods for these maps were not given in the above sources, but they probably originate in the 1980s. These figures continue to be cited in many reports and publications (e.g. Saigal 2003a)]

[f]UNEP/GLAVGIDROMET 1999

[1]This FAOSTAT figure probably represents the area of closed canopy forests alone. 'Other wooded land'; as defined by FAO Forest Resources Assessments (2005 tables) may cover an additional 7 % of the country. The Kazakhstan NAP records these two types of forest as covering a total of 4.6 % of the country, of which just over half is true forest; similar figures are given in data from the Kazakhstan Statistical Agency (2011). Differences between figures, sources and definitions constitute obstacles to the task of measuring change.

[2]This figure includes 1.5 % of usable land (mostly pasture) which has been swamped by drainage water. The remaining 75.3 % of lands subject to degradation of vegetation cover can be broken down as follows: 66.2 % is slightly degraded, 10 % moderately, 0.1 % severely.

[3]The figure of 8.3 % given as a percentage of usable national territory affected by salinization of irrigated land cannot be interpreted as only around 3.5 % of national territory (17,000 km^2) is irrigated, so we have not included this figure in the table. The NEAP states that 68 % of irrigated land is moderately or severely salinized and 28 % is slightly affected.

[4]In a related study (Kharin et al. 1993), out of 214,480 km^2 of rangelands and forests in the Uzbek Aral basin, 100 % were degraded: 58 % slightly, 37 % moderately, 5 % severely.

[5]Of about 220,000 km^2 of pastureland, vegetation on about 63 % has lost 20–40 % of its productivity; vegetation on 10 % of the area has lost more than 40 % of its productivity.

Plate 17.1 Photographs (**a**) *Haloxylon* areas, Moiynkum desert, Kazakhstan © Carol Kerven. (**b**) Tugai forest, Pamir region of Tajikistan. (**c**) Coniferous forests in the Tien Shan mountains. (**d**) Solonchak, central Kazakhstan. (**e**) Orto Syrt—abandoned grazing areas in Kyrgyzstan. (**f**) *Teresken (Ceratoides papposa)* collection, Pamirs. (**g**) The Aral Sea in 1986, 1996 and 2003 (University of Maryland, JAXA). (**h**) Dust storms from the Aral sea, 2010 (NASA) (note the Uzbek-Kazakh border which runs through the sea).

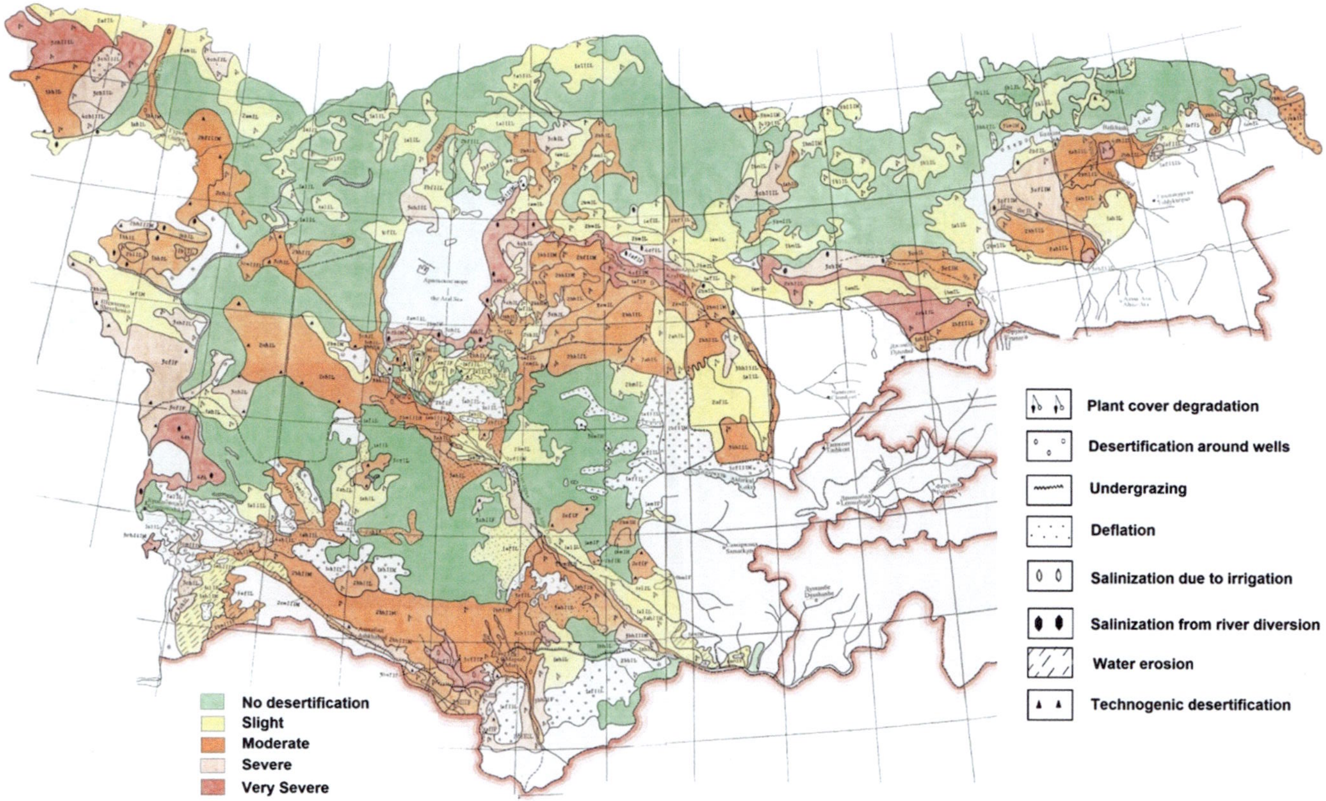

Plate 17.2 Map of human-induced desertification of Arid Lands of the USSR (Adapted from Babaev 1985).

recent research has been able to demonstrate changes in degradation processes associated with current economic and environmental conditions.

17.1 Land Degradation During the Soviet Period

17.1.1 A Regional Overview: Land Degradation Mapping

A major attempt to describe and quantify Soviet-era land degradation processes is the 1985 map of Human-Induced Desertification in Arid Regions of the USSR, which covers Turkmenistan, Uzbekistan and southern Kazakhstan and was produced by researchers at the Institute of Deserts in Turkmenistan (Babaev 1985). The base year for change detection was 1965 and changes were recorded into the mid-1980s. The map is described in detail in a number of publications (Kharin et al. 1986; Kharin and Kiril'tseva 1988; Kharin 2002) and is reproduced here in Plate 17.2.[2] The types of land degradation covered by the map were: *change in vegetation state*; *wind and water erosion, localized loss of vegetation cover caused by activities such as construction, quarrying and road building* (known as 'technogenic degradation') and *salinization of irrigated lands*. Some of the criteria used to measure these processes are given in Box 1.

Box 1. Criteria for production of the 1985 map of desertification of arid lands of the USSR

Degradation was classified into five classes: none, weak, moderate, severe, and very severe. The state of the environment in 1985 was determined by satellite images, whilst for the 1960s aerial photos, thematic maps and questionnaires were used (Kharin and Kiril'tseva 1988). Criteria for measurement of degradation are described in Kharin et al. (1986), Kharin (2002) and Babaev and Kharin (1999).[3] Degradation of vegetation cover was measured by finding actual productivity as a percentage of potential productivity, and species composition indicators. The categories of severe and very severe degradation are characterized by low productivity (<60 and <30 % of potential respectively) and a high proportion of non-palatable species. Areas in this category are relatively small according to some estimations, concentrated mainly around wells and settlements (Kharin et al. 1986). Moderate

[2] A related Desertification Map of the Drylands of Asia (Kharin et al. 1999) was based on a land cover classification using NDVI and surface temperature values from the AVHRR instrument. These values were related to land degradation types using existing desertification maps, which probably explains why this map bears a strong resemblance to the original 1985 map.

[3] Various criteria for measurement of vegetation degradation are given in these sources; the quantitative criteria used here are those listed in Babaev and Kharin (1999) as these seem to be the ones most likely to have been used for the map under discussion.

degradation involves the presence of more or less stable associations that have been productive at 80–90 % of their potential for long periods but include weed species, whilst weak degradation refers to slight changes in species composition only. Wind erosion is measured by the presence of deflation scars and moving sands; and water erosion by the percentage of bare soil, number of rills and gullies per unit area. 'Technogenic' degradation is measured by the disturbance of vegetation (% of total area). In fact, wind, water and technogenic degradation can be considered as types of soil erosion combined with extreme forms of vegetation cover degradation, so separation of these from the broader vegetation cover category must be somewhat artificial. Salinization indicators include both cotton yields and chemical measures of soil and water salinity. In addition to mapping the actual state of desertification, the authors also produced maps of desertification rate and risk (Kharin and Kiril'tseva 1988; Babaev and Kharin 1999).

Remote sensing was used to produce the map, but detailed information on sampling, ground truthing or accuracy assessment could not be found. The definitions of 'potential' productivity and methods for separation of human induced degradation processes from climatic factors are also undescribed in accompanying publications.

The 1985 map is a frequently cited source of key land degradation statistics in the arid and semi-arid lands of Central Asia. Some of these are summarized in Babaeva (1999) and reproduced here in Table 17.1. The 'headline' figure from the map is that 60 % of the arid land area of Central Asia was considered to be some stage of desertification in 1985. All five republics signed the UN Convention to Combat Desertification (UNCCD) and Table 17.1 also presents figures from the National Action Plans (NAPs) and related documents submitted to this body. These figures are a useful starting point for our investigation because they (i) cover the total area of each national republic; (ii) represent officially endorsed or publically acknowledged figures on desertification; and (iii) are usually derived from work by the national scientific community of each republic. Although the NAPs were produced between 2000 and 2005, they are reviewed in this section as many of the statistics they present are based on Soviet-era research. The plans are of varying quality; the Uzbek and Turkmen documents present national statistics with references, the Tajik plan presents quantitative results from local studies without references, whilst the others make mostly general statements with some non-referenced statistics. The next sections discuss some of the figures presented Table 17.1 together with a review of the Soviet-era scientific literature, bearing in mind that some of the Russian sources upon which statistics are based could not be identified or were unobtainable.[4]

[4]We focus primarily on processes affecting degradation of vegetation cover and soil salinization. Wind and water erosion are not treated specifically as these processes act to exacerbate the ploughing, grazing or cutting activities initiated by people.

The figures in Table 17.1 suggest that the most widespread type of desertification in all republics was the degradation of vegetation cover, defined as a decrease in vegetation productivity combined with a reduction in proportion of palatable species. It was considered to be caused primarily by cutting of shrubs for fuel followed closely by overgrazing (Kharin 2002; Zonn et al. 1981). Related factors include "technogenic" disturbances, concentrated along the Karakum canal and the eastern shore of the Caspian sea (Zonn et al. 1981) and conversion of steppe to cropland in Kazakhstan during the Virgin Lands campaign. This led to a loss of topsoil and an increase in dust storms [see references in Pryde (1991)]. Large areas of degraded soils have been recorded in this region and some steppe plant associations were virtually eradicated (Rachkovskaya et al. 1999). An additional consequence was the loss of summer pastures for livestock, increasing grazing pressure elsewhere (Baitulin and Karibaeva 1999). Here we focus in more detail at the literature on impacts of grazing and wood cutting on vegetation cover in the latter Soviet period.

17.1.2 Degradation of Grazing Lands

As mentioned, livestock numbers increased enormously during the Soviet period (for example in Kazakhstan from around 65 million sheep equivalents in 1916 to over 90 million by 1991). Although previously nomadic tribes were settled and the amplitude of long latitudinal livestock migrations decreased, some form of seasonal movements persisted, supported by collective and state farms (Alimaev and Behnke 2008). Vertical transhumance continued in the mountains, and large scale movements of animals were organized between mountains and plains, crossing republic boundaries in many cases. Despite these migrations, the spatial distribution of grazing pressure remained variable.

In the deserts and steppes, grazing damage can be related to spatial patterns of soil type and water availability and was most severe in sandy massifs traditionally used for winter grazing. Kazakh scientists emphasized serious deflation in the Moiynkum and Taukum sandy deserts, amongst others (Dzhanpeisov et al. 1990; Zonov 1974; Zhambakin 1995).[5] Clay deserts were less affected, as these represented transit pastures with few water points. In Kazakhstan, the creation of specialized sheep-raising state farms (*sovkhoz)* in the 1960s concentrated large numbers of animals in pastures which were formerly used briefly during migratory periods, whilst the best summer pastures in the north were ploughed up. This, according to Kazakh scientists, was a cause of degradation across a much broader range of pasture types (Asanov and Alimaev 1990).

---

[5]In contrast, Bizhanova (1999) notes that despite intensive use, a relatively small area of sandy deserts in Kazakhstan could be classified as severely degraded.

Scientists in Kyrgyzstan were also concerned about degradation of vegetation cover, again focusing on overgrazing. It has been estimated that small stock numbers rose from around 1 million head in 1935, to perhaps 10 million in 1987 (Rozanov 1990). The State Institute for Land Management (Giprozem) considered in that in 1985 about 24 % of all pastures were degraded (Kyrgyz Giprozem in Asian Development Bank 2007a); figures in the NAP suggest a total of 74 % of pastures affected; 30 % severely. We shall return to the derivation of these figures in the next section. Most estimates stress that summer pastures were worst affected (e.g. Rozanov 1990). There is also anecdotal evidence to suggest that the pastures were heavily grazed. Accounts of serious pasture damage during the Soviet period were recorded in interviews with herders by the author of this paper (Robinson et al. 2001) and by respondents in a study by Leichti (2012).

The figures provided by Uzbekistan and Tajikistan to the UNCCD also seem to indicate high levels of degradation, although the original scientific studies upon which these are based could not be obtained. Tajik mountains are arid and extremely steep and accordingly the Tajik NAP stresses vegetation degradation combined with serious soil erosion on slopes. Most estimates in the literature concerning Turkmenistan can be traced back to either the 1985 Map of the Present State of Desertification in Turkmenistan or the 1993 Map of Anthropogenic Land Degradation in the Aral Sea Basin, both by the Institute of Deserts. The figures in the Turkmen National Environmental Action Plan (NEAP) in Table 17.1 suggest that around 75 % of the usable area of the republic is affected by degradation of vegetation cover, yet there are significant differences between sources both regarding the total degraded area and the proportion which is severely degraded.[6] This illustrates the importance of documentation of both methodology and definitions of severity class. As we have seen in Box 1, certain criteria used to identify weak or moderately degraded vegetation suggest little reduction in the ability of ecosystems to support livestock; using production-related definitions, such ecosystems would not be described as degraded at all.

Many Soviet-era studies aimed to *describe* the impact of degradation processes on vegetation, without attempting to map them. Long-term studies comparing vegetation at varying distance from wells or between protected and grazed areas were conducted in a number of republics (e.g. Kurochkina and Shabanova 1990; Nechaeva et al. 1979). Such studies are important as they give us some idea of the relationships between grazing pressure and resource condition—for example Nechaeva's study at two wells in the Central Karakum, demonstrates how edible biomass of certain desert associations drops by over 50 % between 1 and 5 km from

[6]According to the source data (Babaev 1996), all usable land in Turkmenistan is degraded in some way. Of this area, around 75 % are classed as having degraded vegetation cover; but only 10 % are said to be moderately or highly degraded, having productivity losses over 25 %. Conversely, another set of figures presented both in Kharin (2002) and elsewhere in the NEAP (which suggest a comparable total area of degraded vegetation) state that 50 % of this area is moderately or severely degraded. Lastly, figures derived from the USSR Arid Lands map suggest that 40 % of arid lands in Turkmenistan have been affected by some form of degradation since the 1960s.

the water source. A focus on forage quality meant that the studies cited here were also able to demonstrate the impact of grazing on the protein and energy content of pastures. Whilst methodologies here are far better documented than those of the maps, they only apply to small geographical areas.

What of the impact of vegetation degradation on the livestock sector itself? In Kazakhstan, Asanov and Alimaev (1990) noted that towards the end of the 1980s, sheep mortality and lambing rates began to be affected. An analysis of Soviet-era livestock statistics for Kazakhstan (Robinson and Milner-Gulland 2003) found a significant positive correlation between death rates and livestock inventories and a negative relationships between animal numbers and both birth rate and live weight. We cannot say for certain whether these decreases in sheep condition were due to increasing competition for natural forage, decreasing quality or quantity of winter food provision, or other reasons. Figures suggest that winter fodder provision kept pace with growing animal numbers, but the cost of meat production more than doubled between 1971 and 1987.

17.1.3 Degradation of Forest Cover

Although forests cover a very small proportion of land, they are disproportionally important to livelihoods as they provide firewood and represent key seasonal grazing resources. The direction and extent of forest cover degradation during the Soviet period is difficult to assess. Both Babaev (1996) and Zonn et al. (1981), suggest that the 1920s were characterized by very high levels of felling in desert areas, leading to the invasion of moving sands on a large scale in the Bukhara region and lower Amu Darya basin. In Kyrgyzstan, deforestation was reported to be massive up to the late 1950s (Asian Development Bank 2007a; Ministry of Agriculture and Water Resources of the Republic of Kyrgyzstan 2000); Tajikistan's forests, which may have covered 25 % of the country at the beginning of the 20th Century, were decimated for cotton production (Kirchhoff and Fabian 2010). All these authors describe how protection and afforestation programmes then led to stabilization and even improvements from the 1950s up to 1990. Most forests came under protection of the State Forestry Department, which was responsible for prevention of illegal cutting. At the same time, incentives for cutting wood must have been considerably reduced as the state provided electricity and cheap coal.

Yet Kharin (2002) recounts that corruption of low paid officials meant that some illegal cutting continued; many *Haloxylon* stands under administration of the forestry departments continued to be grazed, whilst others were transferred as grazing areas to departments of agriculture, which presumably afforded even less protection. A highly critical study by the USSR State Forestry Department (Sukhikh 1991) suggests that both the area and productivity of juniper, walnut, tugai forests and coniferous stands decreased across the region due to heavy grazing. Kharin (2002) presents national statistics for the last six years of the Soviet Union (1983–1989) which suggest negative trends in all republics except Kazakhstan, and a region-wide decrease in forest area of just under 8 % during that period alone. Thus we are left with an impression of both protection and afforestation measures taken by the Soviet administration, mixed with continued illegal cutting and heavy grazing pressure. The overall trends emerging from these conflicting processes appear to be negative, but detailed studies with well-documented methodologies could not be obtained.

17.1.4 Soil Salinization

Other than the forms of vegetation degradation described above, the second major type of desertification in Central Asia is soil salinization, both of irrigated agricultural land and of adjacent rangeland, as salty irrigation water is dumped into poorly drained areas, which develop a system of salt pans or solonchaks (Zonn et al. 1981). During the period from 1965 to 1978 alone, the irrigated area in Turkmenistan and Uzbekistan grew by around 35 % through huge engineering projects such as the irrigation of the previously arid Golodnaya and Karshai steppes in Uzbekistan, and the construction of the Karakum canal in Turkmenistan (Zonn et al. 1981). It also appears that water use increased at a far greater rate than the area of irrigated land (Kolodin and Rabochev 1999).[7]

Almost all irrigation in the region was by open furrow and the preferred crop was cotton, which has high water requirements, yet the level of in-field drainage to remove excess water was low—only 58 % of irrigated lands in Turkmenistan were drained at the beginning of the 1990s (O'Hara 1997). These factors, combined with application of excessive water and seepage from poorly lined irrigation infrastructure, led to rising water tables, waterlogging, and accumulation of salt which rises to the surface through capillary action. Zonn et al. (1981) have estimated that in the 1980s nearly 60 % of irrigated lands in Uzbekistan were affected by salinity to varying degrees. A synthesis of the situation in Turkmenistan at the end of the Soviet Period by O'Hara (1997) gives us some idea of the consequences for agricultural production: in 1992 around 50 % of irrigated land in that republic was affected by medium levels of soil salinity (>8 Ece ds/m[8]). At such levels, yields of almost all crops (especially fruit and vegetables) are reduced. At high salinity levels (>16 Ece ds/m) only a very small selection of crops can be grown (essentially barley, wheat and cotton) and at much reduced yields. Over 15 % of irrigated land in Turkmenistan exhibited these kinds of salinity levels in 1992 and the statistics given here do not include the amount of land which had to be abandoned.

---

[7]In 1960 the area of irrigated lands in Turkmenistan was 496,000 ha with a drainage flow of 0.1 km^3—by 1994 this had climbed to an irrigated area of 1.7 million ha, with a drainage flow of 6.6 km^3.

[8]Electrical conductivity in decisiemens per metre.

17.1.5 The Aral Sea

The loss of the Aral Sea has been described as one of the greatest ecological catastrophes caused by man, but is treated only briefly here as the problem has been extensively documented elsewhere. Prior to the 1960s, although irrigation expansion was large, the sea level (which was then around 50 m above sea level) was not affected, as withdrawals were partially compensated by reduced evaporation and flooding in the delta downstream. Shrinkage started during the 1960s, when this compensation reached its limits and irrigation expanded into more marginal desert areas which required more water per hectare (Micklin 1994). By 1987 water loss was such that the Sea split into two separate lakes known as the small sea (in the north and fed by the Syr Darya) and the large sea (in the south fed by the Amu Darya), and water levels continued to drop—reaching 37 m at the beginning of the 1990s (see Plate 17.1h). By this time, the unique ecosystems of the Amu Darya river delta were practically destroyed and the tugai forests and reed communities had shrunk by 85 %. Irrigated agriculture, practised in the delta for millennia was damaged by salinization (Micklin 1994). The consequences have been severe: the fishing industry, which supported around 60,000 people, was wiped out. The average air temperature rose by around 1.5–2°C whilst humidity levels dropped (Zonn et al. 2009). Dust storms originating in the dry sea bed carried salt deposits as far as 400 km, reducing the fertility of surrounding soils and affecting pastures (see Plate 17.1i). The worst affected area was Karakalpakistan, the Uzbek province in which the southern part of the sea is situated, and in which 96 % of lands are now saline to varying degrees. In this region, which has a population of 1.5 million people, renal ailments related to saline drinking water high concentrations of toxic compounds in breast milk and blood have been recorded (Crighton et al. 2011).

17.1.6 Science and Soviet Policy

The funding allocated to degradation research, publication of maps and the detailed 1981 USSR desertification report to UNEP (Zonn et al. 1981) appear to indicate a willingness by government to acknowledge the issue and to support its researchers to engage with the international scientific community. Yet insistences that desertification was under control, combined with internal censorship (for example of references to the Aral Sea) were also features of the Soviet system (Saiko 1995). Much of the literature thus focusses on solutions, most of which were technical in nature. Strategies employed to mitigate the effects of heavy grazing included massive use of artificial fertilizers and herbicides. Dune fixation methods and other types of sustainable desert improvement were heavily researched, although not always applied over large areas (Zonn et al. 1981; Micklin 1995). A grandiose plan was developed to save the Aral Sea through diversion of water from Siberian rivers, but these ideas were dropped in 1986 amid media criticisms. The new openness which appeared around this time under *glasnost'* allowed public campaigns (for example against 'cotton at any price' in Uzbekistan to flourish) and led to politicization of environmental issues (Pryde 1991). *Goskompriroda* (the state environmental protection agency) was created to enforce compliance with environmental laws and to improve sustainable management of ecosystems, but only in 1990 was a resolution passed which might have given it sweeping powers (Robinson 1989). The presence of both scientists and *Goskompriroda* during the 1991 Forest Department expedition to evaluate the state of forest resources in Central Asia (Sukhikh 1991) illustrates a new relationship between science and policy in which resource management organizations became subject to independent oversight by experts and regulators. However following the collapse of the Soviet Union, widespread impoverishment led to a decrease in public concern about environmental issues and the loss of state funding to study them.

17.2 Drivers of Land Degradation in the Post-soviet Period

The 1990s were a period of severe hardship in Central Asia—the GDP of the new republics shrank abruptly[9] and food security became a priority. Although both collective farm structures and government funding for the agricultural sector continued initially in Turkmenistan and Uzbekistan, levels of support were much lower than before. In Kyrgyzstan, Tajikistan and Kazakhstan subsidies and salaries disappeared, leading to the use of livestock as currency. Collective structures collapsed in some areas; in others they were broken up more slowly as land reform processes introduced individual user rights or even private ownership of land.[10] Former salaried workers suddenly found that they had become peasant farmers, dependant on local soil and water resources for survival. In the 2000s a greater individualisation of farming has also been observed in Uzbekistan and Turkmenistan, although state controls are tighter and security of tenure is low. Some of the new states, which previously benefitted from gas and electricity produced by neighbouring republics, suffered severe shortages as those republics, now sovereign states, cut off supplies. Those changes likely to have the greatest impacts on land use are summarized below, in the following sections the implication of these changes for degradation processes are discussed.

Livestock numbers: Decreases in livestock inventories of between 50 and 70 % were recorded in Kazakhstan, Kyrgyzstan and Tajikistan, where the state withdrew from livestock production. Since around 2000, the livestock sector has recovered to

[9]Year on year GDP change was still negative across Central Asia in 1995, but by 1999 it was positive in all five republics (Babu and Sengupta 2006).

[10]With exceptions including cotton growing areas of Tajikistan and the large arable farming corporations which appeared in arable areas of Kazakhstan.

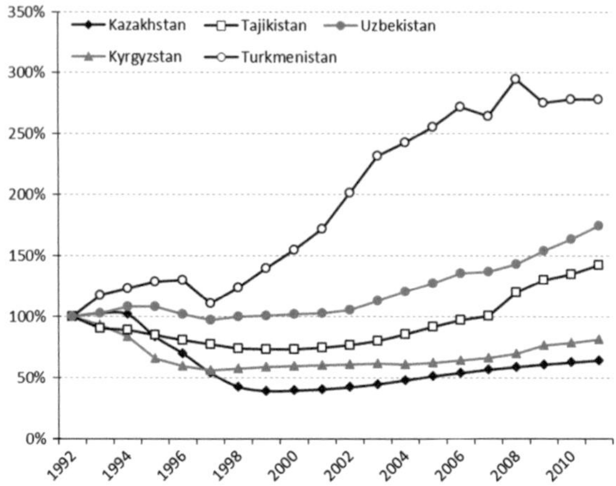

Fig. 17.2 Trends in livestock numbers (sheep equivalents) since 1992. All figures given as percentages of 1992 numbers. *Source* FAOSTAT. *Note* in some republics (namely Turkmenistan) livestock numbers may be overestimated, in Kyrgyzstan inventories are estimated to be underreported by 25–30 % (Bedunah 2006; Robinson 2007).

varying extents (see Fig. 17.2). In Turkmenistan and Uzbekistan initial changes were slight, followed by sharp increases in Turkmenistan, although there has been some controversy over these figures (Suleimenov et al. 2006).

Spatial patterns of pasture use: Trends towards a reduction in livestock mobility and abandonment of remote pastures have been observed in all five republics (Robinson et al. 2012). These are related to collapse of state support for livestock movement and the loss of economies of scale needed for mass migration associated with individualization of livestock ownership (Kerven et al. 2004). These changes were less marked in Turkmenistan, where state ownership of livestock meant that Soviet-era herd sizes and pasture allocations persisted.

Cropping patterns: Food security concerns resulted in a switch from cotton and feed crops, to wheat, a change driven by the state in Turkmenistan and Uzbekistan and by individual farmers in other republics. Subsidized feed from elsewhere in the USSR was also no longer available, resulting in a 'winter fodder crisis' for livestock and drop in animal productivity (Suleimenov et al. 2006; Sedik 2010).[11]

Irrigated land: A virtual cessation of canal maintenance and lack of fuel or spare parts for pumps led to a steep reduction in the proportion of irrigable lands which could actually be watered in some republics. Statistics are difficult to interpret as these often give only the area of irrigable land, but it appears that Turkmenistan, in contrast with other republics, has increased its irrigated area (Lerman and Stanchin

[11]For example, in Kyrgyzstan the area planted to fodder crops dropped from 48 % of total cropped area in 1991 to 18 % in 2003; in Uzbekistan from 25 % of cropped land area in 1991, to 7 % in 2003 (Suleimenov et al. 2006).

2006). The decrease in maintenance also applies to drainage systems, which has consequences for salinization.

Rainfed land: In Tajikistan, land shares allocated to private households were often too small for subsistence requirements; this led to the development of shifting cultivation of rainfed pastures in the mountains. Meanwhile, institutional collapse and fuel shortages led to widespread land abandonment in arable areas of northern Kazakhstan.

Fuel and fertilizer: There has been an increase in extraction of shrubs and trees for fuel, especially in republics such as Tajikistan where electricity supply is now very poor. This also has implications for agriculture—the use of mineral fertilizers in Central Asia declined steeply since 1991, and may have contributed to declining crop yields (in particular cotton) in the 1990s (Goletti and Chabot 2000; Gintzburger et al. 2005). Lack of fertilizer has increased the importance of manure for soil fertility maintenance; but demands for manure as a fuel for heating and cooking have also increased, exacerbating the fertilizer deficit in some areas.

The magnitude of these changes is such that the nature, extent and severity of land degradation is likely to have changed over the last 20 years. The belief that it must be getting worse is almost universal, as exemplified by the following quote: *"There are no modern data on the condition of land, vegetation and water but it is believed that all of these indicators have declined further and that the rate of change is increasing"* (Turkmenistan UNCCD National Working Group 2006). Here we ask whether there really are no recent data which we can use to assess land degradation trends.

17.3 Research in the Post-soviet Period

Unfortunately, one of the consequences of the Soviet collapse was the reduction in budgets for science. In most republics rangeland monitoring systems broke down completely (e.g. Schillhorn-van-Veen et al. 2004), although large scale pasture assessments have recently resumed in Kazakhstan. Many of the best scientists emigrated and salaries fell too low to attract young people, leaving most republics with an acute shortage of specialists. In contrast, scientists from outside the USSR were able to study Central Asian environments for the first time, resulting in a number of collaborative projects. Development assistance enabled the continuation of some national monitoring work.

17.3.1 Rangeland Recovery and Degradation

In Kazakhstan, a number of international scientific projects in the late 1990s documented the retraction of livestock from remote seasonal pastures in Almaty Province and parts of southern and central Kazakhstan. This loss of livestock mobility led to heavy grazing around settlements and barns particularly in winter areas, whilst summer pastures were abandoned (Alimaev et al. 2008; Alimaev and Behnke 2008). Work by Kerven et al. (2006) showed that as households accumulate livestock over time, they usually resume mobile or partially mobile management strategies, resulting in improved livestock performance (measured in terms of sheep live weight) relative to their sedentary neighbours. In some desert areas poor well condition and high repair costs may hamper the return to abandoned pastures.[12]

Concerning vegetation recovery, an expedition to central Kazakhstan in 1998 by the Kazakh Institute of Botany found that areas previously described as degraded were in good condition (Sadvokasov 2000; Robinson 2000). Stogova (1999) found that both vegetation cover and species composition in summer pastures in the Assy Valley of the Tien Shan mountains, monitored from 1979 to 1990 and then in 1999, had recovered. In Almaty oblast, Alimaev (2003) suggested that pasture productivity had increased across a number of pasture types, contrasting data for the years 1974–1986 with 1998–1999. Studies using NDVI and Rainfall Use efficiency (see Box 2) showed similar patterns for the same area (Ellis and Lee 2003; Coughenour 2008). However, no evidence of range degradation or recovery was found in central Kazakhstan using these methods (Coughenour 2008). At the local scale, Karnieli et al. (2008) used Landsat imagery to study vegetation change between 1970 and 2000 around wells on the Ustyurt Plateau and Kyzylkum desert in Kazakhstan. At both sites bare areas, around wells disappeared as rangeland recovery progressed, whilst on the Ustyurt plateau other sources of vegetation removal related to exploitation of gas and oil appeared. Kazakhstan is now funding vegetation assessments once again, but published studies comparing results of current work with past data are not yet available.

Box 2. Rainfall Use Efficiency

Rainfall Use Efficiency (RUE) is the annual aerial biomass production from rainfall produced per mm of annual rainfall or the ability of an ecosystem to produce above ground biomass from rainfall (Le Houérou 1984). It provides us with a way of measuring vegetation change which 'corrects' for inter-annual variation in precipitation. However RUE itself exhibits relationships with rainfall[13] because plants are more efficient at using rainfall to produce biomass in dry years. This compensatory response has been used as an indicator of the sensitivity of vegetation to rainfall, which in turn is related

[12]In Kazakhstan a total of 147 m ha (or 80 %) of rangeland depends on man-made facilities using subsurface water. Most of the boreholes do not function (partially due to disintegration of their construction and partially due to lack of electricity) and about half of shaft-wells are also out of service (Schillhorn-van-Veen et al. 2004).

[13]Le Houérou (1984) suggests that, when *compared between different sites*, RUE is inversely related to aridity since the proportion of inefficient rainfall increases with aridity (i.e. evaporation and runoff are relatively more important). The same has been noted for the steppes of Central Asia (Gilmanov 1995). Conversely, higher RUE values are associated with *lower* rainfall at *single* sites because, at a given location, plants are more efficient at using rainfall in dry years.

to pasture condition. Using such methods Ellis and Lee (2003) and Robinson (2000) found significant differences between sites in Kazakhstan and Ellis and Lee detected change between the periods before and after 1991 in some places.

In Uzbekistan and Turkmenistan, continued engagement of the state in agriculture prevented a collapse in livestock inventories in the 1990s, but as elsewhere limits to private ownership of animals were removed, so inventories rose from a very high base. Yet pasture availability has declined. In Uzbekistan, 40 % of pasture previously belonging to agricultural enterprises has been returned to the state, suggesting possible abandonment (Yusupov et al. 2010). *Dekhkan* farms, a form of household-based holding which own most of the national livestock inventory, do not hold formal entitlement to pastures and must use lands of state enterprises on an informal (and often sedentary) basis; meanwhile many wells have fallen into disrepair effectively taking large areas of pasture out of production (Robinson et al. 2012). In Turkmenistan, many animals continue to be located at desert wells rather than grouped around settlements and mobility is relatively high, but many pastures can only be used if sweet water is trucked in and the cost of this has made it difficult to access some areas (Behnke et al. 2005, 2008; Jumardurdyev 2010). These factors, combined with rising livestock numbers, may explain why a set of five case recent case studies found that many livestock owners are experiencing an acute shortage of usable pastures (Robinson et al. 2015). Yet evidence for recent change in pasture condition is thin; statistics in the country NAPs are based on Soviet-era data. Kapustina (2001) reports a rough halving in vegetation productivity in the Central Kyzylkum between the period 1960–1979 and some unspecified later date (probably in the late 1990s) but it is not clear how these changes relate to changing grazing patterns and no methodological detail is given. In Turkmenistan, a comparison of stocking rates and edible pasture biomass on one former state collective farm estimated that vegetation offtake rates by livestock were well above sustainable levels (Gintzburger et al. 2005). Yet, negative changes in pasture condition over time at this site (and at one other site also) were not detected when trends in Rainfall Use Efficiency between 1989 and 2003 were inspected (Coughenour 2008). A study using spectral unmixing of time series Landsat imagery found increases in vegetation cover in desert pastures of the northern Karakum desert (Kaplan et al. 2006). Another study in the Karakum detected increases in biogenic crusts[14] which are indicators of undergrazing (Orlovsky et al. 2004); and may reflect abandonment of some areas. There is some speculation that these may now be widespread in the Karakum and Kyzylkum deserts, affecting NDVI signals (Lioubimtseva 2007).

In both mountainous republics—Tajikistan and Kyrgyzstan—livestock numbers fell steeply. The concentration of livestock around settlements and in winter areas combined with a reduced use of remoter pastures has been remarked both in

[14]*Karakharsangs*, consisting of the moss *Tortula desertorum*, lichens and cyanobacteria.

Tajikistan (Ludi 2003; Haslinger et al. 2007) and Kyrgyzstan (Farrington 2005; Steimann 2011). Whilst vertical transhumance movements over short distances quickly resumed, longer migrations between provinces are taking longer to re-emerge (Robinson et al. 2010). Reasons for this include the lack of economies of scale required to support movement, as well as pasture access issues and broken infrastructure such as bridges, roads and water supply (Farrington 2005; Robinson 2007). The winter feed deficit is particularly extreme in Tajikistan (Sedik 2010) but is also a problem in Kyrgyzstan and has led to extended grazing periods on autumn pastures. Lack of winter fodder may limit livestock numbers as it did in pre-Soviet times—many households now slaughter animals in the autumn as they cannot feed them through the winter (Ur-Rahim et al. 2014). Yet statistics suggest that overall inventories are still growing in both countries. These observations suggest that vegetation on village pastures and other wintering areas may be heavily affected by grazing, whilst other pastures are improving. We now look at evidence for these patterns.

In Tajikistan, the majority of research has been conducted in the Pamir mountain region, and particularly on the high plateau of the Eastern Pamir (Murghab district). Vanselow et al. (2012a) mapped livestock distributions and forage resources and concluded that, although the pasture resources cover the needs of all livestock at 2012 numbers, the distribution of camps combined with limited daily range of livestock means that some areas are overused and others underused. Due to severely reduced availability of supplementary winter feed, marginal or risky winter pastures with inconsistent weather conditions are no longer used, putting pressure on remaining winter areas; easily accessible pastures are used for longer periods. The productivity of these pastures has been measured over time, but the literature does not include sufficient methodology for evaluation.[15] In the Western Pamir, Breckle and Wucherer (2005) consider that any grazing at all must be categorized as overgrazing and describe negative developments in pasture condition,[16] but it is not clear whether these assessments were made during the Soviet period or subsequently. As expected, the

[15]Aknazarov (2003) presents data for the Eastern Pamir suggesting a steep decline in vegetation productivity between 1965 and 2000, but it is not clear whether the bulk of this change occurred during the Soviet period or afterwards. It is not clear how precipitation is accounted for in the study, which appears to depend on a comparison of two single years. Literature on the Pamir suggests that biomass may fluctuate from year to year by several orders of magnitude. In the Eastern Pamir the coefficient of variation in precipitation was 53.4 % for the years from 1939 until 1953 (Domeisen 2002). In the nearby Western Pamir it was found that estimated that peak annual biomass of *teresken* pasture varied between 240 and 590 kg/ha over a period of just 6 years in a single study area (Agakhanyants and Yusufbekov 1975). This illustrates the significant problem of interpreting studies by Central Asian institutes which are based on changes in vegetation productivity alone.

[16]The authors state that the percentage of ruderal plant species in mountain meadows can reach 50–70 % of biomass whilst cushion thorns (*Acantholimon* spp.) have replaced primary *Artemisia-dominated* plant communities. The proportion of edible biomass reported by Soviet authors was extremely low (around 20 %) and may indicate the result of heavy grazing at that time (Agakhanyants and Yusufbekov 1975). Pasture regeneration times of 20–30 years given by Breckle and Wucherer (2005) combined with recent increases in livestock numbers suggest that these areas may be under heavy pressure.

difference between estimated values of forage supply and livestock demand is much greater on wintering areas than on those used in other seasons (Robinson et al. 2011).

The author of this paper visited pastures at four localities in Kyrgyzstan[17] with scientists from the Kyrgyz Institute of Pasture and Fodder (Robinson et al. 2001). These workers found that the pastures visited were in better condition than any pastures of those types monitored by the Institute during the 1980s, although those close to settlements still showed evidence of heavy grazing. This study was based on expert opinion and the pastures were judged mainly in terms of their ability to support livestock populations. In contrast, Borchardt et al. (2011) employed a botanical and quantitative approach to a study of pastures in another region of Kyrgyzstan. Those communities less exposed to grazing had a higher proportion of endemic and regionally specific species whilst higher grazing pressure was associated with large proportions of ruderal or weedy vegetation. Pastures in that study area were thus botanically affected by grazing, but it is not clear to what extent forage availability for livestock has been affected.

Some research in Kyrgyzstan has focussed on perceptions of degradation—one study undertaken for the Asian Development Bank (ADB) Land Reform Project Bedunah (2006) found that, of 500 herders questioned, the majority believed that there had been no recent deterioration of pasture. Leichti (2012) also found that pasture degradation is not a major concern of its primary users, many of whom believe that overgrazing was serious during the Soviet period but is of little importance today. These impressions seem to be at odds with national monitoring data on pasture condition described in Box 3.

Box 3. Unreliable data, policy making and international assistance—an example from Kyrgyzstan

A number of Kyrgyz institutions, including the State Land Management Agency (Giprozem) and the Institute of Irrigation, generate datasets on pasture condition. The data suffer from inconsistences both between and within organizations (Table 17.2). All data in the table were taken from reports commissioned by international organizations and published in 2007. The figures are bewilderingly variable and also differ from those given in the Kyrgyz NAP shown in Table 17.1 (which reflect figures from the Institute of Irrigation—see Kulov and Zooshev (2007)). Giprozem has also produced time series data on pasture productivity which are widely published in donor reports (see Fig. 17.3)[18]. These figures represent a 26% drop in productivity since 1948 (37% on spring-autumn pastures, 18% on summer pastures and 50% on winter pastures). There is no evidence that they do not correctly

[17]In four districts and two provinces. Each locality corresponded to one *aiyl okmotu*, the smallest administrative unit in Kyrgyzstan. These are equivalent to one Soviet-period sovkhoz and all the pastures used by this entity.

[18]Equivalent figures presented by the Asian Development Bank (2007a) show the same trends, but the absolute biomass numbers are exactly half those provided to this author by Giprozem.

portray the situation of pasture condition in Kyrgyzstan, yet the 50% decrease in livestock numbers during the 1990s might have been expected to lead to recovery, at least on summer pastures. Moreover, it appears that the most recent data point in Fig. 17.3 is in fact based on data for the year 2000 alone (Giprozem, Pers comm.).

For all these datasets, documentation of the number and location of plots, sampling methodology and correction for precipitation and offtake[19] are available only in internal manuals. Data are never presented with levels of methodological detail required for an international scientific publication. Thus, in a circular process, policy is based on evidence that does not meet minimum scientific standards, reinforcing a perception of degradation that is never actually examined rigorously.

Table 17.2 Extent of degradation in 1985 and 2002 as percentage of pasture area

Report	Asian Development Bank (2007a)[a]	World Bank (2007a)	Asian Development Bank (2007b)		USAID (2007)	
Source of data	Giprozem	Giprozem	Institute of Irrigation		Gosregister	
Degradation category	All categories (% of pasture area)	All categories (% of pasture area)	All categories (% of pasture area)	Seriously degraded (% of pasture area)	All categories (% of pasture area)	Seriously degraded (% of pasture area)
Summer pastures	29	35	36	11	–	–
Spring-Autumn pastures	26	47	50	15	–	–
Winter pastures	16	83	70	21	–	–
Total	25	–	49	15	79	30

[a]These figures are also cited in United Nations Development Programme (2007) and World Bank (2007b)

To summarize, a number of post-Soviet studies describe the spatial influence of grazing on vegetation, in particular around wells (Kanchaev et al. 2003; Alimaev 2003; Alimaev et al. 2008; Rajabov 2009) but there are few that convincingly demonstrate change over time. Temporal change is usually investigated by measuring trends in

[19]Biomass in Kyrgyz mountain pastures varies by 2–3 times from year to year depending on rainfall (Korneva 1959), thus fluctuations specific to the year of the survey must be removed. In addition, whilst the effect of grazing in previous seasons on long term productivity will be detected through current edible biomass levels, it is important that biomass has not been removed by animals in the year of the actual cut.

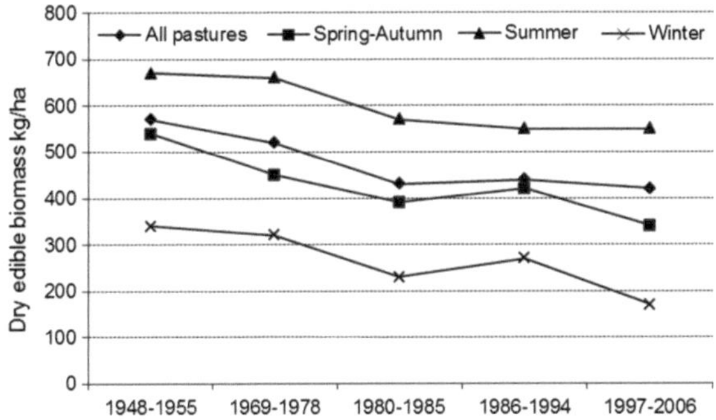

Fig. 17.3 Time series data showing pasture productivity since 1948. *Source* Giprozem.

vegetation productivity, but information on sampling, control of factors other than the degradation processes under investigation, and protocols for measurement of productivity itself are rarely reported with the findings (see also Vanselow 2011). These observations must not be taken to imply that the conclusions of vegetation studies undertaken by Central Asian institutes are incorrect—there is a wealth of data, maps and know-how in the region today which must urgently be preserved and used. One reason for the lack of methodological detail was that some of the documents reviewed here constitute summaries or secondary materials and not the original research papers, or that methodologies are taken from manuals not published along with the findings.

17.3.2 Changes in Cropping Practises

The regeneration of steppe pastures in the southern USSR has been widely documented (Hölzel et al. 2002). In Kazakhstan, by 1998 the sown area of all crops had decreased by 38 % (Suleimenov and Oram 2000). By 2000 12.8 million ha had fallen out of production (Schillhorn-van-Veen et al. 2004). Although this abandonment initially resulted in poorly productive pastures dominated by weeds (Rachkovskaya et al. 1999), if undisturbed, steppe associations similar to those of the original 'virgin' steppe may reappear within 20 years (Rachkovskaya and Bragina 2012). In the meantime, the most productive land will probably be brought back into production: in the provinces of Akmola, Kostanai, and North-Kazakhstan, in which 80 % of Kazakhstan's wheat is produced, agricultural land use increased by about a third between 2003 and 2010 bringing five million ha of cropland back under the plough (Petrick et al. 2012). This new production is conducted largely by agribusiness firms using modern technology such as zero tillage equipment, so it remains to be seen what the environmental impact of these new styles of farming will be.

The opposite trends can be seen in mountainous areas of Central Asia, where many households started to plough pasture on steep rainfed lands, some having mean annual rainfall as low as 200–300 mm. Such cultivation results in very high levels of erosion. The authors of the Tajikistan NAP suggest that slopes greater than 10–12° should be reserved for perennial cultivation such as vineyards, yet many such areas were converted to annual crops such as wheat. Under such conditions, soil losses may reach 175 tonnes/ha (Ministry of Nature Conservation of the Republic of Tajikistan 2001).[20] In Tajikistan, fertilizer is almost never applied to annual crops on rainfed land (Robinson et al. 2008); and the land can only be used in this way for two to three years before it must be abandoned. Even then soil quality remains low, as illustrated in a remote sensing study by Wolfgramm et al. (2007) which measured and mapped levels of soil organic carbon at two sites in Tajikistan. Conversion of pasture to rainfed agriculture should be readily visible using remote sensing, but no studies have yet quantified the areas which were ploughed. There is anecdotal evidence that such land may already have been progressively abandoned as alternative income sources, such as remittances from Russia, increased in importance in the 2000s.

17.3.3 Fuelwood Extraction and Changes in Forest Cover

In the mountainous republics, electricity provision was severely disrupted at the end of the Soviet Union; in much of rural Tajikistan there is little electricity supply in the winter during which temperatures may fall below −20°C (Droux and Hoeck 2004; Hoeck et al. 2007). The resulting fuelwood crisis has been extensively studied and various attempts to quantify its impacts have been made in the Gorno-Badakhshan region of Tajikistan (see Box 4).

Box 4. The *teresken* syndrome

In rural areas of Tajikistan, dried dung and small shrubs known as *teresken* (*Ceratoides papposa)*, mixed with other woody species have become important sources of fuel. In the Pamirs, Breu (2006) estimated that shrubs covered 80 % of energy needs; the unsustainable removal of these shrubs was couched in terms of an ecological catastrophe (e.g. Breckle and Wucherer 2005) and became a focal point for aid and development interventions. Vanselow et al. (2012b) mapped the actual extent of *teresken* pastures using classification of satellite images. They found that seriously degraded areas do not exceed 6 % of the analysed area of more than 8000 km^2 and are limited to land close to main roads and villages. They concluded that the severity of the problem had been somewhat exaggerated (Kraudzun et al. 2014).

[20]To put this figure in context, average rates for soil loss have been estimated at 17 tons/ha/year in the US and Europe, and 30–40 tons/ha/year in Asia, Africa, and South America (Pimental et al. 1995).

The Tajik NAP of 2001 includes some descriptions of forest destruction outside the Pamirs but there are no studies which have actually measured forest losses at a national level. Instead, reports present estimates based on secondary sources. One of these suggests losses of 2 % per year since 1991, a decrease from 3 to 1.75 % total percentage forest cover in the country (Kirchhoff and Fabian 2010). Lower depletion rates have been estimated for Kyrgyzstan (Asian Development Bank 2007a; USAID 2011) but again do not appear to be based on hard data. Elsewhere, estimates of forest loss could not be found, but felling of *Haloxylon* shrubs in Kazakhstan certainly became a problem and was banned in 2002 (Schillhorn-van-Veen et al. 2004). In Turkmenistan gas provision in rural areas is likely to reduce shrub cutting and there is already some evidence of recovery (Kaplan et al. 2006). The FAO global Forest Resources Assessment of 2010 estimated that forest areas across the region have *increased* since 1990, by about 7 % (FAO 2011). The figures are based on government statistics and methodologies are not given. One explanation for the mismatch between these figures and the general perception that forests have deteriorated over the last 20 years, could be that the FAO statistics concern areas designated as forest, and do not actually reflect actual plant cover or wood stocks in those areas; for example in Kyrgyzstan some proportion of the increase is recorded as due to changes in forest land definitions (Burhanov 2005), but FAO figures also show increases in *wood stock* per ha since 1990 for all republics except Tajikistan.[21] Thus, as for the Soviet period, we are again left with a lack of hard evidence to fully evaluate the direction and extent of forest degradation since 1990.

17.3.4 The Aral Sea Since 1991

Plans by the USSR to return water to the sea and stabilize its level at 41 m by 2010 were interrupted by political collapse. After this time, efforts to save the Aral were dependant on cooperation between the new independent states which now share the Aral Sea basin, and to which the irriated agriculture which led to its demise makes and important economic contribution. Today, the small Aral Sea has recovered to some extent due to a dam completed in 2005 which prevents water losses to the larger sea to the south (Zonn et al. 2009). Its depth had risen to 42 m by 2008 and some fishing has again become possible. The large Aral Sea has been more or less abandoned to its fate (see Plate 17.1i); indeed Turkmenistan has even increased its irrigated area and withdrawals from the Amu Darya. This part of the Aral Sea today has a depth of just 29 m and the sea as a whole is just 10 % of its original size. A new desert, the Aral Kum, now covers 2 million hectares in its place (Zonn et al. 2009).

[21]Both excel spreadsheets with country statistics, and national reports are available for the latest (2010) Forest Resources Assessment on http://www.fao.org/forestry/fra/fra2010/en/. The 2005 data were used here because they disaggregate areas into true forests and other wooded areas, which is useful for Central Asia. The 2010 statistical sheets did not include this distinction.

17.3.5 Trends in Salinization Since the End of the Soviet Period

The major driver of recent change outside the Aral Sea area is the deterioration of irrigation canals and drainage systems. In Uzbekistan it was reported that by 2003 50 % of vertical drainage systems were no longer operating (FAO/AGLL 2003) and that repair work was not able to keep pace with requirements (Tashmatov and Tashmatova 2006). Government ministries continued to publish salinization statistics after independence. However, given reduced monitoring capacity and equipment it is not certain how reliable these figures are (FAO/AGLL 2003).[22] Some recent figures are given in Table 17.3, and are all ultimately from government sources. Some are similar to those given in the NAPs or are even taken from them in the case of Turkmenistan.

Given the amount of land which was abandoned after 1991[23] (or brought into production in the case of Turkmenistan), the statistical issues of separating irrigable and irrigated land, and lack of access to primary research data, it is difficult to say whether the figures in the table indicate more or less salinization (in terms of absolute area or proportion of irrigated land) than that recorded for at the end of the Soviet period. There have been reports that deterioration of drainage systems led to an increase in salinization and waterlogging during the 1990s (Bucknell et al. 2003). Although certain figures are contradictory,[24] globally there is little to suggest that salinization has become *less* of a problem since 1991.

17.3.6 Change Detection at the Regional Level

Regional studies tend to be based on remote sensing and use two types of approach: one of these is to classify the data into land cover types and to then examine changes in class between periods. The other approach is to use NDVI as a proxy for Net Primary Productivity (NPP) and to study long term trends, on the assumption

[22]In 2003 a map of soil salinity of Uzbekistan was published using the FAO-LADA methodology (FAO/AGLL 2003). Data from this map suggests the area of salt affected soils in Uzbekistan is 47 % of the total area; about 20 % is characterized with high degree of soil salinization. The worst affected areas are located mainly in the Syrdarya, Djizak and Central Fergana regions of Syr Darya River basin; in the Karakalpakstan and Khorezm regions close to the Aral Sea, over 90 % of land is affected. These figures are not directly comparable with published MAWR data, which are for irrigated land only.

[23]Bucknell et al. (2003) estimate that 600,000 ha of irrigate cropland were abandoned across the region, due to a mixture of water scarcity, lack of inputs and environmental problems such as salinity.

[24]In Kyrgyzstan, figures from the Department of Irrigation given in Kulov and Zhooshov (2007) suggest that the area of salinized land doubled between 1985 and 2002 whilst waterlogged land area increased by 400 %. On the other hand, the 2000 estimate of 12 % for irrigated lands affected by salinity is *lower* than equivalent state figures for 1985 (Braden 1995).

Table 17.3 Salinization of soils in Central Asia

Republic	Area irrigated land (th. ha)	Area affected (th. ha)	% affected	Severity[c]			Source	Year of statistics
				Slight (%)	Moderate (%)	Heavy (%)		
Kyrgyzstan	1077	124	12	–	–	–	Bucknell et al. (2003)	2000
Tajikistan[a]	742	124	17	10	5	2	Toderich et al. (2008)	2004
Kazakhstan[b]	2620	940	36	19	12	5	Toderich et al. (2008)	2005
Uzbekistan	4275	2112	49	31	18	5	FAO/AGLL (2003)	1999
Turkmenistan	1744	1673	96	–	–	–	Bucknell et al. (2003)	2000

[a]Kulyab, Kiurgan Tiube, Sughd and Gissar provinces
[b]Almaty, Dzhambul, Kyzylorda and S. Kazakhstan provinces
[c]The measurement scale here is (in EC, dS/m): non saline < 2–4; slightly 2–4 to 8; moderate 8–16; severe > 16

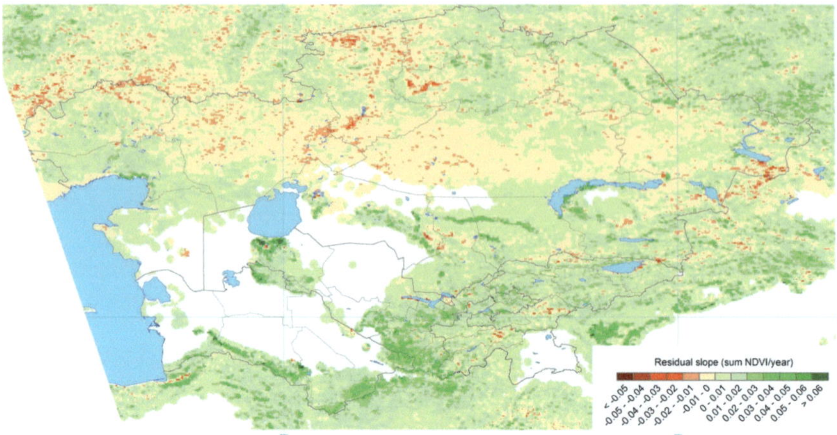

Plate 17.3 GLADA residual trend in annual sum NDVI between 1981 and 2003 (Bai et al. 2008). *Note* Those areas with extremely low rainfall (the Karakum and Kyzykum deserts and Pamir Mountains) are excluded from the dataset, either due to lack of sensitivity of NDVI to change in those areas or because changes in NPP were attributed to precipitation alone.

that change in NPP is a general indicator of a loss of ecosystem function and may result from overgrazing, forest removal or salinization.[25]

The second type of study included the FAO Global Land Degradation Assessment (GLADA) which used the 1981–2003 GIMMS 8 km resolution NDVI dataset to investigate change in NPP since 1981 using indices based on RUE (Bai et al. 2008).[26] Globally, the GLADA study showed very little NPP change in those drylands typically associated with land degradation. The authors suggest this is because "*most areas of historical land degradation have become stable landscapes—with a stubbornly low level of productivity*". But for the Central Asian context GLADA is interesting, because it spans the period of rapid change which we hope to examine.

---

[25]Declining trends in NPP, even allowing for climatic variability, may be unrelated to land degradation (for example a change in crop type, expansion or reduction in irrigated areas and urban growth) but in Central Asia enough is known about changes in land use to allow informed speculation about the likely processes behind trends in NPP.

[26]Firstly those pixels in which annual productivity is related rainfall were identified (in tropical areas rainfall is so high that there is no relationship). Where productivity declined but RUE increased, the decline was attributed to drought and those areas are masked. NDVI trends are presented for the remaining areas as RUE-adjusted NDVI. To correct for relationships between RUE and rainfall itself between sites, statistical associations between observed sum NDVI and rainfall were found. The residuals of sum NDVI (i.e. differences between the observed and predicted sum NDVI) for each pixel were calculated, and the trend of these residuals (RESTREND) was analysed by linear regression. RESTREND points in the same direction as RUE: a negative RESTREND may indicate land degradation. Data are available at http://www.fao.org/geonetwork/srv/en/main.search?any=glada. Because the results from positive and negative RUE-adjusted NPP and RESTREND were very similar, only the latter is presented here.

The map representing residual trend in annual sum NDVI is shown in Plate 17.3. Positive changes are concentrated around rivers, deltas and canals and in mountain ranges such as the Tien Shan and Altai. Increases in primary productivity of irrigated agricultural areas have also been described in Uzbekistan and Turkmenistan by Kariyeva and Leeuwen (2012) and are attributed by these authors to increases in wheat production (negative trends were observed in uncultivated delta and riverine areas). Another explanation given for increases in NDVI at irrigated sites (in Kazakhstan) is altered land management practices linked to privatization (de Beurs and Henebry 2004). These studies tend to find human-induced reasons for change in cultivated areas and climatic ones for other areas.

Other observed changes in NPP visible in Plate 17.2 include negative trends in mixed cropping and grassland areas of northern and central Kazakhstan. A general 'browning' trend in this area has also been observed by a number of authors working with NDVI data and has accelerated over the last decade (de Beurs et al. 2009; de Jong et al. 2012; Mohammat et al. 2013). These authors generally attributed these changes to climatic factors including precipitation decreases (particularly in summer) and a recent cooling trend in spring.

For many rangeland areas outside northern Kazakhstan, trends have not been consistent; several studies found increases in NDVI up to the 1990s followed by declines (Lioubimtseva 2007; Mohammat et al. 2013). It has been speculated that NDVI increases some desert areas during the 1990s may be linked to declining stock numbers combined with appearance of biogenic crusts (Lioubimtseva 2007), but causality could not be demonstrated. Another study attributed increased NDVI in Central Kazakhstan between 1989 and 2003 to changes in precipitation (Coughenour 2008). Over the last decade NDVI has been decreasing over large areas of Central Asia, particularly in western areas of Kazakhstan, Uzbekistan and Turkmenistan, probably due to climatic factors (Mohammat et al. 2013).

Overall, the results of regional remote sensing studies using low resolution imagery remain ambiguous; studies do not always identify the same trends, and explanations remain speculative. There is a discrepancy between the scale at which degradation processes are acting, and the resolution of images used to detect them. Most of the NDVI studies cited above use imagery having a maximum resolution of 250m. But only medium resolution imagery (with a pixel size of 15-30m) can be used to identify and measure specific processes at the local level such as vegetation change around wells and appearance or disappearance of ploughed areas on steep slopes (Karnieli et al. 2008; Wolfgramm et al. 2007). Scaling up of such studies to large areas by mosaicking multiple images is rarely attempted in arid ecosystems (Strand et al. 2007).[27]

[27]Ji (2008) has recommended regional monitoring using MODIS vegetation indices, combined with Landsat-based monitoring of sample sites.

17.3.7 Policy, Funding and Science

In contrast with the latter Soviet period, there are now clear incentives for both countries and individuals to keep land degradation on the agenda. Outside Kazakhstan, government funding now covers only basic running costs of research institutes and minimal salaries. Scientists must look for external funding, both to supplement salaries and produce new work. Around $750 million of official donor assistance were disbursed for sectors related to agricultural land use, water resources and the environment across the region from 2003 to 2012, mostly to Kyrgyzstan, Tajikistan and Uzbekistan (see Fig. 17.4). Funds earmarked for combatting land degradation over this period constituted a significant proportion of international funding. For example, $44 million of Global Environment Facility (GEF) funds (and associated co-financing) were listed for projects targeting land degradation in Kyrgyzstan alone (including those under implementation, completed or approved).[28] Data in Fig. 17.4 suggest that funds targeting environmental research were few, although additional resources for monitoring and institutional capacity building were made available to scientists under other

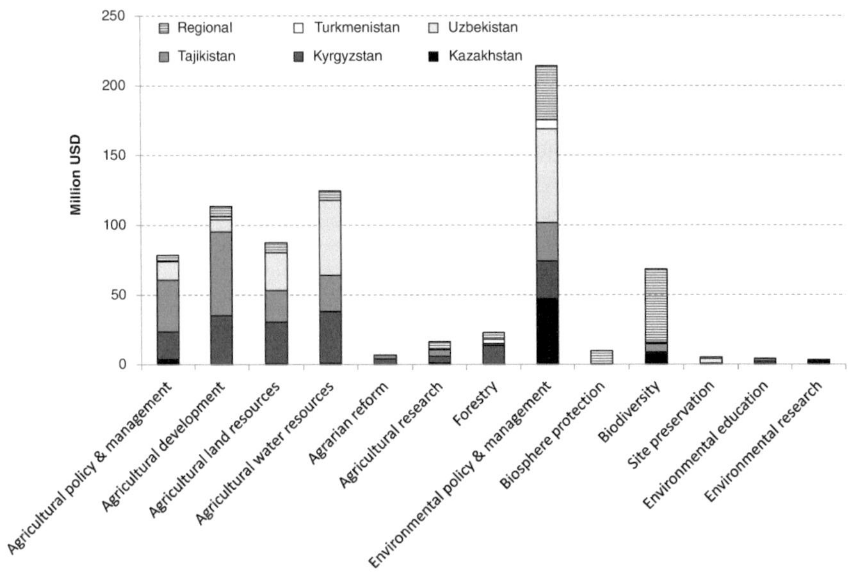

Fig. 17.4 Official development assistance for sectors related to environment and agriculture 2003–2012 Central Asia (*source* OECD Creditor Reporting System).

[28]GEF website: http://www.thegef.org/gef/gef_projects_funding.

categories. Overall, international funding has rarely resulted in improved quality or availability of those primary data which drive the allocation of funds.[29]

The ambiguity and complexity of the evidence for different types of land degradation discussed here are not reflected in the grey literature associated with the dispersal of funds to combat degradation. Soil mining and overgrazing were included amongst the top four problems in a major review of land degradation in Central Asia by ADB (Saigal 2003b). This report stated that "[*pasture*] *lands are now in various stages of land degradation and unable to support livestock production to their full potential*". Yet the evidence reviewed here presents a far more nuanced picture. The GEF website[30] implies that the drop in agricultural yields of 20–30 % across the region since independence is related to land degradation, yet the sudden disappearance of inputs and machinery at the end of the Soviet period must also be partly to blame.

Despite the lack of good evidence for certain types of degradation, attempts to control or ameliorate it may have brought about positive change. Suggested policies to reduce soil mining have included support for arable land reform and input markets, which are important for a broad range of reasons (Saigal 2003b). Few would dispute that irrigation and draining systems are in need of improvement. Ideas such as collaborative forest management and community pasture management (implemented in various forms in a number of republics), have their roots in global experience and arose from a mixture of concerns about environmental degradation, poverty and access to natural resources. On the other hand there are dangers in overstating the issue. As pointed out in Kraudzun et al. (2014), exaggeration of shrub depletion could results in bans, and the loss of a vital resource for people with few alternatives. Central Asian forests are a key resource for livestock in critical seasons and their overuse should be convincingly shown before action is taken to reduce grazing. The perception of widespread environmental damage combined with a penchant for top-down control has resulted in land legislation with clauses which allow land to be seized by governments due to 'environmental degradation'.

17.4 Conclusion

In the last years of the Soviet Union there were serious and systematic efforts to understand, document and map land degradation processes. However, many publications emerging from this work did not include enough methodological detail for others subsequently to fully interpret the results. Upon independence, information coming out of Central Asian institutions was often presented as part of applications for international development assistance. Although sweeping economic change had altered both the

---

[29] In Kyrgyzstan, funding was provided for an SLM information system including collation and digitization of existing data and maps in Russian. However, no published material appears to have emerged from this exercise and, as noted in the project completion document, the system has not made its way into a searchable internet database on the internet (Asian Development Bank 2007b). This example is typical of a number of similar projects in the region.

[30] http://www.thegef.org/gef/projects/CACILM.

drivers and nature of land degradation, many key figures continued to be based on Soviet-era data, presented with even less context or methodology than before. Recent international collaborations have resulted in a number of processes being well-documented locally, but trends on the national and regional scale have been less convincingly demonstrated. Some types of degradation, such as salinization and deforestation, may have worsened in many areas, but actual field data to quantify this are either lacking or presented in reports with no accompanying information on collection methods. Trends in pasture condition appear mixed, with both positive and negative change recorded locally. The assumption that degradation of all types is getting worse across the region is thus not fully supported by the material reviewed here.

17.4.1 The Future

Lastly, what of prospects? These appear to be rather poor for irrigated agriculture. Efforts to rehabilitate the Aral Sea are now confined to measures designed to protect what is left of the Amu Darya delta (Micklin 2007). Globally, it has been found that salinization can only be brought under control by establishment of institutions and pricing systems which encourage efficient use of water (see references in Wichelns 1999); in Central Asian republics the task of creating local water management bodies and financing mechanisms for sustainable irrigation and drainage has only just begun (Tashmatov and Tashmatova 2006). In the early 2000s, the World Bank suggested that World Bank suggested that, whilst irrigation improvements appeared profitable in some areas, over a million people in Uzbekistan and Tajikistan lived in areas where large scale restoration of any kind of irrigation infrastructure, let alone infrastructure which supplied environmentally sustainable services,was unlikely to be economically viable (Bucknell et al. 2003).

Climate change models are also not particularly encouraging. All predict increases in temperature, but there is far less agreement on expected changes in precipitation, with very small increase or decreases predicted by different models. However, decreases in rainfall appear most likely, particularly in spring and summer and in western parts of Central Asia, with possible winter increases, particularly in northern Kazakhstan (Lioubimtseva and Henebry 2009). Although, as we have seen, northern Kazakhstan is currently experiencing NDVI decreases, its cereal producing areas are eventually expected to benefit from warmer springs and increased winter precipitation; Uzbekistan and Turkmenistan, which may become both hotter and drier, are likely to be negatively affected (IPCC 2014).

Irrigated agriculture is largely dependant on water supply from mountain ranges; short term increases in river flow are predicted in some sources as glacier melt accelerates, but over the longer term deficits will emerge as ice cover disappears (Robinson and Engel 2008).[31] Population growth alone is continuously eroding the quantities of water and irrigated land per person (Tashmatov and Tashmatova 2006).

---

[31]Current trends in flow are hard to assess due to high inter-annual variation, but it appears that negative trends in precipitation in the Amu Darya basin since 1978 have been linked to decreased discharge in that river, whilst the Syr Darya has remained stable (Nezlin et al. 2004).

The outlook for rangelands may be somewhat brighter; financial resources are badly needed to rehabilitate desert wells, but less costly changes in management and land tenure arrangements may also make a difference (see review by Robinson et al. 2012). Growth in livestock numbers may reproduce the economics of scale required for re-colonization of remote pastures, taking pressure off heavily stocked areas. However, availability of quality winter pastures will limit growth—if the livestock sector is to flourish, improved winter fodder supplies will be needed, much of which must be produced on irrigated land. The National Communications of Central Asian countries to the UNFCCC all predict negative impacts of higher temperatures on pastures, with the desert belt moving to the north in Kazakhstan, but these changes have yet to be observed.

Acknowledgements I would like to acknowledge Elmar Mamedov, who helped to obtain some of the literature reviewed here, and Carol Kerven for proof reading. The Leverhulme Trust funded my Research Associateship during the period over which this paper was written.

References

Agakhanyants, O. E., & Yusufbekov, K. I. (1975). *Vegetation of the Western Pamirs and experiences in its reconstruction [Rastitel'nost' Zapadnovo Pamira I Opyt ee Rekonstruktsii].* Donish: Dushanbe.

Aknazarov, K. A. (2003). The present situation of pasture land in Eastern Pamir. *Bielefelder Ökologische Beiträge, 18*, 30–32.

Alekseevna, G. T. (2004). *Influence of drought on pasture productivity and the effectiveness of phytomelioration in the plains of Turkmenistan [Vliyanie Zasukhi na Produktivnost' Pasbishch i Effectivnost' ikh Fitomelioratsii v Pavninnom Turkmenistane].* Thesis submitted for title of Candidate of Science, All Russian Scientific Research Institute for Forest Improvement.

Alimaev, I. I. (2003). Transhumant ecosystems: fluctuations in seasonal pasture productivity. In C. Kerven (Ed.), *Prospects for pastoralism in Kazakhstan and Turkmenistan: From state farms to private flocks*. Oxford: Routledge Curzon.

Alimaev, I. I., & Behnke, R. (2008). Ideology, land tenure and livestock mobility in Kazakhstan. In K. Galvin, R. Reid, R. Behnke, & N. Hobbs (Eds.), *Fragmentation in semi-arid and arid landscapes*. Heidelberg, Germany: Springer.

Alimaev, I. I., Kerven, C., Torekhanov, A., Behnke, R., Smailov, K., Yurchenko, V., et al. (2008). The impact of livestock grazing on soils and vegetation around settlements in Southeast Kazakhstan. In R. Behnke (Ed.), *The socio-economic causes and consequences of desertification in Central Asia*. Berlin: Springer.

Asanov, K. A., & Alimaev, I. I. (1990). New forms of organisation and management of arid pastures of Kazakstan. *Problems of Desert Development, 5*, 42–49.

Ash, A. J., & Melvor, J. G. (2005). Constaints to pastoral systems in marginal environments. In J. A. Milne (Ed.), *Pastoral systems in marginal environments*. Wageningen.

Asian Development Bank. (2007a). The Kyrgyz republic natural resource sector study.

Asian Development Bank. (2007b). TA 6357: Central Asian countries initiative for land management mulitcountry partnership framework support project. *Technical Completion Report*.

Babaev, A. G. (1985). *Map of anthropogenic desertification of arid zones of the USSR*. Ashgabat: Institute of Deserts.

Babaev, A. G. (1996). *Problems of arid land development*. Moscow: Moscow University Press.

Babaev, A. M., & Kharin, N. G. (1999). The monitoring and forecast of desertification processes. In A. G. Babaev (Ed.), *Desert problems and desertification in Central Asia*. Berlin: Springer.

Babaeva, T. A. (1999). The mapping of desertification processes. In A. G. Babaev (Ed.), *Desert problems and desertification in Central Asia*. Berlin: Springer.

Babu, S. C., & Sengupta, D. (2006). Policy reforms and agriculture development in Central Asia: An overview of issues and challanges. In S. C. Babu & S. Djalalov (Eds.), *Policy reforms and agriculture development in Central Asia*. Berlin: Springer.

Bai, Z. G., Dent, D. L., Olsson, L., & Schaepman, M. E. (2008). *Global assessment of land degradation and improvement. 1. Identification by remote sensing*. Report 2008/01, ISRIC—World Soil Information, Wageningen. LADA technical Report no. 12.

Baitulin, I. O., & Karibaeva, K. N. (1999). Preconditions and trends of desertification in Kazakhstan [Predposylki i tendentsii opustynivaniya v Kazakhstane]. In I. O. Baitulin, N. P. Ogar & T. M. Bragina (Eds.), *Transformation of natural ecosytems and their components during desertification [Transformatsiya Prirodnykh Ekosistem i ikh Komponentov pri Opustynivanii]*. Almaty.

Bedunah, D. J. (2006). Impact of land reform on Pastureland. Consultant report to ADB under TA 2208.

Behnke, R., Davidson, G., Jabbar, A., & Coughenour, M. (2008). Human and natural factors that influence livestock distributions and rangeland desertification in Turkmenistan. In R. Behnke (Ed.), *The socio-economic causes and consequences of desertification in Central Asia*. Berlin: Springer.

Behnke, R., Jabbar, A., Budanov, A., & Davidson, G. (2005). The administration and practice of leashold pastoralism in Turkmenistan. *Nomadic Peoples, 9*, 147–169.

Bizhanova, G. K. (1999). An estimation of the extent of desertification of psammophytic vegetation of Kazakhstan [Otsenka stepeni opustynivaniya psammofitnovo rastitel'nosti Kazakhstana]. In I. O. Baitulin, N. P. Ogar & T. M. Bragina (Eds.), *Transformation of natural ecosytems and their components during desertification [Transformatsiya Prirodnykh Ekosistem i ikh Komponentov pri Opustynivanii]*. Almaty.

Borchardt, P., Schickhoff, U., Scheitweiler, S., & Kulikov, M. (2011). Mountain pastures and grasslands in the SW Tien Shan, Kyrgyzstan—floristic patterns, environmental gradients, phytogeography and grazing impact. *Journal of Mountain Science, 8*, 363–373.

Braden, K. (1995). Kyrgyzstan. In P. R. Pryde (Ed.), *Environmental resources and constraints in the former Soviet republics*. USA: Westview Press Inc.

Breckle, S. W., & Wucherer, W. (2005). Vegetation of the Pamir (Tajikistan): Land use and desertification problems. In E. M. Spehn, M. Liberman, & C. Körner (Eds.), *Land use change and mountain biodiversity*. UK: Taylor and Francis.

Breu, T. (2006). *Sustainable land management in the Tajik Pamirs: The role of knowledge for sustainable development*. Doctoral Thesis, Bern: University of Berne.

Bucknell, J., Klytchnikova, I., Lampietti, J., Lundell, M., Scatasta, M., & Thurman, M. (2003). Irrigation in Central Asia. Social, economic and environmental considerations. *Europe and Central Asia region environmentally and socially sustainable development*. Washington, DC: World Bank.

Burhanov, A. (2005). *Kyrgyzstan. Global forest resources assessment country reports*. Rome: FAO Forestry Department.

Coughenour, M. (2008). Forage distributions, range condition, and the importance of pastoral movement in Central Asia—a remote sensing study. In R. Behnke (Ed.), *The socio-economic causes and consequences of desertification in Central Asia*. Berlin: Springer.

Crighton, E. J., Barwin, L., Small, I., & Upshur, R. (2011). What have we learned? A review of the literature on children's health and environment in the Aral Sea area. *International Journal of Public Health, 56*, 125–138.

de Beurs, K., & Henebry, G. M. (2004). Land surface phenology, climatic variation, and institutional change: Analyzing agricultural land cover change in Kazakhstan. *Remote Sensing of the Environment, 89*, 497–509.

de Beurs, K. M., Wright, C. K., & Henebry, G. M. (2009). Dual scale trend analysis for evaluating climatic and anthropogenic effects on the vegetated land surface in Russia and Kazakhstan. *Environmental Research Letters, 4*, 1–11.

de Jong, R., Verbeselt, J., Schaepman, M. E., & Bruin, Sd. (2012). Trend changes in global greening and browning: Contribution of short-term trends to longer-term change. *Global Change Biology, 18*, 642–655.

Domeisen, M. (2002). *Marginalised by the impacts of transformation. A study of post-Soviet livestock breeding in the high mountains of the Eastern Pamirs*. Bern: University of Bern.

Droux, R., & Hoeck, T. (2004). *Energy for Gorno-Badakhshan: hydropower and the cultivation of firewood*. Joint Diploma Thesis, Bern: University of Bern.

Dzhanpeisov, R., Alimbaev, A. K., Minyat, V. E., & Smagulov, T. A. (1990). Degradation of soils of mountain and desert pastures in Kazakhstan. *Problems of Desert Development, 4*, 19–26.

Ellis, J., & Lee, R. Y. (2003). Collapse of the Kazakhstan livestock sector. In C. Kerven (Ed.), *Prospects for pastoralism in Kazakhstan and Turkmenistan: From state farms to private flocks*. Oxford: Routledge Curzon.

FAO. (2011). State of the World's forests.

FAO/AGLL. (2003). Land degradation assessment in drylands (LADA), Technical Report, Republic of Uzbekistan. Uzbekistan Ministry of Agriculture and Water Resources.

Farrington, J. D. (2005). De-development in Eastern Kyrgyzstan and persistance of semi-nomadic livestock herding. *Nomadic Peoples, 9*, 171–197.

Fedorovich, B. A. (1973). Natural conditions of the arid zones of the USSR and ways of development of their livestock sectors. [Prirodnye usloviya aridnykh zon SSSR i puti razvitiya v nikh zhivotnovodsva]. In: Abramov, S. M. & Orazov, A. (eds.) *Essays on the agricultural history of the peoples of Central Asia and Kazakhstan [Ocherki po Istorii Khozyaistva Narodov Srednei Asii i Kazakhstana]*. Leningrad: Nauka.

Gilmanov, T. G. (1995). The state of rangeland resources in the newly independent states of the former USSR. In *Fifth international rangelands congress* (pp. 10–13), Salt Lake City, Utah.

Gintzburger, G., Le Houérou, H. N., & Toderich, K. N. (2005). The steppes of Middle Asia: Post-1991 agricultural and rangeland adjustment. *Arid Land Research and Management, 19*, 215–239.

Goletti, F., & Chabot, P. (2000). Food policy research for improving the reform of agricultural input and output markets in Central Asia. In S. Babu & A. Tashmatov (Eds.), *Food policy reforms in Central Asia: Setting the research priorities*. Washington, DC: International Food Policy Research Institute.

Government of Turkmenistan. (2002). National environmental action plan [Natsional'nyi Plan Deisctvii Prezidenta Turkmenistana Po Okhrane Okruzhaiushei Sredy).

Haslinger, A., Breu, T., Hurni, H., & Maselli, D. (2007). Opportunities and risks in reconciling conservation and development in a post-Soviet setting: The example of the Tajik National Park. *International Journal of Biodiversity Science and Management, 3*, 157–169.

Hoeck, T., Droux, R., Breu, T., Hurni, H., & Maselli, D. (2007). Rural energy consumption and land degradation in a post-Soviet setting—an example from the west Pamir mountains in Tajikistan. *Energy for Sustainable Development, XI*, 48–57.

Hölzel, N., Haub, C., Ingelfinger, M. P., Otter, A., & Pilipenko, V. N. (2002). The return of the steppe—large-scale restoration of degraded land in southern Russia during the post-Soviet era. *Journal for Nature Conservation, 10*, 75–85.

IPCC. (2014). Climate change 2014: Impacts, adaptation, and vulnerability. Part B: Regional aspects. Contribution of Working Group II to the Fifth Assessment Report of the Intergovernmental Panel on Climate Change, Cambridge University Press, Cambridge, UK, NY, USA.

Ji, C. Y. (2008). Land degradation in Central Asia. Asian Development Bank, Mandaluyong.

Jumardurdyev, D. (2010). Socio-economic survey of the Karakum Pilot region. Unpublished report for GIZ.

Kanchaev, K., Kerven, C., & Wright, I. A. (2003). The limits of the land: Pasture and water conditions. In C. Kerven (Ed.), *Prospects for Pastoralism in Kazakhstan and Turkmenistan: From state farms to private flocks*. Oxford: Routledge Curzon.
Kapustina, L. (2001). Degradation of the vegetation in the Central Kyzylkym desert (Uzbekistan). In S. W. Breckle, M. Veste, & W. Wucherer (Eds.), *Sustainable land use in deserts*. Berlin: Springer.
Kariyeva, J., & Leeuwen, W. J. D. V. (2012). Phenological dynamics of irrigated and natural drylands in Central Asia before and after the USSR collapse. *Agriculture, Ecosystems and Environment, 162*, 77–89.
Karnieli, A., Gilad, U., Ponzet, M., Svoray, T., Mirzadinov, R., & Fedorina, O. (2008). Assessing land-cover change and degradation in the Central Asian deserts using satellite image processing and geostatistical methods. *Journal of Arid Environments, 72*, 2093–2105.
Kazakhstan Statistical Agency. (2011). Kazakhstan in 2010. Astana.
Kerven, C., Alimaev, I. I., Behnke, R., Davidson, G., Franchois, L., Malmakov, N., et al. (2004). Retraction and explansion of flock mobility in Central Asia: Costs and consequences. *African Journal of Range and Forage Science, 21*, 91–102.
Kerven, C., Alimaev, I. I., Behnke, R., Davidson, G., Malmakov, N., & Smailov, A., et al. (2006). Fragmenting pastoral mobility: Changing grazing patterns in post-Soviet Kazakhstan. *USDA Forest Service Proceedings RMRS-P-39*, 99–110.
Kharin, N. (2002). *Vegetation degradation in Central Asia under the impact of human activities*. New York: Springer.
Kharin, N. G., Kalenov, G. S., & Kiril'tseva, A. A. (1993). *Explanatory note to the map of human-induced land degradation in the Aral Sea basin, scale 1:2,500,000*. Ashgabat: Desert Research Institute.
Kharin, N. G., & Kiril'tseva, A. A. (1988). New data on areas of desertified lands in the arid zone of the USSR. *Problems of Desert Development, 4*, 3–8.
Kharin, N. G., Orlovskii, N. S., Kogai, N. A., & Makulbekova, G. B. (1986). Contemporary status of and prognosis for desertification in the USSR arid zone. *Problems of Desert Development, 5*, 58–68.
Kharin, N. G., Tateishi, R., & Harahsheh, H. (1999). *Degradation of the drylands of Asia & desertification map of the drylands of Asia, scale 1:10,000,000*. Japan: Chiba University.
Kirchhoff, J. F., & Fabian, A. (2010). Forestry sector analysis of the Republic of Tajikstan. German Federal Ministry for Economic Cooperation and Development (GTZ).
Kolodin, M. V., & Rabochev, G. I. (1999). The use of drainage water. In A. G. Babaev (Ed.), *Desert problems and desertification in Central Asia*. Berlin: Springer.
Korneva, I. G. (1959). *Geobotanical studies of the Susamyr valley*. Frunze: Academy of Sciences.
Kraudzun, T., Vanselow, K., & Samimi, C. (2014). Realities and myths of the Teresken syndrome —an evaluation of the exploitation of dwarf shrub resources in the Eastern Pamirs of Tajikistan. *Journal of Environmental Management, 132*, 49–59.
Kulov, K. M., & Zhooshov, P. M. (2007). Desertification processes in Kyrgyzstan [Protsessy opustynivania v Kyrgyzstane]. *Problems of Desert Development, 2*, 7–10.
Kurochkina, L. Y., & Shabanova, L. V. 1990. *Characteristics of pasture zones of Kazakhstan [Kompleksnaya Kharakteristika pastbishch zony Kazakhstana].*Almaty, Nauka.
Le Houérou, H. N. (1984). Rain use efficiency, a unifying concept in arid land ecology. *Journal of Arid Lands, 7*, 213–247.
Lerman, Z., & Stanchin, I. (2006). Agrarian reforms in Turkmenistan. In S. C. Babu & S. Djalalov (Eds.), *Policy reforms and agriculture development in Central Asia*. Berlin: Springer.
Liechti, K. (2012). The meanings of pasture in resource degradation negotiations: Evidence from post-socialist rural Kyrgyzstan. *Mountain Research and Development, 32*, 304–312.
Lioubimtseva, E. (2007). Possible changes in the carbon budget of arid and semi-arid Central Asia inferred from land-use/landcover analyses during 1981–2001. In R. Lal, M. Suleimenov, B. A. Stewart, D. O. Hansen, & P. Doraiswamy (Eds.), *Climate change and terrestrial carbon sequestration in Central Asia*. UK: Taylor and Francis.

Lioubimtseva, E. & Henebry, G. M. (2009). Climate and environmental change in arid Central Asia: Impacts, vulnerability and adaptations. *Journal of Arid Environments, 73*, 963–977.

Ludi, E. (2003). Managing high pastures in the Eastern Pamirs. In T. Breu & H. Hurni (Eds.), *The Tajik Pamirs*. Bern, Switzerland: Centre for Development and Environment (CDE).

Micklin, P. P. (1994). The Aral Sea problem. *Proceedings of the Institution of Civil Engineers—Civil Engineering,* Vol. 102, pp. 114–121.

Micklin, P. (1995). Turkmenistan. In P. R. Pryde (Ed.), *Environmental resources and constraints in the former Soviet republics*. USA: Westview Press Inc.

Micklin, P. (2007). The Aral Sea disaster. *Annual Review of Earth and Planetary Sciences, 35*, 47–72.

Ministry of Agriculture and Water Resources of the Republic of Kyrgyzstan. (2000). *National action plan for combating desertficiation in the Kyrgyz republic [Natsional'nyi Plan Deistvii Po Bor'be s Opustyvaniem V Kyrgyzskoi Respublike]*.

Ministry of Environmental Protection of the Republic of Kazakhstan. (2005). *The program on combating desertification in the Republic of Kazakhstan for (2005–2015)*. Astana.

Ministry of Nature Conservation of the Republic of Tajikistan. (2001). *National action plan for combatting desertification in Tajikistan [Natsional'naya Programma Deistvii Po Bor'be c Opustynivaniem v Tadhzikistane]*. Dushanbe.

Mohammat, A., Wang, X., Xu, X., Peng, L., Yang, Y., Zhang, X., et al. (2013). Drought and spring cooling induced recent decreases in vegetation growth in Inner Asia. *Agricultural and Forest Meteorology, 178*, 21–30.

Nechaeva, N. T., Antonova, K. G., Karshenas, C. D., Mukhammedov, G., & Hurberdiev, M. (1979). *Vegetation productivity in the Central Karakum and implications for user regimes [Produktivnost' rastitel'nosti Tsentral'nykh Karakumov v sviazi s razlichnym rezhimom ispol'zovaniia]*. Moscow: Nauka.

Nezlin, N., Kostianoy, A., & Lebedev, S. (2004). Internannual variations of the discharge of Amu Darya nad Syr Darya estimated from global atmospheric precipitation. *Journal of Marine Systems, 47*, 67–75.

O'Hara, S. L. (1997). Irrigation and land degradation: Implications for agriculture in Turkmenistan, Central Asia. *Journal of Arid Environments, 37*, 137–165.

Orlovsky, L., Dourikov, M., & Babaev, A. (2004). Temporal dynamics and productivity of biogenic soil crusts in the central Karakum desert, Turkmenistan. *Journal of Arid Environments, 56*, 579–601.

Petrick, M., Wandel, J., & Karsten, K. (2012). Rediscovering the virgin lands: Agricultural investment and rural livelihoods in a Eurasian frontier area. *World Development, 43*, 164–179.

Pimental, D., Harvey, C., Resosudarmo, P., Sinclair, K., Kurz, D., McNair, M., et al. (1995). Environmental and economic costs of erosion and conservation benefits. *Science, 267*, 1117–1123.

Pryde, P. R. (1991). *Environmental management in the Soviet union*. Cambridge: Cambridge University Press.

Rachkovskaya, E. I., & Bragina, T. M. (2012). Steppes of Kazakhstan: Diversity and present state. In M. J. A. Werger & M. A. Van Staalduinen (Eds.), *Eurasian steppes. Ecological problems and livelihoods in a changing world*. Berlin: Springer.

Rachkovskaya, E. I., Bragina, T. M., Bragin, E. A., & Evstifeev, I. G. (1999). Influence of ploughing on vegetation cover and wildlife of Kostanai Oblast [Vlianie raspashki zemel' na rastitel'nosti i zhivotnyi mir Kostanaiskoi oblasti].In I. O Baitulin, N. P. Ogar & T. M. Bragina (Eds.), *Transformation of natural ecosytems and their components during desertification [Transformatsiya Prirodnykh Ekosistem i ikh Komponentov pri Opustynivanii]*. Almaty.

Rafikov, V. A. (2007). On the map of desertification of arid zones of Uzbekistan [Karte Opustynivaniya Aridnoi Zony Uzbekistana]. *Problems of Desert Development*, 10–14.

Rajabov, T. (2009). Ecological assessment of spatio-temporal changes of vegetation in response to piosphere effects in semi-arid rangelands of Uzbekistan. *Land Restoration Training Programme Final Project*. Complex Research Institute of Regional Problems, Samarkand Division of the Academy of Sciences of Uzbekistan.

Robinson, N. A. (1989). Soviet environmental protection: The challange for legal studies. *Pace Environmental Law Review, 7*, 117–150.
Robinson, S. (2000). *Pastoralism and land degradation in Kazakhstan*, PhD thesis. Coventry: University of Warwick.
Robinson, S. (2007). *Report of the pasturelands ecologist for ADB agricultural land improvement project*. Asian Development Bank: Kyrgyzstan.
Robinson, S., & Engel, E. (2008). Climate change and land degradation in Central Asia: Scenarios, strategies and funding opportunities. *Report for GTZ CCD.*
Robinson, S., Higginbotham, I., Guenther, T., & Germain, A. (2008). Land reform in Tajikistan: Consequences for tenure security, agricultural productivity and land management practices. In R. Behnke (Ed.), *The socio-economic causes and consequences of desertification in Central Asia*. Amsterdam: IOS and Dordrecht; Heidelberg, Germany: Springer.
Robinson, S., Karasartov, S., Bobukeeva, M., Gorborukova, L. P., Bush, G., & Fitzherbert, A. (2001). Pasture and land tenure in Kyrgyzstan *Report for DFID project 'Sustainable Livelihoods for Livestock Producing Communities'*.
Robinson, S., & Milner-Gulland, E. J. (2003). Political change and factors limiting numbers of wild and domestic ungulates in Kazakhstan. *Human Ecology, 31*, 87–110.
Robinson, S., Safaraliev, G., & Muzofirshoev, N. (2011). Carrying capacity of pasture and fodder resources in the Tajik Pamirs. Rome: FAO.
Robinson, S., Whitton, M., Biber-Klemm, S., & Muzofirshoev, N. (2010). The impact of land reform legislation on pasture tenure in Gorno-Badakhshan: From common resource to private property? *Mountain Research and Development, 30*, 4–13.
Robinson, S., Wiedemann, C., Michel, S., Zhumabayev, Y., & Singh, N. (2012). Pastoral tenure in Central Asia: Theme and variation in the five former Soviet republics. In V. Squires (Ed.), *Rangeland stewardship in Central Asia. Balancing improved livelihoods, biodiversity conservation and land protection*. Berlin: Springer.
Robinson, S., Gardashev, A., Atahanov, G., Akiniyazov, A., Mamedov, E., Abdurasulova, G. & Durikov, M. (2015). Building the resilience of Turkmen pastoralists to environmental variability. *Presented at the Conference on Resilience, Research and Innovation*. Djibouti, 26–28 October 2015.
Rozanov, A. B. (1990). Soil degradation of Tian-Shan high mountain pasture grounds. *Transactions of the 14th International Congress of Soil Science*. Kyoto, Japan.
Kaplan, S. Orlovsky, L., Blumberg, D. G., & Mamedov, E. (2006). Assessment of pasture degradation in Turkmenistan using remote sensing. In M. Ehlers & U. Micheleds (Eds.), *SPIE Proceedings: Remote Sensing for Environmental Monitoring, GIS Applications and Geology IV*. Bellingham: SPIE.
Sadvokasov, R. (2000). The current state of the northern deserts, a case study of Betpak Dala. *Seventh Conference of Young Botanists*. Saint Petersburg.
Saigal, S. (2003a). *Turkmenistan: Issues and approaches to combat desertification. Combating desertification in Asia*. Mandaluyong: Asian Development Bank.
Saigal, S. (2003b). *An overview: A regional synthesis report on issues and approaches to combat desertification*. Mandaluyong: Asian Development Bank.
Saiko, T. A. (1995). Implications of the disintegration of the former Soviet union for desertification control. *Environmental Monitoring and Assessment, 37*, 289–302.
Schillhorn-van-Veen, T. W, Alimaev, I. I., & Utkelov, B. (2004). *Kazakhstan: Rangelands in transition; the resource, the users and sustainable use. World Bank Technical Paper: Europe and Central Asia Environmentally and Socially Sustainable Development Series*. Washington, DC: World Bank.
Sedik, D. (2010). The feed-livestock nexus in Tajikistan: Livestock development policy in transition. Policy Studies on Rural Transition No. 2010-1. FAO Regional Office for Europe and Central Asia.
Sludskii, A. A. (1963). Dzhut in Eurasian steppes and deserts. *Trudii Instituta Zoologii ANKazSSR [Proceedings of the Zoological Institute of the Kazakstan Academy of Sciences]*, Vol. 20, pp. 5–88.

Steimann, B. (2011). *Making a living in uncertainty. Agro-pastoral livelihoods and institutional transformations in post-Soviet rural Kyrgyzstan. Bishkek and Zurich.* University of Zurich.
Stogova, L. L. (1999). Vegetation recovery in high mountain pastures under new socio-economic conditions [Vosstanovlenie rastitel'nosti vysokogornykh pastbish v novykh sotsial'no-ekonomicheskikh usloviakh]. In I. O. Baitulin, N. P. Ogar, & T. M. Bragina (Eds.), *Transformation of natural ecosytems and their components during desertification [Transformatsiya Prirodnykh Ekosistem i ikh Komponentov pri Opustynivanii].* Almaty.
Strand, H., Leimgruber, P., & Mueller, T. (2007). Trends in selected biomes, habitats and ecosystems: Dry and sub-humid lands. In H. Strand, R. Höft, J. Stritthold, L. Miles, N. Horning, E. Fosnight & W. Turner (Eds.), *Sourcebook on remote sensing and biodiversity indicators.* Secretariat of the Convention on Biological Diversity, NASA-NGO Biodiversity Working Group, World Conservation Monitoring Centre of the United Nations Environment Programme.
Sukhikh, V. I. (1991). Problems of forestry in Central Asia and Kazakhstan [Problemy lesnogo khozyaistva v Tsentral'noi Azii i Kazakhstana]. *Lesnoe Khozyaistvo [Forestry], 6,* 2–7.
Suleimenov, M., Iniguez, L., & Musayeva, M. (2006). Policy reforms and livestock developments in Central Asia. In S. C. Babu & S. Djalalov (Eds.), *Policy reforms and agriculture development in Central Asia.* Berlin: Springer.
Suleimenov, M., & Oram, P. (2000). Trends in feed, livestock production and rangelands during the transition period in three Central Asian countries. *Food Policy, 25,* 681–700.
Tashmatov, A. A., & Tashmatova, D. A. (2006). Irrigation systems and sustainable natural resource use under economic reforms in Uzbekistan. In S. C. Babu & S. Djalalov (Eds.), *Policy reforms and agriculture development in Central Asia.*
Toderich, K., Tsukatani, T., Shoaib, I., Massino, I., Wilhelm, M., Yusupov, S., et al. (2008). Extent of salt affected land in Central Asia: Biosaline agriculture and utilization of the salt-affected resources. Kier Discussion Paper Series [Online]
Turkmenistan UNCCD National Working Group. (2006). Central Asian countries initiative for land management: National planning framework (Draft).
United Nations Development Programme (2007). Kyrgyzstan environment and natural resources for sustainable development. Bishkek: In partnership with the Kyrgyz State Agency for Environmental Protection and Forestry.
UNEP/GLAVGIDROMET. (1999). *National action programme to combat desertification in the Republic of Uzbekistan.* Tashkent: Cabinet of Ministers, Main Administrators of Hydrometeorology.
Ur-Rahim, I., Maselli, D., Rueff, H., & Bonfoh, B. (2014). Market access and herder's strategies to pass through lean winter periods in post-Soviet Kyrgyzstan. *APCBEE Procedia, 8,* 323–328.
USAID. (2011). Kyrgyzstan. *USAID country profile: Property rights and resource governance.* USAID.
Vanselow, K. (2011). *The high-mountain pastures of the Eastern Pamirs (Tajikistan): An evaluation of the ecological basis and the pasture potential.* Doctoral thesis, Freidrich-Alexander University.
Vanselow, K., Kraudzun, T., & Samimi, C. (2012a). Grazing practises and pasture tenure in the Eastern Pamirs. *Mountain Research and Development, 32,* 324–336.
Vanselow, K., Kraudzun, T., & Samimi, C. (2012b). Land stewardship in practise: An example from the Eastern Pamirs of Tajikistan. In V. Squires (Ed.), *Rangeland stewardship in Central Asia. Balancing improved livelihoods, biodiversity conservation and land protection.* Berlin: Springer.
Wichelns, D. (1999). An economic model of waterlogging and salinization in arid regions. *Ecological Economics, 30,* 475–491.
Wolfgramm, B., Seiler, B., Kneubühler, M., & Liniger, H. (2007). Spatial assessment of erosion and its impact on soil fertility in the Tajik Foothills. *EARSeL eProceedings 6, 1/2007.*
World Bank. (2007a). Kyrgyz republic country review volume II. Integrating environment into agriculture and forestry. Progress and prospects in Eastern Europe and Central Asia.
World Bank. (2007b). Kyrgyz republic livestock sector review: Embracing the new challenges

Yusupov, Y. B., Lerman, Z., Chertovitskiy, A. S., & Akbarov, O. M. (2010). Livestock production in Uzbekistan: Current state, issues and prospects. Tashkent, Nasaf.

Zhambakin, Z. A. (1995). *Pastures of Kazakhstan [Pastbisha Kazakhstana]*. Almaty, Kainar.

Zonn, I., Glantz, M., Kostianoy, A., & Kosarev, A. (2009). *The Aral Sea encyclopedia*. Berlin: Springer.

Zonn, I. C., Nikolaev, V. N., Orlovskii, N. C., & Svinstov, I. P. (1981). Combating desertification in the USSR. UNEP.

Zonov, A. G. (1974). Wind erosion in grazing lands north of Betpak-dala. *Problems of Desert Development, 3*, 38–42.

Chapter 18
Rangeland Degradation Control in China: A Policy Review

Wenjun Li, Yanbo Li and Gongbuzeren

Abstract In order to prevent rangeland degradation and to recover the ecological services provided by the rangeland ecosystem, China has implemented a series of ecological policies (EPs) since the beginning of the 21st century. Nevertheless, the impacts of these policies are greatly debated in terms of ecosystem, herders' livelihoods, animal husbandry and pastoral social development. This chapter will analyse the narratives employed by different stakeholders, including academic circles, decision-makers and herders, to explain the success or failure of ecological policies for rangelands. For government narratives, we analyse the original policy reports and key relevant government official speeches. We present how the government perceives ecological restoration polices and what narratives are used to portray the impacts on ecological condition and herders' livelihood. Regarding academic narratives, we conduct bibliometric analysis on research articles by academic researchers and published in Chinese academic journals. Some critical findings are summarized. Based on these findings, we will argue that the problems of these policies may be ultimately due to one root cause: failure to understand the linkages and feedbacks between interconnected social and ecological systems (SES). We conclude that the problems of EPs are not the result of improper policy implementation, as most scholars believed, but instead are due to misunderstanding the SES of pastoral areas.

Keywords Rangeland degradation · Retire livestock to restore rangeland · Social ecological system

---

W. Li (✉) · Y. Li · Gongbuzeren
College of Environmental Sciences and Engineering, Peking University, Beijing, China
e-mail: wjlee@pku.edu.cn

R.H. Behnke and M. Mortimore (eds.), *The End of Desertification?*
Springer Earth System Sciences, DOI 10.1007/978-3-642-16014-1_18

18.1 Background

Rangelands cover 41.7 % (400 million ha) of China's total territory, and support around 17 million herders and agro-herders (ECOCAHYB 2011) who depend on natural resources of the ecological systems. Before the 1990s, rangeland use in China was orientated to supporting animal husbandry. Since the 1990s, when Beijing was severely affected by dust storms, the central government has started to adopt different views about the function of rangelands and decision-makers have shifted rangeland services from animal husbandry to provision of ecological services. For instance, new narratives are applied for the rangelands of Inner Mongolia as the "green screen" for north China, whereas the Qinghai-Tibet Plateau is named as the "water tower" for all of China.

In order to prevent rangeland degradation and to recover the ecological services provided by the rangeland ecosystem, China has implemented a series of ecological policies (EPs) since the beginning of the 21st century. The major policy is called "retire livestock to restore rangeland" (RLRR) which started in 2003 and includes fencing rangeland, year-round or seasonal removal of grazing livestock, encouraging intensive animal husbandry instead of extensive grazing, encouraging pastoralists move out from degraded rangelands etc. As year-round grazing bans and seasonal grazing resting greatly constrained livestock husbandry, the main livelihood of herders, the government provided subsidies to the herders to encourage them to switch to practicing intensive animal husbandry based on pen-raising and feeding systems, or to move away from their pastures to engage in alternative livelihoods, such as crop farming or urban wage work. The objectives of RLRR are to reduce overall grazing pressure, because it is believed among decision-makers that overgrazing is the fundamental cause of rangeland degradation. From 2003 to 2011, the Chinese central government invested 2.5 billion USD (1USD = 6.2CNY) in the RLRR programmes (MOA 2012). Under this policy, 26 million ha of rangeland has been fenced to exclude year- round grazing and 28 million ha fenced for seasonal grazing rest (MOA 2012). In 2012, to continue pursuing the above objectives, a follow-up programme "Ecological Subsidy and Award System" was initiated and the central government decided to dramatically increase investment in rangeland restoration, with an annual budget of 2.2 billion USD, which is nearly the total expenditure of the past 8 years. The actual expenditure was 2.4 billion USD in 2012 (MOA 2013), even more than the budget allocation.

Nevertheless, the impacts of these policies are greatly debated. Although many scholars believe that the RLRR programmes have had positive impacts on restoring rangeland ecosystems (Liu et al. 2003; Xin et al. 2005; Gaowa et al. 2006; Chen and Luo 2007; Zhao et al. 2007), others argue that the programmes have generated negative impacts on the larger scale ecosystem (Li et al. 2007; Wang 2009). They point out that grazing pressures are shifted to non-project areas, thus increasing degradation in non-project regions and that long-term grazing exclusion is harmful to vegetation regeneration (Wang 2009; Zhao et al. 2009; Alatandalai et al. 2011; Wang and Qiao 2011). Particularly in terms of impacts on herders' livelihood, many scholars criticize the greatly increased costs of livestock production, because herders have to buy more forage during the periods when grazing is forbidden (Tian 2011; Siyiti et al. 2012; Zhao et al. 2012). Additionally, for those areas where a year-round grazing ban is implemented, the pastoralists were moved out from the rangelands and resettled in towns where they are facing many difficulties in finding alternative sustainable livelihoods (Xue and Jiang 2006; Chen 2008; Nie et al. 2008; Zhao and Jia 2009). However, some other academic positions argue that government subsidies have increased herders' income and along with the shift into intensive animal husbandry and more non-pastoral employment (especially for those who emigrated from pastoral areas into towns), herders' income would be increased in the long-term. The same debates are taking place about the impacts of RLRR programmes on pastoralist culture and social development (Chen and Luo 2007; Chen and Su 2008; Guo and Qia 2008).

This chapter will analyse the narratives employed by different stakeholders, including academic circles, decision-makers and herders, to explain the success or failure of ecological policies for rangelands. Some critical findings will be summarized. Based on these findings, we will argue that the problems with these policies may be ultimately due to one root cause: failure to understand the linkage between social and ecological systems (SES) as interconnected sets with feedback mechanisms. We conclude that the problems of EPs are not the result of improper policy implementation, as most scholars believed, but instead are due to misunderstanding the SES of pastoral areas.

18.2 Methodology

For government narratives, we will analyse the original policy reports as follows: Annual National Grassland Monitoring Reports from 2008, 2009, 2010, 2011, 2012, 2013; Ministry of Agriculture (MOA) reports on rangeland management policies; and key relevant government official speeches. We present how the government perceives ecological restoration polices and what narratives are used to portray the impacts on ecological condition and herders' livelihoods.

Regarding academic narratives, we conduct bibliometric analysis on research articles by academic researchers published in Chinese academic journals. All the relevant and qualified articles are collected from China Academic Network Publishing Database, the largest journal database in China (http://epub.cnki.net/kns/default.htm). We carry out three types of analysis for each paper: impacts analysis, cause-of-failure analysis and reliability-of-conclusion analysis.

In impacts analysis, the impacts of the RLRR policy are categorized into four aspects: (1) rangeland ecosystem; (2) herder income; (3) animal husbandry; and (4) pastoral culture and social development. For each aspect, we categorize the authors' into positive, negative, and ineffective impacts. Positive means that the policy achieved its planned improvement in the above four aspects; negative means that the policy directly caused negative impacts; while ineffective means that the policy failed to reach its expected objectives.

In cause-of-failure analysis, academic narratives about the causes of RLRR policy failures are analysed into two groups: (1) improper policy implementation, where the narratives are convinced the policy was reasonable but attributed their negative impacts or ineffectiveness to flaws in policy design or implementation procedures; and (2) improper policy, where the narratives perceived the policy itself was unreasonable and thus caused failures.

In addition, as the methods used in these research studies were quite diverse and the rigorousness of these methods directly affected the reliability of the study conclusions, we carried out a reliability-of-conclusion analysis for each of the four aspects. Here we mainly focus on the validity of data collection methods and reasoning. If the evidence supporting the conclusion were directly based on primary data from case studies and field samplings, we classify the research as most reliable versus not very reliable for evidence based on secondary data. Finally we classify as not reliable the papers based on no evidence, or lack of data sources and data representativeness.

For herders' narratives, we use data from our household survey in Alax Left Banner (ALB) of Inner Mongolia from 2009 to 2011 (Fig. 18.1). We did fieldwork in the middle and southern area of ALB (the grazing ban area) where the herders lived and also in the resettlement areas (Luanjing Tan, Yaoba Tan and Bayanhuode). We interviewed herder households to understand their perceptions about changes in rangeland brought by the EPs. In each area, the interviews were conducted with randomly selected households. We interviewed 30 households in Bayanhuode from June to July of 2009, and revisited in 2010 and interviewed another 9 households and the government officials. In 2011, we interviewed 16 households in Luanjing Tan and Yaoba Tan, and 17 households in Tengger desert where grazing exclusion was implemented.

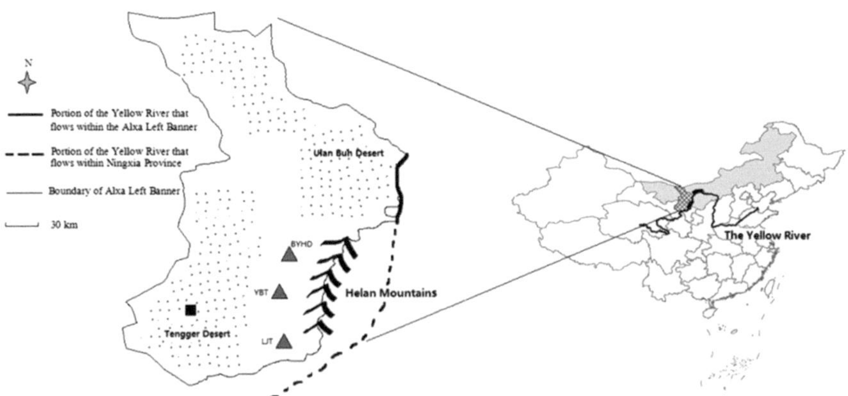

Fig. 18.1 Topographic information on ALB and the location of resettlement sites (LJT-Luanjing Tan, YBT-Yaoba Tan and BYHD-Bayanhuode) and grazing exclusion areas (Tengger desert and Helan mountain).

Table 18.1 Impacts of "retire livestock to restore grassland" (RLRR) on vegetation, from Annual National Grassland Monitoring Reports (2007–2011)

	Compare with non-project area (increased %)			Compare with pre-project condition (increased %)	
Year	Coverage (%)	Height (%)	Above-ground biomass (%)	Coverage (%)	Above-ground biomass (%)
2007	15	47	58	9	26
2008	14	60	68	–	–
2009	12	36	75	6	18
2010	12	38	44	3	8
2011	10	43	50	4	11

Data sources Annual National Grassland Monitoring Report (MOA 2008, 2009, 2010, 2011, 2012)

18.3 Results

18.3.1 Government's Perspectives

According to the "Annual National Grassland Monitoring Report" developed by MOA in recent years (see Table 18.1), the EPs had effectively restored the degraded grassland.[1] Taking the RLRR as an example, the report found that the policy achieved improvements in the aspects of grassland coverage, height and above-ground biomass production in the project areas comparing with non-project areas and pre-project grassland condition (see Table 18.1).

However, as to overall grassland conditions (see Table 18.2), even though EPs have been implemented for years, the MOA acknowledged "the rangeland degradation, desertification, and salinization have not been controlled effectively (MOA 2013)."

According to government perspectives, problems with the RLRR are due to its focus on restoring grassland condition without considering the effects on herders' livelihoods and without providing the means to improve livestock production through artificial grass reseeding and the construction of livestock sheds. Herders

[1]In the national standard (GB 19377–2003), "Parameters for Degradation, Sandification and Salinization of Rangelands", issued by General Administration of Quality Supervision, Inspection and Quarantine of the People's Republic of China (AQSIQ) in 2003, rangeland degradation is defined as a process in which rangeland ecological conditions deteriorated, forage biomass and quality decreased, the utilization performance decreased or was even lost due to adverse natural influences such as drought, sand storms, flooding, salinization, underground water table changes and etc., or by unseasonal animal husbandry such as overgrazing and hay overcutting, or by destructive human activities such as unregulated mining, medicine digging and firewood collecting. In this *Guo Biao* (national standard), the degree of rangeland degradation is classified as 4 levels: non-degraded, minor degraded, moderate degraded and severe degraded, based on monitoring changes in plant community characteristics, plant community composition, indicator species, above-ground biomass and measures of soil erosion.

Table 18.2 Government narratives on the evaluation of overall rangeland conditions

Year	Narratives of overall evaluation
2007	Ecological conditions in project areas have been improved, though overall rangeland conditions are tending to worsen. Degradation is exacerbated in certain areas
2008	Obvious restoration is observed in project areas, though overgrazing is still serious and overexploitation and illegal expropriation of the rangelands, continued. The task of grassland restoration remains a big challenge
2009	The trend of increasing grassland degradation has been controlled to some extent. Obvious improvements are observed in some areas
2010	The trend of increasing grassland degradation has been effectively controlled, although "local conditions improved while overall situation deteriorated". The task of grassland restoration remains a big challenge
2011	Overgrazing is still serious. The overall grassland conditions are still in a tense situation. Grassland restoration projects are getting to a crucial stage

Data source Annual National Grassland Monitoring Report (MOA 2008, 2009, 2010, 2011, 2012)

lack forage supplies and the means to engage in intensive animal husbandry after the implementation of grazing bans and grazing rests, so their long-term livelihoods are in trouble. The majority of the programme areas are located in remote and poor areas. Besides the subsidies from central government, local governments have limited financial capacities to supply the supporting funds needed for the EPs. According to the No. 1 Document of Central Government in 2011(State Council 2011) and Inner Mongolia No. 1 Document of 2010 (IMARG 2010), due to implementation of the grazing ban and grazing rest, the herders have had to sacrifice to achieve grassland protection because of the high costs of alternative systems of livestock management. As a result, herders' income is lower than the average level of farmers' income in crop production areas. Therefore in 2012, the Inner Mongolian Autonomous Region council dramatically increased subsidies to herders when they initiated a new project of the Ecological Subsidy and Award System, a follow-up to RLRR. This policy aims to maintain the ecological programme while improving herder livelihoods through: (1) providing 6 yuan (about \$0.95 USD in 2012) per μ (1 ha = 15 μ) for the year-round "grazing-ban" area; (2) providing 1.5 yuan (about \$0.23USD in 2012) per μ for "forage-livestock balance" area; (3) increasing subsidies for the livestock production system, including livestock breed improvement, artificial grassland improvement, purchase of production tools, and infrastructure improvement; and (4) developing education and providing vocational training to facilitate diversification of herders' livelihoods.

Several key findings can be summarized from the government narratives. First, rangeland degradation in the project areas has been partially controlled and rangelands have been partially restored. Secondly, China's rangeland conditions are continuing to degrade overall. Thirdly, the central government is realizing the problems with previous EPs caused by prioritizing rangeland ecological conditions while failing to consider the financial burdens on herder livelihoods and pastoral production.

18.3.2 Academic Perspectives

The first academic study of EPs was published in 2002 (Liu and Feng 2002), and by the end of 2012 in total 136 such studies were found in the China Academic Network Publishing Database. Among them, 103 papers study policy impacts, while the rest focus on policy design and the implementation process without mentioning the policy impacts. Among the 103 papers, 98 papers discuss ecological impacts, 65 papers discuss the impacts on herder income, 57 papers discuss animal husbandry, and 17 papers are related to cultural and social development.

In terms of project impacts, the majority of papers present positive findings on the ecosystem, while negative findings are dominant on aspects of herder income and pastoral social development (see Fig. 18.2). Regarding ecological impacts, 72 % of the academic perspectives present positive impacts, reflected in the increased height of the vegetation, coverage and biomass production; increased percentage of perennial vegetation; decreased frequency of dust storms; increased biodiversity and lastly, improved water storage functions. However, around 13 % of these papers argue that the projects generated negative impacts on the ecosystem. Their main arguments include: grazing pressures were shifted to non-project areas, thus increasing degradation in non-project regions; planting artificial forage destroyed original natural vegetation and exhausted ground water and soil fertility; and long-term exclusion of grazing is harmful to vegetation regeneration. In addition, around 14 % of the academic papers directly demonstrate that EPs did not achieve the expected objectives, since illegal grazing activities were very common, while grazing bans and grazing rests were not strictly implemented.

Regarding the impacts on herder income, 60 % of the academic papers demonstrate decreasing incomes. Researchers found that in the seasonal grazing rest areas, the costs of livestock production were greatly increased because herders had to buy more cultivated forage during the months when grazing was forbidden. Additionally, in those areas where year-round grazing exclusion was implemented,

Fig. 18.2 Percentages of findings of ecological project impacts on each aspect.

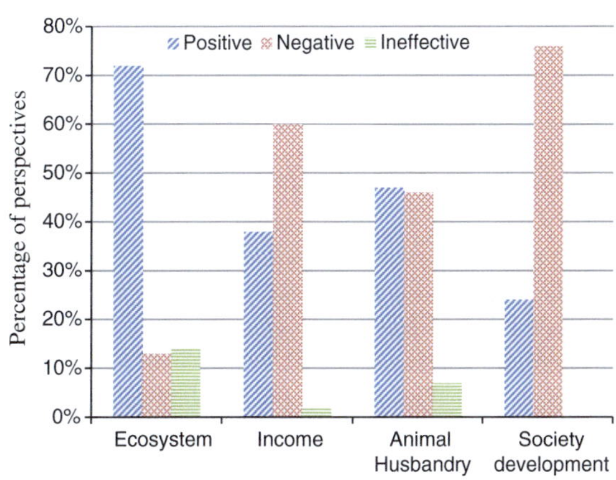

the pastoralists were moved out from the rangeland and resettled in towns, where they faced many difficulties in finding alternative sustainable livelihoods. They merely depended on government subsidies for their support, while living costs were much higher in the towns than previously in the pastoral area. Poverty was increasing among the resettled herders.

Of the academic papers, 38 % demonstrate that EPs have generated positive impacts on herder income. These academic perspectives argue that government subsidies have increased herders' income. Besides, along with shifting into intensive animal husbandry and increasing non-pastoral employment (especially those who emigrated from pastoral areas into towns), it is argued that herders' income should increase in the long-term. In addition, improvement of the natural environment under EPs also helped to improve local people's living conditions.

Regarding animal husbandry, 47 % of the academic papers present positive impacts. These state that EPs promoted shifts from extensive pastoralism into intensive animal husbandry through improvements in productive infrastructure, as well as animal breeding and shelter conditions. On the other hand, the development of intensive livestock feeding systems inevitably increased production costs (increased inputs for forage purchase, labour requirements and buildings), and thus reduced the net benefit from animal husbandry. This is the reason why 46 % of the academic findings state that the EPs generated negative impacts on animal husbandry. In addition, 7 % of the findings show that the policy was not able to achieve the expected objectives because herders encountered diversified challenges of transferring into modern intensive animal husbandry, particularly given the poor conditions of current pastoral production.

On social development, only 24 % of the academic papers hold positive views. These views include the improvement of income and development of intensive animal husbandry which facilitated "harmonious development of pastoral society and reformed the backward traditional concepts of the herders". In addition, the implementation of EPs facilitated a rangeland transfer system, so that the poor could generate income through leasing their pastures. EPs promoted urbanization through resettlement of herders, thus the educational and medical services for people were improved. However, 76 % of the papers argue that EPs generated obvious negative impacts on pastoral society. The increasing conflicts between local governments and herders were the most direct negative impacts, which led to critical challenges for local governance, as herders under EPs were generally involved in illegal grazing to reduce their production costs. Illegal grazing activities also caused conflicts among herders over grazing lands. In addition, grazing bans increased risks for the livelihood of marginalized people (e.g. the poor, the undereducated, elders, women and those living in harsh environments). Due to clear family income reductions after the initiation of grazing bans, the main workforce of pastoral families left the rangeland areas to search for employment, leaving the old and children uncared for at home, which weakened family ties. The shift into intensive animal husbandry and self-interested competition under market conditions affected social relationships and caused changes in cultural customs and identities.

Furthermore, as herders who resettled under EPs were unable to find proper jobs, some of them became involved in criminal and gambling activities.

It is important to note that according to the reliability analysis (Table 18.3), the positive impacts of EPs on the above four aspects—ecological, herder income, animal husbandry and social development—might be overrated. Regarding ecological impacts, 62 % of the papers that provide evidence for positive impacts were based on case studies or direct ecological monitoring. However, among these 71 papers, the majority cited indicators that only reflected short-term changes in vegetation, including aboveground biomass (42 papers), cover (40 papers), height (31 papers) and proportion of perennial grasses in vegetation communities (22 papers). The three former indicators are very sensitive to short-term precipitation changes in drylands, as well as grazing, and thus are unreliable for reflecting the long-term ecological impacts of grazing bans and grazing rests. In addition, as the grazing pressure within programme areas was often transferred to non-programme areas, monitored conditions within the project areas did not necessarily represent the overall grassland condition.

Among the papers that assert the existence of positive EP impact on herder income, less than half (44) studies were based on primary data. In contrast, 70 % of the papers that assert negative impacts and 100 % of papers arguing ineffective impacts are based on primary data from field surveys. Similarly for research on animal husbandry impacts, only 37 % of papers that claim positive impacts were based on case studies and 26 % did not provide enough evidence or presented unrepresentative data. Regarding social development, 100 % of papers arguing for positive impacts supplied no convincing evidence.

The majority of the papers argue that the flaws, failures and emerging issues are the result of the improper implementation of EPs, while few papers considered the possibility that these failures were caused by improper design of the policy itself. Thus, in response to the declining livelihoods and livestock husbandry, most academic studies suggest shifting herders into intensive livestock production systems and/or alternative livelihood approaches, with increased technical inputs and government subsidies, in order to increase pastoral production and income generation.

In general, the bibliometric analysis of academic studies shows that EPs have achieved effective impacts on grassland restoration (at least in the short term within project areas), while generating obvious negative impacts on herder livelihoods and pastoral society. While arguing that EPs have produced positive ecological, these studies focus only on short-term changes in vegetation indicators, and the absence of information on vegetation in non-project areas weakens the reliability of their conclusions. Regarding socio-economic impacts, the academic studies find that negative impacts on herder livelihoods have overridden demonstrated positive impacts, and most studies found serious negative impacts on pastoral society resulting from EPs. According to the results of these academic studies, it can be concluded that EPs prioritized ecological protection and failed to pay enough attention to herders' livelihoods, society and culture. Consequently, even though the policy improved grassland condition in the project areas, it worsened herders' livelihoods, increased social conflict in pastoral areas and weakened pastoral culture.

Table 18.3 Reliability analysis of the academic studies on the impacts of EPs (Percentage of papers that applied different data resources)

Data sources	Ecosystem			Herders' income			Animal husbandry			Social development		
	Positive (%)	Negative (%)	Ineffective (%)	Positive (%)	Negative (%)	Ineffective (%)	Positive (%)	Negative (%)	Ineffective (%)	Positive (%)	Negative (%)	Ineffective (%)
Primary data from case studies and field samplings	60	23	46	44	70	100	37	81	50	0	54	–
Secondary data	35	69	54	20	28	0	37	15	50	0	31	–
No evidence, lack of data sources and/or data representativeness	5	8	0	36	2	0	26	4	0	100	15	–

18.3.3 Herders' Perceptions: Case Study from Inner Mongolia

The Alax Left Banner (ALB) region is located in the west of Inner Mongolia and lies along western Helan mountain (Fig. 18.1). The banner is an administrative region used in Mongolian pastoral areas, similar to a county in the agricultural areas of China. The region covers 80,412 km^2, of which 42.5 % is desert (Chorography Compiling Committee of Alax Left Banner 2000). The climate is extremely arid, with annual precipitation of 64–208 mm, annual evaporation of 2317–3993 mm, and an annual average temperature of 7.4 °C (Chorography Compiling Committee of Alax Left Banner 2000). The banner contains 46,000 km^2 of usable rangeland, mainly steppe desert and desert, but it is reported that 59.6 % of useable rangeland has degraded (Wu et al. 2005). Vegetation coverage has decreased by 30–50 % since the 1950s (Wu et al. 2005), and biomass (dry matter) production decreased by 53.9 %—from 570.03 kg per ha in the 1960s to 262.82 kg per ha in the 1990s (Pei 2011).

This area is sparsely populated. The population was 173,000 in 2012, of which 19 % were Mongolian and 73 % were Han (Statistic Bureau of ALB 2013). About 20,000 people were relying on animal husbandry in 2013 (Zhang et al. 2013). The people in the northern area are mostly Mongolians who raise goats and camels. In the southern part of the banner, Han people predominate, with their main livelihood depending on crop agriculture or a combination of cropping and animal husbandry. There were 1.2 million livestock in the banner in 2013, of which 0.65 million were raised in intensive systems and the others grazing on natural rangeland over most of a year (Wei 2013).

Since 2002, a series of EPs have been implemented in the southern and middle areas of ALB. The basic model for such projects is exclusion of grazing from rangelands and the relocation of local herders into resettlement villages in Luanjing Tan, Yaoba Tan and Bayanhuode development zones. The first two sites are agricultural development areas founded in the 1970s, while the last site is a newly established agricultural zone near Bayanhot, the capital of ALB (Fig. 18.1). In the new settlement, the ex-pastoralists are supposed to adopt crop cultivation, intensive animal husbandry and employment in service sectors. During the process, those pastoralists who formerly were entirely supported by pastoralism receive payments equal to $1198 USD in cash per year per person and in addition the government provides infrastructure in the resettlement villages, including houses, livestock pens, hay fields and irrigation systems. For those agro-pastoralists who formerly depended on crop farming as well as animal husbandry, the government provides only the infrastructure mentioned above, but no annual cash payments. For the older resettled people, males older than 54 years and females older than 49 years receive from the government an annual pension but no cash subsidies. According to the Poverty Alleviation Office of ALB, 7319 working age people were resettled between 2002 and 2010, among whom 6300 switched to crop production or

Table 18.4 Herders' perceptions of effects of grazing ban on rangeland restoration (N = 27)

	Percentage of interviewees	Example of herders' observations
No obvious change	17 (63 %)	Nothing changes after grazing ban, neither better nor worse. This is the characteristic [of ecosystem] in this place. If there's no rain, then no forage. It has nothing to do with grazing
Rangeland degraded after grazing ban	8 (30 %)	Grass would grow better if grazed by goats. It did not grow well after the grazing ban. Now, after the ban, the new grass shoots are white and without the goats grazing, the shoots gradually die
		Grass began to die in the second year after grazing ban. Grass is not as dense as before. If there were livestock grazing, the grass wouldn't be dead block by block like this
		Desertification is more obvious after the grazing ban. Before the ban, we could ride bicycle in the desert steppe, but now, even a motorcycle could not run due to the sand
Vegetation recovered obviously	2 (7 %)	Vegetation in Helan mountain areas is much better after grazing ban. The shrubs are growing up

intensive animal husbandry and 1019 got employment in secondary and tertiary industries.

Regarding the perceptions of herders on changes in rangelands brought about by the EPs, according to 28 herders interviewed, only 7 % reported grass being improved after the imposition of the grazing ban. Of the herders interviewed, 63 % told us that rangeland conditions are much more influenced by rainfall than grazing and therefore the EPs did not have any impact on rangeland restoration. Nearly a third, 30 % of households even believed that rangelands degraded further after grazing bans; Some examples of observations from herders are listed in Table 18.4.

Among reasons given why the EPs had failed to restore rangeland conditions, 65 % of herders pointed out that many forage plants would not grow without grazing, and 60 % of herders believed that the accumulated plant litter due to lack of grazing prevented grass germination or renewal (see Fig. 18.3). Also some herders thought that lack of livestock dung as fertilizer, as well as increased rodent hazards, were reasons causing unexpected ecological changes (Table 18.5).

On the livelihood aspects, we found that crop planting and intensive animal husbandry could not provide feasible livelihoods for the ex-pastoralists. For example, Bayanhuode resettlement village, where government has provided the most financial and material inputs among all the resettlements, received more than 300 in-migration households from 2007 to 2008. The government provided each household with a house, a greenhouse for vegetable planting (Fig. 18.4), and a livestock shelter for keeping animals. Elders reaching the qualified age could get pensions of $92 USD per month. According to the government plan, planting vegetables in the greenhouses would provide subsistence for those in-migrants after

Fig. 18.3 The pictures were taken in July 2010 after 8 years of a grazing ban. The *left* one with *green* and growing vegetation is in a grazing area outside the project fence, while the *right* one with *grey* and withered vegetation is within project area without grazing for 8 years.

Table 18.5 Herders' perceptions about why rangeland failed to restore after grazing ban (N = 20)

	Percentage of interviewees	Example of herders' observations
Lack of compensatory regrowth	13 (65 %)	The length of re-grown shoots is getting shorter after the ban. Some particular species like *R. soongorica* will not grow without grazing
Accumulation of litter prohibits plants from germination or renewal	12 (60 %)	Litter is accumulates without grazing, which prohibits the renewal of the grass. For example, *R. soongorica* could re-grow in the next year if it was grazed, but if it is not grazed this year, it will not grow well next year
Lack of livestock dung as fertilizer	4 (16 %)	Shrubs grow better with manure. There would be no fertilizer without grazing
Increased rodent hazard	2 (10 %)	Rodent-caused disaster increased after grazing exclusion

they give up pastoralism in the rangelands. Nevertheless, the resettled families were pastoralists who used to be engaged in herding animals in rangelands, so the skill of greenhouse cultivation is a big challenge for them. Among the 22 households interviewed, by 2010 after being resettled for 3 years, only 6 households were still engaged in vegetable production in greenhouses, 5 households continued to rely on pastoralism (the older members and children moved to the resettlement village while young family members still herded animals in the pastoral areas), 8 households relied on casual jobs (such as drivers, construction workers etc.) and 3 households had no employment and lived solely on the money from the government. By 2012 when we revisited this site, all the resettled households had given up

Fig. 18.4 The greenhouses provided by EPs in Bayanhuode resettlement village. The picture on the *left* was taken in the summer of 2008 when resettled herders were starting to learn vegetable cultivation, and the picture on the *right* was taken in summer of 2010, showing most greenhouses had been abandoned.

greenhouse vegetable growing, due to water shortages. Water crisis is one of the major threats to crop agriculture in Yaoba Tan and Luanjing Tan as well.

Moving herders away from the rangeland is the government's aim to solve the overgrazing problem. However, the rangeland is now facing other serious over-exploitation of natural resources. Since the 1990s, collecting of wild plants, such as "fat choy", *Cistanche deserticola*, *Cynomorium songaricum* has been destroying rangeland vegetation. In recent years, collecting scorpions for medicinal markets has become a new destructive activity. In 2007, ALB Forestry Agency sanctioned more than 2000 outsiders who came to rangelands to dig *Boschniakia rossica,* a medical plant which parasites *R. soongorica* on its root (Shi 2007). In the summer of 2011, almost one thousand outsiders came to collect scorpions on the rangeland along Helan mountain and Luanjing Tan and killed some 0.2 million scorpions every day (Agency of Agriculture of Alax League 2007). In addition, more and more mineral explorations have taken place in the rangeland after local herders were moved out.

Such activities, generally illegal, are destructive to the ecosystem. As predators on several harmful insects, scorpions play an important role in controlling and preventing insect hazards. Collecting of scorpions threatens ecosystem functions by destroying food webs. Digging plants and exploration of minerals directly disturbs vegetation and soil. Due to the extreme arid climate, growth of shrubs is very slow and, once damaged, it is difficult for the vegetation to recover. Supervision and control of illegal activities is critical to rangeland conservation.

Local herders are not only rangeland users, but also protectors of rangelands for generations. When herders made a living on the rangeland, they were motivated to protect their rangeland. In our survey, we found that herders spontaneously formed patrols to monitor, report and stop the above-mentioned destructive activities. However, after herders left rangelands due to the EPs, the burden of rangeland protection was loaded entirely on government agencies. For example, by June 2011, local governments have punished more than 7000 cases of scorpion collection, with high enforcement costs of vehicles and policemen involved (Agency of Agriculture of Alax League 2007). As the rangeland is extensive in area and the government agencies have human and material resource constraints, such illegal activities could not be effectively controlled. The scorpion collectors always come to rangelands on motorcycles as night falls and have left before dawn. Such high mobility made it rather difficult for government to monitor. Our interviewees reported that "the rangeland was destroyed by people from neighbouring provinces who came to dig *Cong Rong* (*C. deserticola*) and scorpions and sell to the medicine market. These people came in flocks. They dig the rangeland with shovels and even explosives, leaving the land with pits everywhere. In this case, how could the ecosystem be protected as EPs expected?" Even worse, we noticed that due to lack of sustainable livelihoods, some of the ex-pastoralists have come back from resettlement sites to join the scorpion collecting, plant digging and rare stone collecting.

18.4 Discussion and Conclusion

Both government and academic findings show that EPs have had a positive impact on grassland restoration (at least in the short-term within project areas), while generating obvious negative impacts on herder livelihoods, livestock production and pastoral society. Our case study in Alax Left Banner shows that most herders believe rangeland conditions are more influenced by rainfall than grazing, in which case the EPs would have had no impact on rangeland restoration. Some herders even think that the rangeland becomes further degraded after banning grazing, and they consider that the alternative livelihoods and production systems supported by EPs do not work well. The other case studies by our research group in different precipitation regions of Inner Mongolia, covering Hulunbuir (Gu and Li 2013), Xinlingol (Lai 2012), and Chifeng (Wang 2011) have reached the same conclusions as this Alax case study.

Here we are not going to try to decide which conclusions are right or wrong, because different stakeholders have different management goals for the ecosystem and apply different criteria to evaluate ecosystem health. Taking vegetation species as an example, ecologists use dominant species as important indicators to evaluate ecosystem condition, while herders believe that more palatable species, which are not necessarily dominant, indicate a healthier rangeland. Particularly in terms of temporal scale, ecosystem changes should be assessed in the long run and thus it may be too early to judge the effectiveness of China's rangeland EPs.

We need, nonetheless, to analyse the mechanisms of what has already happened and what will likely happen to the target ecosystem following project interventions. A non-equilibrium theory of rangeland functioning could explain some of the processes occurring in China's rangeland (Li and Zhang 2009), particularly in terms of precipitation variability and its impacts on biomass production. But non-equilibrium theory does not fit all cases. In Africa overgrazing might not be the major contribution to rangeland degradation compared with rainfall variability, but in parts of China overgrazing does exist. This is because the social ecological system of pastoral areas in China is no longer closed, particularly after the 1980s with the growth of the market economy. While the number of livestock may still fluctuate in response to variable precipitation, now livestock numbers are also heavily influenced by the availability of forage markets, by the demand for livestock products, and the availability of micro-credit. For example, if the market price of meat is high enough to cover the cost of buying forage from outside the livestock production ecosystem, herders can maintain and even increase their herd sizes in drought years through obtaining micro-credit to purchase forage.

Given the limited applicability of new rangeland ecology in China, here we consider the logic behind the EPs formulation, which may help to understand the current issues in China's rangelands.

It is accepted that government EPs aim to improve ecosystem conditions. The logic behind the projects is that too many people and thus too many livestock have contributed to rangeland degradation and that decreasing the population of both will improve environmental conditions. However, most ecosystems have a long history of human management, as is the case of grazing systems in arid and semi-arid areas of China, where the human system utilizes natural resources and at the same time maintains the ecosystem. In fact, some ecosystems can only be sustained through human management practices (Liu et al. 2007). In such cases the two systems—human and natural resources—are adapted to each other and have evolved into a highly integrated social-ecological system (SES). SES is neither a social system embedded in an ecological system, nor an ecological system embedded in a social system. It has its unique characteristics (Walker et al. 2006), and cannot be managed through focusing solely on an ecological or a social component.

If we treat a pastoral community and the rangelands as an integrated system, it is clear that EPs have driven this system from an extensive grazing state characterized by reciprocal ecological-social feedbacks into another state characterized by isolated social and ecological systems. During this transformation, key processes that control system dynamics have been changed or terminated in ways that may be irreversible without external interventions.

In the Mongolian plateau, the extensive grazing system was formed by nearly a thousand years of co-evolution with its natural conditions, and livestock and vegetation have become mutually interdependent. In this system a degree of grazing disturbance is critical to maintaining the vigor of plants while grazing is also a form of natural resource utilization adapted to the local environment. During this process, local people have developed a body of indigenous ecological knowledge about the interactions between grass, livestock and the abiotic environment, and have managed the rangeland based on this knowledge. They have also protected the rangeland from destructive exploitation by outsiders.

After the EPs were implemented and these herders were moved out, the interactions between the ecological system and the social system were cut off. Removal of grazing disturbance makes plants lose their vigour, and may lead finally to death. Meanwhile, emigration of herders weakened supervision and protection of the rangeland, leaving pasture vulnerable to exploitation by outsiders, which caused serious damage to vegetation and soil. Under the arid climate in ALB, growth of perennial vegetation and the formation of soil are extremely slow processes and once interrupted they are difficult to restore. In addition, when herders moved off the rangeland, they sold their herds for slaughter, reducing the gene pool of livestock, especially that of the unique species of Alax double-hump camel which is highly adapted to the local environment. Finally, pastoralism is founded on indigenous knowledge and when herders gave up pastoralism this knowledge was no longer useful or valued. As time goes by, it will be lost as the carriers of such knowledge disappear. Loss of vegetation, the eroding of the livestock gene pool and the loss of indigenous ecological knowledge will make it nearly impossible to revive the pastoral system.

Depending on rangeland for a living, herders have greater incentives than the government to protect rangeland for sustainable livelihoods. Nevertheless, herders are perceived by EPs as destroyers of the environment and are removed from the rangeland, leaving all the responsibility for protection to the government. Limited by insufficient human and material resources, the government is not able to enforce effective monitoring and, as a result, the rangeland is left without adequate supervision.

Irrigated crop cultivation has further worsened the current water shortage crisis, due to its high dependency on water resources and low water use efficiency, in arid areas where water resources are critical to the maintenance of local ecological and social systems. Over-exploitation of water resources has resulted in a water shortage, a decline in the water table and land salinization in Yaoba Tan, Luanjing Tan and Bayanhot. In some resettlement areas, cultivated sites have been abandoned due to salinization of the land caused by the overwatering of irrigated crops (Water Authority of Alax Left Banner 2004). It therefore seems that crop cultivation is ecologically infeasible in such and arid environment, and is unlikely to generate sustainable livelihoods for resettled herders. Ex-pastoralists, who lack skills for other livelihood activities and have no resources to return to pastoralism are very likely to engage in activities that exploit the environment in destructive ways, creating a vicious feedback loop.[2]

Considering the ecological characteristics of arid rangelands, even in terms of providing the ecosystem services expected by government, it is necessary to maintain a SES based on grazing. Polices that aim to improve ecosystem services through rehabilitation should therefore focus on maintaining pastoralism by enhancing the resilience of grazing systems. At present, pastoralism in China is facing huge internal and external pressures and changes, such as rangeland privatization, decreased livestock mobility (Li and Huntsinger 2011), intensive animal husbandry, decoupling of social systems from ecological systems (Li and Li 2012),

[2]Realizing this problems, local governments are now starting to change strategies. As stated in official documents, the strategy of "pastoralism transfer to agriculture farming" is supposed to be stopped and gradually replaced by "pastoralists transfer to employment in industry or service sectors". But in practice it is not easy to provide appropriate employment opportunities for unskilled ex-pastoralists within a reasonable time.

the influence of commercial markets on traditional knowledge of animal husbandry, and expansion of the mining and manufacturing sectors. All these factors are causing damage and pollution to rangeland, while EPs are moving herders out of the rangeland and pastoralism is declining. There are many challenges to be met if pastoralism is to adjust successfully to these pressures. Taking market influences as an example, as mentioned previously, livestock populations no longer fluctuate naturally to match the variability of precipitation, which may lead to overgrazing on natural pasture. Nevertheless, the response to overgrazing should not simply be the separation of the human system from the ecological system by removing herders from the rangelands, which has been proven not to work. We are not arguing that market forces are bad or should not influence pastoralists, since nobody can stop markets spreading in this globalized world. What we need to explore more fully is how to integrate market mechanisms with traditional society, instead of simply replacing the latter. In sum, our discussion has argued that the problems of EPs are not due to improper implementation of policy, as most scholars have believed, but are instead due to misunderstanding the SES of pastoral areas.

Acknowledgements Our research was supported by the National Natural Science Foundation of China (41171428) and a Ford Foundation grant. Alax Camel Association and Alax White Cashmere Goat Association provided precious assistance in data collection. We thank all the students in our team for their contributions to the case study analysis and for their discussion during our weekly group meeting. Special thanks go to all the hospitable Mongolian herders, who were always patient with our numerous questions. Last but not least, thanks must go to Dr. Roy Behnke and Dr. Carol Kerven for their editing for the manuscript which improves this chapter greatly.

References

Agency of Agriculture of Alax League. (2007). Illegal collecting of scorpions is out of control in Alax League, Inner Mongolia. (Chinese: NeiMengGu Alashan Meng CaoYuan YeSheng XieZi LuanBu LanCai WeiFa XingWei LvJingBuZhi). http://nc.mofcom.gov.cn/articlexw/xw/dsxw/201107/18090226_1.html. Publish date: July 22, 2011, Access date: January 16, 2014.

Alatandalai, Z., Bao J. F., & Dahuerbayaer, G. (2011). Impacts of long-term grazing ban on desert rangeland of Alxa Left Banner (Chinese: ChangQin JinMu Dui Alashan ZuoQi HuangMo CaoYuan De YingXiang). *Inner Mongolia Prataculture [NeiMengGu CaoYe], 23*(1), 56–58.

Chen, J. (2008). Review of canceling herds and returning to grassland policy in Sanjiangyuan Region in Qinghai Province: Based on survey of Maduo County (Chinese: QiangHai Shen SanJiangYuan TuiMuHuanCao He ShengTai YiMing KaoCha: JiYu MaDuo Xian De DiaoCha FenXi). *Mationalities Research in Qinghai [Chinese: QingHai MinZu YanJiu], 1*, 110–115.

Chen, J., & Luo, D. (2007). Survey report on grassland restoration and poverty allivation in Sonid Right Banner, Inner Mongolia (Chinese: NeiMengGu SuNiTe YouQi CaoYuan ShengTai ZhiLi Yu MuQu FanPingKun DiaoCha BaoGao). *Northern Economy, 11*, 29–35.

Chen, J., & Su, Y. L. (2008). Impacts of grazing ban on farmers' production and livelihoods in agro-pastoral area: A survey to 80 households of 4 villages in 2 townships in Yanchi county, Ningxia (Chinese: JinMu Dui NongMu JiaoCuo Dai NongHu ShengChan He ShengJi De

YingXiang: Dui NingXia YanChi Xian 2 Xiang 4 Cun 80 Ge NongHu De DiaoCha). *Agricultural Economic Problems [Chinese: NongYe JingJi WenTi], 6*, 73–79.

Chorography Compiling Committee of Alax Left Banner (Chinese: Alashan ZuoQi Zhi BianZuan WeiYuanHui). (2000). *Chorography of Alax Left Banner (Chinese: Alashan ZuoQi Zhi 2000)*. Huhhot: Inner Mongolia Education Press [Chinese: NeiMengGu JiaoYu ChuBan She], Inner Mongolia, China.

ECOCAHYB (Compiling Committee of Chinese Animal Husbandry Year Book (ZhongGuo XuMuYe NianJian BianZhuan WeiYuanHui). (2011). *Chinese Animal Husbandry Year Book (ZhongGuo XuMuYe NianJian)*. Beijing: Chinese Agricultural Publish.

Gaowa, S. X. W., Li, C.G., & Liu, Z. L. (2006). Effects of grazing rest in spring on natural grassland of Xilingol League (Chinese: XiLin GuoLe Meng TianRan CaoYuan ChunJi XiuMu XiaoGuo JianCe FenXi). *Inner Mongolia Prataculture (Chinese: NeiMengGu CaoYe) Inner Mongolia Prataculture (Chinese: NeiMengGu CaoYe), 18*(3), 42–50.

Gu, Y. C., & Li, W. J. (2013). Research on the grazing ban policy's effect to grassland quality: analysis on the scale of herdersman households. *Acta Scientiarum Naturalium Universitatis Pekinensis, 49*(2), 288–296.

Guo, J. J., & Qia, G. X. (2008). Survey and suggestions on perfecting grazing ban and pen-raising policy (Chinese: GuanYu WanShan CaoYuan "JinMu SheSi" ZhengCe De DiaoCha Yu JianYi). *Journal of Chongqing Institute of Technology (Social Science) (Chinese: ChongQing GongXueYuan XueBao(SheHui KeXue Ban), 22*(12), 1–6.

IMARG (Inner Monogolia Automatous Region Government). (2010). Opinions on facilitating herders income increase by inner mongolia autonomous region government (NeiMengGu ZiZhiQu RenMin ZhengFu GuanYu CuJin ZengJia MuMin ShouRu De ShiShi YiJian). *Document of IMARG (NeiMengGu ZiZhiQu RenMin ZhengFu GongBao)*.

Lai, Y. P. (2012). *Impact of pasture transfer and cooperation management systems on Herders' ability to benefit: A comparative study of pasture re-aggregation strategies after the implementation of grassland contract policy (Chinese: LiuZhuan he HeZuo Duiyu GanHan BanGanHanQu MuMin HuoYi NengLi YingXiang de BiJiao YanJiu-CaoChang ChengBao BeiJing Xia de CaoChang ZhengHe LuJing BiJiao)*. M.S. Thesis, Peking University.

Li, W. Q., Hu, Z. Z., Long, R. J., Gao, X. C., & Li, F. D. (2007). The application effect of the project of restoring grassland from over- grazing in Gansu and the ways for its sustainable development (Chinese: GanSu Sheng TuiMu HuanCao GongCheng ShiSHi JiXiao, CunZai WenTi He DuiCe). *Pratacultural Science (Chinese: CaoYe KeXue), 24*(1), 1–6.

Li, W. J., & Huntsinger, L. (2011). China's grassland contract policy and its impacts on herder ability to benefit in Inner Mongolia: tragic feedbacks. *Ecology and Society, 16*(2), 1.

Li, W. J., & Li, Y. B. (2012). Managing rangeland as a complex system: How government interventions decouples social systems from ecological systems. *Ecology and Society, 17*(1), 9.

Li, W. J., & Zhang, Q. (2009). *Unscrambling the dilemma on rangeland: cognitions on some problems of arid and semi-arid rangeland utilization and management (Chinese: JieDu CaoYuan KunJing: DuiYu GanHan BanGanHan CaoYuan LiYong He GuanLi RuoGan WenTi De RenShi)*. Beijing: Economic Science Press.

Liu, J. G., Dietz, T., Carpenter, S. R., Alberti, M., Folke, C., Moran, E., et al. (2007). Complexity of coupled human and natural systems. *Science, 317*(5844), 1513–1516.

Liu, L., & Feng, W. L. (2002). The necessity of grazing rest in natural rangeland (In Chinese: TianRan CaoYuan JiJieXing XiuMu De BiYaoXing). *Inner Mongolia Prataculture (NeiMengGu CaoYe), 14*(4), 18–19.

Liu, A. J., Xin, Q., Surina Wang, J. J., & Guo, G. F. (2003). Monitoring of the effects of grazing ban and grazing rest on Xilingol Grassland (Chinese: XiLin GuoLe CaoYuan JinMu XiuMu XiaoGuo JianCe YanJiu). *Inner Mongolia Prataculture (Chinese: NeiMengGu CaoYe), 15*(3), 1–4.

MOA (Ministry of Agriculture). (2008). Annual National Grassland Monitoring Report (QuanGuo CaoYuan JianCe BaoGao).

MOA (Ministry of Agriculture). (2009). Annual National Grassland Monitoring Report (QuanGuo CaoYuan JianCe BaoGao).

MOA (Ministry of Agriculture). (2010). Annual National Grassland Monitoring Report (QuanGuo CaoYuan JianCe BaoGao).

MOA (Ministry of Agriculture). (2011). Annual National Grassland Monitoring Report (QuanGuo CaoYuan JianCe BaoGao).

MOA (Ministry of Agriculture). (2012). Annual National Grassland Monitoring Report (QuanGuo CaoYuan JianCe BaoGao).

MOA (Ministry of Agriculture). (2013). Annual National Grassland Monitoring Report (QuanGuo CaoYuan JianCe BaoGao).

Nie, X. M., Shi, H. X., &d Zhao, C. Z. (2008). Investigation of present situation and achievements of grazing withdrawal project in source region of Yellow River (Chinese: HuangHe YuanQu TuiMu HuanCao GongCheng ShiShi XianZhuang Ji JiXiao De DiaoCha YanJiu). *Grassland and Lawn (Chinese: CaoYuan Yu CaoPin), 2*, 59–63.

Pei, H. (2011). *Ecological Characteristics and Environment protections in Alax Desert (Chinese: Alashan HuangMo Qu ShengTai HuanJing TeZheng Yu HuanJing BaoHu)*. Beijing: Meteorology Press (QiXiang ChuBanShe).

Shi, W. S. (2007). Illegal digging of *Boschniakia Rossica* seriously damaged ecosystem (Chinese: LanWa CaoCongRong Zhi Alashan CaoYuan ShengTai Zao YanZhong PoHuai). Legal Daily (Chinese: FaZhi RiBao). http://news.sohu.com/20071113/n253213825.shtml. Publish date: November 13, 2007, Access date: January 16, 2014.

Siyiti, H., Zhu, M. L., & Jiang, Z. Q. (2012). Problems in implementing grassland grazing ban policy and counter measures: Cases of Xinjiang (Chinese: CaoDi JinMu ZhengCe ShiShi Zhong CunZai De WenTi Yu DuiCe JianYi: Yi XinJiang WeiLi). *Agricultural Economic Problems (Chinese: NongYe JingJi WenTi) 3*, 105–109

State Council. (2011). Some opinions of the State Council on promoting good and fast development in pastoral areas (GuoWuYuan GuanYu CuJin MuQu YouHaoYouKuai FaZhan De RuoGan YiJian). http://www.gov.cn/zwgk/2011-08/09/content_1922237.htm. Publish date: August 9, 2011. Access date: November 27, 2012.

Statistic Bureu of Alxa Left Banner. (2013). Bulletin of main statistics of the sixth population census of Alax Left Banner (Chinese: Alashan ZuoQi DiLiuCi RenKou PuCha ZhuYao ShuJu GongBao).

Tian, X. Y. (2011). The influence of policies of returning grazing land to forage land on the nomads' living (Chinese: TuiMu HuanCao ZhengCe Dui WoGuo MuMin ShengHuo De YingXiang). *Chinese Journal o f Grassland (Chinese: ZhongGuo CaoDi XueBao), 33*(4), 1–4.

Walker, B., Gunderson, L., Kinzig, A., Folke, C., Carpenter, S., & Schultz, L. (2006). A handful of heuristics and some propositions for understanding resilience in social-ecological systems. *Ecology and Society, 11*(1), 13.

Wang, L. Q. (2009). Some thoughts on the performance and sustainability of retire livestock restore grassland projects (Chinese: GuanYu TuiMu HuanCao GongChen JiXiao KeChiXu Xing De SiKao). *Journal of Inner Mongolia Agricultural University (Social Science Edition) (Chinese: NeiMengGu NongYe DaXue XueBao (SheHui KeXue Ban)), 11*(2), 116–119.

Wang, Y. H., & Qiao, Y. L. (2011). Problems in implementing retire livestock, restore grassland projects and counter measures (TuiMu HuanCao GongCheng ShiShi Zhong De WenTi Yu DuiCe). *Agricultural Economic Problems (Chinese: NongYe JingJi WenTi), 2*, 99–103.

Water Authority of Alax Left Banner. (2004). Report of irrigation works and water resource survey in Alax Left Banner (Title translated from the Chinese: Alashan ZuoQi Shui ZiYuan Ji Shui GongCheng PuCha BaoGao). Publication restricted.

Wei, B. (2013). Governmental Report of Alax Left Banner in 2013

Wu, Y. J., Zhang, X. Z., Duan, Z. H., Liu, S. J., Ren, Y. L., Luo, H. C., & Xie, S. Z. (2005). Causes of rangeland degradation in Alax and countermeasures for sustainable development (Chinese: Alashan CaoDi TuiHua YuanYin Ji KeChiXu FaZhan DuiCe). *Prataculturul Science (Caoye KeXue), 9*, 20–22

Xin, Q., Liu, A. J., Liu, Y. Z., & Gaowa. (2005). Study on Change Rangeland Vegetation, Change Using MOD IS-NDV I in Xilinguole Grassland (Chinese: YingYong MODIS-NDVI Dui

CaoYuan ZhiBei BianHhua JianCe Yanjiu: Yi XiLin GuoLe Meng WeiLi). *Acta Agrestia Sinica (Chinese: CaoDi XueBao), 13*(Supplement), 15–19.

Xue, F. R., & Jiang, D. M. (2006). Study on problems in implementing retire livestock, restore grassland projects and counter measures (TuiMu HuanCao GongCheng ShiShi Zhong De WenTi Yu DuiCe YanJiu). *Northern Economy (Chinese: BeiFang JingJi), 11*, 77–78.

Zhang, J. L., Bao, G. X., Wu, W. J., & Zhang, Y. C. (2013). Emerging problems and countermeasure in pastoralism of Alax Left Banner (Chinese: Alashan ZuoQi CaoYuan XuMuYe FaZhan XingShi ChuXian De WenTi Ji DuiCe). *Inner Mongolia Forestry Survey and Design (Chinese: NeiMengGu LinYe DiaoCha SheJi), 5*, 120–123

Zhao, C. H., Cao, Z. Z., & Rong, Z. J. (2009). Social-economic influences of retire livestock, restore grassland projects in Alxa Left Banner, Inner Mongolia (Chinese: TuiMu HuanCao GongCheng Dui NeiMengGu ALaShan ZuoQi JingJi SheHui XiaoYi De YingXiang). *Acta Agrestia Sinica (Chinese: CaoDi XueBao), 17*(1), 16–21.

Zhao, C. Z., & Jia, L. H. (2009). Ecological performance and sustainable problems with the grazing forbidden project in the resource regions of the Yellow River (Chinese: HuangHe YuanQu TuiMu HuanCao GongCheng JiXiao Yu WenTi). *Journal of Lanzhou University (Natural Sciences) (Chinese: LanZhou DaXue XueBao (ZiRan KeXue Ban)) 45*(1), 37–41.

Zhao, H., Wang, R. Y., Guo, N., Liu, H. Y., Zhang, K., & Wang, H. L. (2007). Ecology effect of retrieving grassland from grazing on desertif ication grassland vegetation in Anxi (Chinese: TuiMu Dui AnXi HuangMoHua CaoYuan ZhiBei De ShengTai XiaoYing). *Prataculturał Science (Chinese: CaoYe KeXue), 24*(5), 83–88.

Zhao, Y. J., Zhang, Y. Q., Wu, B., Qin, S. G., Shi, H. S., & Zhao, J. H. (2012). Farmers' perceptions of rangeland enclosure policy and influencing factors (NongMu Min Dui JinMu ZhengCe De YiYuan Ji XiangYing YinSu FenXi). *Bulletin of Soil and Water Conservation (Chinese: ShuiTu BaoChi TongBao), 32*(4), 307–311.

Chapter 19
Taking Down the "Great Green Wall": The Science and Policy Discourse of Desertification and Its Control in China

Hong Jiang

Abstract Desertification is a serious problem in China and various research studies and surveys have been conducted to address the issue. The Chinese government has launched the ambitious "Great Green Wall" tree-planting programme to control desertification. But problems abound. This chapter reviews the scientific studies of desertification in China, and highlights their many inconsistencies, connecting them with political control and policy discourses. I point out that the State Forestry Administration, the institution in charge of desertification study and control, has adopted an ineffective tree-focused approach through the "Great Green Wall" programme. Even though the programme has included the establishment of shrubs and other vegetation types in addition to tree planting, the underlying aggressive attitude toward the environment remains, rendering it difficult for the problem of desertification to be addressed fundamentally. This chapter calls for the Chinese government to abandon the dominating relationship with the environment embodied in the "Great Green Wall" programme.

Keywords Desertification · Policy · Science · "Great Green Wall", China

Assessing the current situation of desertification in China is not a simple matter. The following two contradictory stories are typical in describing China's desertification—both are from Minqin county in Gansu province in the Hexi Corridor, northwest China.

One story is about serious and on-going desertification induced by human activities. Situated between the Badain Jaran and Tengger deserts, Minqin is a dryland with 110 mm annual rainfall, and water resources are crucial to its ecological health. Minqin sits at the lower reaches of the Shiyang River, whose water, coming from the melting glacier of the Qilian Mountain to the south, has nurtured much of Minqin's oases. In recent years, water overuse in the upper reaches has resulted in reduced water available to Minqin. To compensate, with government

H. Jiang (✉)
University of Hawaii, Manoa, USA
e-mail: hjiang@hawaii.edu

R.H. Behnke and M. Mortimore (eds.), *The End of Desertification?*
Springer Earth System Sciences, DOI 10.1007/978-3-642-16014-1_19

encouragement, the people of Minqin dug many wells—from zero in 1964 to over 11,000 by 2007—to sustain its agriculture (Zhang et al. 2007a, b). This has led to lowering of the groundwater table from 6 m in 1986 to 20 m in 2006. Natural and planted trees and shrubs have been dying off, the oases are swallowed by sand that advances at 8–10 m each year, and the sand-covered area is as high as 94.5 % (Ni et al. 2013). Oasis expansion into new areas cannot match their reduction from desertification, and increased human activity threatens to worsen the environment (Ning and Yang 2013).

Another, opposing, story line tells of heroic actions in desertification control, with victorious outcomes. The same Minqin has been named a "sand control and tree planting model" and a "tree planting and greenification model" by the Chinese national and Gansu provincial government. Research and news reports (e.g. Jin 2013; Ma and Zhang 2012) claim increasing tree cover in Minqin helped reverse the trend of desertification. Heroic individuals, such as Shi Shuzhu, Chen Yongming and Liu Shizeng, have been touted by the government as "sand control and tree planting" models. In the case of Shi Shuzhu, he has led his Songhe Village to plant over 10,000 mu (or 667 ha) of trees to control sand. "Shi Shuzhu relied on his belief and used his life to build up greenery on the desert, and build a beautiful homeland," said a newspaper report; and his village has been transformed from a "beggar village" to a "rich village" (Tang 2013).

These two conflicting stories from Minqin exemplify the challenges to achieving an accurate understanding of desertification in China. Is desertification expanding or retreating? What are the sources of these conflicting stories? More generally, what is the current trend in desertification, and what has China been doing to control land desertification? This article addresses broad issues implicated in the Minqin stories.

The first part of this chapter assesses scientific research on desertification in China, pointing out many problems and inconsistencies. The second part examines Chinese desertification control efforts—or measures to improve and restore desertified land. Here, I focus on policy discourses and how they trap desertification control into relying on human-dominated measures, especially planting trees through the "Great Green Wall" programme, resulting in wasted resources and continuing desertification. This paper concludes with a call to take down the "Great Green Wall," referring both literally to the tree-planting programme, and symbolically to policy discourses of human domination underlying the programme. Only by shifting to an attitude of respect for both nature and culture can we better understand and address desertification.

19.1 Part 1: Science of Desertification—Problems and Confusions

China faces serious desertification, or dryland degradation—all studies confirm this. But just how severely is China affected? Research from different sources shows a range from 4 to 27.9 % of China's land area as desertified in 2000. How does this huge discrepancy come about? Below I examine two major sources of desertification research—their definitions, academic/institutional lineages, and findings. Problems and confusions emerge, hampering a clear view of desertification in China.

19.1.1 Institute of Desert Research Under the Chinese Academy of Sciences

Prior to 2000, systematic research on desertification was conducted by the Lanzhou Institute of Desert Research under the Chinese Academy of Sciences. Publications from this research have mostly been led by Zhu Zhenda and Wang Tao (e.g. Wang and Zhao 2005; Wang 2008; Zhu 1985, 1989, 1997).

According to Zhu (1985), desertification refers to the conversion from non-desert to desert-like conditions on arid, semi-arid, and sub-humid areas due to a combination of human activities and natural factors. This includes primarily land changes involving sand movement driven by wind. As such, their conception of desertification is narrowly defined (Shi 1983). Later, China joined the United Nations Convention to Combat Desertification (UNCCD) in 1994, and the definitional discrepancy between Chinese studies and UNCCD became clear, as the latter includes all kinds of dryland degradation. To be consistent with the UN definition, Wang (2008) suggested that desertification should include aeolian (wind-driven) desertification, water erosion, and salinization.

In terms of wind-driven desertification, Zhu and colleagues used remote sensing (Thematic Mapper) images and air photos to study its change over time (Wang 2008). They concluded that from the 1950s to the 1970s, the annual (wind-driven) desertification rate was 1,560 km^2. From 1976 to 1988, the annual rate increased to 2,100 km^2. From 1988 to 2000, the annual rate further accelerated to 3,595 km^2. The total area suffering from wind-driven desertification reached 385,700 km^2 in 2000, or 4 % of China's land area. Zhu (1989) added two other types: desertification led by water erosion (394,000 km^2) and salinization and other physiochemical causes (88,000 km^2), and his calculation also included desertification in humid areas.

The Lanzhou Institute of Desert Research was the main institution studying desertification during the period from 1978 to 1999. The institute started in 1959 as a sand-control study team under the Chinese Academy of Sciences, and its main task was to control sand movement along the Baotou-Lanzhou railway passing through the Shapotou sandy area in northwest China's Ningxia region. In 1965, the "sand study unit" moved to Lanzhou, and combined with the "glacier and frozen soil" study unit to form the "Glacier, Frozen Soil, and Desert Study Institute" under the Chinese Academy of Sciences. From 1966 to 1976, the Cultural Revolution stymied research. Then, in 1977, the UNEP Conference on Desertification provided an impetus for the revival of research. The Lanzhou Institute of Desert Research was formed in 1978, with Zhu Zhenda serving as its director. This was when the institute started to study desertification. With its genesis in desert studies, the institute focused on wind-driven desertification (see Wang and Zhao 2005).

By 1999, the institute's focus was combined with cold area studies to form the new Cold and Arid Regions Environmental and Engineering Research Institute. By that time, the mandate of desertification studies had shifted to the State Forestry Administration (SFA).

As Zhu and colleagues were invited to help with sand control in humid areas in southern China, they added the category of "sandification" to their desertification research, referring to sand activation in humid areas. This category has led to much conceptual confusion, as will be discussed later.

19.1.2 The State Forestry Administration (SFA)

When a serious dust storm occurred in 1993, China started to pay more attention to desertification. Signing the UN Convention to Combat Desertification (UNCCD) in 1994, China formed a national committee for desertification control under the SFA (CCICCD 2006). This was when the SFA took over research and control of desertification in China (Wilkening 2006). The SFA completed the first national survey of desertification in 1994. Thereafter, the survey has been conducted once every five years. To align with the UNCCD, these surveys define desertification as "land degradation in arid, semiarid and dry sub-humid areas resulting from various factors including climatic variability and human activities" (Chen and Tang 2005). Conceptually, this definition resembles the expanded version of the definition used by the Lanzhou Institute of Desert Research.

The most recent fourth desertification survey by SFA shows that by 2009, China had 2,623,700 km^2 of desertified land, or 27.33 % of China's land area. Wind-driven desertification made up the majority, 69.82 % of all desertified land. The remaining drivers were identified as water erosion (9.73 %), salinization (6.59 %), and ice melting (13.86 %) (SFA 2011). Here, desertification was classified according to direct physical processes, following previous research from the Lanzhou Institute of Desert Research. Remote sensing images were used for the survey, which involved over 6,000 researchers. The survey concluded that the desertified land area decreased between 2004 and 2009 at the annual rate of 2,491 km^2, signifying a slow shrinkage of desertified land.

The national desertification survey also includes a separate category of land degradation—sandification, referring to land suffering from sandy material on the surface in dry as well as wet (humid) climates. Sandification occurs on natural deserts and sandy lands in humid areas, as well as through sand expansion on drylands. The fourth survey claimed that sandification affected 1,731,100 km^2, or 18.03 % of China's land area in 2009. Of sandified land, moving sand dunes accounted for 23.46 %; semi-fixed sand dunes 10.24 %; fixed sand dunes 16.06 %; exposed sandy land 5.76 %. Other types were sandified cropland (2.58 %), wind-eroded dunes (0.51 %), wind-eroded badland (3.22 %), and sandy desert (38.17 %) (SFA 2011). Table 19.1 summarizes the data from all four national surveys.

Table 19.1 National Desertification Survey results

Surveys	Desertification total area (km^2)	Desertification annual change (km^2)	Sandification total area (km^2)	Sandification annual change (km^2)
1st survey (1994)	2,622,000 (1996)	N/A	1,689,000 (1996)	2,460 (1994-1996)
2nd survey (1999)	2,672,000 (1999)	10,400 (1996–1999)	1,743,100 (1999)	3,436 (1996–1999)
3rd survey (2004)	2,636,200 (2004)	−7,585 (1999–2004)	1,739,700 (2004)	−1,283 (1999–2004)
4th survey (2009)	2,623,700 (2009)	−2,491 (2004–2009)	1,731,100 (2009)	−1,717 (2004–2009)

Sources SFA (2007) for the first three surveys, SFA (2011) for the 4th survey

The SFA surveys suffer from several major problems. In 2009, desertification occurred in 18 provinces/regions, while sandification occurred in 30 provinces/regions. Clearly sandification is not a subcategory of, thus is not entirely included in, desertification. It is not clear how much overlap they have. Since desertification includes Gobi or stony deserts and sandification includes sandy deserts (see SFA 2011), the SFA significantly exaggerated the areas affected. Why did the SFA exaggerate desertification in China? Another question is the reversal in the trend of desertification, which was first reported in the third SFA survey—how can we understand the rather drastic shift in desertification from previous expansion to reduction?

Science and politics are closely linked, and scientific endeavor is often constrained by political and institutional factors (Forsyth 2003; Forsyth and Walker 2008). Desertification study in China is no exception. Since the late 1990s, the SFA has been the institution in charge of not only surveys and research but also of policy and project implementation for desertification control. Seen from this institutional mandate, we may perceive the SFA's incentive to exaggerate desertification data.

Is China winning the battle against desertification? The SFA survey data show that since 1999, China has made a major shift in desertification control, reverting desertified and sandified land. The change from annual increase of 10,400 km^2 between 1996 and 1999 to an annual reduction of 7,585 km^2 from 1999 to 2004 was a major one. Are these numbers credible?

In 2001, China published the Sand Control Law that placed sand control within a policy framework. The SFA (2007) mentioned that the third survey conducted in 2004 "was the first after the Sand Control Law was implemented, and the survey needed to reflect the Law's requirements, as well as considering its continuity." This seems to indicate a connection between the survey and policy implementations. I therefore question the accuracy of SFA's reporting of desertification reversal after 2001, as it may suffer from a similar exaggeration, if in the opposite direction.

19.1.3 Confusion in Concepts and Data

Several major discrepancies and confusions plague desertification research in China.

(a) Conceptual inconsistency

Zhu and colleagues defined desertification as resulting from wind-driven sand movement on dryland. They used the percentage of sand cover (5, 20, and 50 %), sand dune heights, and vegetation cover (30 and 10 %) as criteria or thresholds to assess the severity of desertification (Zhu 1989). The SFA, on the other hand, defines desertification as occurring through wind erosion, water erosion, physical/chemical/biological changes in the soil, and vegetation loss. The severity of desertification is assessed through the addition of scores from two dozen measures in soil, vegetation, and land use/cover types—each according to different physical processes of desertification (SFA 2008). Conceptually Zhu's desertification and SFA's wind-driven desertification should be the same, yet the two assessments adopted different criteria and produced drastically different results. Zhu's desertification was 385,700 km^2 in 2000, while SFA's wind-driven desertification was 1,873,000 km^2 in 1999—a difference of nearly five times.

The definition of "sandification" is even more confusing. Zhu and colleagues defined "sandification" as sand dune movement outside of dryland areas (Wang 2008), which, according to UNCCD, has nothing to do with desertification. The SFA defines sandification based on the movement of sandy material in all types of climate, dry or wet; they classify sandification into moving sand, semi-fixed sandy, fixed sandy land, exposed sandy land, sandy cropland, etc. Vegetation coverage (threshold of 30, 10, and 5 %) is used to classify the severity of sandification.

Not only was sandification data from Zhu's and SFA sources incomparable due to classification differences, even within SFA surveys, the relationship between desertification and sandification is not clear, as was explained earlier. How, then, is sandification constituted in desertification study?

(b) Data comparability and validity

Not only are data from Zhu's and SFA sources incomparable, some have questioned comparability among the different SFA surveys because of methodological inconsistencies (Sun et al. 2012). For example, the first survey used site investigation; the second survey relied on sampled site investigation. The third survey used TM images in combination with site visits. Data generated from the different surveys have systematic differences, and thus cannot really be compared (ibid).

Scholars not only cite the SFA survey data assuming there is full comparability, they also unknowingly compare Zhu's and SFA data in one fell swoop. The well-known scholar of desertification, Wang Tao (see Wang 2008), referred to Zhu's research on the desertification expansion rate from the 1950s to 2000, then stated: "but in recent years, through the joint effort of the government and local people, desertification clearly was reversed, and was reduced at the rate of 1,280 km^2 annually." The reduction rate quoted is reported by SFA for sandification, instead of desertification, in the third SFA survey. This example shows the level of confusion in desertification studies in China.

If desertification denotes changes in the ecosystem on dryland that experiences significant climatic fluctuations, then the SFA's assessment raises further questions. To detect trend change and filter out fluctuations, we need data that spans at least a decade or longer. The SFA's surveys are based on five years, and data can only span three years as the SFA allows the use of data that are two years old. Is this type of study duration meaningful to judge whether desertification has been reversed or not? At best, the SFA's trend assessments are only suggestive.

19.1.4 Causes and Trends in Desertification

Zhu and colleagues from the Lanzhou Institute of Desert Research cited disharmony in human-environment relationship as the cause of desertification (Wang and Zhao 2005). They identified primary drivers and calculated the percentage of desertified land according to each driver in 2000: Over-cultivation in farming accounted for 25.4 %, over-grazing 28.3 %, over-cutting of firewood 31.8 %, water resource misuse and industry/mining-induced destruction 9 %. Natural factors alone driven by wind accounted for only 5.5 %.

The SFA surveys followed the above classification for human drivers of desertification (SFA 2007, 2008, 2011). The Sand Control Law (SFA 2001) is written specifically for sand activation caused "mainly by human irrational activities." In other words, human activities have been identified as the main driver of desertification in China.

These direct "human activities" do not constitute the root causes of environmental change, however. Turner et al. (1990) calls them proximate drivers, which are driven by "underlying causes" such as population increase, technological change, and government policies. Desertification studies in China have often focused on direct drivers, not the underlying ones. There are some exceptions. Many researchers address population growth (see Chen and Tang 2005). Zhu (1989, 1997) pointed to policy errors, such as the "grain first" policy, which, by promoting the conversion of grassland into cropland for grain production, led to grassland degradation. Overall, however, policy critiques have been relatively weak in China's desertification literature.

Some regional studies have teased out the changing role of human activities. For Minqin, the region where the opening stories took place, one study attributed 56.33 % of desertification in 1956–1981 and 66.19 % in 1981–2004 to human factors (Ma et al. 2007). Other studies highlight the role of natural factors (Wang et al. 2007). A study on the Heihe river basin adjacent to Minqin concludes that human activities are the main cause of desertification, while natural factors (climate) are the main reason for desertification reversal (Zhou et al. 2013).

As for trends in desertification, other than the Lanzhou Institute and the SFA, studies covering large areas in China have been scant. According to Shan (2009), a study using a greenness index to measure the total amount of vegetation showed an overall reduction in greenness in northern China between 1981 and 2000. Regional variation existed. The hyper-dry areas remained unchanged; 14 % of the area, mostly in northeast China, saw increased greenness, while 21.36 % of the area, in north and northwest China, saw decreased greenness. This study is consistent with the study of spatial variation by Zhu's colleagues (Wang et al. 2004).

Since 2000, many local assessments follow the story line of "desertification reversal" established by the SFA. "Sand has retreated and people are advancing," as some reports claim (ScienceNet 2013; Ai and Li 2013). Most newspaper articles follow this line, which is understandable given the political control of media in China. Still others point to continuing desertification, with the improved and controlled area exceeded by the newly desertified area (Meng 2011; Zhang et al. 2007a, b). Challenges in assessing the complex process of desertification also play a role (Yang et al. 2005).

In this section, I confined my discussion to scientific assessments of overall desertification in China, and did not address research on biophysical processes of desertification. On that, readers may refer to Chen and Tang (2005), which deems such studies to be inconsistent and short-termed, thus insufficient to guide policies. Field monitoring data come only from several research stations under the Chinese Academy of Sciences, including the Shapotou Experimental Station, Naiman Field Station, and Ordos Sandyland Ecological Station—not enough to uncover the biophysical processes of desertification.

19.2 Part 2: Policy Discourse on Desertification Control

By "desertification control" I refer to human efforts to improve already desertified land. Although Chinese scientists and official documents have pointed out the importance of desertification prevention, China is still spending more resources on reversal and improvement of desertified land, and less on desertification prevention. I will first discuss the general policy discourses in desertification control and then highlight the key anti-desertification programme, the "Great Green Wall," pointing out its fallacies.

19.2.1 Domination Over Nature—Ecological Construction to Guide Desertification Control

Jiang (2006) points out major problems in China's desertification control: lack of attention to protection, too much focus on tree planting, and relying too heavily on human intervention instead of natural recovery. These comments get to the heart of the problems.

Since the 1980s, environmental issues have received attention at various levels of the Chinese government. More corrective programmes have been set up to improve desertified land. While this may be a positive sign, the guiding principle behind these programmes only serves to undermine their effectiveness. The Chinese state policy has championed the rebuilding of ecosystems, under the concept of "ecological construction." What does "ecological construction" mean? Jiang (2010) offers a detailed discussion. To summarize briefly here, ecological construction is based on human control and human efforts, and in practice disregards native environments or species. In other words, the attitude of human domination, perfected by the Communist Party during the Mao era, has continued in some ways under the disguise of environmental improvement.

In around 2006–2007, the term "ecological civilization" became more common, used in conjunction with "ecological construction." This was the period when alarming environmental problems in China drew global attention. Cases such as Lake Tai algae pollution, potential problems with the Three Gorges Dam, increasing sand storms, and expansion of deserts all made international headlines, and China also rose to become the largest emitter of carbon dioxide in 2007. At this point the Chinese government geared up its rhetoric on environmental protection. In the report at the Party's 18th Congress in 2012, the phrase "ecological civilization" was used 15 times.

In desertification control projects, however, the new concept "ecological civilization" has yet to bring about a shift in the human-dominated relationship with the environment. Often, policy discourse has to be placed in the political context in order to understand how the same programme (such as the "Great Green Wall") can be used to serve different discourses (see Table 19.2).

As desertification studies target direct human activities as the main culprit, local people who carry out cropping, grazing, fuelwood collection and other activities have often taken the blame, even though they are the direct victims of desertification. In programmes to control desertification, the state, with its different levels of the government, has been the manager (Chen and Tang 2005). "Desertification control has mainly relied on the government, experts, and other social elites for decision-making, while the local residents are often passively implementing the policies" (Zhang et al. 2007a, b). China's "Sand Control Law" did not give local communities the right to decide on control measures (Wang 2006). Lack of local decision-making and community-based management has led to many harmful decisions and wasteful efforts. Song Jun from the Alashan Ecological Association believes that the Chinese government's environmental protection has rarely been effective and that the main reason is its top-down policy process, with the government as both the rule setter and project executer (Zhou 2006).

To be sure, not all local people are neglected in desertification control. Since the late 1990s, the Chinese government has identified tree planting heroes and heroines —actively promoted their stories nationally to encourage desertification control. They include, for example, Yin Yuzhen, Wang Youde, and Niu Yuqin.

Table 19.2 Desertification control discourses over time

Time period	Desertification control measures/projects	Policy discourses	Political context
1950s–1978s	Fix moving sand along railway lines, plant shrubs, grass and trees to build oasis	Desert reform: Build socialism, transform heaven and earth	China in isolation and Maoist domination over nature
1978s–1990s	"Great Green Wall" afforestation programme, distribution of desertified land for private improvement, sand control	Ecological construction: Revert land degradation and rebuild ecosystems in order to assist economic development	China's drive for rapid economic development—need to strengthen environment as economic base, and to defend ecologically damaging economic policies, and address poverty in drylands
2000–present	"Great Green Wall," conversion of cropland to forest, sand control in Beijing and Tianjin, Sand Control Law, relocate people off desertified land	Ecological civilization and construction: Rebuild ecosystems, sustainable development, carbon sequestration	Economic disparity and need to develop western regions; China's rising global power and becoming the largest CO_2 polluter; global attention to China's environmental issues threatening the CCP's image

The personal stories of these individuals tend to be portrayed in similar ways: Through hard work, they overcame harsh environments; they planted many trees and improved the environment and their livelihood. The story line recounts a typical linear trajectory from poverty to richness, from limitation to liberty, from self to selflessness, during the Mao era. As inspiring as these stories are, it is the vast barrenness surrounding their green forest plots that has supported their tree planting. Dryland ecology dictates that only a small portion of land can succeed in afforestation, given limited water resources (Jiang 2002). These heroic stories can hardly be replicated across the landscape.

Northwest China is home to several major ethnic groups, notably the Mongols, Uyghurs, and Kazakhs. For these nomadic peoples, animal grazing has been their long-held cultural tradition. Government policies have neglected ethnic cultures, exacerbating desertification. In Inner Mongolia, for example, grassland was distributed to households and fenced behind enclosures as part of policy reform (Jiang 2002; Zhou et al. 2007). Grazing mobility was no longer possible and over-grazing became rampant. Liu (2003) believes the grassland contract system and subsequent disappearance of nomadic grazing are the key reasons for grassland degradation. The Mongol's respect for nature (as in shamanism) is dismissed, leading to a "tragedy of the private land" (Zhang 2008).

Desertification control often ran counter to ethnic lifestyles and values. Assessment of desertification control in China is framed by "ecological, economic, and social" considerations, leaving out cultural assessment. Cultural preferences and their connection with the land have been neglected. Williams (2001) points out that the Mongols do not vilify sand, but instead they value sand dunes for providing shelters for the livestock; this view defies the attitude in sand control that sees sand as an enemy or even a "demon" to combat (see Jin 2013). In recent years, "ecological migration," moving people out of seriously desertified areas, has been championed as the major fix for desertification. Ecologist Liu Shurui believes that this practice disregards pastoral practice and has ecological consequences. "Reducing livestock numbers is good, but totally removing them is bad for the pasture." Liu said that without livestock manure as fertilizer, the pasture will be left with only a few species, reducing biodiversity (Zhou 2006). Togochog (2005) from the Southern Mongolian Network considers it a rights violation when pastoral Mongols are moved off the grassland.

To summarize, desertification control in China suffers from three major problems or imbalances: too much focus on human control and disregard for the natural environment; too much focus on elites and government and not on the local people; too much focus on intensive land use and not on pastoral cultural practices. To address these problems in desertification control requires a new way of looking at dryland and the human relationship with it.

19.2.2 An Institutional Trap—the State Forest Administration and Over-Management of Drylands

Why is the SFA, a government agency in charge of forest and trees, designated to lead China's desertification control? Desertification is a land use issue on drylands where forest usually does not grow naturally and it would logically fall under the land use, pastureland, or agricultural administrative branches, which can more easily coordinate prevention before the land is desertified and control after the land is desertified. To understand the role of SFA, we have to go back to 1978, when the North China Afforestation (otherwise known as the "Great Green Wall") programme started.

Prior to 1978, desertification control took two paths. The first was in "campaign" form, such as the "desert transformation" in Inner Mongolia from 1958 to 1965, when shrub and tree planting was promoted on the drylands. The second path revolved around sand control along railways, especially in the Shapotou area along the Baotou-Lanzhou railway (Huang and Wei 2009). Studies prescribed various methods for sand control, including laying straw grids on moving sand, and planting indigenous shrubs such as *Hedysarum scoparium, Caragana internedia*, and *Artemisia ordosica* to fix the sandy land (Wang et al. 2005; Huang and Wei 2009).

In the post-Mao era, to improve the degraded environment in northern China, the Chinese government launched an afforestation programme in 1978. The

Three-North Shelterbelt programme, later dubbed the "Great Green Wall," was to build shelterbelts in the northeast, north and northwest (thus "Three-North") by planting 35.6 million hectares of trees to protect the cropland and grassland from wind and water erosion, increasing forest cover from 5 to 15 % (SFA 2008). The programme covers 42.4 % of China's landmass, and is the largest environmental programme ever launched in China. The State Forestry Bureau was in charge of the programme. The next year, the Agriculture and Forestry Ministry was dismissed, and the State Forest Ministry was established to take over the management of this afforestation programme. In 1997, the ministry was renamed the State Forestry Administration, directly under the State Department.

Given the 73-year plan of the programme, sustained attention in the official media, and funding and policy mandates involving all levels of the government, the "Great Green Wall" programme has become a focal point of ecological improvement. It has also created a discourse of dryland improvement based on tree planting. On the Baidu encyclopedia website (the Chinese equivalent of Wikipedia), the entry on the afforestation programme (called *san bei fang hu lin* in Chinese) pours on lavish praise for its grand-scale planning and achievements. The Baidu article explains that northern China used to be covered by dense forest and lush grassland, and to restore the now desertified and eroded area, a "Great Green Wall" of trees must be built—and is being built.

By the time desertification got on to the government agenda via UNCCD in 1994, the SFA was already deeply engaged in dryland environmental rehabilitation. In 1997, the SFA was elevated in status and funding increased. Between 1998 and 2009, annual financing increased from 13.53 billion yuan (US$1.6 million) to 137.79 billion yuan (US$20.2 million) (Xiao et al. 2010). Now, the SFA is in charge of rehabilitation of most ecosystems in China, from forest to grassland, from wetland to desert.

Since 1997, the SFA has established several institutions on desertification, including the National Bureau to Combat Desertification, the National Desertification Monitoring Center, a National Training Center for Combating Desertification, and a National Research and Development Center for Combating Desertification—all to conduct research and implement policy programmes on desertification in China (Wilkening 2006). In June of 2009, the Institute of Desertification was established by the Chinese Forest Academy, which is under the SFA as well (CCICCD 2006).

The SFA has administered various desertification control programmes, including the "Great Green Wall" programme, the sandification control programme in the vicinity of Beijing and Tianjin, the conversion of cropland into forests (or grassland) programmes, etc. Most of these programmes focus on the planting of trees. China's national plan for sand control also indicates that "planting trees and building forests" are the main methods to be used (Shi 2013; also see Wilkening 2006). In a way, the "Great Green Wall" programme has set the tone for desertification control in China.

On paper, both government planning documents and research publications have pointed out the need to diversify control methods, moving away from the singular tree planting focus. The national sand control plan (2011–2020) states we have to "select methods according to the locale, choose prevention approaches based on the problem, adopt different methods, plant trees and grass in suitable conditions, and leave the land to nature when that is called for" (Shi 2013). Still, the SFA manual on "Great Green Wall" remains focused on trees (SFA 2009). The problem is not the law—the problem lies in the implementation, which is governed by power and interest. Desertification control is trapped in the institutional interest of the SFA, resulting in a narrow method that privileges tree panting—an ineffective approach that in turn helps trap the land in a desertified state, as discussed below.

Other national and global factors have helped strengthen the SFA's role in tree planting. The rampant sand storms in eastern China that began in the 1990s drew national and international attention to desertification and its impact, and government media responded by trumpeting tree planting (e.g. Jin 2013; Chen 2013). Global climate change and carbon sequestration also favor the promotion of tree planting (Gong et al. 2009). The Chinese government continues to champion afforestation as an ideal way to address not only desertification, but also a wide range of ecological and economic issues on drylands.

The SFA is not the only institution that manages the land through a singular approach. With different agencies managing various aspects of the land, institutional mandates have fostered compartmentalized views about the environment. Ecologist Jiang Gaoming described multi-agency land management this way (Jiang 2005): "In our research, we found that on the same piece of pastureland, the SFA dug holes to plant trees, the grassland department air-sowed grass seed, the hydrological department dug a well, and all were happy to conduct their respective monitoring. But all used public funds—tax money from the people. If people did not fix the pastureland, grasses and shrubs would naturally grow on it, and it would need no human improvement." It seems that desertified land has been "over-managed"—not only has overuse led to desertification, over-control for restoration has only exacerbated desertification, as will be discussed later. Institutional mandates and government approaches to the environment threaten to trap the land in a desertified condition.

19.2.3 The "Great Green Wall" to Control Desertification—What's Science Got to Do with It?

While the Three-North Shelterbelt, or the "Great Green Wall" programme, was launched before desertification control programmes officially started, its original intent and later national policy planning have both made it the most sustained desertification control programme in China. Scheduled to last for 73 years ending in 2050, the programme aims to control land degradation and desertification. A swath 4,480 km long and 560–1,460 km wide is included in the programme, affecting a population of several hundred million. Much of programme area, except for the eastern part, is dryland where forest is not the principal natural vegetation. This section examines the programme's ecological problems, as well as the role of politics in keeping it going.

Scientific Studies of Dryland Tree Planting: Scientific research has warned us about the ecological problems associated with afforestation in areas not previously forested for generations. For example, the grass-to-forest transition can lead to the loss of biodiversity (Ginsberg 2006; Hawlena and Bouskila 2006; Oxbrough et al. 2006), decline in water recharge (Van Der Salm et al. 2006), and loss in soil moisture and quality (Farley 2007; Nosetto et al. 2005). Farley et al. (2005) conducted a global study of 26 dryland sites to show that afforestation results in a 30–100 % reduction in annual water runoff, thus causing or intensifying water shortage.

Tree planting on drylands has also been questioned in China. As early as the 1980s, Huang (1981, 1982) cautioned that the function of forests as a means for conserving water was exaggerated, and since then empirical studies in northern China have pointed out the problems of soil desiccation and groundwater depletion caused by aggressive land improvement projects and tree planting (Chen et al. 2005; Jiang 2002; Yang et al. 2004). Cao et al. (2010) point out that dryland precipitation cannot support trees on a large scale and afforestation has exacerbated environmental degradation. Single species forestry adds to the problems. For example, in Ningxia, 70 % of the trees planted are poplar and willow trees. In 2000, one billion poplar trees were lost to a disease (*Anoplophora*), wiping out 20 years of planting effort (Yu et al. 2008). Single species forests help increase government figures for trees planted, but they fail to improve the environment; thus they have been called "green deserts" (Jiang 2008a).

Afforestation is costly, and tree survival rates are low. Of all the trees planted on China's drylands since 1949, only 15 % have survived (Cao 2008). Persistence in the tree planting programme has meant high costs, including the future generation's precious water resources (Jiang 2005). In Minqin, the location of the stories that opened this chapter, due to worsened water condition, tree planting costs have increased over tenfold since the 1980s and many planted trees have died (Fan et al. 2005). In general, it costs 2,250–15,000 yuan ($368–$2,450)/ha to plant trees. The central government only put in a small amount of the cost, about 5 % in the 1990s (Weng 2000). The rest came from the local government and local people.

I should point out that in areas with annual precipitation over 380–400 mm, natural forest exists and tree planting is more suitable. This is why China's "Great Green Wall" project has seen better results in the eastern part of north China (Shan 2009). In areas with less than 380 mm of annual rainfall, tree planting can achieve limited success if groundwater is shallow and replenished from runoff. Otherwise, trees are only marginally suitable and have to be selected carefully (Chen and Cai 2003). Even as trees increase, the ecosystem may continue to worsen. In Hexi corridor of which Minqin is a part, during 1990–2007, trees increased (Pang et al. 2012), but desertification continued (Zhang et al. 2007a, b; Fan et al. 2005).

Dryland tree planting has created astounding failures in the world. The most famous has to be Stalin's "Great Plan for Transformation of Nature" in 1948, in which over 30,000 km^2 of trees were planted on the steppe; by the 1960s only 2 % remained (Jiang 2006). In North Africa, tree planting programmes such as the "Green Dam" and "Green Wall" have only yielded limited success. In 1971, Algeria planned to plant 3 million hectare of trees in a belt 20 km wide and 1500 km long. By 2003 only 100,000 ha had been planted, mainly single species stands of *Pinus halepensis* (Belaaz 2003).

Why and How Does the Chinese State Continue to Promote Trees? Despite known ecological problems with afforestation, the Chinese government continues to push ahead with the "Great Green Wall" programme. Shrubs and grass have been included in the programme description, but they have played only minor roles. By 2012, the programme had reportedly increased tree cover from 5 to 12.4 % in the programme area, with the cumulative tree planting area reaching 26.47 million hectare (gov.cn website 2013). Currently, the programme has entered its fifth stage (2011–2020), which aims to plant 9.884 million hectare of forest, with funding of more than 90 billion yuan (US$14.6 billion) (Yan et al. 2012). This is a major increase in funding compared with 4 billion total during the initial 28 years of the programme (Du 2006). Over time, the afforestation programme has become increasingly significant as official rhetoric attaches more meaning to it, such as ecological security, sustainable development (Qi et al. 2009) and carbon sequestration (Gong et al. 2009). Countless media and scientific reports in China have valorized the "Great Green Wall" programme for its ecological, social, and economic achievements. Many do point out problems in the programme, but more as comments supporting its improvement. Criticisms that challenge the programme's very viability, albeit sharp, have been far and few between (see e.g., Jiang 2006, 2007, 2008a; Wang 2007; Weng 2000).

To be fair, the state's glorification of the "Great Green Wall" is partly justified by localized successful stories of tree planting in key project areas and experimental sites, which account for less than 10 % of the area in northern China (Liu and Cao 2005). For example, the Ordos region of Inner Mongolia, a key project site, is reported to have increased vegetation cover from 30 % to over 75 % from 2000 to 2008 through the planting of trees and shrubs (Li et al. 2008). Trees do add up, although they do not necessarily equal improvement of desertified land. Increased tree cover in 10 % of the project area can hardly attest to the programme's success, nor can it explain why many Chinese scientists, journalists, and the public have participated in perpetuating the state-sponsored "Great Green Wall" programme, drowning out fundamental criticisms.

"Forest cover is like GDP—it is a measure of political achievement," commented Meng (2011), even though tree planting often does not serve the livelihood needs of the local people (Zhang et al. 2007a, b). A party head from Shanxi Province once proclaimed, "planting trees is to plant humanity, plant history, plant political achievements" (Geng 2012) and this statement was quoted in praise by Central Chinese TV. One can see how entrenched the rhetoric of tree planting has become. Not only does the SFA have staff at each level of the government,

government leaders also see tree cover as a measure of political performance. Over-reporting of tree cover is common (see Jiang 2006; Meng 2011). In Alashan Right Banner at the edge of Badain Jaran Desert, tree cover reportedly increased by 5.6 times between 2000 and 2009. Added to the reported figure was the area with shrubs. The SFA's forest resource technical guide specified that for drylands with less than 400 mm annual precipitation, protective shrubs with cover of over 30 %, as well as economic shrubs, can be treated as forests (SFA 2009; Meng 2011). No wonder the number of reported trees added up quickly.

Following the predominantly positive reports about the "Great Green Wall" by the Chinese official media and the SFA, a wide range of publications lauded the programme's achievements and the hardworking spirit of tree planting heroes (Liu et al. 2009). The online Baidu encyclopedia states that in scale and achievements, China's "Great Green Wall" programme trumps Stalin's nature transformation programme, Roosevelt's shelterbelt programme, and northern Africa's "Green Dam" project (for academic writing championing the same, see Jiang et al. 2013). Al Gore is reported to have commented that China plants 2.5 times more trees than the total number of trees planted in the rest of the world each year (Moxley 2010). Quoting from the SFA, the UN book on "Desertification" cited the "Great Green Wall" as one of the best examples of desertification control in the world (UNCCD 2011, 16).

The short attention span of politics helps perpetuate dryland tree planting. When investment and climatic conditions are favorable, planted trees can survive on dryland for an initial number of years. But over time, some of these trees shrivel under harsh environmental conditions and eventually wither away. For example, in the 1970s, "the initially successful afforestation that was used to restore the environment of the Mu Us Sandland in northern Shanxi Province was used as a model for the rest of China, but 30 years later, most of the planted trees had died" (Cao et al. 2010). Thirty years later, however, the political leaders responsible for the programme have long reaped their rewards; problems of nature (climate or disease) can always be blamed for the death of trees by the current leadership. The fundamental problems of the afforestation programme are not addressed.

Voices from scientists and within the SFA—If the "Great Green Wall" project largely violates ecological principles, why have scientists not risen up against the programme? One needs only to turn to the media control and internet censorship Chinese people have to face today to understand the risks people may perceive when they criticize government policies. The effect of the silencing of scholars during Mao's anti-rightest movement in 1957 can still be felt in China, as generations of scholars have learned to protect themselves through self-censorship, abandoning the age-old Confucian tradition of scholarly responsibility for the nation.

Institutionalization of science and scientists plays a role. Many researchers from the SFA have helped defend the positive effects of tree planting (e.g. Zhang et al. 2007a, b). Those questioning the programme are represented by ecologists and geographers. To highlight a few dissenters, ecologists Jiang Gaoming and Cao Shixiong have both voiced fundamental concerns about the afforestation programme. In a blog article titled "Speak the truth to authorities," Jiang (2007) told the story of his being excluded from academic meetings sponsored by the SFA after criticizing the tree planting programme. "Grassland does not have trees, and this is natural law," Jiang commented, "Ecologists often take this position firmly in private discussions. But strangely, in front of reporters, especially facing the camera of the Chinese Central Television, their voices suddenly become weak. No one wants to speak the truth and offend the SFA" (Jiang 2008b). Cao, a forest ecologist, has published in English several summaries questioning the effect of dryland afforestation (Cao 2008). Huang Bingwei, seen as a founding father of modern physical geography in China, stated in the 1980s: "In dry areas, planted trees cannot survive, and water is being wasted. Sometimes the more you plant the more destructive it is." But he later lamented, after his articles were attacked: "[I can] at best write a few small articles; there is no place to publish them, and no one would listen" (Ge 2005). Zheng Du, another physical geographer, cautions that tree planting on drylands does not fit with geographic zones (Zheng 2006).

Most scientists, however, offer "self-disciplined" criticisms that only help advance the state's tree planting projects. They point out problems such as a lack of funding, the need for better maintenance, and the need for more diversification of tree species (e.g. SDRC 2010). These suggestions serve to prolong the fundamentally flawed afforestation programme with more investment. Others direct their criticism at the local people and local leaders, citing their greed for money as the reason why the problematic tree planting programme has been perpetuated (e.g., Ge 2005). These publications have yet to explore deeper into the deep rooted problems of state policies and institutional politics.

There are plenty of examples of scientists' warm support for the programme. I quote just one enthusiastic remark: "The tenth five-year plan, the fourth phase of the Three North Project … is driven by the bringing forth of new ideas in ecological systems and science, and it also gives a full return to investment and guarantees the strong development of forestry with the help of laws and science. Over these five years, the Three North Project has used 4,147 million yuan (US$545.6 million) of investment, created 2,840 million yuan (US$373.7 million) of ecological benefits, and 8,060 million yuan (US$1,060.5 million) of economic benefits. And it also lays a solid foundation for building a harmonious society and promoting the construction of a socialist new countryside" (Zhang et al. 2007a, b).

As a government agency, the SFA is responsible for creating much of the discourse supporting the "Great Green Wall" programme, which is echoed by many scientists, journalists and the general public. Yet within the SFA, the lower-level staff, those working closely on the land, often question the tree planting programme, but only privately. Wang Man, Forest Bureau head in in Zhangbei of Hebei province, said, "Planted trees absolutely cannot be guaranteed to survive; they must be replanted in 3–4 years. At the maximum 2 in 5 trees can survive the first year. Continuous replanting requires more input" (Weng 2000). Gao Yuchuan, Forest Bureau head, Jingbian County, Shaanxi, commented, "Planting trees for 10 years is not as good as enclosing [the grassland] for one year" (Ding 2006, 40).

To summarize, the "Great Green Wall" tree planting programme, China's largest and longest-lasting effort to fix desertification, represents an astounding case of an ecological mismatch. Plenty of scientific studies, both in China and abroad, have uncovered the ecological problems of dryland tree planting. But the programme continues to be pushed along due to a policy discourse of human control over the environment and institutional politics involving the SFA. The "Great Green Wall" project exemplifies the fundamental problems of China's desertification control programmes.

19.3 Conclusion—Taking Down the "Great Green Wall"?

In this article I have examined the state of scientific studies of desertification in China, and demonstrated how policy discourses in desertification control have disregarded the dryland environment and displayed a tenacious tendency to remake the ecosystem and to report on positive results. I started with two opposing stories from Minqin, the first of an alarming situation of continued desertification, and second of the successful control of desertification through heroic human effort. Let me turn back to each story and put it in the context of what we have learned in this chapter.

The first story shows that China faces serious challenges from desertification. Even as the wetter part of the drylands has seen reductions in desertification, even as model communities have successfully beaten back the sand, desertification remains a serious problem. Owing to inconsistency in concepts and discrepancies in research methods, however, much confusion still exists, preventing us from charting out a clear trajectory of change for desertification in China. While past research by the Lanzhou Institute of Desert Research and ongoing surveys by the SFA can be used as important references, we need more and unbiased studies on desertification in China.

If the first Minqin story represents alarming desertification, the second story demonstrates the official approach to desertification control—relying on tree planting, human effort, and positive reporting. Localized successes are often emphasized, at the risk of obscuring the overall reality. Tree planting has been bound up together with desertification control even in areas as dry as Minqin, testifying to the paucity of control measures. The State Forestry Administration, the government agency in charge of desertification control, has continued to champion tree planting as a key approach to stop the rolling sands on drylands where trees do not naturally grow, and continued to promote artificial improvement in ecosystems where researchers have found natural recovery much more effective than aggressive planting. The top-down approach represented by the domination of the SFA in environmental programmes traps desertification control in a situation more severe than that of the "emperor's new clothes," where an innocent voice of truth could wake up the nation.

If there is one message this chapter can offer for China's desertification control, it is: Take down the "Great Green Wall" afforestation programme. This suggestion has several layers of meaning. First, the "Great Green Wall" project has not worked well for most of the drylands, where shrubs, grass, or bare sand are the native and natural land covers, instead of trees. Second, for China to stop the afforestation programme, the institutional approach to ecological rehabilitation has to be replaced by a more comprehensive one, and science has to play a more important role in guiding policies. Third—and perhaps more important as it underlies the first and second—is to change the attitude of human control and replace it with a genuine respect for nature and culture. Let me explain several key issues related to this suggestion of a "paradigm" shift.

If the "Great Green Wall" programme is to be taken down, what measures can be used to replace tree planting? Scientific research has demonstrated that natural recovery is much more effective in restoring degraded drylands (Jiang 2006). Other methods, proven to be successful, include laying straw grids on sand dunes, and planting native shrubs and grasses in areas that are not too dry. Some of the experiments on specialized industries for drylands—fruit trees and herbal medicine—may be expanded if conditions are suitable. When moving people off the pastureland, both the grassland-grazing symbiotic relationship and the cultural preferences of pastoral peoples must be considered, and the government should stop imposing on people the values of urban and agricultural life from the outside. Cultural survival and economic and environmental survival often go hand in hand.

Trees may be planted, but only at a small scale, more for lifestyle than for livelihood. A new generation of tree-planting heroes have started to chart this path. Zhang Yinglong and Guo Zhihui from Shenmu county, Shaanxi province are both educated and well-connected. Both made money in other sectors and used their savings to plant trees. Zhang Yinglong articulated a connection with nature that goes beyond livelihood dependence to a spiritual connection (Yang 2007). He established an Ecological Association and mobilized people to join him in planting trees on his contracted sandy land of 430,000 mu (or 28,667 ha). The spiritual and educational value of his effort sets Zhang apart from other tree-planting heroes, some of whom have found themselves strapped when they rely on the planted trees to make a living. This new type of tree planting applies to a small number of people who have financial resources. This also fits with the reality that drylands can only support trees in limited locations.

State control through top-down governance, as represented by the "Great Green Wall" programme, also has to change if desertification control is to be effective. Liu (2005a) suggests that ecological construction should not be solely controlled by the government. Liu (2005b) proposes that the government invest directly in individual people, not through the top-down command chain of government whereby funds are eaten up at each level on the way down, and funding allocations cannot be guaranteed. Sand control costs will be much less. Individual "capable persons" (such as Yin Yuzhen and Wang Minghai) spent less than 1/5 of the funds for sand control compared to that spent by the government. Cutting out the government chain also means releasing desertification control from the SFA trap.

Ultimately, taking down the "Great Green Wall" means replacing an aggressive attitude toward nature with a non-aggressive approach, fostering a respect for nature and culture and a harmonious relationship between the two. Doing so requires learning from nature and from local people. A political system that does not care about what people want and what nature needs is ineffective at best and brutal at worst. In no way am I saying attitude change is easy or desertification control is not challenging; political change is even more complicated. Yet if the current trends continue, the Chinese government will continue to undermine the very environment it claims to improve. The ills of human over-control are widespread around the modern world, yet not only can science offer better ways to address desertification, the Chinese cultural tradition also provides us with the resources to nurture responsibility and care (through Confucianism) and non-aggressive actions (through Daoism). Instead of "controlling nature," we need to "follow nature" (Li 2004, 7), treat the land in holistic ways, and "return magic to nature" (Zhang 2008). This includes, ultimately, allowing sand the freedom to roll: In certain locations it is the moving sands that, by allowing rainfall to infiltrate into the ground to replenish the groundwater, nurture the land of green around them.

Acknowledgements I wish to thank the book editor, Roy Behnke, and contributor Lynn Hunsinger, for their encouragements, comments and editing of this chapter. All errors remain my own.

References

Ai, F., & Li, Z. (2013). *China builds "Green Wall" with hard work, so "sand is in retreat and people are advancing"*. Xinhua Net (Ningxia), August 19. http://www.nxnews.net/wj/system/2013/09/03/010941528.shtml Accessed September 27, 2014. (in Chinese).

Belaaz, M. (2003). Le barrage vert en tant que patrimoine naturel national et moyen de lutte contre la désertification. In *Proceedings of the XII World Forestry Congress, Quebec, Canada*, September 21–28, 2003.

Cao, S. (2008). Why large-scale afforestation efforts in China have failed to solve the desertification problem. *Environmental Science & Technology*, 1826–1831.

Cao, S., Tian, T., Chen, L., Dong, X., Yu, X., & Wang, G. (2010). Damage caused to the environment by reforestation policies in arid and semi-arid areas of China. *Ambio, 39*, 279–283.

CCICCD (China National Committee for the Implementation of the UNCCD). (2006). China national report on the implementation of the United Nation Convention to Combat Desertification. Beijing. (in Chinese).

Chen, S. (2013). Reporters witness million-tree planting plan, as volunteers planted 2,000 trees on the grassland. *Youth Daily*, May 7. (in Chinese).

Chen, Y., & Cai, Q. (2003). Current condition, reasons and strategy of desertification in Hunshandake Sandy Land north of Beijing. *Progress in Geography, 22*(4), 353–359. (in Chinese).

Chen, Y., & Tang, H. (2005). Desertification in North China: background, anthropogenic impacts and failures in combating it. *Land Degradation & Development, 16*, 367–376.

Chen, H., Wang, K., & Shao, M. (2005). A review on the effect of vegetation rehabilitation on the deep soil layer on the Loess Plateau. *Scientia Silvae Sinicae*, *41*(4), 155–161. (in Chinese).

Ding, M. (2006). Planting poplar and willow trees cannot stop Three-North from deteriorating. *Liaowang News Weekly*, 39–41, August 7. (in Chinese).
Du, J. (2006). Using 28 years and 4 billion yuan to build Three-North Shelterbelt to benefit the society. *People's Daily*, June 18. (in Chinese).
Fan, S., Ma, Y., & Zhou, L. (2005). Farmers' response to ecological control strategies in recent years in Minqin oasis, Gansu. *Journal of Desert Research, 25*(3), 397–403. (in Chinese).
Farley, K. A. (2007). Grasslands to tree plantations: Forest transition in the Andes of Ecuador. *Annals of the Association of American Geographers, 97*(4), 755–771.
Farley, K. A., Jackson, R. B., & Jobbagy, E. G. (2005). Effects of afforestation on water yield: A global synthesis with implications for policy. *Global Change Biology, 11*(10), 1565–1576.
Forsyth, T. (2003). *Critical political ecology: The politics of environmental science*. London and New York: Routledge.
Forsyth, T., & Walker, A. (2008). *Forest guardian, forest destroyers: The politics of environmental knowledge in northern Thailand*. Seattle: University of Washington Press.
Ge, J. (2005). Why invest tens of billions to the "ineffective" Three-North Shelterbelt? *Oriental Morning News*, July 28. (in Chinese).
Geng, G. (2012). Green monument erected on China's land. *Green China TV*, October 9, 2012. http://www.greenchina.tv/news-4052.xhtml. Accessed September 27, 2014. (in Chinese).
Ginsberg, P. (2006). Restoring biodiversity to pine afforestations in Israel. *Journal for Nature Conservation, 14*(3–4), 207–216.
Gong, W., Li, J., He, Y., Ding, F., Yao, Y., Zhang, H., & Sun, Y. (2009). Develop forestry for carbon sequestration, promote the construction of North China Shelterbelt. *Chinese Journal of Ecology, 28*(9), 1691–1695. (in Chinese).
Gov.cn website. (2013). *Three-North Shelterbelt programme reaching a "difficult" stage*. August 11, 2013. http://www.gov.cn/jrzg/2013-08/11/content_2464941.htm. Accessed September 27, 2014. (in Chinese).
Hawlena, D., & Bouskila, A. (2006). Land management practices for combating desertification cause species replacement of desert lizards. *Journal of Applied Ecology, 43*(4), 701–709.
Huang, B. (1981). Accurately assess the role of forest. *Geographical Knowledge, 1*, 9. (in Chinese).
Huang, B. (1982). Reiterate the role of reforest. *Geographical Knowledge, 2–4*, 10. (in Chinese).
Huang, R., & Wei, Z. (2009). Miracles created along the desert Railway. *Modern Science, 3*, 22–24. (in Chinese).
Jiang, H. (2002). Culture, ecology, and nature's changing balance: Sandification on Mu Us Sandy Land, Inner Mongolia, China. In J. F. Reynolds & D. M. Stafford Smith (Eds.), *Global desertification: Do humans cause deserts?* (pp. 181–196). Berlin: Dahlem University Press.
Jiang, G. (2005). Why the more China controls desertification the more desertified land there is? *Environment and Innovation, 9*, 18. (in Chinese).
Jiang, G. (2006). Three mistakes in desertification control in China. *People's Daily*, March 30, p. 16 (in Chinese).
Jiang, G. (2007). Speaking the truth to authority. March 31, http://www.sciencenet.cn/blog/user_content.aspx?id=1027. Accessed September 27, 2014. (in Chinese).
Jiang, G. (2008a). *China's "green deserts"*. January 10. http://www.sciencenet.cn/blog/user_content.aspx?id=14245. Accessed September 27, 2014. (in Chinese).
Jiang, G. (2008b). *Use nature's force to recover nature*. Beijing: Hydrological and Hydro-power Press. (in Chinese).
Jiang, H. (2010). Desertification in China: Problems with policies and perceptions. In J. J. Kassiola & S. Guo (Eds.). *China's environmental crisis* (pp. 13–40). London: Palgrave Macmillan.
Jiang, F., Yu, Z., Zeng, D., & Zhu, J. (2013). North China shelterbelt programme calls for ecological civilization. *Protection Forest Science and Technology, 5*, 1–3. (in Chinese).
Jin, F. (2013). Sand control in Minqin—Let green hold up new hope. *Gangsu Net*, March 11. http://roll.sohu.com/20130311/n368372804.shtml Accessed September 27, 2014. (in Chinese).
Li, Y. (2004). Ecological environment, cultural values and sustainable development. *Journal of Guangxi Ethnic College, 4*, 12–17. (in Chinese).

Li, B., Pan, Z., Hu, Y., & Liu, Z. (2008). Model in implementing scientific development in western China: Deciphering the Ordos model. *Shijian (Practice)*, *9*, 24–26. (in Chinese).
Liu, S. (2003). Political system and cultural homogenization are key factors in large-scale grassland degradation in Inner Mongolia. Friends of Nature (website), April 7. (in Chinese).
Liu, M. (2005a). Ecological construction should not be controlled by the government. *Northern Economy, 6*, 31. (in Chinese).
Liu, M. (2005b). Strategy of "giving sand control to able people" is a must. *Northern Economy, 8*, 21–23. (in Chinese).
Liu, B., Gong, W., Gong, W., & Song, L. (2009). Opportunities and challenges faced by Three-North Shelterbelt forestry construction. *Chinese Journal of Ecology, 28*(9), 1679–1683. (in Chinese).
Liu, J., & Cao, X. (Interview with Jiang Gaoming). (2005). Build forest and control sand—How much benefit is driving behind the wasted effort. *Southern Weekend*, July 22. (in Chinese).
Ma, Y., Fan, S., Zhou, L., Dong, Z., Zhang, K., & Feng, J. (2007). Temporal change of driving factors in the course of land desertification in arid region of North China: The case of Minqin county. *Environmental Geology, 51*, 999–1008. (in Chinese).
Ma, S., & Zhang, Y. (2012). On-site report of sand control in Minqin. *Gansu Daily*, December 7. (in Chinese).
Meng, S. (2011). China's forestry fudge? *Chinadialogue.net*. https://www.chinadialogue.net/article/show/single/en/4259-China-s-forestry-fudge. Accessed September 27, 2014.
Moxley, M. (2010). China's "Great Green Wall" grows in climate fight. *The Guardian*, September 23.
Ni, G., Zheng, F., Ding, D., & Yu, Z. (2013). Oasis agriculture, tragedy of the commons, and land desertification: Example of Minqin county, Gansu. *Journal of Northwest University (Social Science Edition), 13*(3), 12–16. (in Chinese).
Ning, L., & Yang, J. (2013). Remote sensing analysis of desertification in Minqin County. *Journal of Hunan University of Science & Technology (Natural Science Edition), 28*(2), 115–119. (in Chinese).
Nosetto, M. D., Jobbagy, E. G., & Paruelo, J. M. (2005). Land-use change and water losses: The case of grassland afforestation across a soil textural gradient in central Argentina. *Global Change Biology, 11*(7), 1101–1117.
Oxbrough, A. G., Gittings, T., O'Halloran, J., Giller, P. S., & Kelly, T. C. (2006). The initial effects of afforestation on the ground-dwelling spider fauna of Irish peatlands and grasslands. *Forest Ecology and Management, 237*(1–3), 478–491.
Pang, G., Dong, X., Song, X., Lim, B., & Ma, R. (2012). Monitoring of protective forestry in Hexi Corridor using remote sensing since the construction of "Three-North" shelterbelt forestry. *Journal of Desert Research, 32*(2), 539–544. (in Chinese).
Qi, L., Su, Z., & Liu. N. (2009). Ceremony of 30-year of Three-North Shelterbelt programme held in Beijing. *Green Times*, November 20. (in Chinese).
ScienceNet. (2013). *China leads the world in having sand retreat and people advance.* August 7. http://news.sciencenet.cn/htmlnews/2013/8/280948.shtm. Accessed September 27, 2014. (in Chinese).
SDRC (State Development and Reform Commission). (2010). Integrated improvement planning of the Loess Plateau region (2010–2030). (in Chinese).
SFA. (2008). *Technical stipulation of the 4th national desertification and sandification survey.* Beijing: SFA. (in Chinese).
SFA. (2011). *A bulletin of status quo of desertification and sandification in China.* http://www.forestry.gov.cn/uploadfile/main/2011-1/file/2011-1-5-59315b03587b4d7793d5d9c3aae7ca86.pdf Accessed September 27, 2014.
SFA (Guo, C., Ed.). 2007. *China's desertification and sandification.* http://www.cctv.com/docu/special/C19398/20071109/105716.shtml. Accessed September 27, 2014. (in Chinese).
SFA (Institute of Forest Inventory, Planning and Design under SFA). (2009). Technical manual for the planning of Three-North Shelterbelt system, 5th stage project. (in Chinese).

SFA (State Administration of Forestry). (2001). *Sand control law*. Beijing: China Forestry Publisher House. (in Chinese).

Shan, Z. (2009). Greenness: A new concept everyone should know. *Friends of China Newsletter*. http://www.fon.org.cn/content.php?aid=11116 Accessed September 27, 2014. (in Chinese).

Shi, P. (1983). Explaining the concept of desertification. *World Desert Studies, 4*, 26–28. (in Chinese).

Shi, Y. (2013). Ten-year plan of sand control, stopping the desert to add colors to the beautiful China. *Green Times*, March 31. (in Chinese).

Sun, W., Huo, R., & Tang, X. (2012). Dynamic analysis techniques of desertification and sandification survey data. *Journal of Beijing Forestry University, 34*(5), 105–112. (in Chinese).

Tang, C. (2013). Sand control hero Shi Shuzhu will spend lifetime to control sand. *Gansu Net*, March 13, 2013. (in Chinese).

Togochog, E. (2005). Ecological immigration and human rights in Inner Mongolia. Southern Mongolian Network (website).

Turner, B. L, I. I., Clark, W. C., Kates, R. W., Richards, J. F., Mathews, J. T., & Meyer, W. B. (1990). *The Earth as transformed by human action: Global and regional changes in the biosphere over the past 300 years*. Cambridge: Cambridge University Press.

UNCCD (UN Convention to Combat Desertification). (2011). Desertification: a visual synthesis. Bonn: UNCCD Secretariat Publication.

Van Der Salm, C., Wieggers, R., Van Den Toorn, A., Denier Van Der Gon, H., & Bleeker, A. (2006). The effect of afforestation on water recharge and nitrogen leaching in The Netherlands. *Forest Ecology and Management, 221*(1–3), 170–182.

Wang, H. (2006). Comparison of UNCCD and China's Sand Control Law. In *China Proceedings of Law Association Environment and Resources Law Study Annual Meeting*. http://www.riel.whu.edu.cn/article.asp?id=28623. Accessed September 27, 2014. (in Chinese).

Wang, L. (2007). Tree planting benefit is a sensitive topic. *Science Times*, August 12. (in Chinese).

Wang, T. (2008). Desertification change and its control strategies in North China in the last 50 years. *Journal of Yunnan Normal University (Philosophy and Social Science Edition), 40* (3), 23–30. (in Chinese).

Wang, X., Eerdun, H., Zhou, Z., & Liu, X. (2007). Significance of variations in the wind energy environment over the past 50 years with respect to dune activity and desertification in arid and semiarid northern China. *Geomorphology, 86*, 252–266.

Wang, T., Wei, W., Xue, X., Sun, Q., Zhang, W., & Han, Z. (2004). Spatial-temporal change of desertification in northern China in the last 50 years. *Acta Geographica Sinica, 59*(2), 203–212. (in Chinese).

Wang, T., & Zhao, H. L. (2005). Fifty years of desert science in China. *Journal of Desert Research, 25*(2), 145–165. (in Chinese).

Weng, B. (2000). "The lonely wall": Reflection on the Three-North Shelterbelt programme after 20 years. *Window of Southern Wind, 8*, 14–25. (in Chinese).

Wilkening, K. (2006). Dragon dust: Atmospheric science and cooperation on desertification in the Asia and Pacific Region. *Journal of East Asian Studies, 6*, 433–461.

Williams, D. M. (2001). *Beyond the Great Wall: Environment, identity, and development on the Chinese grasslands of Inner Mongolia*. Stanford, CA: Stanford University Press.

Xiao, W., Dai, G., & Zhang, S. (2010). China's strategy and financing for forestry sustainable development. United Nations Forum on Forests, August 16. www.un.org/esa/forests/pdf/aheg/aheg1/China_case_study.pdf Accessed September 27, 2014.

Yan, B., Pang, S., Lu, X., & Li, Q. (2012). Green shelter to block wind and sand (p. 1). *People's Daily* (Overseas Edition), August 28. (in Chinese).

Yang, T. (2007). Zhang Yinlong: For the green in my heart. December 13, 2007, cnwest.com. (in Chinese).

Yang, J., Liang, Z., Han, R., Sun, Q., & Cui, L. (2004). Water use efficiency and water consumption characteristics of poplar under soil drought conditions. *Acta Phytoecologica Sinica, 28*(5), 630–636. (in Chinese).

Yang, X., Zhang, K., Jia, B., & Ci, L. (2005). Desertification assessment in China: An overview. *Journal of Arid Environments, 63*, 517–531.

Yu, L., Li, W., & He, J. (2008). Problems and benefits of the Three-North Shelterbelt project in Ningxia Autonomous Region. *Journal of Northwest Forestry University, 23*(4), 228–232. (in Chinese).

Zhang, W. (2008). Anthropological study of grassland desertification: Using the example of B Gacha at the northern edge of the Mu Us Sandy Land. *Society, 28*(4), 189–205. (in Chinese).

Zhang, W., Yang, J., & Zhang, Y. (2007a). Three-North Shelterbelt project growth analysis and assessment during the tenth five-year period. *Forestry Economics, 10*, 48–51. (in Chinese).

Zhang, J., Zhang, B., & Zhang, H. (2007b). Reflection on desertification control in Shiyang River watershed in the last 50 years. *Ecological Environment*, 394–397. (in Chinese).

Zheng, D. (2006). Dryland ecological construction should follow geographical zone and laws. *Science Times*, September 11. (in Chinese).

Zhou, J. (2006). Ecological refugees in Alashan. *Chinadialogue.net*. www.chinadialogue.net/article/show/single/ch/605-Alashan-s-environmental-refugees. Accessed September 27, 2014.

Zhou, L., Ma, Y., & Ma, S. (2007). On grains and cropland-to-forestry (grassland) conversion in Northern China's mixed wind-water erosion area at the agriculture-pasture transitional zone. *Journal of Desert Research, 27*(4), 552–557. (in Chinese).

Zhou, W., Sun, Z., Li, J., Gang, C., & Zhang, C. (2013). Desertification dynamic and the relative roles of climate change and human activities in desertification in the Heihe River Basin based on NPP. *Journal of Arid Land, 5*(4), 465–479.

Zhu, Z. (1985). The current condition and trend of desertification in northern China. *Journal of Desert Research, 3*, 3–11. (in Chinese).

Zhu, Z. (1989). Concept, causes and control of land desertification in China. *Quaternary Sciences, 2*, 145–154. (in Chinese).

Zhu, Z. (1997). Global change and desertification. *Earth Science Frontiers, China University of Geosciences, 4*(1–2), 213–219. (in Chinese).

Part IV
Overview

Chapter 20
Desertification: Reflections on the Mirage

Mark Stafford Smith

Abstract This chapter concludes the book with a reflection on its key messages. Prior chapters add up to a strong denunciation of the simplistic but far reaching notion of desertification as promulgated over the past century. It continues to be enlisted to support policies that disempower dryland peoples around the world. There remains a question, though, of how much the misconceived notion of desertification drove these outcomes in sub-Saharan Africa, as opposed to colonization and then globalization being the underlying forces which simply found desertification a convenient crutch for actions that would have happened regardless. Today, we have a better understanding of mobility, equilibrium (and other) dynamics, the effectiveness and limitations of local institutions, the diversity of dryland systems, and accelerating global change. These combine to provide a set of more nuanced recommendations for the future. It still matters to understand causation, however, as these recommendations themselves could be undermined by the extraordinary power of narrative and narrative of power. A new and appealing narrative of drylands is needed, that is strong enough to rebut (or evade) the apparent inevitability of centralized power dynamics, and to respond to the inexorable realities of global change.

20.1 Introduction

The chapters of this book tell a remarkable and rather uncomfortable story. Centuries of scientific interpretation funnelled into an overstated narrative of desertification, which may have mobilized some supportive investment in the Sahel but at the same time imposed disruptive, even deadly, centralized interventions that damaged local institutions and culture; challenged in the past three decades by a counter-narrative that was at times equally simplistic in idealising nomadism and downplaying any desertification (whatever that exactly meant), partly in a bid to

M. Stafford Smith (✉)
CSIRO Land and Water, PO Box 1700, Canberra, ACT 2601, Australia
e-mail: mark.staffordsmith@csiro.au

R.H. Behnke and M. Mortimore (eds.), *The End of Desertification?*
Springer Earth System Sciences, DOI 10.1007/978-3-642-16014-1_20

swing the pendulum towards more local self-determination. At the end of this period, the book emphatically shows that the reality is less black and white but also explores issues that make it hard to work out this reality. Indeed one underlying issue, mostly somewhat subliminal in this book, is whether its focus is strictly on mobile pastoralists themselves or the systems in which they are embedded, given that the latter have become inexorably more complex, connected and mixed with sedentary activities over the decades.

Desertification as enshrined in political processes over the last half century is clearly shown to be a mirage here. But of course mirages are real phenomena, just not the ones that they appear to be, and they can influence behaviour catastrophically; so too is desertification, as the record here shows. And if you march across the desert through one mirage, as often as not another appears ahead of you.

Parenthetically, I should declare my relationship to these issues. My first published paper was on fieldwork in Niger, but my career has been based in Australia and mostly in the rangelands around Alice Springs. I have repeatedly dipped into issues of rangelands management in Africa, the US, Asia and South America, but my prejudices are clearly coloured by my Australian experience. In the absence of an Australian case study in the latter chapters of this book, the editors have urged me to bring some of that into this concluding chapter.

My colleagues and I were invited to contribute a chapter to the book (Behnke et al. 1993) that originally staked out the controversy that the present book seeks to bookend; we did so with some bemusement (Stafford Smith and Pickup 1993) as the controversy seemed to us then to be based on a failure to recognize the complexity of rangelands that was being explored in Australia at the time. The present volume provides a much more holistic view of how those scientific issues were embedded in the narratives and policy dynamics of local, national and above all international institutions, issues which have played out in Australia too in the intervening time, albeit in different ways.

In 2001, we engaged again with this challenge through a Dahlem conference attended by several of the present authors (Reynolds and Stafford Smith 2002); and from this emerged a Science paper in 2007 that proposed the Drylands Development Paradigm ('DDP'—Reynolds et al. 2007). This was not a theory, but rather a set of diagnostic propositions; these aimed to empower those debating dryland issues to ensure that they engaged with the requisite complexity of drylands. In brief precis, they urged any analysis of drylands to systematically consider the linked social and environmental system, key slow variables and their thresholds, the crucially cross-scale nature of drylands, and the vital role that mental models play as simplified narratives that drive decision-making in the social-ecological system. All these issues emerge repeatedly in this book. Both 'DDP' publications concluded that 'desertification' was useless as an operational term, and, given how embedded it is in international conventions, should at most be reserved as a generalized regional concept.

This book, as cogently summarized by the editors in the Introduction, goes further to explain why the concept of desertification should be well past its use-by-date. However, it remains embedded in the UN Convention for Combatting

Desertification (UNCCD), and has now been re-vitalized in Goal 15 of the newly adopted Sustainable Development Goals (UN 2015). So is there a new set of more contemporary issues, which could become the next mirage? Here I want to work towards these.

20.2 The Making of the Mirage

The foregoing chapters of this book add up to a most insightful set of contributions, that are well synthesized by Behnke and Mortimore (Chap. 1). I will not repeat this richness, but highlight a few key issues that underpin my own interpretation.

The story provided by these chapters is deeply nested spatially and institutionally, from local case studies in the Sahel and elsewhere, embedded in tales of changing provincial and national programmes and policy, to the international dimension of aid, trade and international agreements (particularly UNCCD). It is also nested in time from the longer term history of how the narrative of desertification developed over centuries, to the changes of the past half century, down to the impacts of short term policy changes, and, in recent decades, the accelerating implications of population growth and global environmental change. Through it all comes the pervasive sense that there is a diversity of systems so that any simplistic one-size-fits-all recommendation is bound to be wrong in most specific places and times. No wonder it is hard for any administrator, however well-meaning, to readily absorb the full complexity of the system and then apply the understanding in an often short-term posting to some dryland task!

Here are some threads that particularly strike me in the making of the mirage.

20.2.1 Origins of the Concept of Desertification in the Sahel

Davis (Chap. 8) and Toulmin and Brock (Chap. 2) together document how 'desiccation theory'—the idea that denuded landscapes will dry out—pervaded French colonial control in Africa and elsewhere, and eventually resonated well with perceptions that the desert was advancing; and how these crisis narratives in turn led to the promulgation of the concept of desertification well beyond the francophone sphere of interest. As a consequence, well-meaning (but often distantly managed) interventions to intensify agriculture, as well as those driven more by less well-meaning self-interest among local power elites, enlisted the concept to impose change on drylands and their traditional local institutions; these usually marginalized pastoralists, poor farmers and communal interests in the process.

It is worth considering the extent to which this specific history is a symptom of underlying causes rather than a cause in its own right, however. In other parts of the world on different timings, similar disruptions to local institutions and management systems were occurring, whether in Central Asia during the soviet era (Robinson,

Chap. 17; Alimaev and Behnke 2008), or China with its tree planting since the 1950s (Jiang, Chap. 19), or Argentina with its settlement of Patagonia around 1900 (Oliva et al., Chap. 13). Indeed, this history helps me understand why the concept of desertification never resonated in Australia: although white settlers displaced many nomadic Aboriginal populations in the late 1800s (as was also happening in Patagonia at the time), this was mostly rationalized by the (supposed) absence of land use, not its misuse. By the time the concept of desertification was being codified in the UNCCD, the debate on degradation in dryland Australia was focused on the perception of widespread damage being caused by the white settlers themselves, rather than some notion of 'troublesome nomads' (who had in fact by that time recently achieved the right to vote…). Worldwide, this was an era of increased connectivity and globalization, with governance and market power becoming increasingly remote from most drylands; this trend was spearheaded by French and English colonialization in the 1800s, but replicated by other powers taking greater central control over drylands whether within nations or across empires. The desertification narrative may have been the specific tool to support control in the context of globalization in the Sahel, whilst other narratives did so elsewhere in pre-UNCCD times.

20.2.2 Technology and Monitoring

It is also worth reflecting on the double-edged role of technological developments in these trends. The Sahelian drought and famine of 1983–85 was perhaps the first major event to capture a widespread public imagination in the West through events such as LiveAid, thanks to the increasingly live television coverage. This increasing connectedness of the world no doubt enhanced the sense of crisis and reached people who would never have been aware of the 'unfolding tragedy' even a decade earlier, let alone a century (never mind that this awareness may have been too little, too late, and that the multiple phases of famine may have been caused as much by government action as by climate). The continued development of communications through to today's mobile phones was instrumental in globalization and control, as noted above, but has, of course, also opened up opportunities for diversification, and for pastoralists to engage more knowledgeably with markets (Hiernaux et al., Chap. 6), highlighting that connectedness has both positive and negative aspects.

Indeed, satellite technology was not only speeding up global communications connectivity; around this time it was also starting to probe the promise of universal monitoring. The first Landsat satellite was launched in 1972, and the first Advanced Very High Resolution Radiometer (AVHRR) carrying satellite in 1978. The 1980s were the era of experimentation with what these sensors could offer. 'Very high resolution' in AVHRR was in fact resolution of no better than 1.1 km but global coverage could be obtained monthly.

In the first draft of this chapter, I made an unguarded assertion here that was reasonably questioned by reviewers. This made me go back to literature of the

1980s; among other sources I looked through the entire proceedings of the International Rangelands Congresses in 1984 (Adelaide) and 1991 (Montpellier) to revisit the flavour of debate at the time. In 1984, most of the discussion of remote sensing was still about methodology; there were a few reports of long-term trends in vegetation, but these were mostly from air photos or ground surveys. The flavour of preconceptions is epitomized by this quote regarding ground and aerial photo records between 1955 and 1981 in Upper Volta (recalling that this was towards the end of the decade of Sahelian drought):

> …the degradation of the vegetal cover has been increasing…the production of forage in the native pastures has been decreasing greatly. These changes depend mostly on overgrazing, recent drought being only an aggravating factor (Toutain 1984).

The contemporary level of remote sensing had also already debunked the idea of a moving front of desertification, noting that 'desertification' occurred well away from the desert boundary where production was less rainfall-limited and then patchily, at least in the eastern Sahel (Olsson 1984). Several studies did document detectable and credible declines in vegetation productivity, though other authors were starting to distinguish ecological condition from what might matter for production for different purposes. By 1991, the papers on satellite remote sensing had moved from methods to starting to deliver an understanding of spatially disaggregated trends (e.g. Kharin 1993). In addition, the idea that the interpretation of 'degradation' might be in the eye of beholder has taken a firm grip (e.g. Danckwerts and Adams 1993; Maxwell 1993; Teer 1993), and aspects of the multi-scaled nature of the social-ecological systems were being explored (e.g. Balent and Stafford Smith 1993) with implications for how research should be done.

We now understand the local and remote drivers of regional climate variability and change much better (Giannini, Chap. 10). But given these developments in the 1980s, it is peculiar that there remain challenges in reconciling the interpretation of remote sensing with ground data, such that Hermann and Sop (Chap. 5) still feel the need to observe, "With few exceptions, regional-scale trends of vegetation productivity in the Sahel, observed from satellite data, and local-scale changes in the vegetation cover on the ground have so far mostly been addressed separately and by different researchers. The findings from those two perspectives have led to differing and sometimes conflicting conclusions on the nature and extent of degradation/desertification in the Sahel." 'Desiccation theory' and its initial descendants had been based on biased local observations. Remote sensing held out the promise of universal observations that should have resolved these biases. But, perversely, simplistic interpretation insufficiently linked to ground observations with a strong typology of landscapes and social systems seems to have still not re-written the narrative of desertification.

Interestingly in the Australia the first users of remote sensing to attempt to measure dryland land condition and trend—Barney Foran, Geoff Pickup and Gary Bastin in particular but with numerous successors—lived and worked in the rangelands that they were aiming to monitor, and consequently could see the limitations of their analyses first hand (e.g. Foran 1987; Bastin et al. 1993b; Pickup et al. 1994). Proportionately, they expended great effort on ground-truthing and on new models for interpreting the meaning of the satellite signals in complex landscapes; they had to respond to the observed reality that reduced productivity could mean lower green cover in one landscape element where the basal area of perennial grasses had been grazed out, but increased green cover in a neighbouring element where unpalatable shrubs had invaded due to the same grazing and lack of fire. By the early 1990s, as the desertification narrative was being questioned in the Sahel, Pickup and Bastin (e.g. Bastin et al. 1993a; Pickup et al. 1994) had developed a grazing gradient model that accounted for these landscape differences and clearly extracted a land management signal from natural landscape variability, identifying different forms of degradation. Before long the results were starting to be applied in a three-way research-policy-pastoralist dialogue to assist local management (e.g. Bastin et al. 1996). Whilst the same models are not readily transferrable to more densely settled areas (though see Bastin et al. 2014 more recently), the principle of having a very close relationship between remote sensing and ground data should be unremarkable, as Hermann and Sop (Chap. 5) point out ("although the need for field data appears self-evident….") and is indeed recorded in many sub-Saharan Africa studies since the 1980s. In Australia these sources of data were deeply integrated across nested scales in the Australian Collaborative Rangelands Information System (ACRIS) (Bastin et al. 2009). Looking forwards, one must ask why this did not appear to happen widely in the Sahel.

20.2.3 Understanding Local Institutions

A part of the initial challenge to the desertification narrative was from researchers working locally and observing the frequent effectiveness of local institutions for managing natural resources, apparently continuing to be productive despite claims of degradation, and being radically disrupted by interventions from above. Just as early descriptions of environmental changes were local, biased and over-generalized, so too were some interpretations of this new wave of institutional analyses. Some commentaries reacted to the dominant desertification narrative that "everywhere was degrading" with an equally implausible narrative that local institutions could solve everything and "nowhere is degrading"; I distinctly recall sitting in some workshops in Africa in the early 1990s where a few social scientists were proposing that degradation was entirely a political construct. Both threads are still visible in public narratives but most oversimplifications have been laid to rest over the past 2 decades. As Toulmin and Brock (Chap. 2) say, "solutions to the environmental and economic problems faced by drylands systems…need to be more firmly rooted in a nuanced understanding of ecological changes and the links between climate, vegetation and people. They must also involve a shift in power to local people…"

Indeed, several chapters document the frequent benefits of empowered local institutions, whilst also noting their limitations in the face of rapid changes wrought by more global forces. Boubacar (Chap. 7) describes the decentralization of power to local communities and institutions in establishing the Régénération Naturel Assisté (RNA) programme in Niger in the 1980s, the successes of which are now readily visible from remote sensing. He also notes that the increasing individualisation in production systems since has triggered a crisis in family models. Coppock (Chap. 12) documents a series of central policy changes in Ethiopia which have permitted more or less local determination in the subhumid Borana region, but generally reflect the reality that "pastoralists have never been a priority for any Ethiopian government". Toulmin and Brock (Chap. 2) outline 5 cases of the "tug-of-war between those who advocate a rooted decentralized approach, drawing on local knowledge and perspectives, and those who assert their power and expertise through planning large-scale schemes and mobilizing big investments." They note how this is often a highly politicized process, framed by power relations at different scales; but that there is evidence of what works, one need being "a shift of power to local people and removing attempts at bureaucratic control". Importantly this requires a marrying of modern science and technical solutions with local knowledge systems, often indigenous, and recognising local land tenure and access rights.

There is little doubt that the desertification narrative supported too much intervention and external control of drylands; but the best outcomes are likely to emerge from some dynamic partnership between levels of governance, not by expecting any one level to deliver everything. In fact, summarising the outcomes of a session at the IVth International Rangelands Congress in Montpelier in 1991, Balent and Stafford Smith (1993) discussed how, through its 1980s LandCare programme, Australia had embarked on constructing a community-level of governance that had never existed in its colonial history, at the same time as other parts of the world were seeing that level fragmented and lost. Though elements persist today in Australia, enabling a far more effective community engagement of individual farmers with their peers and other stakeholders in regional issues than had been enacted previously, this level has since had a patchy history (Curtis and Lockwood 2000; Curtis et al. 2014) thanks to the ever-centralizing tendencies of government. Thus a key question is not only how to establish the local institutions described by Mortimore (Chap. 3), Boubacar (Chap. 7) and others, but also how to ensure that they persist in the face of cross-scale power relationships, given that we now have three decades of experience in some cases.

20.2.4 Population and Global Environmental Change

Debate about how much of the changes detected from satellite were induced locally by management on the ground as opposed to how much was the influence of regional climate now seems reasonably well resolved, at least at this regional scale.

Giannini (Chap. 10) summarizes the latest understanding from climate modelling, indicating that major drivers for late-20th century drought in the Sahel arose from regional climate variability. There is evidence that local land surface-atmosphere interactions play some subsidiary amplifying role, but human-induced land cover change could simply not occur at the scale needed to be the main driver for regional drought. This is not, of course, to say there was no land use change or land degradation occurring at all—ground-based monitoring in various regions confirms this, though not nearly as universally as the desertification narrative would have it (Hiernaux et al., Mortimore, Herrmann and Sop, Chaps. 6, 3 and 5; not to mention parallels in the case studies from other parts of the world; these can be found as far back as the 1984 International Rangelands Congress—e.g. Olsson (1984), Skovlin (1984), though causation was already being debated—e.g. Cassanelli (1984), Mortimore (1984)). Giannini (Chap. 10) also shows that there is increasing evidence that Sahel drought may have been exacerbated by greenhouse gas emissions —that is, there is some anthropogenic signal but it is global not local. Toulmin and Brock (Chap. 2) argue that, ironically, the same climate understanding that eventually showed that the simplistic desertification notion of an advancing desert front in the Sahel was false is now being harnessed in a new crisis narrative of desertification driven by climate change, though at least one aspect of this—a causal link to conflict—is largely debunked for the Sahel by Benjaminsen (Chap. 4).

Climate is only one form of global change that is emphasized in this book in ways that were not so evident 30 years ago, however. Aside from the forces from globalization already noted above, many chapters return to the themes of increasing population (local in the drylands but in a broader context), urbanization as an opportunity and challenge to dryland populations, and occasionally to other challenges such as limits to the water cycle and availability of nutrients such as phosphorus; and these themes are echoed from other parts of the world. Authors repeatedly note that, while more empowering of local institutions is required, simply returning to some past condition is not an option. In some regions the population balance has been dramatically altered through in-migration, particularly of farmers; in others, the population has simply increased past the level at which traditional practices can support everyone on the basis of past social models. For example, Boubacar (Chap. 7) notes that, despite their effectiveness, "RNA techniques…will be of limited value in addressing the worrying demographic trends in Niger…[now] increasing by 3.9 % each year…"; and in the Borana Plateau "the regional forage base is simply insufficient to support the current densities of people" (Coppock, Chap. 12).

Change is endemic. Mortimore (Chap. 3) thus argues for a resilience paradigm as guide for policies that build up local adaptive capacity to cope with change. The resilience literature is very clear in emphasizing the need for multi-scaled analyses of such interventions (e.g. Anderies et al. 2006), as was the Dryland Development Paradigm (Stafford Smith and Reynolds 2002; Reynolds et al. 2007—Proposition 4).

20.2.5 Narratives and Mental Models

One of the strongest points in the current book is its open treatment of the importance of narrative, an issue which, like power, is not commonly taught to natural scientists. Not only is the power of narratives recognized in many chapters, but Huntsinger (Chap. 11) explicitly analyses the nature of simple narratives in driving policy in the USA, and Shanahan (Chap. 15) reports studies of media representations in Kenya, India and China, showing how these reinforce the different narratives in each country. Whilst one should clearly (the lesson of this whole debate!) be cautious of unitary causes, Huntsinger makes a powerful case that simple narratives triumph over more realistic representations of real complexity, and Shanahan supports this by showing that reporters even self-censor to fit with black and white tales that help to sell news.

Huntsinger's analysis talks of narratives at two levels—more broadly as to the history of US rangelands, and more narrowly with regard to the underpinning scientific theory. The former emphasizes, as she says, the failure to privatize, lack of accountability, the influence of outside investment and ignorance or discounting of carrying capacity, but neglects wider issues of failure to support local management institutions, loss of indigenous management and distant lobbying influences on Congress. These issues partially enlisted the science: the theory for most of the 20th century was the Clementsian model of climax succession which we now understand "simply does not fit most rangelands outside of mesic Midwestern grasslands" (Huntsinger, Chap. 11). This has since been replaced by the multi-equilibrial concept of states and transitions, drawing on the idea of non-equilibrium systems through the original landmark publication of Westoby et al. (1989). State and transitions have themselves been critiqued as an underpinning theory as opposed to useful mental model to inform management (Stafford Smith and Pickup 1993; Stafford Smith 1996), but have driven major investment in detailed mapping and management in the SW US (Bestelmeyer et al. 2009, 2013) and elsewhere, and have come to underpin much other rangeland professionals' thinking. In the more public discourse, however, Huntsinger suggests they lack "the directionality and purpose of a good story—lacking a moral compass as it were....things just happen. Within this narrative vacuum, the concept of Holistic Resource Management has flourished" despite the absence of any evidence of its underpinning ecological reality. We will return to the challenges of representing the complexities of reality, but Huntsinger's tale says much about the impact of readily grasped mental models in research as well as society.

In the African context, Ellis and Swift (1988) suggested that drylands with higher levels of interannual variability functioned differently to those that were more stable, proposing a coefficient of variance in annual rainfall of 33 % as a rough boundary. They proposed that grazing pressure may be sufficiently decoupled from vegetation responses in the more variable systems that grazing is not a primary cause of degradation, and indeed that such systems may be essentially immune to degradation. Further analysis honed up the value and limitations of this proposal, most clearly explicated by Illius and O'Connor (1999a, b). They showed how spatial and temporal variability in resources lead to a more complex but analytically tractable story which should lead, as they said, to the point where "Rather than ignoring degradation, policy-makers and ecologists should seek to identify the characteristics of grazing systems that predispose some systems toward degradation, while others appear to be resistant." (Illius and O'Connor 1999a, abstract). They also helpfully clarify the difference between non-equilbrial and disequilibrial systems which I will not elaborate here except to note that the term non-equilibrium is often loosely applied to systems which are really just disequilibrial; and that understanding spatial heterogeneity and key resources often lies at the base of defining truly non-equilibrium relations (Illius and O'Connor 1999a). Simple theory breaks down when the system is supported with additional externally-derived inputs that can substitute for these key resources.

Despite these efforts to resolve the 'equilibrium/non-equilibrium' debate (e.g. Reynolds and Stafford Smith 2002; Vetter 2005), it continues in some quarters (e.g. Zemmrich 2007; Derry and Boone 2010; von Wehrden et al. 2012). For example, von Wehrden et al. (2012) provide some useful recent evidence that systems with a high CV of inter-annual rainfall variability (>33 %) may be dominated by different processes, though they note that it stands up only for their 'zonal degradation' category, that is, lands away from key spatially differentiated resources or water, and for a definition of degradation that relates to significant soil degradation in particular. This picks up on Illius and O'Connor's earlier predictions, above.

A related narrative is the uncritical primacy of pastoral mobility (which has some echoes of the 'noble savage'). There is plenty of evidence that in an unconstrained and unsubsidized system, mobility is a powerful strategy for managing variability. By unconstrained, I mean that pastoralists have access to more-or-less any landscape component; and by unsubsidized I mean that there are minimal inputs (whether of fodders, labour or money) from outside so that grazing resources must be used within the biophysical and other capacities available in the region. The effectiveness of mobility is not universal, depending in reasonably predictable (but not fully explored) ways on the spatial arrangement and temporal autocorrelation of resource availability (Illius and O'Connor 1999a; McAllister 2012); many other studies have shown that spatial strategies for moving or divesting stock also interact with alternative stocking strategies over time as well as the targeted product—meat, wool, milk or animals numbers (Foran and Stafford Smith 1991; Stafford Smith and Foran 1992; Illius et al. 1998; Campbell et al. 2006; Hunt et al. 2014 among many others, including Oliva et al., Chap. 13).

But few systems remain either unconstrained or unsubsidized in recent decades. Alimaev and Behnke (2008) (echoed by Robinson, Chap. 17 here) provide a classic description of how increasing fodder inputs during winters in Kazakhstan during the Soviet era enabled much greater livestock production with reduced mobility for a period, but at the collapse of the Soviet Union's support the numbers of stock crashed. This illustrates two points: externally-derived inputs can replace mobility, and may enable a greater productivity to be obtained from a region than mobility, probably at lower net cost, in both cases crucially assuming the inputs can be maintained. This is a cross-scale issue (Phoenix, Arizona, and the State of Israel are extreme dryland examples from outside the pastoral domain of survival on external inputs—Stafford Smith 2008). Behnke (2008) discusses the trends towards landscape fragmentation in different drylands, noting how the political economy drives the closure of more productive parts of the landscape, thus possibly removing key assets from mobile pastoralists. In systems where these key assets can persistently produce more resources under more intensive agriculture than under low intensity grazing, the net production of the region may well increase. This process has underlain agricultural intensification in mesic systems; it is more risky in drylands since agriculture is not always resilient in marginal environments. Whether mobility is the best option today depends on context. And so a key task becomes to clarify which systems can tolerate constraints from intensification or subsidies persistently, and which cannot; and then to facilitate different developmental strategies in different contexts.

Whilst range management was already moving from a steady state conceptualisation to ecosystem management, all these developments are now taking it further to resilience-based management (Reynolds and Stafford Smith 2002; Mortimore, Chap. 3), which, in the words of Bestelmeyer and Briske (2012), embraces the inevitability of change and emphasizes that management should seek to guide change to benefit society (see their Table 2 for the characteristics of the different approaches). Is the emergence of a narrative based around resilience sufficient? A rather uncritical and normative framing of resilience is certainly a growing fad in the UN bodies at present, but there are many examples of dryland system which need transformation, not increased resilience of the current state. Thus resilience alone is not the narrative—it requires an added sense of direction or purpose.

20.3 Behind the Mirage

Behnke and Mortimore (Chap. 1) review many aspects of the mirage. Here let me pick up some of the bigger picture in and beyond the Sahel.

My reading is that there has been a global move from drylands that were quite independent to ones that were increasingly managed by distant bureaucracies and elites, on diverse timings but in general disrupting local institutions and mobility. This occurred for many different reasons in detail but reflected the growing connectedness of the world over the past two or three centuries, originally through colonial interventions but since the 1950s mainly in autonomous nation states increasingly linked into a global trade system. In the Sahel and some other places, these interventions were assisted by the crisis narrative of desertification; in other places there were other narratives. Desertification was the one most institutionalized at a global level in the form of UNCCD, although this has been rather feebly reinventing itself as time has passed by.

In most places, these forces, coupled with growing local population pressure, also disrupted mobility which had been a widespread strategy for dealing with variability. However, the importance of this strategy varies greatly with the nature and level of biophysical variability that is experienced, so settled agriculture or forestry worked well in some places (in general, higher rainfall or lower variability, coupled with certain soil types), but not in others. In places where settled agriculture did not work well (or where it could not be sustained without major external inputs of fodder, e.g. in parts of Central Asia) there has generally been detectable, long-term land degradation of some sort, at least in some key landscape elements. In early years poor definitions and limited datasets meant that it was hard to distinguish real degradation from variation driven by natural (or at least externally imposed) climatic variability and change; real degradation was occurring mainly in systems which are, technically, disequilibrial rather than non-equilibrial. However, where settled agriculture can work, it seems that local institutions can usually self-organize to maintain land productivity, given the time and support (or lack of interruption) to establish those institutions, coupled with, sometimes, inputs of technical knowledge. Research and technology played both positive and negative roles in relation to power politics through this process, though not really as a cause in itself. Better interdisciplinary research integration earlier could have saved much argument.

In the past two decades, the analysis of how all these factors have intersected with power and politics has become sophisticated, especially in Africa but also elsewhere, and the more extreme black and white stories of researchers, at least, have become more nuanced and convergent, recognising the complexity of forces at play across multiple scales in dynamic and variable dryland social-ecological systems. But the simple narratives and associated debates persist powerfully.

Finally it is worth noting that we now recognize the 'Great Acceleration' since about 1950 in the many drivers and consequences of global change (Steffen et al. 2007). Of course some of these drivers were well-recognized in the 1970s and 1980s (for example, frequently in the pages of the International Rangelands Congress already cited), but only recently has research been able to document just how rapidly pressures from population, consumption and resulting changes such as climate change have been accelerating during precisely the period in which all of these debates have taken place. The baseline against which disrupted traditional arrangements might have been judged was itself changing drastically.

By the 1990s, the debate became split between two positions that were reinforced ideologically, one being that degradation was widespread, real and largely caused by local overuse, sometimes acknowledging cross-scale institutional causal factors too but seeing the solutions as top-down control; and the other arguing that true degradation was negligible as productivity continued despite this, and that local suffering was largely due to centralized interference with local institutions and power balances, seeing the solution as empowering local communities and minimising external control. These are caricatures, of course, but, transferred to the media and policy domains, the former narrative supported the reality of desertification, whereas the latter saw this as a matter of power relations that were devastating local culture. One saw local people creating desertification and causing their own suffering; the other saw desertification as a politically inspired notion that permitted distant power elites to inflict local suffering. In their black and white extremes, one saw dryland inhabitants as villains, the other as victims; one saw desertification everywhere, the other nowhere.

As with most black and white dichotomies in the real world, neither was right. These views had far-reaching, often negative, consequences for local people in terms of the policies and interventions supported through international aid and national political processes. This occurred not only in the individual expressions of ideological extremes, but also in some countries in the flip-flopping between policies (cf. Coppock, Chap. 12) which added major social variability (Stafford Smith 2008) to whatever biophysical uncertainty local people may already have had to endure.

20.4 Past the Mirage

More positively, a clear general recipe for the future of drylands also emerges from these pages, which I would summarize (based particularly on Toulmin and Brock, Mortimore, Hiernaux et al., Coppock—Chaps. 2, 3, 6 and 12—and lessons from the non-African case studies) as:

1. Actively strengthen local governance institutions within a sensitized cross-scale political economy that provides support and not bureaucratic control, and do this with long-term stability to allow local systems to be worked through.
2. In the face of growing local populations (and climate change), accept that it is usually no longer an option for mobile pastoralism to be the only livelihood that supports everyone.
3. Where agriculture is viable, accept more-or-less loss of mobility; where it is not, encourage the maintenance of pastoral mobility (which includes helping it to connect to markets, etc).
4. Given that, in most places, there will be too many people for agriculture or pastoralism to support everyone, support livelihood diversification and people to migrate in orderly and voluntary ways.
5. In developing these options, insist on inclusive processes, where technical research, if it is needed, is married (in Toulmin and Brock's term) with local knowledge, and the solutions often require engagement across scales of governance.
6. Above all, one size does not fit all: recognize that drylands are diverse biophysically, socially and culturally, and in their cross-scale interactions, so that the degree and nature of implementation of each of the above needs to be context sensitive.

These points imply that the major exercise for drylands researchers ought to be to tease out what sorts of approaches are more or less likely to work in what contexts, in order to inform well-meaning, facilitatory policy at the appropriate level of governance. The idea of distinguishing non-equilibrial regions was one of these (Ellis and Swift 1988) but was clearly an exploratory blunt instrument. More comprehensive classifications are needed which are facilitatory not prescriptive.

Given that the drylands occupy around two fifths of the earth's land surface, it is unimaginable that they are not differentiated in important ways across the planet. Despite the fact that everyone recognizes this when challenged, at all scales we persist with promulgating simplistic 'universal truths'. One of the most useful but as yet incomplete contributions to this debate is the gradual development of a typology of drylands as complex social ecological systems, which genuinely combines social and policy aspects that help to say what sort of goals and players are likely to dominate in different places, and how these intersect with attributes of the biophysical system. Early classifications were climatic or soil driven and very much biophysical only, or based only on social systems. The first real attempt to bring these aspects together was Geist and Lambin (2004) who reported a typology of degradation causes and outcomes based on Geist (2004). More recently, syndrome approaches have been also taken more regionally (e.g. Maru and Chewings 2008; Stellmes et al. 2013), but Sietz et al. (2011) have been the first to systematize this at a global level. These need further development, drawing on key slow variables and in a way that is nested across scales to permit much more sophistication in getting beyond universal shopping lists of problems and solutions. Behnke and Mortimore (Chap. 1) review other evidence that could contribute to this as yet very incomplete synthesis.

However, power and policy processes remain a critical concern, to which we now return.

20.5 Avoiding the Next Mirage?

The six point general recipe above makes sense and can be suitably elaborated. In fact these points, like many aspects of this book, resonate closely with the diagnostics that were postulated as the Dryland Development 'Paradigm' by Reynolds et al. (2007)—the drylands must be seen deeply as co-evolving social and ecological systems; there are slow variables in both biophysical and social domains that are critical to observe and that may have thresholds; many critical processes, both social and environmental, cannot be understood without analyses at multiple scales; and the mental models that underlie local management and policy decision-making form the fundamental linkage between ecological and social processes. But such an analysis begs new questions that arise from recognising the importance of power and narrative, and of underlying forces of change.

In short, the six points may be well intended, but without a straightforward narrative, a means of dealing with distant centres of power, and attention to major global trends, is it all wishful thinking, another mirage approaching? I have more questions than answers, but, at the risk of creating gross generalizations, these suggest some priority questions for those who care about the drylands.

20.5.1 The Telling Narrative

The message that comes through more strongly in this book than in previous volumes on the future of the drylands is the importance of narrative. Narrative will be put to work in political (and research) contexts, and is a necessary if not sufficient contributor to a more positive future for drylands. The simple crisis narrative of desertification has long passed its use-by-date; if it was ever really true, we know from other areas of endeavour, such as climate change and even weight loss, that perpetually emphasizing the problem serves to further disempower and marginalize. Drylands tend to carry a deficit narrative, despite their many positive stories (think of amazing people, plants, animals, geology, institutions, religions and so on). We need a narrative that presages solutions and empowerment, more than just resilience, to encompass a path to the future. The scientific realities, whether it is states and transitions or 'non-equilibrial dynamics' or other expressions of the complexity of the system, lack Huntsinger's moral compass. A narrative that simply says "empower locals because they know best" is likely to fail too, because locals require active support across scales in the face of power issues and global change (see Boubacar, Chap. 7, and the following points).

In remote Australia we sought a positive narrative under the rubric of 'desert knowledge', underpinned by the recognition that remote areas have their own strengths that can be sold to the world thanks to globalization, as long as globalisation doesn't roll over those areas first (Stafford Smith et al. 2008; Stafford Smith and Cribb 2009). It is not clear that this is working as yet, nor that it necessarily translates easily to other drylands.

But it clarifies the key question—***what is the high level, positive, and probably value-laden narrative about the future of drylands across the world which would override the impacts of power dynamics and global change?***

20.5.2 Power to the powerless

If the narrative is challenging, the power dynamics issue is acute. Though this is not universally true, drylands in general are remote from national or international centres of power, and politically marginalized (Reynolds et al. 2007; Stafford Smith 2008). It is wishful thinking to say that those centres of power should pay continuous and sympathetic attention to their dry hinterlands, however populated—their focus will inevitably be on where most people live. Powerful forces are required to counter the centralizing tendencies of government, even when they are accidental, let alone deliberate. Keeley and Scoones (2004) argue for more analysis of the policy process to understand how knowledge and power interact to create particular policy solutions and implementation strategies. This is indeed needed, but the question is, what then?

In caricature, the relationship between centres of power and drylands falls into one of three categories—rape and pillage, well-intentioned but poorly understood intervention, or benign neglect. The desert knowledge narrative argues that remote peoples need to understand how their system works, and then take advantage of the periods of benign neglect to establish local institutions that are able to promote change opportunistically in the periods of intervention; in these they should encourage actions that build resilience to the next period of neglect (or indeed rape and pillage) (e.g. in Australia, see Walker 2015). These actions may include devolved funding and governance arrangements, and effective marketing systems; they will also involve the active creation of supportive networks in the centres of power (and above them, where possible) through the conscious organization of the diaspora that is common from remote regions. Behind these suggestions is the assumption that, if regions can create a long enough breathing space of stability, then local institutions have a good chance of working through intra-regional conflicts; of course, this may not be so if there are entrenched (resilient) power imbalances. Indeed, the foregoing may be a view coloured by living in a pluralist democracy, where these interventions are hard enough; they may be hopeless in totalitarian or oligocratic system, or even those that are just deeply prejudiced.

So, this is an oversimplification that cannot be generalized, and may be academic for regions in the grip of war, but if this issue is not tackled head on, it is not clear that there can ever be any resolution for local people. In any case it highlights the deep need for cross-scale solutions. It also serves to pose the more general question —***recognising the inevitability of centralized power dynamics, how can drylands organize themselves and create cross-scale alliances to build desirable resilience?***

20.5.3 The Reality of Global Change

Answering the foregoing questions would be hard enough in a stationary world. But we do not have that luxury, and whilst there are lessons to be had from the institutions and management systems of previous eras, dryland solutions must look to a non-stationary future, with increasing populations in most regions, and with the superposition of global environmental change of which climate change is the harbinger. As ever, a non-deficit view sees opportunity in this. There will be more labour; some of that labour will emigrate and can be the kernel of national and even global networks; globalization and technologies open up many new opportunities, providing these are grasped before the accompanying risks flood in; even climate change will create new livelihood opportunities. Just as Sahelian pastoralists track markets on their mobiles phones (Hiernaux et al.), so too do Aboriginal inhabitants of the most remote communities, like Tjuntjuntjara in the Great Victorian Desert, sell art direct into New York.

As we have repeatedly asserted, dryland social ecological systems are diverse. In general, the powerless do not benefit from a narrative of dependence. However, in some regions drylands will need support and subsidy from higher levels of governance to grasp these opportunities and avoid the risks of global change. This requires imaginative navigation of the issues of power. However, it raises a third question—***how can different drylands best be given support to handle the pressures of global change as they play out in local populations and environments, without creating dependency?***

20.6 Conclusion

I have not written much about the word desertification here, despite the title of the book. In the concluding chapter to the Dahlem Conference publication in 2002 (Stafford Smith and Reynolds 2002), we suggested that desertification as a term was embedded in the UNCCD and so would persist; but that it was only useful as a loose term for the broader scale, emergent outcomes of degradation, and we stressed that at more local scales it is essential to specify what factors are degraded and in whose eyes (pp. 406, 413). The term, in fact, has never caught on in Australia, which has always talked (if imperfectly) of specific *types* of degradation. This book has undoubtedly documented the desirability of an end to the desertification crisis narrative. However, it is unlikely that the term itself will vanish soon from the lexicon of international conventions, given its re-emergence in the Sustainable Development Goals—it is mentioned three times in the outcome document, as well in the title of Goal 15 and in target 15.3: "By 2030, combat desertification, restore degraded land and soil, including land affected by desertification, drought and floods, and strive to achieve a land degradation-neutral world".

So perhaps another aim is in order. If we could cease to use the term desertification for anything operational at a local level, and we could deploy a new and more positive narrative about drylands, a new role could eventually evolve for the UNCCD. After all, it is, at present, the nearest thing to an opportunity for a supra-national alliance among the dryland regions (though currently controlled by national governments, not their dryland regions, of course). It would be helpful if the UNCCD became a convention for supporting the successful wellbeing of drylands rather than one founded on their problems, if it could help to steer and keep in check the cross-scale forces that lie at the root of the suffering of drylands people over the past century or so. Could it become a force for answering the foregoing three questions? Probably not with the current political economy and framing, and the UN's history of imperfect action (e.g. MacDonald, 1986 in an early and prescient analysis). But, then, could the more inclusive and universal nature of the Sustainable Development Goals provide a new framing?

In writing this closing chapter, the editors challenged me to ask the most pertinent question: "in the end, have 30 years of debate actually helped the dryland people of sub-Saharan Africa and their livelihoods?"

It is a sobering question to contemplate.

If this debate had not occurred, punctuated most notably by the Behnke et al. (1993) book, the pressures for centralized intervention may have gone unchallenged: not that the challenge has removed them. Overall I am left with a sense of astonishing lack of nimbleness in the research community in the last decades of the 20th Century—most of the issues raised in the present book were already being discussed in the 1984 International Rangeland Congress, and were apparently quite mainstream by the 1991 Congress. Yet there seems to have been very poor cross-fertilization of ideas among different strands of work, across disciplines, between regions and across cultures, to the detriment of progress in a part of the world that is most susceptible to intermittent and unintegrated interventions. Even today, this book assembles a first class set of case studies from across the world; most of the key findings were clear a decade ago, yet debate continues in the literature, as recorded here. Meanwhile many dryland regions continue to experience drastic swings in policy, whether in Niger, Ethiopia, or China, with local consequences described herein. The debate has probably helped some researchers' careers, it may have helped aid donors gradually recognize the importance of local institutions, it has probably had limited impact on distant national policy paradigms; yet, whilst having been necessary, it has seen further decades of local suffering in many regions. These indictments should not be forgotten, and should drive reduced fragmentation in the desert research community which remains diverse and poorly connected. The research community at least should not continue to feed misdirected policy that continues to emerge in other regions (Behnke and Mortimore Chap. 1)

My revisit of the papers from the International Rangeland Congresses of 1984 and 1991 makes me feel that there has been a profound failure of the research community to get to grips with the complexity of dryland social-ecological systems across the world, to provide clarity and (reasonable) consensus on some agreed universal aspects of their functioning, and to delineate other aspects that need clear contextualization and must not be generalized across systems. The universal aspects should then underpin a powerful common narrative that drives policy in a way that is likely to deliver good outcomes for local and regional populations; and the differentiated aspects require a well-thought through, multi-scale, social and physical typology through which differences and similarities can be legitimately asserted. These two elements would not, of course, remove the structural and motivational effects of local and globalized power but might at least reduce the unintended complicity of research with these.

But turning to the more positive narrative needed for the future, the full story is complex, as befits a complex system in many locations that experience relatively high variability, both biophysical and social. As Scoones (2004) and Stafford Smith and Cribb (2009) have independently observed (see also Brooks et al. 2009), global change increasingly means that we live in a world where knowledge about the future is uncertain, and predictability and control are false hopes ("a *mirage* of surety and precision", Scoones actually says!). We can learn a great deal from settings where uncertainty has always been part of day-to-day life and survival.

This 'desert knowledge' is becoming increasingly important for all the world's inhabitants in the face of global change. Indeed these lands, along with mountains and the cryosphere, are the sentinel regions of the world in the face of change, and the potential source of critical wisdom in responding to that change.

Let us find a positive narrative for the drylands.

References

Alimaev, I. I., & Behnke, R. H. J. (2008). Ideology, land tenure and livestock mobility in Kazakhstan. In K. A. Galvin, R. S. Reid, R. H. J.Behnke & N. T. Hobbs (Eds.), *Fragmentation in semi-arid and arid landscapes: Consequences for human and natural systems*. Dordrecht: Springer.

Anderies, J. M., Walker, B. H., & Kinzig, A. P. (2006). Fifteen weddings and a funeral: Case studies and resilience-based management. *Ecological Society, 11*, 21. [online]. URL: http://www.ecologyandsociety.org/vol11/iss21/art21/.

Balent, G., & Stafford Smith, D. M. (1993). Conceptual model for evaluating the consequences of management practices on the use of pastoral resources. In *Session Summary, Proceedings of IVth International Rangelands Congress*, Montpellier (pp. 1158–1164), April 1991.

Bastin, G., Denham, R., Scarth, P., Sparrow, A., & Chewings, V. (2014). Remotely-sensed analysis of ground-cover change in Queensland's rangelands, 1988–2005. *The Rangeland Journal, 36*, 191–204.

Bastin, G. N., Pickup, G., Chewings, V. C., Pearce, G. (1993a). Land degradation assessment in central Australia using a grazing gradient method. *The Rangeland Journal, 15*, 190–216.

Bastin, G. N., Pickup, G., & Stanes, A. (1996). Estimating landscape resilience from satellite data and its application to pastoral land management. *The Rangeland Journal, 18*, 118–135.

Bastin, G.N., Sparrow, A. D., & Pearce, G. (1993b). Grazing gradients in central Australian rangelands: Ground verification of remote sensing-based approaches. *The Rangeland Journal, 15*, 217–233.

Bastin, G. N., Stafford Smith, D. M., Watson, I. W., & Fisher, A. (2009). The Australian Collaborative Rangelands Information System: Preparing for a climate of change. *The Rangeland Journal, 31*, 111–125.

Behnke, R. H., Scoones, I., & Kerven, C. (Eds.). (1993). Range ecology at disequilibrium: New models of natural variability and pastoral adaptation in African Savannas. London: Overseas Development Institute.

Behnke, R. H. J. (2008). The drivers of fragmentation in arid and semi-arid landscapes. In K. A. Galvin, R. S. Reid, R. H. J.Behnke & N. T. Hobbs (Eds.), *Fragmentation in semi-arid and arid landscapes: Consequences for human and natural systems*. Berlin: Springer.

Bestelmeyer, B. T., & Briske, D. D. (2012). Grand challenges for resilience-based Management of Rangelands. *Rangeland Ecology & Management, 65*, 654–663.

Bestelmeyer, B. T., Duniway, M. C., James, D. K., Burkett, L. M., & Havstad, K. M. (2013). A test of critical thresholds and their indicators in a desertification-prone ecosystem: More resilience than we thought. *Ecology Letters, 16*, 339–345.

Bestelmeyer, B. T., Tugel, A. J., Peacock, G. L., Jr., Robinett, D. G., Shaver, P. L., Brown, J. R., Herrick, J. E., et al. (2009). State-and-transition models for heterogeneous landscapes: A strategy for development and application. *Rangeland Ecology & Management, 62*, 1–15.

Brooks, N., Grist, N., & Brown, K. (2009). Development futures in the context of climate change: Challenging the present and learning from the past. *Development Policy Review, 27*, 741–765.

Campbell, B.M., Gordon, I.J., Luckert, M.K., Petheram, L., Vetter, S., 2006. In search of optimal stocking regimes in semi-arid grazing lands: One size does not fit all. Ecol. Econ. 60, 75-85.

Cassanelli, L. V. (1984). Historical perspectives on pastoral development: Examples from the Somali and the Maasai, eastern Africa. In *Second International Rangeland Congress*, Adelaide, May 1984. CSIRO Wildlife and Rangelands Research, p. n/a (1 pp).

Curtis, A., & Lockwood, M. (2000). Landcare and catchment management in Australia: Lessons for state-sponsored community participation. *Society & Natural Resources, 13*, 61–73.

Curtis, A., Ross, H., Marshall, G. R., Baldwin, C., Cavaye, J., Freeman, C., Carr, A., et al. (2014). The great experiment with devolved NRM governance: Lessons from community engagement in Australia and New Zealand since the 1980s. *Australasian Journal of Environmental Management, 21*, 175–199.

Danckwerts, J. E., & Adams, K. M. (1993). Dynamics of rangelands ecosystems. In *Session Summary, Proceedings of IVth International Rangelands Congress,* Montpellier, April 1991 (pp. 1066–1069).

Derry, J. F., & Boone, R. B. (2010). Grazing systems are a result of equilibrium and non-equilibrium dynamics. *Journal of Arid Environments, 74*, 307–309.

Ellis, J., & Swift, D. (1988). Stability of African pastoral ecosystems: alternative paradigms and implications for development. *Journal of Range Management, 41*, 450–459.

Foran, B. D. (1987). Detection of yearly cover change with Landsat MSS on pastoral landscapes in Central Australia. *Remote Sensing of Environment, 23*, 333–350.

Foran, B. D., & Stafford Smith, D. M. (1991). Risk, biology and drought management strategies for cattle stations in Central Australia. *Journal of Environmental Management, 33*, 17–33.

Geist, H. J. (2004). *The causes and progression of desertification*. Burlington, VT: Ashgate.

Geist, H. J., & Lambin, E. F. (2004). Dynamic causal patterns of desertification. *Bioscience, 54*, 817–829.

Hunt, L. P., McIvor, J. G., Grice, A. C., & Bray, S. G. (2014). Principles and guidelines for managing cattle grazing in the grazing lands of northern Australia: Stocking rates, pasture resting, prescribed fire, paddock size and water points—a review. *The Rangeland Journal, 36*, 105–119.

Illius, A. W., Derry, J. F., & Gordon, I. J. (1998). Evaluation of strategies for tracking climatic variation in semi-arid grazing systems. *Agricultural Systems, 57*, 381–398.

Illius, A. W., & O'Connor, T. G. (1999a). On the relevance of nonequilibrium concepts to arid and semiarid grazing systems. *Ecological Applications, 9*, 798–813.

Illius, A. W., & O'Connor, T. G. (1999b). When is grazing a major determinant of rangeland condition and productivity?

Keeley, J., & Scoones, I. (2004). Understanding policy processes in Ethiopia: A response. *Journal of Modern African Studies, 42*, 149–153.

Kharin, N. G. (1993). Regional and local aspects of range inventory and monitoring. In *Session Summary, Proceedings of IVth International Rangelands Congress*, Montpellier, April 1991 (pp. 1087–1089).

MacDonald, L. H. (1986). *Natural Resources Development in the Sahel: The role of the United Nations System*. Tokyo: The United Nations University.

Maru, Y. T., & Chewings, V. H. (2008). How can we identify socio-regions in the rangelands of Australia? *The Rangeland Journal, 30*, 45–53.

Maxwell, T. J. (1993). Diagnosis and improvement methods in range utilization systems. In *Session Summary, Proceedings of IVth International Rangelands Congress*, Montpellier, April 1991 (pp. 1147–1154).

McAllister, R. R. J. (2012). Livestock mobility in arid and semiarid Australia: Escaping variability in space. *The Rangeland Journal, 34*, 139–147.

Mortimore, M. (1984). Monitoring and management of range resources in areas prone to desertification in northern Nigeria. In *Second International Rangeland Congress*, Adelaide, May 1984. CSIRO Wildlife and Rangelands Research, p. n/a (1 pp).

Olsson, L. (1984). Monitoring the impact of rain fed agriculture and grazing on desertification in the central Sudan. In *Second International Rangeland Congress*, Adelaide, May 1984. CSIRO Wildlife and Rangelands Research, p. n/a (1 pp).

Pickup, G., Bastin, G. N., & Chewings, V. H. (1994). Remote-sensing-based condition assessment for nonequilibrium rangelands under large-scale commercial grazing. *Ecological Applications, 4*, 497–517.

Reynolds, J. F., & Stafford Smith, D. M. (Eds.). (2002). Global desertification: Do humans cause deserts? Dahlem Workshop Report 88. Berlin: Dahlem University Press.

Reynolds, J. F., Stafford Smith, D. M., Lambin, E. F., Turner, B. L., II., Mortimore, M., Batterbury, S. P.J ., Downing, T. E., et al. (2007). Global desertification: Building a science for dryland development. *Science, 316*, 847–851.

Scoones, I. (2004). Climate change and the challenge of non-equilibrium thinking. *IDS Bulletin Institute of Development Studies, 35*, 114–+.

Sietz, D., Luedeke, M. K. B., & Walther, C. (2011). Categorisation of typical vulnerability patterns in global drylands. *Global Environmental Change-Human Policy Dimensions, 21*, 431–440.

Skovlin, J. M. (1984). Long-term trends from selected range sites throughout Kenya. In *Second International Rangeland Congress*, Adelaide, May 1984. CSIRO Wildlife and Rangelands Research, p. n/a (1 pp).

Stafford Smith, D. M., & Foran, B. (1992). An approach to assessing the economic risk of different drought management tactics on a South Australian pastoral sheep station. *Agricultural Systems, 39*, 83–105.

Stafford Smith, D. M., & Pickup, G. (1993). Out of Africa, looking in: Understanding vegetation change and its implications for management in Australian rangelands. In R. H. Behnke, I. Scoones & C. Kerven (Eds.), *Range ecology at disequilibrium: New models of natural variability and pastoral adaptation in African Savannas* (pp. 196–226). London: Overseas Development Institute.

Stafford Smith, D. M., & Reynolds, J. F. (2002). Desertification: A new paradigm for an old problem. In J. F. Reynolds & D. M. Stafford Smith (Eds.), G*lobal desertification: Do Humans cause deserts?* (pp. 403–424) Dahlem Workshop Report 88. Berlin: Dahlem University Press.

Stafford Smith, M. (1996). Management of rangelands: Paradigms at their limits (Chapter 12). In J. Hodgson & A. Illius (Eds.), *The ecology and management of grazing systems* (pp. 325–357). Wallingford, UK: CAB International.

Stafford Smith, M. (2008). The 'desert syndrome'—Causally-linked factors that characterise outback Australia. *The Rangeland Journal, 30*, 3–14.

Stafford Smith, M., & Cribb, J. (2009). Dry times: A blueprint for a red land. Melbourne: CSIRO Publishing.

Stafford Smith, M., James, C., McGregor, M., & Ferguson, J. (2008). Guest editorial: Building a science of desert living. *The Rangeland Journal, 30*, 1–2.

Steffen, W., Crutzen, P. J., & McNeill, J. R. (2007). The anthropocene: Are humans now overwhelming the great forces of nature. *Ambio, 36*, 614–621.

Stellmes, M., Roder, A., Udelhoven, T., & Hill, J. (2013). Mapping syndromes of land change in Spain with remote sensing time series, demographic and climatic data. *Land Use Policy, 30*, 685–702.

Teer, J. G. (1993. Performance and productivity of domestic and wild herbivores. In *Session Summary, Proceedings of IVth International Rangelands Congress*, Montpellier (pp. 1129–1133), April 1991.

Toutain, B. (1984). Recent vegetation changes and degradation in some Sahelian pastoral ecosystems of Western Africa. In *Second International Rangeland Congress*, Adelaide, May 1984. CSIRO Wildlife and Rangelands Research, p. n/a (1 pp).

UN. (2015). *Transforming our world: The 2030 agenda for sustainable development*. New York: United Nations.

Vetter, S. (2005). Rangelands at equilibrium and non-equilibrium: Recent developments in the debate. *Journal of Arid Environments, 62*, 321–341.

von Wehrden, H., Hanspach, J., Kaczensky, P., Fischer, J., & Wesche, K. (2012). Global assessment of the non-equilibrium concept in rangelands. *Ecological Applications, 22*, 393–399.

Walker, B.W. (2015). Radicalising the rangelands: disruptive change or progressive policy? *The Rangeland Journal*, *37*, 631–635.

Westoby, M., Walker, B., & Noy-Meir, I. (1989). Opportunistic management for rangelands not at equilibrium. *Journal of Range Management, 42*, 266–274.

Zemmrich, A. (2007). Equilibrium or non-equilibrium ecosystems? Scaling effects in rangeland ecosystems of Western Mongolia. *Erdkunde*, *61*, 335–343.

MIX
Papier aus verantwortungsvollen Quellen
Paper from responsible sources
FSC® C105338

If you have any concerns about our products,
you can contact us on
ProductSafety@springernature.com

In case Publisher is established outside the EU,
the EU authorized representative is:
Springer Nature Customer Service Center GmbH
Europaplatz 3, 69115 Heidelberg, Germany

Printed by Libri Plureos GmbH
in Hamburg, Germany